Graph-Theoretic Problems and Their New Applications

Graph-Theoretic Problems and Their New Applications

Special Issue Editor

Frank Werner

MDPI • Basel • Beijing • Wuhan • Barcelona • Belgrade • Manchester • Tokyo • Cluj • Tianjin

Special Issue Editor
Frank Werner
Otto-von-Guericke-Universität Magdeburg
Germany

Editorial Office
MDPI
St. Alban-Anlage 66
4052 Basel, Switzerland

This is a reprint of articles from the Special Issue published online in the open access journal *Mathematics* (ISSN 2227-7390) (available at: https://www.mdpi.com/journal/mathematics/special_issues/gtptna).

For citation purposes, cite each article independently as indicated on the article page online and as indicated below:

LastName, A.A.; LastName, B.B.; LastName, C.C. Article Title. *Journal Name* **Year**, *Article Number*, Page Range.

ISBN 978-3-03928-798-7 (Pbk)
ISBN 978-3-03928-799-4 (PDF)

Contents

About the Special Issue Editor

Frank Werner studied Mathematics from 1975 to 1980 and graduated from the Technical University Magdeburg (Germany) with honors. He defended his Ph.D. thesis on the solution of special scheduling problems in 1984 summa cum laude and his habilitation thesis in 1989. In 1992, he received a grant from the Alexander von Humboldt Foundation. Currently, he works as an Extraordinary Professor at the Faculty of Mathematics of the Otto von Guericke University Magdeburg (Germany). He is the author or editor of six books, among them a textbook "Mathematics of Economics and Business", and he has published more than 280 papers in international journals. He is on the Editorial Board of 17 journals; in particular, he is the Editor-in-Chief of Algorithms and an Associate Editor of the International Journal of Production Research and Journal of Scheduling. He has been a member of the Program Committee of more than 80 international conferences. His research interests include operations research, combinatorial optimization, and scheduling.

Preface to "Graph-Theoretic Problems and Their New Applications"

Graph Theory is an important area of Applied Mathematics with a broad spectrum of applications in many fields. This book results from a special issue entitled 'Graph-Theoretic Problems and Their New Applications'. In the Call for Papers for this issue, I asked for submissions presenting new and innovative approaches for traditional graph-theoretic problems as well as for new applications of graph theory in emerging fields, such as network security, computer science and data analysis, bioinformatics, operations research, engineering and manufacturing, physics and chemistry, linguistics, or social sciences.

In response to the Call for Papers for this issue, we had an enormous resonance, and altogether 151 submissions have been received among which finally 20 papers have been accepted, all of which are of high quality, reflecting the great interest in the area of Graph Theory. This corresponds to an acceptance rate of 13.2%. The authors of these accepted publications come from 13 different countries: USA, China, Pakistan, India, Iran, Marocco, Slovenia, United Arab Emirates, Oman, Spain, Mexico, Serbia, and Belarus, where most authors are from the first two countries. All submissions have been reviewed, as a rule, by at least three experts in the field of Graph Theory. The articles in this book cover a broad spectrum of graph theory, e.g., topological indices, domination in graphs, neutrosophic graphs or mixed graphs. This book contains one survey article by Sotskov and 19 further articles. Subsequently, the articles are briefly discussed according to the sequence in this book. We hope that the readers will find interesting theoretical ideas in this special issue and that researchers will find new inspirations for future works.

In the first article, Sotskov gives a detailed review about mixed graph colorings in relation to scheduling problems with minimizing the makespan. Such a mixed graph contains both directed arcs and undirected edges. He presents known results for two types of vertex colorings, referring to the chromatic number and the strict chromatic number of a graph, respectively, and he also reviews the complexity of these problems. Then he discusses in detail how the results for mixed graph colorings can be used for job shop scheduling problems with unit processing times as well as general shop scheduling problems. Further separate sections deal with colorings of arcs and edges of a mixed graph as well as with non-strict colorings of a mixed graph.

The second article by Tilley is related to the 4-color theorem which has been proven by showing that a minimal counterexample does not exist. Here the author proves that a minimum counterexample must also satisfy a particular coloring property which he denotes as Kempe-Locking one. However, the main intention of this paper is not an alternative proof of the 4-color theorem but an exploratory paper aimed at gaining a deeper understanding of why the 4-color theorem is true and a new approach to understand why planar graphs are 4-colorable by investigating whether the connectivity and coloring properties are compatible.

The third article by Zhang et al. considers so-called generalized hypergraphs H denoted as r-uniform if all the hyperedges have the same cardinality r. Such a graph is called a generalized hypertree GHT, if after removing any hyperedge E, $GHT - E$ has exactly k components with $2 \leq k \leq r$. Focusing first on the case $k = 2$, they determine bounds on the number of edges. In particular, the authors show that an r-uniform generalized GHT on n vertices has at least $2n/(r+1)$ edges and at most $n - r + 1$ edges if $r \geq 3, n \geq 3$ and that the lower and upper bounds on the edge number are tight. Finally, the case of a fixed value $k \leq r - 1$ is also discussed.

The next article by Wang et al. deals with the matrix $A_\alpha(G) = \alpha D(G)+(1-\alpha)A(G)$ with $\alpha \in [0, 1]$, introduced by Nikiforov, where $A(G)$ is the adjacent matrix and $D(G)$ is the resulting diagonal matrix of the degrees of a graph G. They determine the graphs with largest $A_\alpha(G)$-spectral radius with fixed vertex or edge connectivity. In addition, the corresponding extremal graphs are given and equations satisfying the $A_\alpha(G)$-spectral radius are derived.

In the fifth article, Jan et al. deal with fuzzy graphs. The goal of this paper is to show that there are some serious flaws in the existing definitions of several root-level generalized fuzzy graph structures with the help of some counterexamples. To achieve this, first, we aim to improve the existing definition for interval-valued fuzzy graphs, interval-valued intuitionistic fuzzy graphs and their complements. The authors also point out that a single-valued neutrosophic graph is not well defined in the literature by illustrative examples and present then a new definition and an application of such graphs in decision making.

The next paper by Wang et al. deals with the k-rainbow domination number which is the minimum weight of a k-rainbow dominating function. In particular, they determine this domination number of the grid graph $P_3 \square P_n$ for $k \in \{2, 3, 4\}$ and all n, where P_m is a path of order m.

Ye et al. consider the Cartesian product of directed circles C_m and C_n of length m and n, respectively ($n \geq m \geq 3$). In this paper, the authors extend the known results from the literature for m up to 21. They also give the exact values of the domination numbers for n up to 31.

Then Xu et al. deal with a particular topological index, namely with the distance degree index introduced by Dobrynin and Kochetova. Topological indices can be used e.g., for predicting physical, chemical, or pharmaceutical properties of organic molecules and chemical compounds. The authors derive expressions for the distance degree index for a variety of graphs, namely for a line graph, a subdivision graph, a vertex-semitotal graph, an edge-semitotal graph, a total graph, and a paraline graph.

The next article by Liu et al. considers the normalized Laplacian which plays an important role when studying the structural properties of non-regular networks. They determine the normalized Laplacian spectrum of a linear heptagonal network by a decomposition theorem for the normalized Laplacian matrix and elementary operations. In addition, the authors derive explicit formulas for the degree-Kirchhoff index and the number of spanning trees with respect to a linear heptagonal network. Here the authors use the relationships between the roots and coefficients.

In the next article, Wang et al. consider another distance-based topological index, namely the Padmakar-Ivan (PI) index. They obtain results for this index from trees to recursively clustered trees, the so-called k-trees. Moreover, tight upper bounds of such indices for k-trees are obtained by recursive relationships, and also the corresponding extremal graphs are given. In addition, the PI values of some classes of k-trees are derived and compared.

Then Liu et al. deal with several topological indices. In particular, they derive expressions for reformulated Zagreb indices of some derived graphs, such as the complement graph, the line graph, the subdivision graph, the edge-semitotal graph, the vertex-semitotal graph, the total graph and the paraline graph of a graph.

In another article, Liu et al. use the edge isoperimetric problem to determine the exact wirelengths of embedding an enhanced hypercube into windmill and necklace graphs by partitioning the edge set of the host graph. The results obtained in this paper may have a great impact on parallel computing systems.

Yang et al. consider the subtree problem of so-called fan graphs, wheel graphs and also graphs obtained from partitioning wheel graphs under dynamic evolution. The enumeration of these subtree numbers is done through so-called subtree generation functions of graphs. In particular, they study extremal graphs, subtree fitting problems and subtree densitiy behaviors of the graphs under consideration.

Then Huang et al. deal with the idea of regularity in neutrosophic graph theory. They describe the utility of a regular neutrosophic graph and a bipartite neutrosophic graph to model an assignment problem, a road transport network, and a social network. Neutrosophic graphs are a useful concept to cope with uncertainty resulting from the inconsistent or indeterminate information in real-world problems. In particular, a regular neutrosophic graph, a star neutrosophic graph, a regular complete neutrosophic graph, a complete bipartite neutrosophic graph and a regular strong neutrosophic graph are introduced. The authors prove some properties of these graphs. Moreover, the concept of an m-highly irregular neutrosophic graph on cycle and path graphs is introduced. The definition of busy and free nodes in a regular neutrosophic graph is also presented. In addition, some properties of complement and isomorphic regular neutrosophic graphs are also given.

The article by Hui et al. derives necessary and sufficient conditions for the graph join of a cycle with m vertices and a path with n vertices to be induced matching-extendable and bipartite-matching extendable, respectively. A graph G is called induced matching extendable, if every induced matching in this graph is included in a perfect matching of G. Similarly, a graph is bipartite matching extendable if every bipartite matching is included in a perfect matching. The paper finishes with some suggestions for future work, e.g., to investigate the relationships between k-extendable and forbidden subgraphs of a graph.

In the next article, Falcon et al. derive some results on graph theory in the context of molecular processes occurring during the S-phase of a mitotic cell cycle. After presenting some basic concepts on genetics, genetic algebras, evolution algebras, graph theory, and isotopisms of algebras, they introduce a total-colored graph that can be associated with any given evolution algebra over a finite field. Finally, the existence of a faithful functor between both considered categories of evolution algebras and their total-colored graphs is shown.

The article by Carbollosa et al. introduces the f-index and the f-polynomial of a graph. Using this polynomial of several topological indices, they study relations e.g., of the inverse degree index, the generalized first Zagreb index, and sum lordeg indices. They obtain inequalities involving the f-polynomial of many graph operations including the corona product graph, the join graph, and line graph and the Mycielskian graph. This leads to new inequalities for the topological indices considered.

The article by De la Sen et al. considers so-called $(s-q)$-graphic contraction mappings in b-metric like spaces. Their approach is used to show that a Picard sequence is Cauchy in the context of a b-metric like space which generalizes known results from the literature. The obtained results are illustrated by some examples.

Then Marappan et al. deal with the asymptotic analysis of several evolutionary operators (mutations and crossovers) for finding the chromatic number of a graph which is the minimum number of colors necessary to color the vertices of a graph such that no adjacent vertices have the same color. The selection of an appropriate operator has a great influence on finding good bounds for the chromatic number as well as on the achievement of a faster convergence with a smaller population size. In addition, necessary and sufficient conditions for the global convergence of evolutionary algorithms have been derived. Finally, the stochastic convergence of some recently suggested evolutionary operators is investigated.

In the last article, Worawannotai et al. consider particular domination games. Such a game is played by two players, namely the Dominator and the Staller, which alternatively choose vertices until all vertices are dominated. They study a version of a domination game, where the set of chosen vertices is always independent. This game turns out to be a competition-independence game, which is played by a Diminisher and a Sweller, who want to construct a maximal independent set M: however, while the Diminisher tries to minimize $|M|$, the Sweller wishes to maximize $|M|$. In this paper, the authors check whether some well-known results for domination games also hold for such competition-independence games and describe a family of graphs for which many parameters are equal.

Frank Werner
Special Issue Editor

mathematics

MDPI

Review

Mixed Graph Colorings: A Historical Review

Yuri N. Sotskov

United Institute of Informatics Problems, National Academy of Sciences of Belarus, Surganova Street 6, 220012 Minsk, Belarus; sotskov48@mail.ru; Tel.: +375-17-284-2120

Received: 31 January 2020; Accepted: 2 March 2020; Published: 9 March 2020

Abstract: This paper presents a historical review and recent developments in mixed graph colorings in the light of scheduling problems with the makespan criterion. A mixed graph contains both a set of arcs and a set of edges. Two types of colorings of the vertices of the mixed graph and one coloring of the arcs and edges of the mixed graph have been considered in the literature. The unit-time scheduling problem with the makespan criterion may be interpreted as an optimal coloring of the vertices of a mixed graph, where the number of used colors is minimum. Complexity results for optimal colorings of the mixed graph are systematized. The published algorithms for finding optimal mixed graph colorings are briefly surveyed. Two new colorings of a mixed graph are introduced.

Keywords: mixed graph; vertex coloring; chromatic number; edge coloring; chromatic index; chromatic polynomial; unit-time scheduling; makespan criterion

---

1. Introduction

Let $G = (V, A, E)$ denote a finite mixed graph with a non-empty set $V = \{v_1, v_2, \dots, v_n\}$ of n vertices, a set A of (directed) arcs, and a set E of (undirected) edges. It is assumed that the mixed graph $G = (V, A, E)$ contains no multiple arcs, no multiple edges, and no loops. Arc $(v_i, v_j) \in A$ denotes the ordered pair of vertices $v_i \in V$ and $v_j \in V$. Edge $[v_p, v_q] \in E$ denotes the unordered pair of vertices $v_p \in V$ and $v_q \in V$. If $A = \emptyset$, we have a graph $G = (V, \emptyset, E)$. If $E = \emptyset$, we have a digraph $G = (V, A, \emptyset)$. In 1976 [1], a mixed graph coloring was introduced for the first time as follows.

Definition 1. *An integer-valued function $c : V \to \{1, 2, \dots, t\}$ is a coloring (called c-coloring) of the mixed graph $G = (V, A, E)$ if non-strict inequality*

$$c(v_i) \leq c(v_j) \tag{1}$$

holds for each arc $(v_i, v_j) \in A$, and $c(v_p) \neq c(v_q)$ for each edge $(v_p, v_q) \in E$. A c-coloring is optimal if it uses a minimum possible number $\chi(G)$ of different colors $c(v_i) \in \{1, 2, \dots, t\}$, such a minimum number $\chi(G)$ being called a chromatic number of the mixed graph G.

A mixed graph $G = (V, A, E)$ is t-colorable if there exists a c-coloring with t different colors for the mixed graph G. If $A = \emptyset$, then a c-coloring is a usual coloring of the vertices of the graph $G = (V, \emptyset, E)$. Finding an optimal coloring of a mixed graph $G = (V, A, E)$ is NP-hard even if $A = \emptyset$ [2]. It should be noted that paper [1] was published in Russian along with other papers [3–9] published before 1997. In 1997 [10], another mixed graph coloring (we call it a strict mixed graph coloring) was introduced as follows.

Definition 2. *An integer-valued function $c_< : V \to \{1, 2, \dots, t\}$ is a coloring (called $c_<$-coloring) of the mixed graph $G = (V, A, E)$ if strict inequality*

$$c_<(v_i) < c_<(v_j) \tag{2}$$

holds for each arc $(v_i, v_j) \in A$*, and* $c_<(v_p) \neq c_<(v_q)$ *for each edge* $(v_p, v_q) \in E$*. A* $c_<$*-coloring is optimal if it uses a minimum possible number* $\chi_<(G)$ *of different colors* $c_<(v_i) \in \{1, 2, \ldots, t\}$*, such a minimum number* $\chi_<(G)$ *being called a strict chromatic number of the mixed graph* G*.*

A mixed graph $G = (V, A, E)$ is $t_<$-colorable if there exists a $c_<$-coloring with t different colors for the mixed graph G.

Obviously, one can use a c-coloring (Definition 1) instead of a $c_<$-coloring (Definition 2) in a special case of the mixed graph $G = (V, A, E)$ such that the implication in Equation (3) holds for each arc $(v_i, v_j) \in A$

$$(v_i, v_j) \in A \Rightarrow [v_i, v_j] \in E \tag{3}$$

Remark 1. *A* $c_<$*-coloring of the mixed graph G is a special case of a c-coloring, if each inclusion* $(v_i, v_j) \in A$ *implies the inclusion* $[v_i, v_j] \in E$ *in the mixed graph* $G = (V, A, E)$ *to be colored.*

It is required to use more general c-colorings for some applications of mixed graph colorings in planning and scheduling. On the other hand, for some applications, it is sufficient to consider a special $c_<$-coloring. Therefore, we present the known results for c-colorings and $c_<$-colorings separately provided that the published result is not identical for both colorings of the vertices of a mixed graph.

In [11], a coloring of arcs and edges in the mixed graph $G = (V, A, E)$ was determined as follows.

It is required to color arcs A and edges E in the mixed graph $G = (V, A, E)$ *in such a way that any two adjacent edges in the graph* $(V, \emptyset, E)$ *get different colors, and for any two adjacent arcs* $(v_i, v_j) \in A$ *and* $(v_p, v_q) \in A$ *forming a path* (v_i, v_j, v_p, v_q) *in the digraph* $(V, A, \emptyset)$*, the color of arc* (v_i, v_j) *must be less than the color of arc* (v_p, v_q)*.*

Such a coloring of arcs and edges in the mixed graph $G = (V, A, E)$ can be treated as a $c_<$-coloring of a special mixed graph (called a mixed line graph) generated from the mixed graph G as follows.

Definition 3. *For a given mixed graph* $G = (V, A, E)$*, we determine its mixed line graph* $L(G) = (A \cup E, A_{A \cup E}, E_{A \cup E})$ *as a mixed graph having vertex set* $A \cup E$*, arcs* $(e_{ij}, e_{jk}) \in A_{A \cup E}$ *connecting all pairs of arcs* $e_{ij} := (v_i, v_j) \in A$ *and* $e_{jk} := (v_j, v_k) \in A$*, and edge set* $E_{A \cup E}$ *connecting all the remaining pairs of elements of the set* $A \cup E$*, which share at least one vertex of the set* V*.*

The coloring of arcs and edges in the mixed graph $G = (V, A, E)$ is a $c_<$-coloring of vertices in the mixed line graph $L(G) = (A \cup E, A_{A \cup E}, E_{A \cup E})$, and vice versa. Therefore, one can use the following definition for the $c_<$-coloring of arcs and edges in the mixed graph $G = (V, A, E)$ [11].

Definition 4. *Let an integer-valued function* $c_< : \{A \cup E\} \rightarrow \{1, 2, \ldots, t\}$ *be a* $c_<$*-coloring of the mixed line graph* $L(G) = (A \cup E, A_{A \cup E}, E_{A \cup E})$*, i.e., strict inequality*

$$c_<(e_{ij}) < c_<(e_{jk}) \tag{4}$$

holds for each arc $(e_{ij}, e_{jk}) \in A_{A \cup E}$*, and* $c_<(e_{pq}) \neq c_<(e_{qr})$ *for each edge* $[e_{pq}, e_{qr}] \in E_{A \cup E}$*. A* $c_<$*-coloring of the vertices of the mixed line graph* $L(G)$ *is called an edge coloring of the mixed graph* $G = (V, A, E)$*. An edge coloring is optimal if it uses a minimum possible number* $\chi'(G)$ *of different colors* $c_<(e_{ij}) \in \{1, 2, \ldots, t\}$*, such a minimum number* $\chi'(G)$ *being called a chromatic index of the mixed graph* G*.*

For each type of colorings, the following questions have to be studied.

(a) *Existence:* Does a coloring exist for the given mixed graph?
(b) *Optimization:* How should an optimal coloring of the given mixed graph be found?
(c) *Enumeration:* How should all colorings existing for the given mixed graph be constructed?

From an answer to Question (c), one can directly obtain answers to both Questions (a) and (b). However, in practice, it is possible to construct all colorings existing for the mixed graph $G = (V, A, E)$

only if the order $n = |V|$ of the mixed graph G is rather small. Otherwise, instead of Question (c), one can study the following questions.

(d) *Counting and Estimation:* How should a cardinality of the set of all colorings existing for the given mixed graph be determined (or estimated)?

The rest of this paper is organized as follows. The results published for the c-coloring of the mixed graph G are described in Section 2, where the following decision problem $C(G, p)$ is considered.

Problem ($C(G, p)$). *Given a mixed graph $G = (V, A, E)$ and an integer $p \geq 1$, find out whether the mixed graph G admits a c-coloring using at most p different colors $c(v_i)$.*

Section 3 contains the results published for the $c_<$-coloring of the mixed graph G with the following decision problem $C_<(G, p)$.

Problem ($C_<(G, p)$). *Given a mixed graph $G = (V, A, E)$ and an integer $p \geq 1$, find out whether the mixed graph G admits a $c_<$-coloring using at most p different colors $c_<(v_i)$.*

Three tables with the results published in the OR literature are presented in Section 4. In Section 5, we show how a unit-time scheduling problem with the makespan criterion may be interpreted as an optimal coloring of the mixed graph. Section 6 contains a few results published for the edge coloring of the mixed graph. In Section 7, we introduce new types of colorings of the mixed graphs. The paper is concluded in Section 8. Throughout the paper, we use the terminology from [12,13] for graph theory and that from [14,15] for scheduling theory.

2. Mixed Graph Colorings

In Sections 2 and 3, we present known results for two types of mixed graph colorings (c-coloring in Section 2 and $c_<$-coloring in Section 3) in the order of their publications without repetitions. If a result was first published in a weak form and then was published in a stronger form, we present both results in this survey with indicating years of their publications.

Remark 2. *If a "positive result" is proven for a c-coloring (e.g., a polynomial algorithm is derived), it remains correct for a $c_<$-coloring for a special mixed graph $G = (V, A, E)$, where the implication in Equation (3) holds for each arc $(v_i, v_j) \in A$ (see Remark 1). On the other hand, a "positive result" proven for a $c_<$-coloring may remain unproven (open) for a c-coloring. If NP-hardness is proven for $c_<$-colorings of some class of mixed graphs, then NP-hardness remains correct for c-colorings of the same class of mixed graphs.*

The following criterion for existing a c-coloring of the mixed graph is proven in [1].

Theorem 1. *A c-coloring of the mixed graph $G = (V, A, E)$ exists if and only if the digraph $(V, A, \emptyset)$ has no circuit containing some adjacent vertices in the graph $(V, \emptyset, E)$.*

In the proof of Theorem 1, it is shown how to construct a c-coloring of the mixed graph $G = (V, A, E)$ provided that such a coloring exists.

Let $f(G, t)$ denote a number of all different c-colorings with colors $c(v_i) \in \{1, 2, \ldots, t\}$. If $A = \emptyset$, then $f(G, t)$ is a chromatic polynomial of the graph $G = (V, \emptyset, E)$ [12,13,16]. If $E = \emptyset$, then $f(G, t)$ is a chromatic polynomial of the digraph $G = (V, A, \emptyset)$ [17]. In [1,18], it is shown that $f(G, t)$ is a chromatic polynomial of t for the mixed graph $G = (V, A, E)$ with $A \neq \emptyset \neq E$.

In the c-coloring of the t-colorable mixed graph $G = (V, A, E)$, all vertices on a circuit in the digraph $G = (V, A, \emptyset)$ must have the same color from set $\{1, 2, \ldots, t\}$. Let $_{\{v_i, v_j\}}G$ denote a mixed graph obtained from the mixed graph $G = (V, A, E)$ as a result of identifying vertices $v_i \in V$ and

$v_j \in V$ along with identifying multiple edges, multiple arcs, and deleting loops, if these multiple edges, arcs, or loops arise in the mixed graph obtained due to identifying vertices v_i and v_j in $G = (V, A, E)$.

The above vertex identification may be generalized on a set N of the vertex pairs $\{v_i, v_j\}$. Let $_NG = (_NV,_NA,_NE)$ denote a mixed graph obtained from the mixed graph $G = (V, A, E)$ as a result of successive identifying vertices v_i and v_j for each pair of vertices $\{v_i, v_j\} \in N$.

In [1], Lemma 1 and Theorems 2 and 3 have been proven.

Lemma 1. *If vertices v_i and v_j are not adjacent in the graph $(V, \emptyset, E)$, then*

$$f(G,t) = f((V, A, E \cup \{[v_i, v_j]\}), t) + f((_{\{v_i,v_j\}}G), t) \tag{5}$$

Theorem 2. *If $M \subseteq E$ and graph $(V, \emptyset, M)$ has no cycle, then*

$$f(G,t) = \sum(-1)^{n-|_NV|} f(_N(V, A, E \setminus M), t), \tag{6}$$

where the summation is realized for all subsets $N \subseteq M$ such that the graph $(V, \emptyset, N)$ has no chain connecting adjacent vertices in the graph $(V, \emptyset, E \setminus M)$.

Let $\Pi(V, A, E)$ denote a set of all circuit-free digraphs generated by the mixed graph $G = (V, A, E)$ as a result of substituting each edge $[v_i, v_j] \in E$ by one of the arcs, either (v_i, v_j) or (v_j, v_i). The cardinality of set $\Pi(V, A, E)$ is denoted by $\pi(V, A, E) = |\Pi(V, A, E)|$.

Theorem 3. *Let $E \cap M = \emptyset$ and the graph $(V, \emptyset, E \cup M)$ is complete. Then,*

$$f(G,t) = \sum \pi(_N(V, A, E \cup M)) \begin{pmatrix} t \\ |_NV| \end{pmatrix}, \tag{7}$$

where the summation is realized for all subsets $N \subseteq M$ such that labeled mixed graphs $_NG$ are different and there is no chain in the graph $(V, \emptyset, N)$ between vertices, which are adjacent in the graph $(V, \emptyset, E)$.

Using Theorem 3, the coefficient of t^n and that of t^{n-1} in the chromatic polynomial $f(G,t)$ for the mixed graph G have been calculated in [1]. It is also proven that the sum Σ of all coefficients of the chromatic polynomial $f(G,t)$ is equal to zero, if $E \neq \emptyset$ and $\Sigma = 1$, if $E = \emptyset$.

In [19], a reciprocity theorem for the chromatic polynomials $f(G,t)$ is established based on order polynomials of partially ordered sets due to giving interpretations of evaluations at negative integers.

In [20], it is shown that the chromatic polynomial $f(G,t)$ of any mixed graph $G = (V, A, E)$ can be reduced to a linear combination of the chromatic polynomials $f(G,t)$ of simpler mixed graphs G such as trees. The reciprocity theorem for chromatic polynomials $f(G,t)$ has been investigated from a standpoint of inside-put polytopes and partially ordered sets.

In [7], the recurrent functions were determined for calculating several lower bounds on the minimum number of colors used in the c-coloring of the mixed graph $G = (V, A, E)$. These bounds were used for calculating lower bounds on the chromatic number $\chi(G)$ [7,8]. Several lower and upper bounds on the chromatic number $\chi(G)$ have been proven in [21]. Some of these bounds are tight.

Different algorithms for mixed graph colorings were developed and tested in [22–29]. In [25], a branch-and-bound algorithm was developed for calculating the chromatic number $\chi(G)$ and the strict chromatic number $\chi_<(G)$. This algorithm is based on the conflict resolution strategy with adding appropriate arcs to the mixed graph $G = (V, A, E)$ in order to resolve essential conflicts of the vertices, which may be colored by the same color. Computational results for randomly generated mixed graphs of the orders $n \leq 150$ showed that the developed algorithm outperforms the branch-and-bound algorithm described in [10] in cases of sufficiently large values of the strict chromatic numbers $\chi_<(G)$.

In [18], it is shown that a large class of scheduling problems induce mixed graph collorings (either c-colorings or $c_<$-colorings). Three algorithms for mixed graph colorings were coded in FORTRAN and tested on PC 486 for coloring randomly generated mixed graphs with the orders $n \leq 100$.

The algorithms proposed in [26,27] were modified in [29] in order to restrict the computer memory used in the branch-and-bound algorithm. The reported computational results on the benchmark instances showed that the modified algorithms are more efficient in terms of the number of optimal colorings constructed and sizes of the search trees.

The degree of vertex $v_i \in V$, denoted by $d_G(v_i)$, is the number of edges and arcs incident to vertex v_i. In [21], it is shown how to find the chromatic numbers $\chi(G)$ and optimal c-colorings for the following simple classes of mixed graphs.

Theorem 4. *Let $G = (V, A, E)$ be a mixed tree, where $E \neq \emptyset$. Then, $\chi(G) = 2$.*

Theorem 5. *Let $G = (V, A, E)$ be a chordless mixed cycle. Then, $\chi(G) = 2$.*

In [21], it is shown that the decision problem $C(G, p)$ with a fixed integer p may be polynomially solved for the following two classes of mixed graphs.

Theorem 6. *The problem $C(G, p)$ is polynomially solvable if $G = (V, A, E)$ is a partial mixed k-tree for a fixed integer k.*

Theorem 7. *The problem $C(G, 2)$ is polynomially solvable.*

In the proof of Theorem 7, it is shown that the problem $C(G, 2)$ may be (polynomially) reduced to the following k-satisfiability problem k-SAT with $k = 2$ that is known to be polynomially solvable [2].

Problem (k-SAT). *Given a set U of Boolean variables and a collection C of clauses over U, each clause containing $k \geq 1$ Boolean variables, find out whether there is a truth assignment to the Boolean variables that satisfies all clauses in C.*

The following complexity results (NP-completeness) for c-colorings have been proven in [21].

Theorem 8. *The decision problem $C(G, 3)$ is NP-complete even if $G = (V, A, E)$ is a planar bipartite mixed graph with the maximum degree 4.*

In the proof of Theorem 8, it is shown that the NP-complete decision problem $C_<(G, 3)$ is polynomially reduced to the decision problem $C(G, 3)$. In Section 3, we present Theorem 17 claiming that the decision problem $C_<(G, 3)$ is NP-complete if $G = (V, A, E)$ is a planar bipartite mixed graph with the maximum degree equal to 3.

Theorem 9. *The decision problem $C(G, 3)$ is NP-complete even if $G = (V, A, E)$ is a bipartite mixed graph with the maximum degree 3.*

In the proof of Theorems 9, it is shown that the problem $C_<(G, 3)$ is polynomially reduced to the problem $C(G, 3)$. In [30], it is proven that the problem $C_<(G, 3)$ is NP-complete if $G = (V, A, E)$ is a bipartite mixed graph with the maximum degree 3 (see Theorem 18 in Section 3).

The following claim is proven in [31].

Theorem 10. *The decision problem $C(G, 3)$ is NP-complete even if $G = (V, A, E)$ is a cubic planar bipartite mixed graph.*

In the proof of Theorem 10, it is shown that the problem $C_<(G,3)$ is polynomially reduced to the problem $C(G,3)$. In Section 3, Theorem 20 is presented, where it is established that the problem $C(G,3)$ is NP-complete if $G = (V,A,E)$ is a cubic planar bipartite mixed graph.

The above NP-completeness result is the best possible. Indeed, the problem $C(G,2)$ is polynomially solvable. Furthermore, a mixed graph having the maximum degree 2 consists of a family of disjoint mixed chains and mixed cycles. In [21], it is proven that an optimal c-coloring of a mixed cycle can be constructed in polynomial time. An optimal c-coloring of a mixed chain is trivial.

3. Strict Mixed Graph Colorings

In this section, we consider $c_<$-colorings of the mixed graph $G = (V,A,E)$. Due to Remark 1, Theorem 1 may be rewritten for a strict mixed graph coloring as follows.

Theorem 11. *A $c_<$-coloring for the mixed graph $G = (V,A,E)$ exists if and only if the digraph $(V,A,\emptyset)$ has no circuit.*

Algorithms for calculating and estimating the value of $\pi(V,A,E)$ used in the equality in Equation (7) and algorithms for constructing set $\Pi(V,A,E)$ of the circuit-free digraphs generated by the mixed graph $G(V,A,E)$ are described in [4,5], where the following claim is proven.

Lemma 2. *If vertices v_i and v_j are not adjacent in the graph $(V,\emptyset,E)$, then*

$$\pi(G) = \pi(V,A,E \cup [v_i,v_j]) - \pi({}_{v_i,v_j}G) \tag{8}$$

Using Lemma 2 and numbering $E = \bigcup_{k=1}^{|E|}[v_i,v_j]$ of the edges, the following equality is obtained:

$$\pi(G) = \pi(V,A,\emptyset) + \sum_{[v_i,v_j]_r \in E} \pi({}_{v_i,v_j}(V,A,E \setminus \bigcup_{m=1}^{r-1}[v_p,v_q]_m) \tag{9}$$

The value of $\pi(V.\emptyset,E)$ was investigated in [17]. The formulas analogous to Equations (5)–(7) presented in Section 2 for the value of $f(G,t)$ were proven for the value of $\pi(G)$ in [5].

The following claim has been proven in [10].

Theorem 12. *If mixed graph $G = (V,A,E)$ is a nontrivial mixed tree, then an optimal $c_<$-coloring for the mixed graph G may be constructed in $O(n^2)$ time.*

The result of Theorem 12 was strengthened in [32], where it was proven that an optimal $c_<$-coloring of the nontrivial mixed graph G may be constructed in $O(n)$ time.

Let V_o denote a set of vertices, $V_o \subseteq V$, which are incident to at least one arc in the mixed graph $G = (V,A,E)$. We denote by $G(V_o)$ the mixed subgraph of the mixed graph G induced by the vertex set V_o in the digraph $G^o = (V_o,A,\emptyset)$. Let $n(G^o)$ denote a number of vertices on the longest path in the mixed graph $G(V_o)$. Notice that the length of a longest path in the mixed graph $G(V_o)$ is equal to $n(G^o) - 1$. The in-rank of vertex $v_i \in V$, denoted by $in(v_i)$, is the length of a longest path in the digraph $G^o = (V_o,A,\emptyset)$ ending at vertex v_i. The out-rank of vertex $v_i \in V$, denoted by $out(v_i)$, is the length of a longest path in the digraph $G^o = (V_o,A,\emptyset)$ starting at vertex v_i. If vertex v_i is not incident to any arc from set A, then $in(v_i) = 0 = out(v_i)$. The length of a longest path in the mixed graph G is equal to $\max_{v_i \in V}\{in(v_i) + out(v_i)\}$. The above parameters can be determined if digraph $G^o = (V_o,A,\emptyset)$ has no circuit. Let $n(P)$ denote a number of vertices on path P in the digraph $(V,A,\emptyset)$.

The following two claims have been proven in [30].

Theorem 13. *Let mixed graph $G = (V,A,E)$ have the following properties:*

(1) *For each vertex $v_i \in V$, there exists a vertex $v_j \in V$ such that $(v_i, v_j) \in A$ or $(x_j, x_i) \in A$.*
(2) *For each maximal path P in the digraph $(V, A, \emptyset)$, either $n(P) = \chi_<(G^o)$ or $n(P) = \chi_<(G^o) - 1$.*

Then, deciding whether equality $\chi_<(G) = \chi_<(G^o)$ holds or inequality $\chi_<(G) > \chi_<(G^o)$ holds can be done in polynomial time.

The proof of Theorem 13 is based on transformation of the considered problem into the 2-SAT problem. Corollary 1 follows from the proof of Theorem 13.

Corollary 1. *Let mixed graph $G = (V, A, E)$ have the following properties:*

(1) *For each vertex $v_i \in V$, there exists a vertex $v_j \in V$ such that $(v_i, v_j) \in A$ or $(x_j, x_i) \in A$.*
(2) $\chi_<(G^o) \leq \chi_<(G) \leq \chi_<(G^o) + 1$.
(3) *For each maximal path P in the digraph $(V, A, \emptyset)$, either $n(P) = \chi_<(G^o)$ or $n(P) = \chi_<(G^o) - 1$.*

Then, the strict chromatic number $\chi_<(G)$ can be determined in polynomial time.

Several upper bounds on the strict chromatic number $\chi_<(G)$ have been proven in [10], where it was shown that inequalities

$$n(G^o) \leq \chi_<(G) \leq n(G^o) + 1 \tag{10}$$

hold for the bipartite mixed graph $G = (V_1 \cup V_2, A, E)$, where $V = V_1 \cup V_2$ and neither set V_1 nor set V_2 has adjacent vertices in the mixed graph $G = (V_1 \cup V_2, A, E)$.

The following decision problem (called a precoloring extension) was used for proving several results for calculating a value of the strict chromatic number $\chi_<(G)$.

Problem (PrExt(G, q)). *Given an integer $q \geq 1$ and a graph $G = (V, \emptyset, E)$ some of whose vertices are colored using at most q colors, find out whether this coloring of the subset of vertices can be extended to a coloring of all vertices of the graph G using at most q colors.*

In [33], it was proven that the problem PrExt$(G, 2)$ is polynomially solvable for a bipartite graph $G = (V, \emptyset, E)$, for a split graph G, and for a complement G of the bipartite graph. In [34], it was proven that the problem PrExt(G, q) is polynomially solvable for a cograph $G = (V, \emptyset, E)$. In [35], it was proven that the problem PrExt(G, q) is polynomially solvable for the graph $G = (V, \emptyset, E)$ with maximum degree 3. In [21], the following claim was proven.

Theorem 14. *The decision problem $C_<(G, n(G^o))$ is polynomially solvable if:*

(1) *every vertex in the digraph $G^o = (V_o, A, \emptyset)$ is on a path of length $n(G^o) - 1$; and*
(2) *the problem PrExt$(G^*, n(G^o))$ is polynomially solvable, where the graph G^* is obtained by transforming each arc in the mixed graph G into an edge, which is incident to the same vertices.*

In [30], it is shown that the decision problem $C_<(G, 3)$ is polynomially reduced to the decision problem PrExt$(G, 2)$, i.e., the following claim is proven.

Theorem 15. *The problem $C_<(G, 2)$ is polynomially solvable if $G = (V, A, E)$ is a bipartite mixed graph.*

Polynomial algorithms were developed for the class of k-trees defined recursively as follows.

A k-tree on k vertices consists of a clique on k vertices (called a k-clique). Given a k-tree T_n on n vertices, one can construct a k-tree on $k + 1$ vertices by adjoining a new vertex v_{n+1} to the k-tree T_n, which is made adjacent to each vertex of some k-clique existing in the k-tree T_n and nonadjacent to all the remaining $n - k$ vertices in this k-tree. The mixed graph G is called a partial k-tree if G is a subgraph of a k-tree. The following theorem is proven in [30].

Theorem 16. *The decision problem $C_<(G, p)$ is polynomially solvable if:*

(1) *$G = (V, A, E)$ is a bipartite partial mixed k-tree, where k is fixed; and*
(2) *for each maximal path p in the mixed graph G, either equality $n(P) = p$ or equality $n(P) = p - 1$ holds.*

In the proof of Theorem 16, it is shown that the considered problem is polynomially reduced to the problem PrExt$(G, 2)$. From Theorem 16, the following claim is obtained.

Corollary 2. *The decision problem $C_<(G, 3)$ is polynomially solvable if $G = (V, A, E)$ is a bipartite partial mixed k-tree, where k is fixed.*

In [36], it is proven that the problem PrExt$(G, 3)$ is NP-complete for a planar bipartite graph $G = (V, \emptyset, E)$. Based on the reduction of the NP-complete problem PrExt$(G, 3)$ to the decision problem $C_<(G, 3)$ with a planar mixed graph $G = (V, A, E)$, the following claim is proven in [30].

Theorem 17. *The decision problem $C_<(G, 3)$ is NP-complete even if $G = (V, A, E)$ is a planar bipartite mixed graph.*

The following claim is also proven in [30].

Theorem 18. *The decision problem $C_<(G, 3)$ is NP-complete even if $G = (V, A, E)$ is a bipartite mixed graph with the maximum degree 3.*

In the proof of Theorem 18, it is shown that the problem 3-SAT, which is NP-complete [37], is polynomially reduced to the problem $C(G, 3)$. In [21], Theorem 17 was strengthened as follows.

Theorem 19. *The decision problem $C_<(G, 3)$ is NP-complete even if:*

(1) *$G = (V, A, E)$ is a planar bipartite mixed graph with a maximum degree 4; and*
(2) *each vertex $v_i \in V$, which is incident to an arc, has a maximum degree 2.*

The following claim is proven in [31].

Theorem 20. *The decision problem $C_<(G, 3)$ is NP-complete if $G = (V, A, E)$ is a cubic planar bipartite mixed graph.*

In the proof of Theorem 20, it is shown that the following problem $LiCol(G)$ is polynomially reduced to the decision problem $C(G, 3)$ considered in Theorem 20.

Problem ($LiCol(G)$)**.** *Given a graph $G = (V, \emptyset, E)$ together with sets of feasible colors $L(v_i)$ for all vertices $v_i \in V$, find out whether the graph $G = (V, \emptyset, E)$ admits a vertex coloring (i.e., adjacent vertices get different colors) such that every vertex $v_i \in V$ is colored with a feasible color from the given set $L(v_i)$.*

In [35], it is proven that the problem $LiCol(G)$ is NP-complete if the total number of available colors is equal to 3, $|L(v_i)| = 3$, and if graph $G = (V, \emptyset, E)$ is a cubic planar bipartite mixed graph.

It should be noted that the NP-completeness proven in Theorem 20 is best possible in the sense that the problem $C_<(G, 3)$ for the mixed graph $G = (V, A, E)$ having a maximum degree 2 and the problem $C_<(G, 2)$ are both polynomially solvable. In [21], Lemma 3 was proven in order to establish several upper bounds on the strict chromatic number $\chi_<(G)$.

Lemma 3. *Let $G^o = (V_1 \cup V_2, A, \emptyset)$ be a bipartite digraph. Assume that all paths of length $n(G^o) - 1$ start at the same vertex set, say V_1. Then, it is possible to find a $c_<$-coloring of the digraph G^o with the number of colors $n(G^o)$ such that all vertices in set V_1 have even colors, and all vertices in set V_2 have odd colors.*

Using Lemma 3, the following two theorems have been proven in [21].

Theorem 21. *Let $G = (V_1 \cup V_2, A, E)$ be a bipartite mixed graph. Assume that all paths of length $n(G^o) - 1$ start at the same vertex set, say V_1. Then, it is possible to find a $c_<$-coloring of mixed graph G^o with the number of colors $n(G^o)$ such that all vertices in set V_1 have even colors, and all vertices in set V_2 have odd colors.*

Theorem 22. *Let $G = (V_1 \cup V_2, A, E)$ be a bipartite mixed graph. Then, equality $\chi_<(G) = n(G^o)$ holds if and only if all paths of length $n(G^o) - 1$ start in the same vertex set V_i, where $i \in \{1, 2\}$.*

Theorem 23. *Let $G = (V, A, E)$ be a mixed graph such that non-strict inequality $\chi_<(G(V_o)) \leq n(G^o) + 1$ holds for the mixed graph $G(V_o)$. Suppose also that the following inequality holds:*

$$\max_{G' \subseteq G} \{ \min_{v_i \in G'} d_{G'}(v_i) \} \leq n(G^o), \tag{11}$$

where G' is a subgraph of the mixed graph G containing vertex set V_o. Then, the non-strict inequality $\chi_< G) \leq n(G^o) + 1$ holds.

The following two claims follow from Theorem 23 and the inequalities in Equation (10).

Corollary 3. *Let $G = (V, A, E)$ be a mixed graph such that $G(V_o)$ is a bipartite mixed graph and the inequality in Equation (11) holds for the subgraph G' of the mixed graph G containing vertex set V_o. Then, non-strict inequality $\chi_<(G) \leq n(G^o) + 1$ holds.*

Corollary 4. *Let $G = (V, A, E)$ be a mixed graph such that odd cycle C in mixed graph G contains at least one vertex, which is not incident to any arc, and the inequality in Equation (11) holds for the subgraph G' of the mixed graph G containing vertex set V_o. Then, the non-strict inequality $\chi_< G) \leq n(G^o) + 1$ holds.*

In [38], it is shown that the problem $C_<(G, p)$ is polynomially solvable for a series parallel mixed graph G. In other words, an exact algorithm for solving the problem $C_<(G, p)$ is developed, where G is a partial 2-tree. The complexity of this algorithm is $O(n^{3.376} \log n)$.

In [21], the following Theorem 24 has been proven, strengthening Theorem 16 proven in [31].

Theorem 24. *The decision problem $C_<(G, p)$ is polynomially solvable if $G = (V, A, E)$ is a bipartite partial mixed k-tree, where k is fixed.*

In the proof of Theorem 24, it is shown that the problem $C_<(G, p)$, where $G = (V, A, E)$ is a bipartite partial mixed k-tree with a fixed k, is polynomially reduced to the problem $LiCol(G)$, which is known to be solvable in $O(n^{k+2})$ time for a partial k-tree [34]. The exact algorithm developed for solving the problem $C_<(G, p)$, where $G = (V, A, E)$ is a bipartite partial mixed k-tree with a fixed k, has the complexity $O(n^{2k+4}|A|^{k+2})$.

Algorithms for constructing an optimal $c_<$-coloring of the mixed graph $G = (V, A, E)$ have been derived in [39,40]. In [40], it is shown that an optimal $c_<$-coloring may be constructed for the mixed graph $G = (V, A, E)$ in $O(|E|^2 \cdot 2^{|E|} + |E| \cdot |A|)$ time provided that inequality $|E| < |A|$ holds.

The following polynomially solvable case for an optimal $c_<$-coloring was discovered in [40].

Theorem 25. *The decision problem $C_<(G, p)$ is solvable in $O(|A| \cdot |E|)$ time if edge set E of the mixed graph $G = (V, A, E)$ has no redundant edges and form a clique $(V, \emptyset, E)$.*

In [28], an algorithm based on the mixed integer linear programming and a tabu search algorithm were developed for constricting heuristic $c_<$-colorings of the mixed graph $G = (V, A, E)$ and calculating upper bounds on the strict chromatic number $\chi_<(G)$. The performances of the proposed heuristic algorithms were evaluated through several benchmark instances. It was shown that the developed tabu search algorithm outperforms the mixed integer linear programming algorithm.

4. Tables with Results Published on Vertex Colorings of the Mixed Graphs

This section contains three tables of the results published on c-colorings of the vertices of mixed graphs (Tables 1 and 3) and on $c_<$-colorings of the vertices of mixed graphs (Tables 2 and 3).

In Table 2, property (N) indicated in Column 3 is described in the corresponding theorem (corollary) indicated in Column 5 on the same line of Table 2. The sense of other columns are determined in their titles in Tables 1–3.

Table 1. The complexity of optimal c-colorings of the mixed graphs.

	Decision Problem	$\chi(G)$	Properties of the Mixed Graph G	Complexity Status	Theorem	Reference	Year
1	$C(G,p)$	$\chi(G)=2$	G is a mixed tree	$O(1)$	Theorem 4	[21]	2008
2	$C(G,p)$	$\chi(G)=2$	G is a chordless mixed cycle	$O(1)$	Theorem 5	[21]	2008
3	$C(G,p)$	$\chi(G)$	G is a partial mixed k-tree (k is fixed)	$\mathcal{P}$	Theorem 6	[21]	2008
4	$C(G,2)$	$\chi(G)=2$		$\mathcal{P}$	Theorem 7	[21]	2008
5	$C(G,3)$	$\chi(G)=3$	G is a bipartite mixed graph with maximum degree 3	NP-complete	Theorem 8	[30]	2007
6	$C(G,3)$	$\chi(G)=3$	G is a planar bipartite mixed graph with maximum degree 4	NP-complete	Theorem 9	[21]	2008
7	$C(G,3)$	$\chi(G)=3$	G is a cubic planar bipartite mixed graph	NP-complete	Theorem 10	[31]	2010

Table 2. The complexity of optimal $c_<$-colorings of the mixed graphs.

	Decision Problem	$\chi_<(G)$	Properties of the Mixed Graph G	Complexity Status	Theorem, Corollary	Reference	Year
	1	2	3	4	5	6	7
1	$C_<(G,p)$	$\chi_<(G)=\chi_<(G^0)$	Properties (1) and (2)	$\mathcal{P}$	Theorem 13	[30]	2007
2	$C_<(G,p)$	$\chi_<(G)$	Properties (1)–(3)	$\mathcal{P}$	Corollary 1	[30]	2007
3	$C_<(G,n(G^0))$	$\chi_<(G)=n(G^0)$	Properties (1) and (2)	$\mathcal{P}$	Theorem 14	[21]	2008
4	$C_<(G,2)$	$\chi_<(G)=2$	G is a bipartite mixed graph	$\mathcal{P}$	Theorem 15	[30]	2007
5	$C_<(G,p)$	$\chi_<(G)$	Properties (1) and (2)	$\mathcal{P}$	Theorem 16	[30]	2007
6	$C_<(G,p)$	$\chi_<(G)$	Properties (1) and (2)	$O(n^{2k+4}\|A\|^{k+2})$	Theorem 24	[21]	2008
7	$C_<(G,3)$	$\chi_<(G)=3$	G is a bipartite partial mixed k-tree (k is fixed)	$\mathcal{P}$	Corollary 2	[30]	2007
8	$C_<(G,p)$	$\chi_<(G)=3$	G is a series parallel mixed graph	$O(n^{3.376}\log n)$		[38]	2008
9	$C_<(G,3)$	$\chi_<(G)=3$	G is a planar bipartite mixed graph	NP-complete	Theorem 17	[30]	2007
10	$C_<(G,3)$	$\chi_<(G)=3$	G is a bipartite mixed graph with maximum degree 3	NP-complete	Theorem 18	[30]	2007
11	$C_<(G,3)$	$\chi_<(G)=3$	Properties (1) and (2)	NP-complete	Theorem 19	[21]	2008
12	$C_<(G,3)$	$\chi_<(G)=3$	G is a cubic planar bipartite mixed graph	NP-complete	Theorem 20	[31]	2010

Table 3. Existence, estimation and enumeration of c-colorings and $c_<$-colorings of the mixed graphs.

	Problems and Notations	Formulas	Sections	Theorems, Corollaries, Lemmas	References	Years
	1	2	3	4	5	6
1	Existence of c-coloring		Section 2	Theorem 1	[1]	1976
2	Existence of $c_<$-coloring		Section 3	Theorem 11	[1,10]	1976, 1997
3	Counting of c-colorings $f(G,t)$ $\pi(V,A,E)$	Equation (5) Equation (6), Equation (7)	Section 2 Section 2 Section 2	Lemma 1 Theorems 2 and 3	[1] [1] [1,3,5]	1976 1976 1970, 1976
4	Counting of $c_<$-colorings $f(G,t)$ $\pi(V,A,E)$	 Equation (8, 9)	Section 3 Section 3 Section 3	 Lemma 2	[17] [5] [3,4]	1973 1976 1970, 1974
5	Enumeration of c-colorings $\Pi(V,A,E)$		Section 2 Section 5		[5,41] [6,42]	1976, 1975 1980, 1981
6	Enumeration of $c_<$-colorings		Sections 2		[5,42] [41]	1976, 1980 1975
7	Estimation of the number of c-colorings		Sections 2		[5,42] [3,41]	1976, 1980 1970, 1975
8	Estimation of the number of $c_<$-colorings		Sections 2		[5,42] [41]	1976, 1980 1975
9	Calculation of the chromatic number $\chi(G)$		Section 2		[9] [27]	1993 2006
10	Calculation of the strict chromatic number $\chi_<(G)$		Section 3 Section 3 Section 2 Section 2 Sections 2 and 5		[10] [22–24] [18,25,26] [27] [28,29,39,40]	1997 2000 2001, 2002 2006 2014, 2017, 2019
11	Bounds on the chromatic number $\chi(G)$		Section 2 Section 2		[7,8] [14] [21]	1982 1994 2008
12	Bounds on the strict chromatic number $\chi_<(G)$	Equation (10)	Section 3 Section 3 Section 3 Section 3	 Theorem 13 Theorems 22 and 23 Corollaries 3, 4 	[10] [30] [21] [40]	1997 2007 2008 2019

5. Several Scheduling Problems as Mixed Graph Colorings

The minimization of the maximum completion time (i.e., the length of a schedule) of n partially ordered operations $V = \{v_1, v_2, \ldots, v_n\}$ with unit processing times $p_i = 1$, where $i \in \{1, 2, \ldots, n\}$, (or what means the same, with equal processing times) of operations V may be interpreted as an optimal $c_<$-coloring of the mixed graph $G = (V, A, E)$, where V is a set of operations, arc set A determines precedence constraints, and edge set E determines capacity constraints [14,41,43–45].

5.1. *Strict Mixed Graph Colorings and Job-Shop Scheduling Problems*

Due to Definition 2, a $c_<$-coloring $c_< : V \to \{1, 2, \ldots, t\}$ of a mixed graph G determines a feasible assignment of operations $V = \{v_1, v_2, \ldots, v_n\}$ to the following set of unit-time intervals:

$$[0,1], (1,2], (2,3], \ldots, (t-1,t]. \tag{12}$$

An optimal $c_<$-coloring $c_< : V \to \{1, 2, \ldots, \chi_<(G)\}$ of the mixed graph G determines an assignment of operations V to a minimum number of unit-time intervals:

$$[0,1], (1,2], (2,3], \ldots, (\chi_<(G)-1, \chi_<(G)]. \tag{13}$$

The assignment of operations V to the unit-time intervals (13) is makespan optimal, i.e., it determines an optimal schedule of operations V, the length of which is equal to the strict chromatic number $\chi_<(G)$. In this section, we consider an optimal $c_<$-coloring of the mixed graph G, which corresponds to a unit-time, minimum-length, job-shop scheduling problem denoted by $J|p_i = 1|C_{max}$

in the three-field classification $\alpha|\beta|\gamma$, where α specifies the machine environment, β specifies the job characteristics, and γ denotes the objective criterion [46]. An optimal $c_<$-coloring of the mixed graph G determines a makespan optimal schedule, i.e., $\gamma = C_{\max} := \max\{C_1, C_2, \ldots, C_j\}$, where C_i denotes the completion time of the job J_i. Using the graph terminology, we have to assume that the mixed graph $G = (V, A, E)$ under consideration in this section has the following two mandatory properties.

Property 1. *The partition $(V, \emptyset, E) = (V_1, \emptyset, E_1) \cup (V_2, \emptyset, E_2) \cup \ldots \cup (V_m, \emptyset, E_m)$ holds, where subgraph $(V_k, \emptyset, E_k)$ is a clique for each $k \in \{1, 2, \ldots, m\}$ and equality $V_k \cap V_l = \emptyset$ holds for $k \neq l$.*

Property 2. *The digraph $(V, A, \emptyset)$ has no transitive arcs and the partition $(V, A, \emptyset) = (V^{(1)}, A^{(1)}, \emptyset) \cup (V^{(2)}, A^{(2)}, \emptyset) \cup \ldots \cup (V^{(j)}, A^{(j)}, \emptyset)$ holds, where digraph $(V^{(k)}, A^{(k)}, \emptyset)$ is a path $(v_{k_1}, v_{k_2}, \ldots, v_{k_{r_k}})$ for each $k \in \{1, 2, \ldots, j\}$ and equality $V^{(k)} \cap V^{(l)} = \emptyset$ holds for $k \neq l$.*

Property 1 (Property 2, respectively) means that the subgraph $(V, \emptyset, E)$ of a mixed graph G is a union of disjoint cliques (the subgraph $(V, A, \emptyset)$ is a union of disjoint paths). In the job-shop scheduling problem $J||C_{\max}$, the number m and number j denote the cardinality of the machine set $M = \{M_1, M_2, \ldots, M_m\}$ and the cardinality of the job set $J = \{J_1, J_2, \ldots, J_j\}$, respectively.

From Property 2, it follows that, if inclusion $v_i \in V^{(k)}$ holds, then operation v_i is a part of the job $J_k \in J$, and vice versa (Definition 2). Each job $J_k \in J$ consists of a set $V^{(k)}$ of linearly ordered operations, i.e., job J_k is represented as a path $(v_{k_1}, v_{k_2}, \ldots, v_{k_{r_k}})$ in the digraph $(V, A, \emptyset)$. All operations $V^{(k)}$ have to be processed in the order determined by machine route or the path $(v_{k_1}, v_{k_2}, \ldots, v_{k_{r_k}})$.

From Property 1, it follows that, if inclusion $v_i \in V_k$ holds, then operation v_i has to be processed on machine $M_k \in M$. Due to Definition 2, Property 1 means that each machine $M_k \in M$ can process at most one operation within a unit-time interval from the set (12).

Properties 1 and 2 determine usual assumptions used in scheduling theory in terms of graph theory. There exists a one-to-one correspondence between all $c_<$-colorings of the mixed graph $G = (V, A, E)$ complying with Properties 1 and 2, and all semi-active schedules [14,15] existing for the problem $J|p_i = 1|C_{max}$. A schedule is called semi-active if no job (operation) can be processed earlier without changing the processing order or violating the given constraints.

We next present the correspondence of the used terms:

$$\{\text{vertex } v_i \in V\} \leftrightarrow \{\text{operation } v_i \in V\};$$

$$\{\text{set of vertices on the path } (V^{(k)}, A^{(k)}, \emptyset)\} \leftrightarrow \{\text{set of operations of the job } J_k \in J\};$$

$$\{\text{set of vertices of the clique } (V_k, \emptyset, E_k)\} \leftrightarrow \{\text{set of operations processed on machine } M_k \in M\};$$

$$\{c_<\text{-coloring of the mixed graph } G\} \leftrightarrow \{\text{semi-active schedule for the problem } J|p_i = 1|C_{\max}\};$$

$$\{\text{optimal } c_<\text{-coloring using } \chi_<(G) \text{ colors}\} \leftrightarrow \{\text{optimal schedule for the problem } J|p_i = 1|C_{\max}\}; \text{ and}$$

$$\{\text{strict chromatic number } \chi_<(G)\} \leftrightarrow \{\text{optimal value of the makespan } C_{\max}\}.$$

In [39], the scheduling problem $J|p_i = 1|C_{\max}$ is presented as finding an optimal $c_<$-coloring of a special mixed graph satisfying Properties 1 and 2. The lower and upper bounds on the strict chromatic number $\chi_<(G)$ have been proven. A tabu search algorithm using a dynamic neighborhood structure was adapted for solving large benchmark instances heuristically. Computational experiments were conducted to estimate the efficiency of the proposed algorithm.

Most results observed in this section have been proven in [23,24,26]. The complexity of an optimal $c_<$-coloring for special cases of mixed graphs follow from those for the problem $J|p_i = 1|C_{max}$. The corresponding references are given in Tables 4 and 5, where it is assumed that all mixed graphs $G = (V, A, E)$ satisfy both Properties 1 and 2. Along with the mandatory Properties 1 and 2, we consider Property 3, which means that any two sequential operations of the same job $J_k \in J$ in the problem $J|p_i = 1|C_{max}$ have to be processed on different machines from the set M, i.e., "machine repetition" in processing two sequential operations of the same job is not allowed. Taking into account that the digraph $(V, A, \emptyset)$ has no transitive arcs (Property 2), we present Property 3 as follows.

Property 3. *The subgraph $(V_k, A_k, \emptyset)$ of the digraph $(V, A, \emptyset)$ is empty for each $k \in \{1, 2, \dots, m\}$, i.e., equality $A_k = \emptyset$ holds.*

The notation $J|p_i = 1|C_{max}$ is used if the mixed graph G has Property 3. If machine repetition in processing a job is allowed, the notation $J|p_i = 1, rep|C_{max}$ is used. The problem $J|p_i = 1|C_{max}$ is a special case of the problem $J|p_i = 1, rep|C_{max}$, which is equivalent to the problem $J||p_i|, pmtn|C_{max}$ with integer processing times and allowed preemptions of an operation. Property 3 influences the complexity of a scheduling problem [47,48]. An example of such an influence was given in [49], where it was proven that the job-shop problem $J2|j = 3, p_i = 1, rep|C_{max}$ is NP-hard, while in [50] polynomial algorithms for the corresponding job-shop problem without machine repetition have been derived. Column 3 in Table 4 and Column 5 in Table 5 are used to indicate whether the mixed graph G has Property 3 (in this case, the column contains 'yes') or not (in this case, the column contains 'no').

In Table 4, we present complexity results for an optimal $c_<$-coloring of a mixed graph G when the strict chromatic number $\chi_<(G)$ is small. More precisely, the recognition of inequality $\chi_<(G) \leq l$ is considered with a fixed positive integer l (Column 2 in Table 4). Testing inequality $\chi_<(G) \leq 2$ is a trivial problem when either Property 3 holds or not. Indeed, equality $\chi_<(G) = 1$ holds if and only if $E = \emptyset$ and $A = \emptyset$. A simple criterion for equality $\chi_<(G) = 2$ is given in Lemma 4, as proven in [26].

Table 4. The complexity of a mixed graph coloring with short paths and small cliques.

	$\chi(G) \leq l$	Property 3	Complexity Status	References	Years
1	$l = 3$	yes	$O(n)$	[51,52] [26]	1976, 1997, 2001
2	$l = 3$	no	$O(n)$	[51,52] [26]	1976, 1997, 2001
3	$l = 4$	yes	NP-complete	[51]	1997

Lemma 4. *Equality $\chi_<(G) = 2$ holds if and only if*

(1) $|A| + |E| \geq 1$,
(2) $\max_{J_k \in J} |V^{(k)}| \leq 2$,
(3) $\max_{M_k \in M} |V_k| \leq 2$,
(4) *there are no two paths $(v_{k_1}, v_{k_{r_k}})$ and $(v_{s_1}, v_{s_{r_s}})$ such that $[v_{k_1}, v_{s_1}] \in E$ or $[v_{k_{r_k}}, v_{s_{r_s}}] \in E$.*

In [51], it is proven that the problem of deciding if there is an optimal schedule for the problem $J||p_i||C_{max}$ with a length of at most 3 can be reduced to the 2-SAT problem in $O(n)$ time. Since the problem $J|p_i = 1|C_{max}$ is a special case of the problem $J||p_i||C_{max}$ and taking into account that the 2-SAT problem can be solved in $O(n)$ time [52], we conclude that the recognition of inequality $\chi_<(G) \leq 3$ can be done in $O(n)$ time if Property 3 holds for the mixed graph G (Row 1 in Table 4).

Using the polynomial reduction from [51], it is shown in [26] that inequality $\chi_<(G) \leq 3$ holds if Property 3 does not hold (Row 2 in Table 4). Using the polynomial reduction similar to the one described in [51], it is shown in [26] that deciding if there is a schedule for the problem $J||p_i||C_{max}$ with length 3 can be reduced to the problem 2-SAT.

Obviously, for a $c_<$-coloring $c_< : V \to \{1, 2, 3\}$ (if any) only paths of the length of at most 3 and cliques of the cardinality of at most 3 are allowed. In [26], an $O(n)$-algorithm was developed based on the algorithm developed in [52] for solving the problem 2-SAT. It was shown that the logical formula constructed by $O(n)$-algorithm is satisfiable if and only if $\chi_<(G) = 3$. To test inequality $\chi_<(G) \leq 3$ when Property 3 does not hold takes $O(n)$ time (Row 2 in Table 4).

In [51], it is proven that deciding if there is an optimal schedule for the problem $J||[p_i]||C_{max}$ with a length of at most 4 is NP-complete. More precisely, a polynomial reduction was constructed from the restricted version of the 3-SAT problem (which is NP-complete) to the problem $J|p_i = 1|C_{max}$, which is a special case of the problem $J||[p_i]||C_{max}$ with integer processing times of operations (Row 3 in Table 4).

Since inequality $\chi_<(G) < l$ implies both inequalities $|V^{(k)}| < l$ and $|V_t| < l$, small values of the strict chromatic number $\chi_<(G)$ may be possible only for a mixed graph G with short paths $(V^{(k)}, A^{(k)}, \emptyset), k \in \{1, 2, \ldots, j\}$, and small cliques $(V_t, \emptyset, E_t), t \in \{1, 2, \ldots, m\}$.

As follows from Table 4, the boundary between polynomially solvable and NP-complete problems of testing inequality $\chi_<(G) \leq l$ is between $l = 3$ and $l = 4$.

The recognition of inequality $\chi_<(V, \emptyset, E) \leq 3$ is an NP-complete problem [2], while the recognition of inequality $\chi_<(V, \emptyset, E) \leq 2$ may be done in polynomial time since inequality $\chi_<(V, \emptyset, E) \leq 2$ holds if and only if the graph $(V, \emptyset, E)$ has no cycle with odd length.

In [53,54], it was proven that the problem $J2|p_i = 1, rep|C_{max}$ is NP-hard (Row 1 in Table 5). In [55], it was proven that the problem $J3|p_i = 1|C_{max}$ is NP-hard (Row 2 in Table 5). In [50,56], an $O(n)$-algorithm has been developed for the problem $J|p_i = 1|C_{max}$ (Row 3 in Table 5)).

We next observe the complexity of an optimal $c_<$-coloring of a mixed graph G with Properties 1 and 2. If $j = n$, then due to Property 2, we obtain $(V, A, \emptyset) = (\{v_1\}, \emptyset, \emptyset) \cup (\{v_2\}, \emptyset, \emptyset) \cup \ldots \cup (\{v_n\}, \emptyset, \emptyset)$, i.e., the set of arcs A is empty and $G = (V, \emptyset, E)$. Due to Property 1, the strict chromatic number $\chi_<(G)$ is equal to the maximum size of a clique in the graph $G = (V, \emptyset, E)$, i.e., $\chi_<(G) = \max_{k=1}^{m} |V_k|$.

If the input data include a list of adjacent vertices for each vertex $v_i \in V$, then the calculation of the strict chromatic number $\chi_<(G)$ takes $O(n)$ time. If the input data include the sets $V_1, V_2, \ldots, V_m$ of vertices, then the calculation of the strict chromatic number $\chi_<(G)$ takes $O(m)$ time.

If $m = n$ in Property 1, then $M = \{M_1, M_2, \ldots, M_n\}$ and each operation $v_i \in V$ has to be processed on a separate machine $M_t \in M$. Therefore, we have $G = (V, A, \emptyset)$ and the strict chromatic number $\chi_<(G)$ is equal to the maximum length $r = \max_{k=1}^{j} |V^{(k)}|$ of a path in the digraph $G = (V, A, \emptyset)$, i.e., $\chi_<(G) = r = \max_{k=1}^{j} |V^{(k)}|$.

If the input data include the sets $V^{(1)}, V^{(2)}, \ldots, V^{(j)}$ of vertices, then the calculation of the strict chromatic number $\chi_<(G)$ takes $O(j)$ time, otherwise it takes $O(n)$ time.

Summarizing, we conclude that, if $m = n$ or $j = n$, the strict chromatic number $\chi_<(G)$ can be found in $O(n)$ time (Rows 4 and 5 in Table 5).

For the case $j = 2$, there are polynomial algorithms based on the geometrical approach [57–59]. In [54,60], the geometric $O(r^2 \log r)$-algorithms were developed for the job-shop problem $J|j = 2|\Phi$ with two jobs, real processing times and any regular criterion Φ.

If all processing times p_i are integers, then the problem $J|j = 2, [p_i]|C_{max}$ is equivalent to the problem $J|j = 2, p_i = 1, rep|C_{max}$, in which p_i unit-time operations correspond to one operation with integer processing time equal to p_i in the problem $J|j = 2|C_{max}$. In [26], it is shown how to improve the geometrical $O(r^2 \log r)$-algorithm developed in [54,60] for the case of unit-time operations.

Table 5. The complexity of a mixed graph coloring with long paths or large cliques.

	Scheduling Problem	Number of Paths	Number of Cliques	Property 3	Complexity	References	Years
1	$J2\|p_i = 1, rep\|C_{max}$	j	2	no	NP-hard	[14,53,54]	1990, 1991, 1994
2	$J3\|p_i = 1\|C_{max}$	j	3	yes	NP-hard	[55]	1979
3	$J2\|p_i = 1\|C_{max}$	j	2	yes	$O(n)$	[50,56]	1982, 1985
4	$Jm\|j = n, p_i = 1\|C_{max}$	n	m	yes	$O(m)$ or $O(n)$	[26]	2001
5	$Jn\|p_i = 1\|C_{max}$	j	n	yes	$O(j)$ or $O(n)$	[26]	2001
6	$J\|j = k, p_i = 1\|C_{max}$	k	m	yes	$O(r^k)$	[26]	2001
7	$J\|j = 2, p_i = 1, rep\|C_{max}$	2	m	no	$O(r)$ or $O(r^2)$	[26,54,60]	1985, 1991, 2001
8	$J\|j = 2, p_i = 1\|C_{max}$	2	m	yes	$O(r)$	[26,54,60]	1985, 1991, 2001

5.2. Mixed Graph Colorings and General Shop Scheduling Problems

There are other applications of mixed graph colorings for solving real-life optimization and enumeration problems [20,61–67]. Some of these applications are described in this section.

In [20], it was shown that modeling of metabolic pathways in biology and a process management in operating systems may be based on mixed graph colorings. In particular, it was demonstrated how the chromatic polynomial $f(G, T)$ described in Section 2 may be used for solving these problems.

The following school timetabling problem was considered in [67]. One has to arrange unit-time intervals at which a set of lectures has to be given provided that lectures $v_i \in V$ and $v_j \in V$ cannot be held at some unit-time intervals since there may be students who wish to attend both of them. Such a restriction on the pair of lectures v_i and v_j may be given by edge $[v_i, v_j] \in E$ and this scheduling problem is equivalent to the coloring of the vertices of the graph $G = (V, \emptyset, E)$. There are often more similar restrictions generated by students and staff requirements, which have to be taken into consideration in finding a satisfactory timetable of the lectures. In [67], a coloring of the vertices of the graph $(V, \emptyset, E)$ is considered such that forbidden colors are given for some vertices from set V.

Note that an arc $(v_p, v_q) \in A$ may arise in the mixed graph $G = (V, A, E)$ to be colored if lecture v_p must proceed lecture v_q in the desired timetable of the lectures.

A class of so-called general shop scheduling problems was determined in [14,15], where a mixed (disjunctive) graph $G = (V, A, E)$ was used for presenting an input data for a general shop scheduling problem. Any semi-active schedule existing for the general shop scheduling problem may be determined by a specific digraph generated by the mixed graph G. Algorithms for enumerating semi-active schedules generated by the mixed graph G are developed in [42,68].

Several algorithms with different asymptotic complexities were developed in [42,61,62,66,69–71] for solving the general shop scheduling problems based on the mixed graph models.

The general shop scheduling problem $G||F_{max}$ to minimize a maximum penalty $F_{max} := \max_{J_k \in J} \phi_k(C_k)$ is investigated in [69]. The input data for the problem $G||F_{max}$ is presented by a weighted mixed graph $G_w = (V, A_w, E_w)$, where set V is a set of operations, each arc $(v_i, v_j) \in A_w$ has a weight $w_{ij} \geq 0$ and each edge $[v_p, v_q] \in E_w$ has a pair of weights $w_{pq} \geq 0$ and $w_{qp} \geq 0$.

There exists a one-to-one correspondence between a set of all semi-active schedules existing for the problem $G||F_{max}$ and a set $\Pi(V, A_w, E_w)$ of all circuit-free digraphs generated by the weighted mixed graph $G_w = (V, A_w, E_w)$ as a result of substituting each edge $[v_p, v_q] \in E$ by one of the weighted arc either arc (v_p, v_q) with weight w_{pq} or arc (v_q, v_p) with weight w_{pq}. Using the weighted mixed graph G_w, in [69], it was proven that a solution of the problem $G||F_{max}$ is reduced to solving several problems k-SAT. Due to this reduction, a polynomial algorithm is developed for a special case of the problem $G||F_{max}$ based on the $O(n)$-algorithm available for solving the problem 2-SAT [52].

We assume that equality $\phi_k(C_k) = C_k$ holds for each job $J_k \in J$, inclusion $w_{ij} \in \{0, 1\}$ holds for each arc $(v_i, v_j) \in A_w$, and both inclusions $w_{pq} \in \{0, 1\}$ and $w_{pq} \in \{0, 1\}$ hold for each edge

$[v_p, v_q] \in E_w$. It is clear that equalities $\phi_k(C_k) = C_k$, $J_k \in J$ imply that $F_{max} = C_{max}$ and this special case $G||C_{max}$ of the general shop scheduling problem $G||F_{max}$ is equivalent to the problem of finding an optimal c-coloring of the mixed graph $G = (V, A, E)$, which is obtained from the weighted mixed graph $G_w = (V, A_w, E_w)$ as follows. If $w_{ij} = 0$, then $(v_i, v_j) \in A$. If $w_{ij} = 1$, then $(v_i, v_j) \in A$ and $[v_i, v_j] \in E$. In other words, set A of arcs in the obtained mixed graph $G = (V, A, E)$ coincides with set A_w of the same arcs without weights. Set E of edges in the mixed graph $G = (V, A, E)$ is a union of set E_w of edges without weights and the set $\{[v_i, v_j] : (v_i, v_j) \in A_w, w_{ij} = 1\}$ of the edges generated by arcs $(v_i, v_j) \in A_w$ with $w_{ij} = 1$.

In [70], a more complicated general shop scheduling problem was considered. The input data for this general shop scheduling problem is presented by the weighted mixed multigraph $G^*_w = (V, A_w, E^*_w)$, where set V is a set of operations, each arc $(v_i, v_j) \in A_w$ has a weight w_{ij}, and each edge $[v_p, v_q]^k \in E^*_w$ has a pair of weights w^k_{pq} and w^k_{qp}. Note that weight w_{ij} prescribed to arc $(v_i, v_j) \in A_w$, and weights w^k_{pq} and w^k_{qp} prescribed to edge $[v_p, v_q]^k \in E^*_w$ may be arbitrary real numbers, a negative weight of an arc or edge being also possible. The mixed multigraph $G^*_w = (V, A_w, E^*_w)$ contains no multiple arcs and no loops, while multigraph G^*_w may contain multiple edges $[v_p, v_q]^k \in E^*_w$.

Let $\Pi(V, A_w, E^*_w)$ denote a set of all digraphs generated by the mixed multigraph $G^*_w = (V, A_w, E^*_w)$ as a result of substituting each edge $[v_i, v_j]^k \in E^*_w$ by one of the weighted arc either arc $(v_p, v_q)^k$ with weight w^k_{pq} or arc $(v_q, v_p)^k$ with weight w^k_{pq} along with successive deleting all multiple arcs except a single arc (v_p, v_q) with the largest weight incident to the same ordered vertices v_p, v_q. In [44], the following theorem was proven.

Theorem 26. *Let equality $E^*_w = \emptyset$ hold. Then, a schedule admissible for the weighted digraph $G^*_w = (V, A_w, \emptyset)$ exists if and only if the weighted digraph $G^*_w = (V, A_w, \emptyset)$ has no circuit with a strictly positive weight.*

We next consider a weighted mixed multigraph $G^*_w = (V, A_w, E^*_w)$, where $A_w \neq \emptyset$ and $E^*_w \neq \emptyset$. Let H denote a set of all circuits in the weighted digraph $G^*_w = (V, A_w, \emptyset)$. Let H^* denote a subset of set H containing all circuits with strictly positive weights. The following theorem was proven in [70].

Theorem 27. *For existing a schedule admissible for the weighted multigraph $G^*_w = (V, A_w, E^*_w)$, it is necessary that $H^* = \emptyset$ and sufficient that $H = \emptyset$.*

In [71], the following general shop scheduling problem $G||C_{max}$ was considered. The input data for this general shop scheduling problem is presented by the weighted mixed graph $G_w = (V, A_w, E_w)$. Let $\Pi(V, A_w, E_w)$ denote a set of all digraphs generated by the weighted mixed graph $G_w = (V, A_w, E_w)$ as a result of substituting each edge $[v_p, v_q] \in E$ by one of the weighted arc either arc (v_p, v_q) with weight w_{pq} or arc (v_q, v_p) with weight w_{pq}. Let $\Pi^*(V, A_w, E_w)$ denote a subset of set $\Pi(V, A_w, E_w)$ containing all digraphs without circuits with strictly positive weights. There exists a one-to-one correspondence between all semi-active schedules existing for the problem $G||C_{max}$ and all digraphs from set $\Pi^*(V, A_w, E_w)$. In [71], the following theorem was proven.

Theorem 28. *If $H^* = \emptyset$, a problem of testing equality $\Pi(V, A_w, E_w) = \Pi^*(V, A_w, E_w)$ is co-NP-complete.*

The following theorem was proven in [70].

Theorem 29. *If $H^* = \emptyset$ and $H \neq \emptyset$, then a decision problem of testing whether a schedule admissible for the weighted mixed graph $G_w = (V, A_w, E_w)$ exists is an NP-complete problem in a strong sense even if there exists only one negative weight prescribed to an arc from the set A_w.*

In the proof of Theorem 29, it is shown that the flow-shop scheduling problem $F3||C_{\max}$ to minimize a schedule length $C_{\max}$ for processing n jobs on three different machines with identical machine routes for all jobs J is polynomially reduced to the problem determined in Theorem 29. In [72], it is proven that the flow-shop scheduling problem $F3||C_{\max}$ is NP-hard in the strong sense. If conditions of Theorem 29 do not hold, then testing whether a schedule admissible for the weighted mixed graph $G^*_w = (V, A_w, E^*_w)$ exists may be realized using a polynomial algorithm. The asymptotic complexities of such algorithms for different special cases of the problem are determined in [70].

Concluding this section, we note that most scheduling problems with equal processing times of the jobs (operations) may be interpreted as some types of optimal colorings of the vertices or edges and arcs of the mixed graphs with special structures. The main restriction for such interpretations of the scheduling problems is the prohibition of operation preemptions. Furthermore, an objective criterion must be either the minimization of makespan C_{max} or the minimization of maximal lateness $L_{max} = \max\{C_k - D_k : J_k \in J\}$, where D_k denotes the due date given for the job J_k. Such scheduling problems have been considered in papers [73–79], among many others.

6. Colorings of Arcs and Edges of the Mixed Graph

Theorem 11 and Definition 4 imply that an edge coloring exists for the mixed graph $G = (V, A, E)$ if and only if the digraph $(V, A, \emptyset)$ has no circuit.

The following decision problem is connected with an optimal edge coloring of the mixed graph.

Problem ($C^e(G, p)$). *Given an integer $p \geq 1$ and a mixed graph $G = (V, A, E)$ without circuits in $(V, A, \emptyset)$, find out whether the mixed graph G admits an edge coloring using at most p different colors $c_<(e_{ij}) \in \{1, 2, \ldots, p\}$.*

Let $l(G)$ denote the number of arcs on a longest path in the digraph $(V, A, \emptyset)$ and $\Delta(G)$ denote the maximum degree $d_G(v_i)$ of a vertex $v_i \in V$ in the mixed graph $G = (V, A, E)$. The following lower bound (Lemma 5) and upper bound (Lemma 6) on the value of $\chi'(G)$ have been proven in [11].

Lemma 5. *Let digraph (V, A, E) have no circuit. Then, $\chi'(G) \geq \max\{l(G), \Delta(G)\}$.*

Lemma 6. *Let digraph (V, A, E) have no circuit. Then,*

$$\chi'(G) \leq \begin{cases} l(G)[\Delta(G) - 1] + 1, & \text{if } l(G) \geq 2, \\ \Delta(G) + 1, & \text{if } l(G) \leq 1. \end{cases} \quad (14)$$

If inequality $l(G) \leq 1$ holds, the proof of the bound in Equation (14) follows from Vizing's theorem [80], since at most $(\Delta(G) + 1)$ colors are needed for edge coloring of G. The whole upper bound in Equation (14) is tight even if mixed graphs G are trees with arbitrary values of $\Delta(G)$ and $l(G) \geq 2$.

Theorems 30 and 31 were proven in [11]. The proof of Theorem 30 uses a polynomial algorithm developed in [81] for edge coloring of the graph, which is a star.

Theorem 30. *The problem $C^e(G, p)$ can be solved in polynomial time if $G = (V, A, E)$ is a mixed tree.*

Theorem 31. *The decision problem $C^e(G, \Delta(G) = l(G))$ is NP-complete even if $G = (V, A, E)$ is a bipartite outerplanar mixed graph.*

The proof of Theorem 31 is based on a polynomial reduction to the decision problem $C^e(G, \Delta(G) = l(G))$ from the precoloring extension problem on a graph. The latter problem is NP-complete even for bipartite outerplanar graphs, as proven in [82]. The proof of Theorem 31 holds when the number of allowed colors is unbounded.

For the case of a constant number of colors, one can provide a polynomial algorithm for any mixed partial k-tree by adapting algorithm described in [81]. In [11], the following two theorems have been proven for edge colorings of the digraphs $G = (V, A, \emptyset)$. A mixed graph $G = (V, A, E)$ is $(k, k+1)$-regular if every vertex $v_i \in V$ has a degree $d_G(v_i)$ of either k or $(k+1)$.

Theorem 32. *The decision problem $C^e(G, 5)$ is NP-complete if $G = (V, A, E)$ is a $(2,3)$-regular bipartite digraph, $l(G) = 3$, and $E = \emptyset$.*

The proof of Theorem 32 is based on a polynomial reduction to the considered problem $C^e(G, 5)$ from the edge coloring problem for a (3)-regular graph. In [83], it is proven that the latter problem is NP-complete.

Theorem 33. *The decision problem $C^e(G, 5)$ is NP-complete if $G = (V, A, E)$ is a cubic bipartite digraph, $l(G) = 3$, and $E = \emptyset$.*

The proof of Theorem 33 is based on the NP-completeness (Theorem 32) of the problem $C^e(G, 5)$ with a $(2,3)$-regular bipartite digraph $G = (V, A, \emptyset)$. The following theorem was also proven in [11].

Theorem 34. *The decision problem $C^e(G, 3)$ is NP-complete even if $G = (V, A, E)$ is restricted to be a cubic planar bipartite mixed graph, $l(G) = 2$, and all paths in the digraph $(V, A, \emptyset)$ are vertex disjoint.*

The proof of Theorem 34 is based on a polynomial reduction to the considered problem $C^e(G, 3)$ from the NP-complete precoloring extension problem for the bipartite outerplanar graph [82].

As demonstrated in [84,85], colorings of edges in the graph may be used to model a certain job-shop scheduling system consisting of unit-time jobs assigned to specific pairs of machines. In the case of the mixed graph $G = (V, A, E)$, it is convenient to look upon arc $(v_i, v_j) \in A$ as a unit-time data transmission from machine v_i to machine v_j requiring the cooperation of machines v_i and v_j, which cannot simultaneously process other jobs. For such a job-shop scheduling system, coloring of the arcs of the mixed graph $G = (V, A, E)$ corresponds to a schedule such that each vertex first receives input data from all incoming arcs, next uses all the collected data for local computations, and finally sends the output data along outgoing arcs. The edges of the mixed graph $G = (V, A, E)$, which appear in some scheduling applications, correspond to possibly unrelated two-machine jobs processed in the job-shop scheduling system, such as mutual self-testing of machines.

In [86], the following edge coloring problem was considered. It is necessary to color each edge in one color and each arc in two colors, such that the color of the first half of an arc is smaller than the color of the second half. The colors used at the same vertex must all be different. A bound for the minimum number of colors needed for such colorings is obtained in [86]. For the graph $G = (V, A, \emptyset)$, a polynomial algorithm for such coloring G with a minimum number of colors is developed in [86].

7. Non-Strict Colorings of Mixed Graphs

We next introduce two types of coloring of the mixed graph $G = (V, A, E)$. A $c_{\leq}$-coloring of the vertices of the mixed graph G is introduced in Section 7.1 and a $c^e_{\leq}$-coloring of the arcs and edges of the mixed graph G is introduced in Section 7.2.

7.1. A Non-Strict Coloring of the Mixed Graph

Definition 5. *An integer-valued function $c_{\leq} : V \to \{1, 2, \ldots, t\}$ is a coloring (called $c_{\leq}$-coloring) of the mixed graph $G = (V, A, E)$ if non-strict inequality*

$$c_{\leq}(v_i) \leq c_{\leq}(v_j) \tag{15}$$

holds for each arc $(v_i, v_j) \in A$, and the following condition

$$c_{\leq}(v_p) \neq c_{\leq}(v_q) \tag{16}$$

holds for each edge $(v_p, v_q) \in E$ *such that* $(v_p, v_q) \notin A$ *and* $(v_i, v_j) \notin A$. *A* $c_{\leq}$*-coloring is optimal if it uses a minimum possible number* $\chi_{\leq}(G)$ *of different colors* $c_{\leq}(v_i) \in \{1, 2, \ldots, t\}$, *such a minimum number* $\chi_{\leq}(G)$ *being called a non-strict chromatic number of the mixed graph* G.

Obviously, one can use a c-coloring (Definition 1) instead of a $c_{\leq}$-coloring (Definition 5) in the special case of the mixed graph such that the implication in Equation (17) holds for each arc $(v_i, v_j) \in A$

$$(v_i, v_j) \in A \Rightarrow [v_i, v_j] \notin E \tag{17}$$

Remark 3. *A* $c_{\leq}$*-coloring of the vertices of the mixed graph* $G = (V, A, E)$ *is a special case of a c-coloring provided that the implication in Equation (17) holds for each arc* $(v_i, v_j) \in A$.

Due to Remark 3, for using a c-coloring instead of a $c_{\leq}$-coloring, it is sufficient to delete all edges $[v_i, v_j]$ from set E such that vertices v_i and v_j are adjacent in the digraph $(V, A, \emptyset)$. However, for some application of $c_{\leq}$-coloring it is useful to color vertices of a general mixed graph $G = (V, A, E)$, where the implication in Equation (17) does not hold for some arcs $(v_i, v_j) \in A$. Definition 5, which is applicable to any mixed graph $G = (V, A, E)$, may have a sense in some applications.

The following decision problem $C_{\leq}(G, p)$ arises for the $c_{\leq}$-coloring of a mixed graph G.

Problem ($C_{\leq}(G, p)$). *Given a mixed graph* $G = (V, A, E)$ *and an integer* $p \geq 1$, *find out whether the mixed graph* G *admits a* $c_{\leq}$*-coloring using at most* p *colors* $c_{\leq}(v_i) \in \{1, 2, \ldots, p\}$.

Most results (but not all) presented in Section 2 remain correct for non-strict $c_{\leq}$-colorings of vertices of the mixed graph.

7.2. A Non-Strict Edge Coloring of the Mixed Graph

We define a non-strict $c_{\leq}$-coloring of arcs and edges of the mixed graph $G = (V, A, E)$ as follows.

Definition 6. *It is necessary to color all arcs* A *and all edges* E *in the mixed graph* $G = (V, A, E)$ *in such a way that any two adjacent edges in the graph* $(V, \emptyset, E)$ *get different colors from set* $\{1, 2, \ldots, t\}$ *and for any two adjacent arcs* $(v_i, v_j) \in A$ *and* $(v_p, v_q) \in A$ *forming a path* (v_i, v_j, v_p, v_q) *in the digraph* $(V, A, \emptyset)$ *the color of arc* (v_i, v_j) *must equal or be less than the maximum color of arc* (v_p, v_q) *and edge* $[v_p, v_q]$ *(if any). An edge* $c_{\leq}$*-coloring is optimal if it uses a minimum number* $\chi'_{\leq}(G)$ *of different colors, such a number* $\chi'_{\leq}(G)$ *being called a non-strict chromatic index of the mixed graph* G.

The following decision problem is connected with the optimal edge $c_{\leq}$-coloring of the mixed graph $G = (V, A, E)$.

Problem ($C^e_{\leq}(G, p)$). *Given an integer* $p \geq 1$ *and a mixed graph* $G = (V, A, E)$, *find out whether the mixed graph* G *admits an edge* $c_{\leq}$*-coloring using at most* p *different colors* $c^e(e_{ij}) \in \{1, 2, \ldots, p\}$.

8. Concluding Remarks

Sections 2–4 contain a review of known results related to the two types of vertex coloring of the mixed graph. One of the reasons for introducing both types of coloring of the vertices of the mixed graph (c-coloring and $c_<$-coloring) is connected with searching a common approach for solving two types of scheduling problems arising in real-world scheduling systems.

In Section 5, it is demonstrated how $c_<$-colorings may be used for solving unit-time job-shop scheduling problems with the makespan criterion. Although we used both graph terminology and scheduling terminology for the problems under consideration in Sections 2–5, it is possible to describe most presented results using either only graph terminology or only scheduling terminology. The published results on a coloring of arcs and edges of the mixed graphs are reviewed in Section 6.

Some results on mixed graph colorings and their applications for solving scheduling problems have only been published in Russian and are not widely known elsewhere. These results are described in Sections 2–6.

As it mentioned in the Abstract, we review mixed graph colorings in the light of scheduling problems. Thus, many closed results are not presented in Sections 2–7 since we do not find their interpretations in scheduling or production planning. In particular, this survey does not cover results published on the homomorphism of the colored mixed graphs introduced in [87] and surveyed in [88].

In Section 7, we introduce new types of colorings of vertices, arcs, and edges of the mixed graphs since they may be applied for optimal scheduling. These types of colorings and their applications in scheduling and production planning may be considered as subjects of future research.

Further research is also needed to extend the complexity results to some other classes of mixed graphs. It is important to find the borders between NP-hard and polynomially solvable classes of t-colorable mixed graphs. It would be interesting to analyze colorings of mixed graphs $G = (V, A, E)$ containing digraphs $(V, A, \emptyset)$ with a special structure. It is also interesting to ask about the complexity of the coloring problems when the numbers of colors are restricted for some vertices (arcs and edges).

It would be worth developing exact and approximate (as well as heuristic) algorithms for coloring small and medium (as well as large) mixed graphs. For testing exact, approximate, and heuristic algorithms and software, it is necessary to determine sets of the benchmark instances.

Funding: This research received no external funding.

Acknowledgments: The author would like to thank anonymous reviewers for their insightful comments on the early draft of this paper.

Conflicts of Interest: The author declares no conflict of interest.

Abbreviations

Notations used in Definitions 1–6

$\chi(G)$	Chromatic number of mixed graph $G = (V, A, E)$, i.e., a minimum number of different colors $c(v_i) \in \{1, 2, \ldots, t\}$ in the c-coloring of the vertices in G	Definition 1
$\chi_<(G)$	Strict chromatic number of mixed graph $G = (V, A, E)$, i.e., a minimum number of different colors $c(v_i) \in \{1, 2, \ldots, t\}$ in the $c_<$-coloring of the vertices in G	Definition 2
$L(G)$	Mixed line graph $(A \cup E, A_{A\cup E}, E_{A\cup E}) = L(G)$ for mixed graph $G = (V, A, E)$, where arcs $(e_{ij}, e_{jk}) \in A_{A\cup E}$ connect all pairs of arcs $e_{ij} = (v_i, v_j) \in A$ and $e_{jk} = (v_j, v_k) \in A$ and edge set $E_{A\cup E}$ connect all the remaining pairs of elements of the set $A \cup E$, which share at least one vertex of the set V	Definition 3
$\chi'_<(G)$	Chromatic index of the mixed graph $G = (V, A, E)$, i.e., a minimum number of colors in the $c_<$-coloring of the arcs and edges in G (a minimum number of colors $c_<(e_{ij})$ in the $c_<$-coloring of the vertices $A \cup E$ in line graph $L(G) = (A \cup E, A_{A\cup E}, E_{A\cup E})$ for mixed graph G)	Definition 4
$\chi_\leq(G)$	Non-strict chromatic number of the mixed graph $G = (V, A, E)$, i.e., a minimum number of colors $c_\leq(v_i)$ in the $c_\leq$-coloring of the vertices in G	Definition 5
$\chi'_\leq(G)$	Non-strict chromatic index of mixed graph $G = (V, A, E)$, i.e., a minimum number of colors $c_\leq(e_{ij})$ in the $c_\leq$-coloring of the arcs and edges in G	Definition 6

References

1. Sotskov, Y.N.; Tanaev, V.S. A chromatic polynomial of a mixed graph. *Vestsi Akademii Navuk BSSR Seryya Fizika-Matematychnykh Navuk* **1976**, *6*, 20–23. (In Russian)
2. Karp, R.M. Reducibility among combinatorial problems. In *Complexity of Computer Computations*; Miller, R.E., Thatcher, J.W., Eds.; Plenum Press: New York, NY, USA, 1972; pp. 85–103.
3. Lambin, N.V.; Tanaev, V.S. On a circuit-free orientation of mixed graphs. *Dokladi Akademii Navuk BSSR* **1970**, *14*, 780–781. (In Russian)
4. Sotskov, Y.N.; Tanaev, V.S. On an enumeration of circuit-free digraphs generated by a mixed graph. *Vestsi Akademii Navuk BSSR Seryya Fizika-Matematychnykh Navuk* **1974**, *2*, 16–21. (In Russian)

5. Sotskov, Y.N.; Tanaev, V.S. On one approach to the enumeration of circuit-free digraphs generated by a mixed graph. *Vestsi Akademii Navuk BSSR Seryya Fizika-Matematychnykh Navuk* **1976**, *5*, 99–102. (In Russian)
6. Sotskov, Y.N. On an orientation of edges of the mixed graph. *Vestsi Akademii Navuk BSSR Seryya Fizika-Matematychnykh Navuk* **1981**, *5*, 22–24. (In Russian)
7. Levin, G.M. On a bound on chromatic characterizations of the mixed graph. *Vestsi Akademii Navuk BSSR Seryya Fizika-Matematychnykh Navuk* **1982**, *1*, 17–20. (In Russian)
8. Levin, G.M.; Sotskov, Y.N.; Tanaev, V.S. On the coloring of vertices of a mixed graph. In *The Second All-Union Symposium of the Academy of Sciences of the USSR "Methods and Programs for Solving Optimization Problems on Graphs and Networks"*; Part 2 "Theory and Algorithms"; The Siberian Branch of the Academy of Sciences of the USSR: Novosibirsk, Russia, 1982; pp. 87–90. (In Russian)
9. Klimova, L.I.; Sotskov, Y.N. Coloring of vertices of a mixed graph. In *Algorithms for Solving Optimization Problems*; Institute of Engineering Cybernetics: Minsk, Belarus, 1993; pp. 90–96. (In Russian)
10. Hansen, P.; Kuplinsky, J.; de Werra, D. Mixed graph colorings. *Math. Methods Oper. Res.* **1997**, *45*, 145–160. [CrossRef]
11. Furmanczyk, H.; Kosowski, A.; Ries, B.; Zylinski, P. Mixed graph edge coloring. *Discret. Math.* **2009**, *309*, 4027–4036. [CrossRef]
12. Harary, F. *Graph Theory*; Addison-Wesley: Reading, MA, USA, 1969.
13. Thulasiraman, K.; Swamy, M.N.S. *Graphs: Theory and Algorithms*; John Wiley & Sons, Inc.: Toronto, ON, Canada, 1992.
14. Tanaev, V.S.; Sotskov, Y.N.; Strusevich, V.A. *Scheduling Theory: Multi-Stage Systems*; Kluwer Academic Publishers: Dordrecht, The Netherlands, 1994.
15. Brucker, P. *Scheduling Algorithms*; Springer: Berlin, Germany, 1995.
16. Christifides, N. *Graph Theory: Algorithmic Approach*; Academic Press: New York, NY, USA, 1975.
17. Stanley, R.P. Acyclic orientations of graphs. *Discret. Math.* **1973**, *5*, 171–178. [CrossRef]
18. Sotskov, Y.N.; Tanaev, V.S; Werner, F. Scheduling problems and mixed graph colorings. *Optimization* **2002**, *51*, 597–624. [CrossRef]
19. Beck, M.; Blado, D.; Crawford, J.; Jean-Louis, T.; Young, M. On weak chromatic polynomials of mixed graphs. *Graphs Comb.* **2015**, *1*, 91–98. [CrossRef]
20. Beck, M.; Blado, D.; Crawford, J.; Jean-Louis, T.; Young, M. Mixed Graph Colorings. In Proceedings of the Science National Conference 2012 Society for Advancement of Hispanics/Chicanos and Native Americans in Science National Conference, Seattle, WA, USA, 11–14 October 2012.
21. Ries, B.; de Werra, D. On two coloring problems in mixed graphs. *Eur. J. Comb.* **2008**, *29*, 712–725. [CrossRef]
22. Andreev, G.V.; Sotskov, Y.N.; Werner, F. Branch and bound method for mixed graph coloring scheduling. In Proceedings of the 16th International Conference on CAD/CAM, Robotics and Factories of the Future, CARS and FDF, Trinidad and Tobago, 2000; Volume 1, pp. 1–8. Available online: https://www.researchgate.net/publication/230725053_A_Branch_and_Bound_Method_for_Mixed_Graph_Coloring_and_Scheduling (accessed on 9 January 2020).
23. Sotskov, Y.N. Scheduling via mixed graph coloring. In *Proceedings of the Symposium on Operations Research, Magdeburg, Germany, 1–3 September 1999*; Inderfurth, K., Schwodiauer, G., Domschke, W., Juhnke, F., Kleinschmidt, P., Wascher, G., Eds.; Springer: Berlin, Germany, 2000; pp. 414–418.
24. Sotskov, Y.N.; Dolgui, A.; Werner, F. Optimal coloring of the vertices from the union of paths and cliques. In Proceedings of the International Workshop on Discrete Optimization Methods in Scheduling and Computer-Aided Design, Minsk, Belarus, 5–6 September 2000; Proth, J.-M., Tanaev, V.S., Eds.; pp. 93–98.
25. Andreev, G.V.; Sotskov, Y.N. Colorings of the vertices of a mixed graph via the branch and bound method. *Vestsi Akademii Navuk BSSR Seryya Fizika-Matematychnykh Navuk* **2001**, *1*, 124–129. (In Russian)
26. Sotskov, Y.N.; Dolgui, A.; Werner, F. Mixed graph coloring for unit-time job-shop scheduling. *Int. J. Math. Algorithms* **2001**, *2*, 289–323.
27. Al-Anzi, F.S.; Sotskov, Y.N.; Allahverdi, A.; Andreev, G.V. Using mixed graph coloring to minimize total completion time in job shop scheduling. *Appl. Math. Comput.* **2006**, *182*, 1137–1148. [CrossRef]
28. Kouider, A.; Ait Haddadene, H.; Ourari, S.; Oulamara, A. Mixed integer linear programs and tabu search approach to solve mixed graph coloring for unit-time job shop scheduling. In Proceedings of the 11th Annual IEEE International Conference on Automation Science and Engineering (IEEE CASE 2015), Gothenburg, Sweden, 24–28 August 2015; doi:10.1109/CoASE.2015.7294257. [CrossRef]

29. Kouider, A.; Ait Haddadene, H.; Oulamara, A. On minimization of memory usage in branch-and-bound algorithm for the mixed graph coloring: Application to the unit-time job shop scheduling. *Comput. Oper. Res.* **2019**, *4967*, 1001–1008.
30. Ries, B. Coloring some classes of mixed graphs. *Discret. Appl. Math.* **2007**, *155*, 1–6. [CrossRef]
31. Ries, B. Complexity of two coloring problems in cubic planar bipartite mixed graphs. *Discret. Appl. Math.* **2010**, *158*, 292–296. [CrossRef]
32. Furmanczyk, H.; Kosowski, A.; Zylinski, P. A note on mixed tree coloring. *Inf. Process. Lett.* **2008**, *106*, 133–135. [CrossRef]
33. Huijter, M.; Tuza, Z.S. Precoloring extension II Graphs classes related o bipartite graphs. *Acta Math. Univ. Commenianae* **1993**, *LXII 1*, 1–11.
34. Jansen, K.; Scheffler, P. Generalized coloring for tree-like graphs. In *Lecture Notes in Computer Science*; Springer Science+Business Media: Berlin, Germany, 1993; Volume 657, pp. 50–59.
35. Chlebik, M.; Chlebikova, J. Hard coloring problems in low degree planar bipartite graphs. *Discret. Appl. Math.* **2006**, *154*, 1960–1965. [CrossRef]
36. Kratochvil, J. Precoloring extension with fixed color bound. *Acta Math. Univ. Commenianae* **1994**, *LXIII 1*, 139–153.
37. Garey, M.R.; Johnson, D.S. *Computer and Inractabulity, a Guide to the Theory of NP-Completeness*; Freemam: New York, NY, USA, 1979.
38. Furmanczyk, H.; Kosowski, A.; Zylinski, P. Scheduling with precedence constraints: Mixed graph coloring in series-parallel graphs. In *Lecture Notes in Computer Science*; Wyrzykowski, R., Dongarra, J., Karczewski, K., Wasniewski, J., Eds.; Springer: Berlin, Germany, 2008; Volume 4967, pp. 1001–1008.
39. Kouider, A.; Ait Haddadene, H.; Ouran, S.; Oulamara, A. Mixed graph coloring for unit-time job shop scheduling. *Int. J. Prod. Res.* **2017**, *55*, 1720–1729. [CrossRef]
40. Damaschke, P. Parameterized mixed graph coloring. *J. Comb. Optim.* **2019**, *38*, 326–374. [CrossRef]
41. Tanaev, V.S.; Shkurba, V.V. *Introduction to Scheduling Theory*; Nauka Publishers: Moscow, USSR, 1975. (In Russian)
42. Sotskov, Y.N. An enumeration of schedules generated by a mixed graph. *Vestsi Akademii Navuk BSSR Seryya Fizika-Matematychnykh Navuk* **1980**, *5*, 132–133. (In Russian)
43. Balas, E. Machine-sequencing via disjunctive graphs: An implicit enumeration algorihm. *Oper. Res.* **1969**, *17*, 941–957. [CrossRef]
44. Sussmann, B. Scheduling problems with interval disjunctions. *Math. Methods Oper. Res.* **1972**, *16*, 165–178.
45. Pinedo, M. *Scheduling: Theory, Algorithms, and Systems*; Prentice-Hall: Englewood Cliffs, NJ, USA, 2002.
46. Graham, R.L.; Lawler, E.L.; Lenstra, J.K.; Rinnooy Kan, A.H.G. Optimization and approximation in deterministic sequencing and scheduling. *Ann. Discret. Appl. Math.* **1979**, *5*, 287–326.
47. Lenstra, J.K. The mystical power of twoness: In memoriam Eugene L. Lawler. *J. Sched.* **1998**, *1*, 3–14. [CrossRef]
48. Brucker, P.; Sotskov, Y.N.; Werner, F. Complexity of shop-scheduling problems with fixed number of jobs: A survey. *Math. Methods Oper. Res.* **2007**, *65*, 461–481. [CrossRef]
49. Brucker, P.; Kravchenko, S.A.; Sotskov, Y.N. Preemptive job-shop scheduling problems with a fixed number of jobs. *Math. Methods Oper. Res.* **1999**, *49*, 41–76.
50. Timkovsky, V.G. Polynomial-time algorithm for the Lenstra-Rinnooy Kan two-machine scheduling problem. *Kibernetika* **1985**, *2*, 109–111. (In Russian)
51. Williamson, D.P.; Hall, L.A.; Hoogeveen, J.A.; Hurkens, C.A.J.; Lenstra, J.K.; Sevast'janov, S.V.; Shmoys, D.B. Short shop schedules. *Oper. Res.* **1997**, *45*, 288–294. [CrossRef]
52. Even, S.; Itai, A.; Shamir, A. On the complexity of timetable and multicommodity flow problems. *SIAM J. Comput.* **1976**, *5*, 691–703. [CrossRef]
53. Sotskov, Y.N. The complexity of optimal processing three jobs. *Kibernetika* **1990**, *5*, 74–78. (In Russian)
54. Sotskov, Y.N. The complexity of shop-scheduling problems with two or three jobs. *Eur. J. Oper. Res.* **1991**, *53*, 326–336. [CrossRef]
55. Lenstra, J.K.; Rinnooy Kan, A.H.G. Computational complexity of discrete optimization problems. *Ann. Discret. Math.* **1979**, *4*, 121–140.
56. Hefetz, N.; Adiri, I. An efficient optimal algorithm for the two-machine, unit-time, job-shop, schedule-length, problem. *Math. Oper. Res.* **1982**, *7*, 354–360. [CrossRef]

57. Brucker, P. An efficient algorithm for the job-shop problem with two jobs. *Computing* **1988**, *40*, 353–359. [CrossRef]
58. Hardgrave, W.W.; Nemhauser, G.L. A geometric model and graphical algorithm for a sequencing problem. *Oper. Res.* **1963**, *11*, 889–900. [CrossRef]
59. Szwarc, W. Solution of the Akers-Friedman scheduling problem. *Oper. Res.* **1960**, *8*, 782–788. [CrossRef]
60. Sotskov, Y.N. Optimal scheduling two jobs with regular criterion. *Design Processes Automating*; Institute of Engineering Cybernetics: Minsk, Belarus, 1985; pp. 86–95 (In Russian)
61. Sotskov, Y.N. Software for production scheduling based on the mixed (multi)graph approach. *Comput. Control Eng. J.* **1996**, *7*, 240–246. [CrossRef]
62. Sotskov, Y.N. Mixed multigraph approach to scheduling jobs on machines of different types. *Optimization* **1997**, *42*, 245–280. [CrossRef]
63. De Werra, D. Restricted coloring models for timetabling. *Discret. Math.* **1997**, *165/166*, 161–170. [CrossRef]
64. De Werra, D. On a multiconstrained model for chromatic scheduling. *Discret. Appl. Math.* **1999**, *94*, 171–180. [CrossRef]
65. De Werra, D.; Hertz, A. Chromatic scheduling. In *Topics in Chromatic Graph Theory*; Cambridge University Press: Cambridge, UK, 2015; pp. 255–276.
66. Kruger, K.; Sotskov, Y.N.; Werner, F. Heuristic for generalized shop scheduling problems based on decomposition. *Int. J. Prod. Res.* **1998**, *36*, 3013–3033. [CrossRef]
67. Kubale, M. Interval vertex-coloring of a graph with forbidden cilors. *Discret. Appl. Math.* **1989**, *74*, 125–136.
68. Sotskov, Y.N. An optimal schedule of a set of jobs generated by the mixed graph. *Vestsi Akademii Navuk BSSR Seryya Fizika-Matematychnykh Navuk* **1977**, *4*, 133. (In Russian).
69. Sotskov, Y.N. Schedules on the mixed graphs with a restricted maximum penalty. *Vestsi Akademii Navuk BSSR Seryya Fizika-Matematychnykh Navuk* **1980**, *2*, 37–42. (In Russian).
70. Sotskov, Y.N.; Tanaev, V.S. A construction of schedules admissible with respect to the mixed multigraph. *Vestsi Akademii Navuk BSSR Seryya Fizika-Matematychnykh Navuk* **1989**, *4*, 94–98. (In Russian)
71. Sotskov, Y.N. On the complexity of constructing schedules admissible with respect to a mixed graph. *Vestsi Akademii Navuk BSSR Seryya Fizika-Matematychnykh Navuk* **1991**, *1*, 121. (In Russian)
72. Garey, M.R.; Johnson, D.S.; Sethi, R. The complexity of flowshop and jobshop scheduling. *Math. Oper. Res.* **1976**, *1*, 117–129. [CrossRef]
73. Brucker, P.; Garey, M.R.; Johnson, D.S. Scheduling equal-length tasks under tree-like precedence constraints to minimize maximum lateness. *Math. Oper. Res.* **1977**, *2*, 275–284. [CrossRef]
74. Baptiste, P. Polynomial time algorithms for minimizing the weighted number of late jobs on a single machine with equal processing times. *J. Sched.* **1999**, *2*, 245–252. [CrossRef]
75. Garey, M.R.; Johnson, D.S.; Simons, B.B.; Tarjan, R.E. Scheduling unit-time tasks with arbitrary release times and deadlines. *J. Comput.* **1981**, *10*, 256–269. [CrossRef]
76. Kravchenko, S.A. On the complexity of minimizing the number of late jobs in unit time open shop. *Discret. Appl. Math.* **2000**, *100*, 127–132. [CrossRef]
77. Baptiste, P. Ten notes on equal-processing-time scheduling. *4OR* **2004**, *2*, 111–127. [CrossRef]
78. Baptiste, P.; Brucker, P. Scheduling equal processing time jobs: A survey. In *Handbook of Scheduling: Algorithms, Models, and Performance Analysis*; Leung Y.T., Ed.; CRC Press LLC: Boca Raton, FL, USA, 2004; pp. 78–96.
79. Chrobak, M.; Durr, C.; Jawor, W.; Kowalik, L.; Kurowski, M. A note on scheduling equal-length jobs to maximize throughput. *J. Sched.* **2006**, *9*, 71–73. [CrossRef]
80. Vizing, V.G. On an estimate of the chromatic class of a p-graph. *Discret. Anal.* **1964**, *3*, 25–30. (In Russian)
81. Zhou, X.; Nishizeki, T. Algorithm for the cost edge-coloring of trees. *Lect. Notes Comput. Sci.* **2001**, *2108*, 288–297.
82. Marx, D. NP-completeness of list coloring and precoloring extension on the edges of planar graphs. *J. Graph Theory* **2005**, *49*, 313–324. [CrossRef]
83. Holyer, I. The NP-completeness of edge coloring. *SIAM J. Comput.* **1981**, *10*, 718–720. [CrossRef]
84. Kubale, M. *Introduction to Computational Complexity and Algorithmic*; GTN: Gdansk, Poland, 1998.
85. Furmanczyk, H.; Kubale, M. Scheduling of unit-length jobs with bipartite incompatibility graphs on four uniform machines. *Bull. Polish Acad. Sci. Tech. Sci.* **2017**, *65*, 29–34. [CrossRef]
86. Mel'nikov, L.S.; Vizing, V.G. The edge chromatic number of a directed/mixed multigraph. *J. Graph Theory* **1999**, *31*, 267–273. [CrossRef]

87. Nesetril, J.; Raspaud, A. Colored homomorphisms of colored mixed graphs. *J. Comb. Theory Ser. B* **2000**, *1*, 147–155.
88. Sopena, E. Homomorphisms and colorings of oriented graphs: An updated survey. *Discret. Math.* **2016**, *339*, 1993–2005. [CrossRef]

mathematics

MDPI

Article

Kempe-Locking Configurations

James Tilley

61 Meeting House Road, Bedford Corners, NY 10549, USA; jimtilley@optonline.net; Tel.: +1-914-242-9081

Received: 27 October 2018; Accepted: 4 December 2018; Published: 7 December 2018

Abstract: The 4-color theorem was proved by showing that a minimum counterexample cannot exist. Birkhoff demonstrated that a minimum counterexample must be internally 6-connected. We show that a minimum counterexample must also satisfy a coloring property that we call Kempe-locking. The novel idea explored in this article is that the connectivity and coloring properties are incompatible. We describe a methodology for analyzing whether an arbitrary planar triangulation is Kempe-locked. We provide a heuristic argument that a fundamental Kempe-locking configuration must be of low order and then perform a systematic search through isomorphism classes for such configurations. All Kempe-locked triangulations that we discovered have two features in common: (1) they are Kempe-locked with respect to only a single edge, say xy, and (2) they have a Birkhoff diamond with endpoints x and y as a subgraph. On the strength of our investigations, we formulate a plausible conjecture that the Birkhoff diamond is the only fundamental Kempe-locking configuration. If true, this would establish that the connectivity and coloring properties of a minimum counterexample are indeed incompatible. It would also imply the appealing conclusion that the Birkhoff diamond configuration alone is responsible for the 4-colorability of planar triangulations.

Keywords: graph coloring; Kempe chain; Kempe-locking; Birkhoff diamond

MSC: 05C15

---

1. Terminology

We use the standard terminology of graph theory: the basic terms vertex, edge, face, simple graph, planar graph, subgraph, induced subgraph, path, cycle, connectivity, adjacent vertices, order, vertex degree, and minimum degree are all assumed to be known (refer to the introductory text by Chartrand and Zhang [1]). This article is primarily about the coloring of planar graphs. We restrict our consideration to coloring the vertices of simple, connected, planar graphs, and, as is usual, to *proper* colorings of such graphs, colorings in which no two adjacent vertices have the same color. The 4-color theorem states that it takes no more than four distinct colors to color any simple planar graph properly (see [2]). In studying the 4-color problem, we need consider only planar graphs that are *triangulations*, graphs in which all faces are delineated by three edges. That is because any planar graph that is not a triangulation can be turned into a triangulation by inserting edges until all faces are triangular. After properly coloring the triangulation, the inserted edges can be deleted, leaving the original graph with a proper coloring. If all faces of a planar graph except one are triangular, the graph is referred to as a *near-triangulation*. Near-triangulations in which the only non-triangular face is four-sided are particularly important in this article; such a near-triangulation can be derived by deleting any edge in a triangulation. As is customary, we use the term *configuration* for a near-triangulation whose non-triangular face is the infinite face.

2. Introduction

This article is not about finding an alternative proof of the 4-color theorem; instead, it is an exploratory paper aimed at gaining a deeper understanding of why the 4-color theorem is true, what it is about a planar graph that guarantees that it can be 4-colored. We set about this task by focusing on the properties of a minimum counterexample. The two we single out for thorough examination relate to *connectivity* and *coloring*. These are the subjects of Sections 3 and 4. A new approach to understanding why planar graphs are 4-colorable is to investigate whether the connectivity and coloring properties of a minimum counterexample are compatible. We shall see that this approach bears fruit. Our analysis suggests strongly that the two properties are not compatible. An important contribution this article makes to the literature on planar graph coloring is the notion of Kempe-locking, defined in Section 4. If an edge xy is deleted from a planar triangulation T, a near-triangulation G_{xy} results. Consider any 4-coloring of G_{xy} in which vertices x and y have the same color. If T is Kempe-locked with respect to xy, then any sequence of interchanges of colors on Kempe chains leaves x and y still sharing the same color. Thus, Kempe interchanges applied to G_{xy} with x and y colored the same cannot result in a 4-coloring of G_{xy} in which the deleted edge xy can be replaced in order to achieve a 4-coloring of T. In Section 4, we prove that a minimum counterexample to the 4-color theorem must be Kempe-locked with respect to every one of its edges. Each edge in a minimum counterexample must have this coloring property.

In Section 5, we present examples of Kempe-locked triangulations. In Section 6, we present a methodology to determine if a triangulation is Kempe-locked and define two key terms, *Kempe-locking configuration* and *fundamental Kempe-locking configuration*. We then look back briefly to Section 5 to identify the pertinent Kempe-locking configurations for the various examples. In Section 7, we report the results of an extensive search through all isomorphism classes of planar triangulations of low order. Several Kempe-locked triangulations were discovered, each Kempe-locked with respect to only a single edge. Call that edge xy. We find that each such triangulation has a Birkhoff diamond subgraph with endpoints x and y. Our search confirms that the Birkhoff diamond is a fundamental Kempe-locking configuration and strongly suggests that it is the only fundamental Kempe-locking configuration. Section 8 presents an even more fundamental graph—a triangulation of order 9 that gives rise to the Birkhoff diamond. It is constituted as a single vertex within a pentagon within a triangle.

Section 9 concludes the article. It emphasizes the likely singular role of the Birkhoff diamond in the matter of establishing the incompatibility of the connectivity and coloring properties of a minimum counterexample and thus in the matter of the 4-colorability of planar triangulations. It also describes future directions for this research, not the least of which is finding collaborators with substantially greater computing power to extend the search for Kempe-locked triangulations.

Although our primary concern is the theoretical question of 4-colorability of planar graphs, it should be noted that graph coloring finds application more generally to many real-world problems, including the scheduling of sports events, determining examination timetables, and arranging seating plans. Graph coloring also underpins the creation of Sudoku puzzles. While new techniques, such as those discussed in this article, will likely find their way into problems involving complex networks, it is not our intent to explore such applications here.

3. Connectivity Property

A graph G is *connected* if there is a path joining every pair of vertices. It is said to be *k-connected* if it has more than k vertices and remains connected whenever fewer than k vertices are deleted. Therefore, for example, if a graph has 4-cycle with vertices both inside and outside the cycle, then deleting the vertices of the 4-cycle (and all their edges) disconnects the graph into the subgraph that lies outside the 4-cycle and the subgraph that lies inside the 4-cycle. Such a graph might be 4-connected but it cannot be 5-connected. A *separating k-cycle* is a k-cycle in a graph such that there are one or more vertices in the graph both inside and outside the cycle. Thus, a graph with a separating k-cycle can be at most k-connected. A 5-connected triangulation is minimum degree 5 and has no separating 3-cycles or 4-cycles. However, in a 5-connected triangulation, any separating 5-cycle is permitted. Suppose, instead, we consider triangulations for which no separating 5-cycle has two or more vertices both inside and outside the cycle. Then, each separating 5-cycle will have a single vertex either inside or outside the cycle. In other words, each 5-cycle will "enclose" a vertex of degree 5. If that is the case, then the triangulation is said to be *internally 6-connected*. Internally 6-connected triangulations form a proper subset of 5-connected triangulations.

Property 1. ***(Connectivity)*** *A planar triangulation is said to be internally 6-connected if it is minimum degree 5 and has no cycle of length 5 or less for which there are two or more vertices both inside and outside the cycle.*

Every existing proof of the 4-color theorem has been based on showing that a *minimum counterexample* cannot exist. If a counterexample does exist, then there must be one or more smallest counterexamples, where by "smallest" we mean the counterexample(s) having the lowest order (fewest vertices). In a landmark paper in 1913, Birkhoff [3] showed that a minimum counterexample to the 4-color theorem must be internally 6-connected. (That was not a term of his choosing; it arrived on the graph theory scene much later.) In that paper, Birkhoff also introduced the important notion of reducibility. Suppose that an arbitrary planar triangulation T contains a configuration C as a subgraph and let $G = T - C$ denote the near-triangulation formed by removing C from T. The configuration C is said to be *reducible* if, whenever G is 4-colorable, T is also 4-colorable. Thus, a planar triangulation that is a minimum counterexample to the 4-color theorem cannot contain a reducible configuration.

The Appel–Haken team [4,5] and the Robertson–Sanders–Seymour–Thomas team [6] each proved the 4-color theorem by constructing an unavoidable set of reducible configurations (1482 such configurations for [4,5] and 633 for [6]). The term *unavoidable* means that every internally 6-connected triangulation contains at least one of the configurations from the set as a subgraph—the set cannot be avoided. Then, because each configuration in the set is reducible, no internally 6-connected triangulation can be a minimum counterexample.

4. Coloring Property

We follow the standard practice of indicating the color of a vertex by using one of the integers $1, 2, 3, 4, \ldots$. If the first k integers are used to color the vertices of a graph, we say that the graph is *k-colored*. It is important to understand what we mean by a *distinct coloring* of G. Another way to think of a k-coloring of G is that the set of all vertices in G has been partitioned into k independent sets, each set assigned a distinct color from 1 to k. An *independent set* of vertices is one in which no two vertices are adjacent. What matters is the partition (what vertices belong to what independent set); it does not matter how we assign colors to the various independent sets. Thus, a mere permutation of colors does not lead to a distinctly different coloring of G.

A useful tool in graph coloring is the *Kempe chain*, named after the British mathematician whose attempt at proving the 4-color theorem failed [7]. In a given coloring of a graph G, a Kempe chain is a maximal, connected, induced subgraph of G whose vertices use only two colors, let us say colors i and j. An i-j Kempe chain is "maximal" in that every vertex adjacent to, but not in, the chain uses a color other than i or j. In the six-panel figure that accompanies the first example presented in Section 5, there is a single 1–2 Kempe chain in each of the four uppermost graphs and two distinct 1–2 Kempe chains in each of the two lowermost graphs. It is customary to consider a vertex colored i not adjacent to any vertex colored j a *short i-j chain* and likewise a vertex colored j not adjacent to any vertex colored i.

Kempe chains are particularly useful in "navigating" among the various distinct 4-colorings of a graph G because interchanging colors on a chain—that is, interchanging the color labels i and j for all vertices constituting an i-j chain—leaves G properly colored. Interchanging colors on an i-j Kempe chain does not lead to a distinctly different coloring of G if there is only one i-j chain. However, for example, if there are three distinct i-j chains, then interchanging colors on any one or on any pair of those chains does lead to a distinctly different coloring of G. A straightforward description of using Kempe chains to color and recolor graphs can be found in [8].

The idea that underpins our different approach to the question of 4-colorability derives from a graph-coloring problem that is the subject of another article [9], a coloring problem posed for planar graphs that are near-triangulations in which the single non-triangular face is quadrilateral. Although we do not pursue that coloring problem per se in this article, we do focus on near-triangulations with a 4-face and find it useful to establish a few conventions regarding the drawing and vertex labeling of such graphs. In this article, we always draw them with the 4-face as the infinite (exterior) region oriented as shown in Figure 1 and we always label the *boundary cycle* of the 4-face $uxvy$ with u the bottom vertex, x the left-hand vertex, v the top vertex, and y the right-hand vertex. Then x and y constitute one pair of *opposite vertices* and u and v the other pair. We use the notation G_{xy} to refer to the near-triangulation that results when the edge xy is deleted from a planar triangulation T. Such a near-triangulation can always be drawn as depicted in Figure 1.

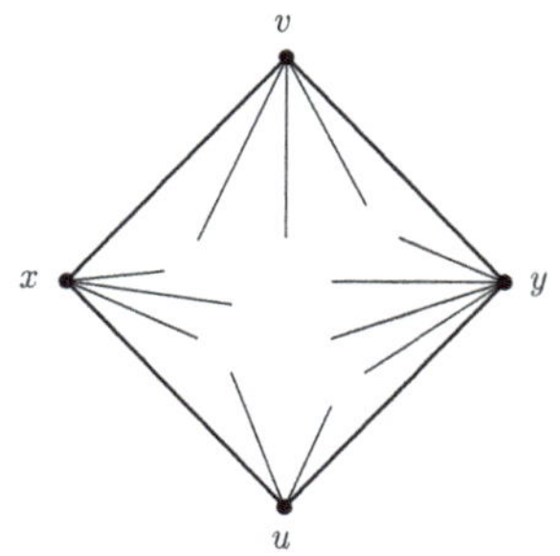

Figure 1. The schematic representation of a near-triangulation with an exterior 4-face. The interior of the graph includes vertices and edges that form triangular faces.

Property 2. ***(Coloring)*** *A planar triangulation T is Kempe-locked with respect to the edge xy if, in every 4-coloring of G_{xy} in which the colors of x and y are the same, there are precisely three distinct Kempe chains that include both x and y.*

Kempe-locking is a property associated with an edge in a triangulation. A triangulation can be Kempe-locked with respect to one particular edge and not with respect to any of its other edges. If T is Kempe-locked with respect to the edge xy, then given a 4-coloring of G_{xy} in which x and y are both colored 1, there must be 1–2, 1–3, and 1–4 Kempe chains including both x and y. Interchanging colors on any of those chains leaves x and y sharing the same color. Moreover, for any such 4-coloring, because T is Kempe-locked, interchanging colors on any Kempe chain that does not include x and y must lead to a 4-coloring in which there are 1–2, 1–3, and 1–4 Kempe chains that include both x and y. Thus, if interchanging colors on Kempe chains is the only method of recoloring available, G_{xy} is "locked into" a state in which x and y have the same color.

The notion of Kempe-locking should not be confused with the matter of entangled Kempe chains, the condition that afflicted Kempe's attempted proof of the 4-color theorem (see [2]). If an i-j Kempe chain and a j-k Kempe chain, $i \neq k$, intersect at one or more vertices colored j, then the two chains are said to be *entangled*. By this definition, there are entangled Kempe chains in every 4-coloring of every triangulation. (Consider any vertex v colored j in a triangulation on at least four vertices; v is adjacent to at least three other vertices that form a cycle enclosing v. Because that cycle must involve at least two colors different from j, say i and k, there will be i-j and j-k Kempe chains entangled at v.) In contrast, Kempe-locking turns out to be rare—it is a coloring property that characterizes not merely a pair of Kempe chains, but an entire triangulation. It refers to a coloring condition shared by all members of a particular set of 4-colorings of the near-triangulation derived by deleting a specified edge from its parent triangulation.

To demonstrate that a minimum counterexample has Property 2, we first need to describe the notion of an *edge contraction*, also known as *vertex identification*. Adjacent vertices x and y are said to be *identified*, or equivalently, the edge xy is said to be *contracted*, if x and y are coalesced into a single vertex w, with all edges formerly incident to x and y now incident to w. The act of contracting xy makes a smaller graph; it decreases the order of the graph by one.

Theorem 1. *A minimum counterexample to the 4-color theorem is Kempe-locked with respect to every one of its edges.*

Proof. Let T be a minimum counterexample to the 4-color theorem. Contract an arbitrary edge xy by coalescing x and y into w. This edge contraction yields a new triangulation T' with one fewer vertex than T. Because T is assumed to be a minimum counterexample, T' can be 4-colored. Then, when w is split apart into the original x and y, but without replacing the edge xy, we obtain a near-triangulation G_{xy} in which the colors of x and y are the same.

Now suppose that T is not Kempe-locked with respect to the edge xy. Then there must be a 4-coloring of G_{xy} in which x and y are colored the same and in which there is a Kempe chain that includes x but does not include y. Interchanging colors on such a chain results in a 4-colored G_{xy} with the color of x not the same as the color of y. In this 4-coloring, the edge xy can be inserted to yield a 4-coloring for T, in contradiction to the assumption that T is a minimum counterexample. □

In contrast to the situation of a minimum counterexample, no 4-colorable planar triangulation can be Kempe-locked with respect to every one of its edges. The proof of that result can be found in Appendix A.

The key point of this article is that Property 1 and Property 2—connectivity and coloring—are likely incompatible. If that could be proved, then the 4-color theorem would be proved because a minimum counterexample, unable to exhibit both properties, could not exist. We capture this idea in the form of a plausible conjecture, one likely to be true but beyond difficult to prove.

Conjecture 1. *No planar triangulation that is Kempe-locked with respect to an edge is internally 6-connected.*

To investigate Conjecture 1, we set about trying to find planar triangulations that are Kempe-locked with respect to some edge. Obviously, we need to consider only internally 6-connected triangulations; however, to find as many Kempe-locked triangulations as possible so that we can learn what structural properties they might have in common, we cast a wider net by searching for any that are at least 4-connected. Because we are studying the properties of a minimum counterexample to the 4-color theorem, we can omit 3-connected triangulations from consideration by virtue of the following argument. Any 3-connected planar triangulation T on more than four vertices must have a separating triangle. Assume that T is a minimum counterexample. Then, the vertex sets of both (1) the induced proper subgraph of T consisting of the separating triangle and everything outside it and (2) the induced proper subgraph of T consisting of the separating triangle and everything inside it can be 4-colored in such a way that the separating triangle is colored identically in each subgraph (through a permutation of colors, as necessary). Thus, T can be 4-colored, a contradiction.

5. Examples of Kempe-Locked Triangulations

Figure 2 shows the six distinct 4-colorings of a particular near-triangulation G_{xy} of order 12. These are the only 4-colorings in which x and y (the leftmost and rightmost vertices on the exterior 4-face) are colored the same. This can be seen by noting, without loss of generality, that x and y can be colored 1 and v (the topmost vertex on the exterior 4-face) colored 2. Then, the other two vertices adjacent to v must be colored 3 and 4. Without loss of generality, the vertex adjacent to both x and v can be colored 3 and the vertex adjacent to both y and v can be colored 4. There are seven vertices remaining to be colored. Noting the color possibilities for each remaining vertex and then successively trying each, one quickly discovers that there are only six distinct 4-colorings of G_{xy}. As can be seen from Figure 2, in each of the 4-colorings there are 1–2, 1–3, and 1–4 Kempe chains that include both x and y. Thus, the triangulation T formed by joining x and y is Kempe-locked with respect to the edge

xy. T does not contradict Conjecture 1 because the bottommost and topmost vertices on the 4-face (u and v, respectively) have degree 4. Indeed, T is 4-connected.

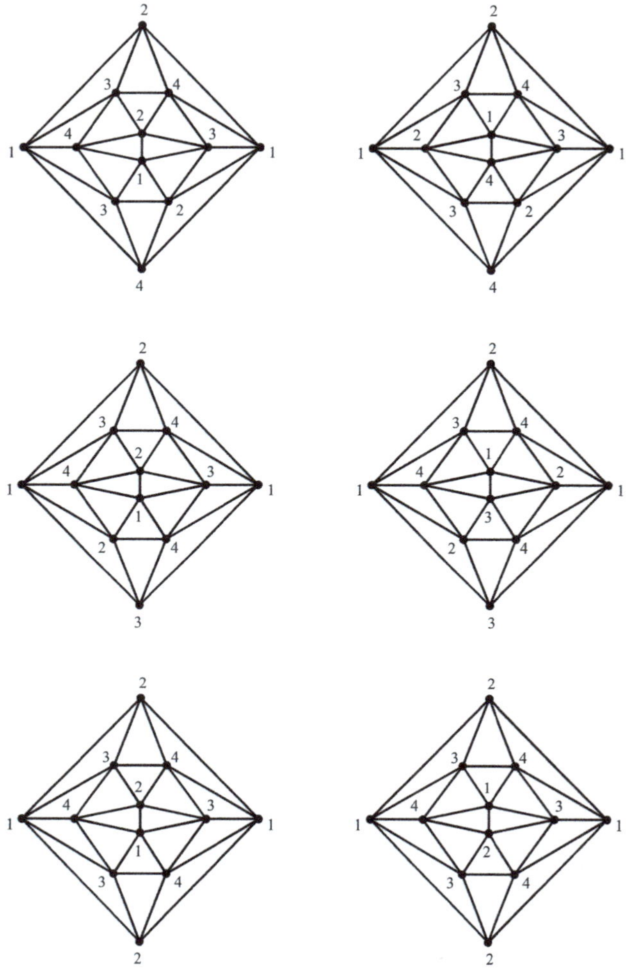

Figure 2. A near-triangulation of order 12 with all six of its distinct 4-colorings in which the colors of x and y are the same, without loss of generality color 1.

We note that the near-triangulation shown in Figure 2 becomes the icosahedron when u and v, instead of x and y, are joined. In that situation, we refer to the near-triangulation as G_{uv} because it is the edge uv that is deleted from the icosahedron to form the near-triangulation. In neither of the bottom two 4-colorings of G_{uv} in Figure 2, in which u and v and both colored 2, are there 2–3 or 2–4 Kempe chains that include both u and v. Because (u, v) can be taken to represent any pair of adjacent vertices in the highly symmetric and regular icosahedron, we conclude that the icosahedron is not Kempe-locked with respect to any of its edges.

Figure 3 shows a particular induced subgraph of order 10 of the near-triangulation of order 12 in Figure 2. That subgraph, a configuration, is known as the *Birkhoff diamond*, named for the mathematician who first highlighted its importance in the 4-color problem [3]: it was the very first configuration to be proved reducible; its presence in a triangulation guarantees that the triangulation cannot be a minimum counterexample. Ironically, we will see that the Birkhoff diamond makes its appearance in our analysis in an opposite way—namely, that it likely must be present for a triangulation to be Kempe-locked with respect to an edge and thus possibly a candidate to be a minimum counterexample. Our systematic search among 4-connected and 5-connected triangulations, which we shall describe in Section 7, uncovered a number of Kempe-locked triangulations. All share a common feature: one or more Birkhoff diamond subgraphs with endpoints x and y.

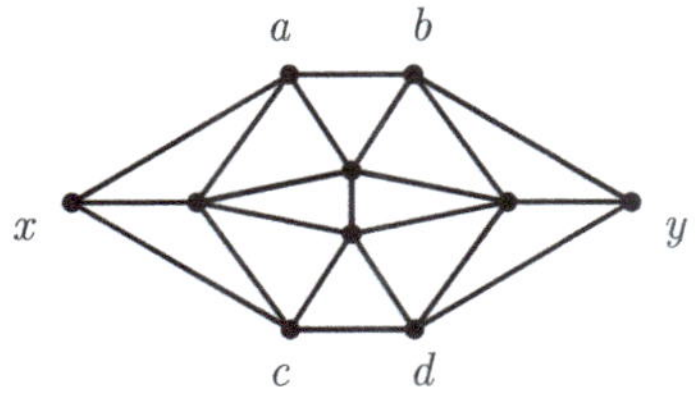

Figure 3. The Birkhoff diamond, a configuration of order 10, with its left- and right-hand endpoints labeled x and y, respectively, and its topmost and bottommost edges labeled ab and cd, respectively.

Four additional examples of Kempe-locked triangulations (once x and y are joined) are illustrated in Figure 4. It is not obvious by inspection that any of these triangulations is Kempe-locked. Using a computer program to determine all possible 4-colorings of each triangulation quickly establishes the Kempe-locking. Note that each of the four triangulations features a Birkhoff diamond subgraph with endpoints x and y. The graph in the lower right panel suggests that by including an arbitrary number of Birkhoff diamonds with endpoints x and y in an appropriate way in a triangulation that is Kempe-locked with respect to xy, we can obtain an infinite number of Kempe-locked triangulations. This surmise becomes evidently correct by applying the type of analysis discussed in Section 6 to a vertical stack of Birkhoff diamonds with appropriate intervening configurations. However, we should note that not all triangulations containing one or more Birkhoff diamonds with endpoints x and y are Kempe-locked—we shall have more to say on this point in the following Section 6.

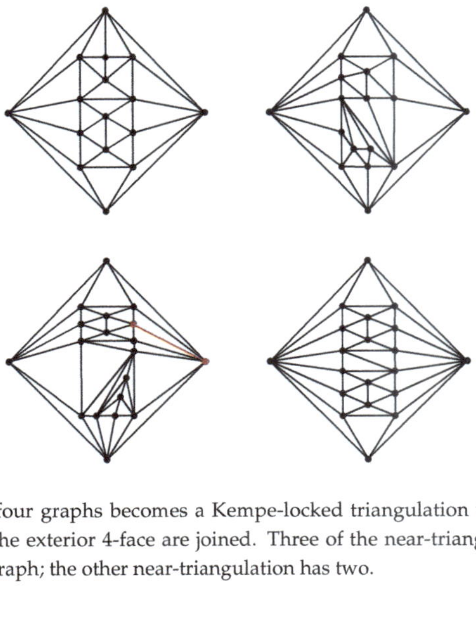

Figure 4. Each of the four graphs becomes a Kempe-locked triangulation when the leftmost and rightmost vertices on the exterior 4-face are joined. Three of the near-triangulations have a single Birkhoff diamond subgraph; the other near-triangulation has two.

6. Kempe-Locking Configurations

As is our established convention, we let G_{xy} be the near-triangulation obtained by deleting the edge xy in a triangulation T. Figure 5 depicts the general structure of such a G_{xy}. Every thick solid line represents one or more edges. Any oval, together with the thick solid lines that join it to endpoints x and y, represents a configuration, which, because all the configurations we are interested in include both x and y, we shall refer to as an *xy-configuration*. Figure 5 shows five such xy-configurations but there can be any (non-negative) number of them. The bottom boundary of each xy-configuration in the vertical stack always coincides with the top boundary of the next lower xy-configuration, if there is one. Because xy-configurations can be combined into a single larger xy-configuration or deconstructed into smaller xy-configurations, any given G_{xy} can be represented in multiple ways as a vertical stack of xy-configurations. However, for a given purpose, there is often a choice that renders the desired analysis more straightforward. For example, because we have now observed the importance of the Birkhoff diamond, the near-triangulation shown in Figure 2 is best displayed in the form of Figure 5 with a single xy-configuration, a Birkhoff diamond: v is joined to that configuration by two edges, and likewise for u, and that configuration includes three edges emanating from each of x and y. The near-triangulation in the lower right of Figure 4 can be put into the general form of Figure 5 most usefully with three xy-configurations, two Birkhoff diamonds with an xy-configuration between them. The middle configuration has order 6—endpoint vertices x and y and four others, two on the top boundary, one adjacent to x, and one adjacent to y, and the same for the two on the bottom boundary, one adjacent to x and one adjacent to y.

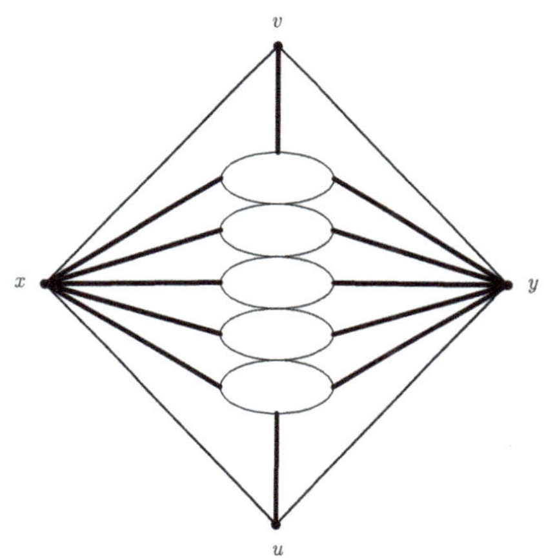

Figure 5. The general structure of a near-triangulation G_{xy}.

This manner of representing G_{xy} allows us to think somewhat differently about what it takes for T to be Kempe-locked with respect to the edge xy. The definition states that in every 4-coloring of G_{xy} in which the colors of x and y are the same there must be three Kempe chains that include x and y. Without loss of generality, let the common color of x and y be 1 and the color of v be 2. Then the three required chains are colored 1–2, 1–3, and 1–4. We are interested in finding an equivalent condition that might be of greater utility in understanding what is necessary to create a Kempe-locked triangulation. We proceed by considering two cases:

1. If the color of u is 2, then there cannot be either a 2–3 path or a 2–4 path between v and u because the first would have to intersect the 1–4 Kempe chain that includes both x and y and the second would have to intersect the 1–3 Kempe chain that includes both x and y, neither of which is possible.

2. If the color of u is not 2, we can choose it to be 3 without loss of generality. There cannot be a 2–3 path between v and u because it cannot intersect the 1–4 Kempe chain that includes both x and y. Obviously there cannot be a 2–4 path between v and u when v is colored 2 and u is colored 3.

We have just shown that the existence of three Kempe chains including both x and y implies the non-existence of any 2-color paths between v and u using colors different from the common color of x and y. The converse is proved in [9], thus establishing the truth of the following Theorem 2.

Theorem 2. *Let T be a planar triangulation and G_{xy} the near-triangulation derived from it by deleting the edge xy. Let $uxvy$ be the 4-cycle delineating the 4-face of G_{xy}. Then T is Kempe-locked with respect to the edge xy if and only if, in every 4-coloring of G_{xy} in which the colors of x and y are the same, there is no 2-color path between v and u that uses colors different from the common color of x and y.*

Consider Figure 5 with x and y both colored 1 and v colored 2. If a 2–3 path between v and u were to exist, it would have to pass through every xy-configuration in the vertical stack on its way from v to u. Thus, there would have to exist a 4-coloring for each xy-configuration that transmits the 2–3 path from the configuration's top boundary to its bottom boundary and such 4-colorings for the various xy-configurations would have to match along their coincident boundaries. Each xy-configuration would have to be able to take the 2–3 path presented to it as input at its top boundary and carry it through to the bottom boundary, that output then passed along as input to the next xy-configuration. Thus, there might exist certain xy-configurations for which this "passing along" cannot be assured or which make it impossible to "complete" the path at vertex u and achieve a proper coloring of G_{xy}. Such a configuration potentially enables a Kempe-locked triangulation. The Birkhoff diamond is one of these special configurations.

To understand how the Birkhoff diamond can prevent a 2-color path between v and u, we analyze all possible colorings of Figure 3 when x and y are both colored 1. Because the vertices a, b, c, and d are all adjacent to either x or y, they must take their colors from {2,3,4}. There are six possible pairings of colors for either (a,b) or (c,d): (2,3), (2,4), (3,2), (3,4), (4,2), and (4,3). The various proper colorings of the Birkhoff diamond subject to this constraint are not provided here—they are easily determined. Instead, we summarize the results in terms of the possible transitions from the colors of the (a,b) pair along the top boundary of the Birkhoff diamond to the colors of the (c,d) pair along the bottom boundary. We can represent the permitted transitions between (a,b) and (c,d) as a *bipartite graph*, a graph in which each vertex is assigned to one of two independent sets and the only edges in the graph thus occur between a vertex in one set and a vertex in the other. Figure 6 shows such a graph for the Birkhoff diamond with both x and y constrained to have color 1. The possibilities for (a,b) appear as the top set of six vertices and the possibilities for (c,d) appear as the bottom set of six vertices. An edge exists between a vertex in the top set and a vertex in the bottom set if and only if there exists a proper 4-coloring of the Birkhoff diamond for that pairing of top and bottom vertices. We have depicted the various edges in Figure 6 as thin or thick solid lines or as dotted or dashed lines. A thin solid line indicates the presence of both a 2–3 path and a 2–4 path between (a,b) and (c,d), a thick solid line indicates the absence of both a 2–3 path and a 2–4 path, a dotted line the presence of a 2–3 path only, and a dashed line the presence of a 2–4 path only.

We can use Figure 6 to provide a straightforward demonstration that the triangulation giving rise to the near-triangulation depicted in Figure 2 is Kempe-locked with respect to the edge xy. With x and y both colored 1 and v colored 2, the only possibilities for (a,b) are (3,4) and (4,3). Consider the case (3,4). From Figure 6, we see that the only possibilities for (c,d) are (2,4), (3,2), and (3,4). (This conclusion is confirmed by inspection of Figure 2.) There is only a 2–4 path between (3,4) and (2,4) and it cannot be completed because u must then take color 3 for a proper coloring. There is only a 2–3 path between (3,4) and (3,2) and it cannot be completed because u must then take color 4 for a proper coloring. Finally, there is neither a 2–3 nor a 2–4 path between (3,4) and (3,4). Thus, for the near-triangulation G_{xy} in Figure 2, neither a 2–3 nor a 2–4 path between v and u can exist if (a,b) is (3,4). Interchanging colors 3 and 4 throughout G_{xy} leads to the same conclusion when (a, b) is (4,3). The bipartite graph in Figure 6 explains the Birkhoff diamond's ability to enable a Kempe-locked triangulation. With only minor modifications, this methodology can be used to show that each near-triangulation in Figure 4 derives from a Kempe-locked triangulation.

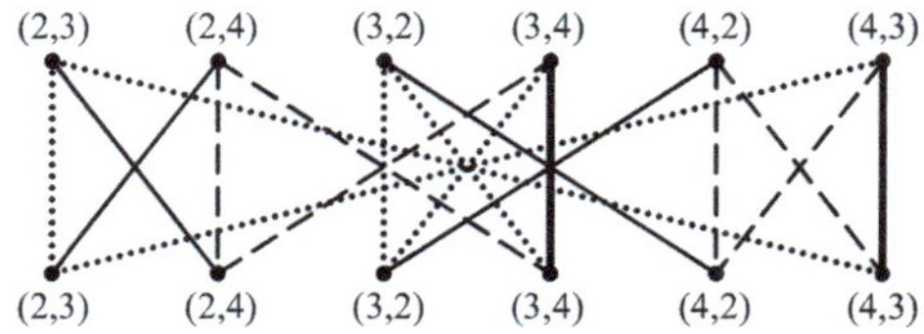

Figure 6. A bipartite graph that shows for the Birkhoff diamond (see Figure 3) the possible transitions from the input pair of colors (a, b) at the top of the graph to the output pair of colors (c, d) at the bottom of the graph.

Appealing to the bipartite methodology may seem like overkill. Why not just use Figure 2 to prove that there is no 2–3 or 2–4 path between v and u? For such an uncomplicated situation, the bipartite analysis is indeed unnecessary, but it becomes highly useful in more complicated situations in which it is not feasible, either manually or with the aid of a computer, to list all possible distinct 4-colorings in which x and y are both colored 1. In such situations, representing a near-triangulation G_{xy} as a stack of xy-configurations and then analyzing each configuration in the stack separately permits us to determine more easily whether a 2–3 or 2–4 path can be transmitted from v to u. Moreover, Figure 5, combined with the bipartite analysis, sheds light on how an xy-configuration can prevent the transmission or completion of a 2-color path, whereas simply analyzing a long list of all distinct colorings of a candidate near-triangulation does not.

In Section 5, we stated that not all triangulations that include a Birkhoff diamond subgraph with endpoints x and y are Kempe-locked. The bipartite methodology can be used to understand why this statement is true. From Figure 6, we see that all coloring states of the top edge (a, b) other than (3,4) and (4,3) can transmit both 2–3 and 2-4 paths through the diamond to its bottom edge (c, d). Thus, many near-triangulations G_{xy} having a Birkhoff diamond subgraph with endpoints x and y will not be Kempe-locked. The mere presence of a Birkhoff diamond is not sufficient—if there is any chance for the triangulation T of which the Birkhoff diamond is a subgraph to be Kempe-locked, the Birkhoff diamond has to appear in just the right way relative to the other xy-configurations in the vertical stack. As an example of this, take the graph in the lower right panel of Figure 4 and eliminate the middle xy-configuration entirely; squeeze the two Birkhoff diamonds together so that the bottom boundary of the upper diamond coincides with the top boundary of the lower diamond, then apply Figure 6 twice to verify that the triangulation with x and y joined is not Kempe-locked with respect to xy.

The foregoing discussion motivates the following definitions of Kempe-locking configurations and fundamental Kempe-locking configurations, those particular xy-configurations that enable a triangulation T to be Kempe-locked with respect to an edge xy.

Definition 1. *Let T be a planar triangulation that is at least 4-connected and G_{xy} the near-triangulation derived from it by deleting the edge xy. Let $uxvy$ be the 4-cycle delineating the 4-face of G_{xy}. If T is Kempe-locked with respect to the edge xy, then the xy-configuration K_{xy} induced by all vertices of G_{xy} other than u and v is the Kempe-locking configuration for T.*

Definition 2. *Let K_{xy} be the Kempe-locking configuration for a planar triangulation T (that is at least 4-connected) with an edge xy. Then K_{xy} is said to be a fundamental Kempe-locking configuration if it contains* *no proper subgraph K'_{xy} that is the Kempe-locking configuration for some planar triangulation T' (that is at least 4-connected) with an edge xy.*

On the basis of these definitions, we see readily that none of the Kempe-locking configurations for the graphs in Figure 4 is fundamental, since, in each case, the Kempe-locking configuration is obtained by deleting vertices v and u and their incident edges. We note that each such Kempe-locking configuration has a Birkhoff diamond with endpoints x and y as a proper subgraph and the Birkhoff diamond is the Kempe-locking configuration for the triangulation obtained by joining x and y in the near-triangulation shown in Figure 2. We will see in Section 7 that the Birkhoff diamond is actually a fundamental Kempe-locking configuration.

7. The Search for Fundamental Kempe-Locking Configurations

The approach we adopted in searching for fundamental Kempe-locking configurations is analogous to that of experimental physicists in their search for a new elementary particle with specified properties. Physicists confine their explorations to collision events involving total energy in an interval sufficient to bring the sought-after particle into being. Similarly, we explored planar triangulations of orders in which fundamental Kempe-locking configurations would be expected to be found. The best chance to discover those configurations would seem to occur when there is a small number of distinct 4-colorings of a near-triangulation G_{xy} with x and y the same color, thus admitting the possibility that every one of those few colorings will feature three Kempe chains that include both x and y. Equivalently (by means of Theorem 2), we note that in a G_{xy} of large order with x and y colored the same there are sufficiently many possibilities for 2-color paths from v to u using colors different from the common color of x and y to render it a virtual certainty that at least one such 2-color path will form part of a proper coloring of G_{xy}. Quite simply, the combinatorial possibilities for coloring a high-order planar triangulation militate against the existence of a high-order fundamental Kempe-locking configuration.

To assure that we did not miss any low-order triangulations, we generated the full set of 8044 isomorphism classes for 4-connected triangulations of orders 6–15 and the full set of 9733 isomorphism classes for 5-connected triangulations of orders 12–24 (see [10,11]). Because Kempe-locking is associated with an edge of a triangulation, we tested every edge separately in a triangulation from each of the isomorphism classes. Referring to the edge being tested as xy, we deleted the edge and then determined whether there are any 4-colorings with x and y colored the same but with fewer than three Kempe chains that include both x and y. It turns out to be rare for this not to be the case. The only Kempe-locked triangulations we encountered were 4-connected; there were none at all among 5-connected triangulations (and hence none at all among internally 6-connected triangulations). There are no Kempe-locked triangulations of order less than 12 and a single 4-connected Kempe-locked triangulation of order 12, the one illustrated in Figure 2 with the leftmost and rightmost vertices on the 4-face joined. Thus, the Birkhoff diamond is a fundamental Kempe-locking configuration. There are no Kempe-locked triangulations of order 13, and one 4-connected Kempe-locked triangulation for each of orders 14 and 15, each of those featuring a Birkhoff diamond with endpoints x and y and each Kempe-locked with respect to only a single edge.

In an expanded search for fundamental Kempe-locking configurations, we examined 4-connected triangulations of orders 16–20. Because the number of isomorphism classes grows rapidly with increasing order (from 30,926 at order 16 to 24,649,284 at order 20—refer to [11]) and because the number of edges in a triangulation increases with increasing order (a planar triangulation of order n has $3n - 6$ edges—see [1]), we soon ran into computation-time limitations imposed by our laptop computer. After deciding to limit aggregate computer execution time to several weeks instead of several months, we proceeded in the expanded search by examining all 30,926 isomorphism classes of order 16 and all 158,428 isomorphism classes of order 17, but only 100,000 randomly generated non-isomorphic triangulations for each order from 18 through 20.

For orders 16 and 17, we discovered 8 and 14 non-isomorphic triangulations, respectively, that are Kempe-locked, all with respect to a single edge, call it xy in each case. Each of those 4-connected Kempe-locked triangulations features a Birkhoff diamond with x and y as endpoints. Figure 4 shows the G_{xy} derived from two such Kempe-locked triangulations. In the random samples of 100,000 triangulations each for orders 18–20, we discovered additional non-isomorphic Kempe-locked triangulations: 10 of order 18, 8 of order 19, and 5 of order 20, all locked with respect to a single edge xy and all featuring a Birkhoff diamond configuration with x and y as endpoints. Figure 4 also shows near-triangulations G_{xy} arising from Kempe-locked triangulations of orders 19 and 20. It is an open question whether there are any triangulations that are at least 4-connected and Kempe-locked with respect to more than a single edge.

Let us take stock of the results of our search for fundamental Kempe-locking configurations. We found no 5-connected (and hence no internally 6-connected) Kempe-locked triangulations of orders 12–24. The only fundamental Kempe-locking configuration we found is the Birkhoff diamond of order 10. We have shown that there are no fundamental Kempe-locking configurations of orders 9 or lower or between 11 and 15, inclusive, and a sample of 100,000 non-isomorphic 4-connected triangulations each for orders 18–20 turned up no fundamental Kempe-locking configurations of orders 16–18. We conclude that the "experimental" case is strong that the Birkhoff diamond is the only fundamental Kempe-locking configuration.

Conjecture 2. *The Birkhoff diamond is the sole fundamental Kempe-locking configuration.*

Conjecture 2 implies Conjecture 1. To see this, refer to the vertex labeling for the Birkhoff diamond given in Figure 3. We observe that a triangulation T that is Kempe-locked with respect to an edge xy by means of a Birkhoff diamond subgraph with endpoints x and y is at most 4-connected because both $xaby$ and $xcdy$ are separating 4-cycles. Hence, all triangulations we discovered that exhibit Property 2 (Kempe-locking) are 4-connected and cannot satisfy Property 1 (internal 6-connectivity). Moreover, our computer search through low-order isomorphism classes of 5-connected triangulations found no Kempe-locked triangulations. Hence, every triangulation that satisfied Property 1 (internally 6-connected triangulations are a subset of all 5-connected triangulations—refer to the discussion in the first paragraph of Section 3) did not satisfy Property 2. To sum up, the "experimental" results presented and discussed in this article (strongly) suggest that Property 1 and Property 2 are incompatible.

Furthermore, if Conjecture 2 is true, then a minimum counterexample simply cannot be constructed because the vertices x and y serving as endpoints for any edge xy that is Kempe-locked must have degree at least 6. Because the signature feature of a Birkhoff diamond is its four central vertices of degree 5, none of the five edges involving pairs of those four central vertices can be Kempe-locked and, by Theorem 1, a counterexample cannot exist. Conjecture 2 alone has sufficient power to imply 4-colorability of planar graphs.

8. An Even More Fundamental Graph

Based on the arguments and analysis presented in this article, it is highly likely that the Birkhoff diamond serves an essential role in creating a Kempe-locked triangulation. Thus, it is illuminating to consider an even more fundamental graph from which the Birkhoff diamond naturally arises. Key to the notion of Kempe-locking with respect to a deleted edge xy is that the vertices x and y have the same color and that they cannot be made to differ in color through any sequence of Kempe exchanges. Consequently, we need to analyze all possible 4-colorings of Figure 3 in which x and y have the same color. That set consists of the six distinct 4-colorings shown in Figure 2 as restricted to the Birkhoff diamond subgraph. In turn, we note that that particular set of six distinct 4-colorings is identical to the set of all possible 4-colorings of the triangulation of order 9 formed by coalescing x and y into a single vertex of degree 6. That triangulation can be drawn as depicted in Figure 7—a triangle enclosing a pentagon enclosing a single vertex. It is quite satisfying to think that such an elegant graph might lie at the heart of why a minimum counterexample to the 4-color theorem does not exist, at the heart of why all planar triangulations must be 4-colorable.

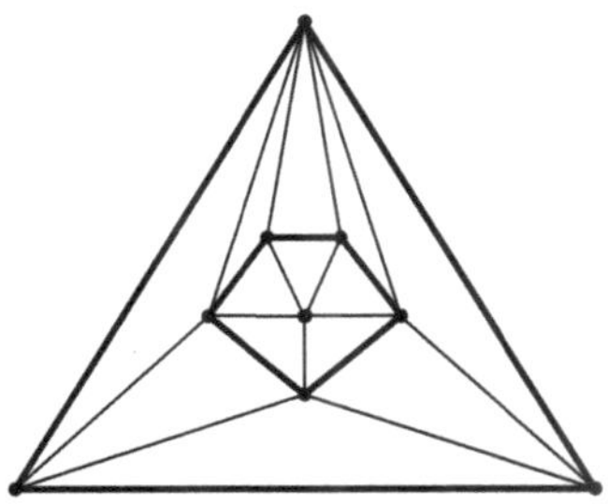

Figure 7. This 4-connected triangulation of order 9 is likely the key to why a minimum counterexample to the 4-color theorem does not exist. The Birkhoff diamond of order 10 results when the topmost vertex is split into two vertices, each of degree 3.

9. Conclusions and Future Directions

We have introduced the key notion of Kempe-locking and have used it to establish the coloring property that a minimum counterexample to the 4-color theorem must satisfy with respect to each of its edges. After an exhaustive search through low-order isomorphism classes of 4-connected and 5-connected triangulations, we found no 5-connected triangulations that are Kempe-locked. The only Kempe-locked triangulations we found are Kempe-locked with respect to only a single edge, far from what is necessary for a minimum counterexample. Moreover, each discovered Kempe-locked triangulation, Kempe-locked with respect to an edge xy, features a Birkhoff diamond with endpoints x and y. The mere presence of a Birkhoff diamond subgraph renders the triangulation at most 4-connected. These results strongly suggest that the connectivity and coloring properties of a minimum counterexample are incompatible. They also point to the likely singular role that the Birkhoff diamond and its progenitor, the single-vertex-within-a-pentagon-within-a-triangle graph, play in the matter of the 4-colorability of planar graphs.

What future directions does this work indicate? First, since the matter of graph coloring has applications to various problems in graph theory, the notion of Kempe-locking may find practical application beyond its theoretical use in this article. Second, with respect to further experimental studies, it would be helpful if researchers with substantially more powerful computing resources than a laptop computer could complete the search for Kempe-locked configurations among the isomorphism classes for 4-connected triangulations of orders 18, 19, and 20, and perhaps for even higher orders. Such extensions of the study reported here would either solidify support for or refute Conjecture 2. Because Conjecture 2 implies Conjecture 1 and lends itself to an appealing explanation for why a planar graph must be 4-colorable, it is the conjecture on which to focus. Even better than further experimental studies would be a direct proof of Conjecture 2, but that may well be a more difficult task than was proving the 4-color theorem.

Funding: This research received no external funding.

Acknowledgments: The author would like to thank one of the reviewers for expressing the view that the near-triangulation method presented in this article is promising, with potential applications to the calculation of spanning trees in some graphs. In particular, the reviewer mentioned work on the number of spanning trees, Laplacian eigenvalues, and the Laplacian Estrada index of subdivided-line graphs.

Conflicts of Interest: The author declares no conflict of interest.

Appendix A

In this appendix, we prove that no 4-colorable planar triangulation can be Kempe-locked with respect to every one of its edges. We begin with a few preliminary demonstrations.

Theorem A1. *If T is a 4-colorable planar triangulation that is Kempe-locked with respect to the edge xy, then every 4-coloring of T has a 2-color cycle including the edge xy.*

Proof. Suppose there is a 4-coloring χ of a planar triangulation T that is Kempe-locked with respect to xy in which there is *not* a 2-colored cycle including both x and y. Without loss of generality, let the colors of x and y in χ be 1 and 2, respectively. Then, in the near-triangulation G_{xy} formed by deleting the edge xy, the coloring χ (as it applies to G_{xy}) can be transformed into a coloring χ' by performing a 1–2 interchange of colors on the Kempe chain that includes y but does not include x. In χ', both x and y have color 1. However, by the statement of Property 2, T then cannot be Kempe-locked, contrary to our initial supposition. □

Lemma A1. *Every 2-color cycle in a planar triangulation must enclose at least one vertex.*

Proof. Any cycle with a fully-triangulated interior and no interior vertices requires at least three colors for a proper vertex-coloring because any triangle demands three colors. □

Theorem A2. *No 4-colorable planar triangulation can be Kempe-locked with respect to every one of its edges.*

Proof. Suppose that T is a 4-colorable planar triangulation that is Kempe-locked with respect to every one of its edges. Let xy be one such edge and let T be given a 4-coloring in which, without loss of generality, the colors of x and y are 1 and 2, respectively. By Theorem A1, there is in T a 2-color cycle C using the colors 1 and 2 that includes xy. By Lemma A1, C must enclose at least one vertex and because T is a triangulation we know that there is such an enclosed vertex u that is adjacent to both x and y. Without loss of generality, u has color 3. Any vertex adjacent to u must be have color 1, 2, or 4. If none has color 4, then, without loss of generality, we can consider the cycle C to enclose only u. If there is a vertex v colored 4 adjacent to u, then because T is Kempe-locked with respect to uv, by Theorem A1 there must be a 2-color cycle enclosed by C that uses the colors 3 and 4 and includes the edge uv. By Lemma A1, this 3–4 cycle must enclose at least one vertex. Because T has finite order and because T is assumed to be Kempe-locked with respect to every one of its edges, such "telescoping" of 2-color cycles must terminate at a 2-color cycle that encloses a single vertex. Thus, without loss of generality, we can assume that such a cycle is, in fact, the starting cycle C colored 1–2 that includes xy and encloses a single vertex u colored 3. We now drop the vertex labels x and y.

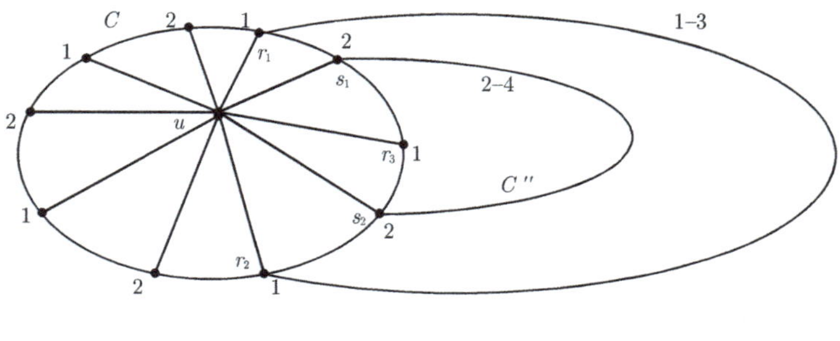

Figure A1. The "telescoping" process used to prove Theorem A2.

Let r_1 be a vertex in C colored 1. By assumption, T is Kempe-locked with respect to the edge ur_1. Thus, there must be a vertex $r_2 \neq r_1$ in C also colored 1 so that the path r_1ur_2 is part of a 2-color cycle

C' using the colors 1 and 3. See Figure A1. Recolor u from 3 to 4. Let s_1 be a vertex colored 2 in C that lies between r_1 and r_2 and is enclosed by C' and suppose it is the only vertex lying between r_1 and r_2 that is enclosed by C'. Then a 2-color cycle using the colors 2 and 4 that includes the edge us_1 cannot exist and T cannot be Kempe-locked with respect to us_1. However, if there is a vertex $s_2 \neq s_1$ colored 2 lying between r_1 and r_2, then by Theorem A1 a 2-color cycle C'' using the colors 2 and 4 exists and C'' is essentially enclosed by C', the two cycles touching only at u. Recolor u from 4 to 3 and let r_3 be a vertex colored 1 that lies between s_1 and s_2 and is enclosed by both C' and C''. If r_3 is the only vertex lying between s_1 and s_2 that is enclosed by C'', then the edge ur_3 cannot be Kempe-locked. However, if there is a vertex $r_4 \neq r_3$ colored 1 lying between s_1 and s_2, then by Theorem A1 there must be a 2-color cycle including the path r_3ur_4 that uses the colors 1 and 3. Because the order of T is finite, this "telescoping" process of switching the color of u between 3 and 4 and finding "nested" cycles touching only at u cannot continue indefinitely; at some point, there will be either a single vertex in C colored 1 between vertices in C colored 2 (with u colored 3) or a single vertex in C colored 2 between vertices in C colored 1 (with u colored 4). In either case, the edge containing u and that single vertex will be one for which T is not Kempe-locked, contradicting the initial supposition. □

References

1. Chartrand, G.; Zhang, P. *A First Course in Graph Theory*; Dover Publications, Inc.: Mineola, NY, USA, 2012.
2. Wilson, R. *Four Colors Suffice*; Princeton University Press: Princeton, NJ, USA, 2002.
3. Birkhoff, G. The reducibility of maps. *Am. J. Math.* **1913**, *70*, 429–490. [CrossRef]
4. Appel, K.; Haken, W. Every planar map is four colorable, Part I: Discharging. *Ill. J. Math.* **1977**, *21*, 429–490.
5. Appel, K.; Haken, W.; Koch, J. Every planar map is four colorable, Part II: Reducibility. *Ill. J. Math.* **1977**, *21*, 491–567.
6. Robertson, N.; Sanders, P.; Seymour, P.; Thomas, R. The four-color theorem. *J. Comb. Theory B* **1997**, *70*, 2–44. [CrossRef]
7. Kempe, A. On the geographical problem of four colours. *Am. J. Math.* **1879**, *2*, 193–200. [CrossRef]
8. Tilley, J. Using Kempe exchanges to disentangle Kempe chains. *Math. Intell.* **2018**, *40*, 50–54. [CrossRef]
9. Tilley, J. The a-graph coloring problem. *Discret. Appl. Math.* **2017**, *217*, 304–317. [CrossRef]
10. Brinkmann, G.; McKay, B. Construction of planar triangulations with minimum degree 5. *Discret. Math.* **2005**, *301*, 147–163. [CrossRef]
11. Brinkmann, G.; McKay, B. Fast generation of planar graphs. *MATCH Commun. Math. Comput. Chem.* **2007**, *58*, 323–357.

mathematics

MDPI

Article

The Bounds of the Edge Number in Generalized Hypertrees

Ke Zhang [1,2,3], Haixing Zhao [1,2,3,*], Zhonglin Ye [2,3,4], Yu Zhu [1,2,3] and Liang Wei [5]

1 School of Computer, Qinghai Normal University, Xining 810008, China; hbsanli@163.com (K.Z.); sdcxzhuyu@163.com (Y.Z.)
2 Key Laboratory of Tibetan Information Processing and Machine Translation in QH, Xining 810008, China; zhonglin_ye@foxmail.com
3 Key Laboratory of the Education Ministry for Tibetan Information Processing, Xining 810008, China
4 School of Computer Science, Shannxi Normal university, Xi'an 710062, China
5 School of mathematics and statistics, Qinghai Normal University, Xining 810008, China; weiliang@qhnu.edu.cn
* Correspondence: h.x.zhao@163.com

Received: 11 November 2018; Accepted: 15 December 2018; Published: 20 December 2018

Abstract: A hypergraph $H = (V, \varepsilon)$ is a pair consisting of a vertex set V, and a set ε of subsets (the hyperedges of H) of V. A hypergraph H is r-uniform if all the hyperedges of H have the same cardinality r. Let H be an r-uniform hypergraph, we generalize the concept of trees for r-uniform hypergraphs. We say that an r-uniform hypergraph H is a generalized hypertree (GHT) if H is disconnected after removing any hyperedge E, and the number of components of $GHT - E$ is a fixed value k $(2 \leq k \leq r)$. We focus on the case that $GHT - E$ has exactly two components. An edge-minimal GHT is a GHT whose edge set is minimal with respect to inclusion. After considering these definitions, we show that an r-uniform GHT on n vertices has at least $2n/(r+1)$ edges and it has at most $n - r + 1$ edges if $r \geq 3$ and $n \geq 3$, and the lower and upper bounds on the edge number are sharp. We then discuss the case that $GHT - E$ has exactly k $(2 \leq k \leq r-1)$ components.

Keywords: hypergraph; generalized hypertree; bound; component

1. Introduction

Graph theory is a powerful tool in exploring various network types, and in many areas of application, such as communication, biology, and chemistry. A key limit of graphs is that edges can only model the pairwise relations among the objects. This limitation stands out in many real complex systems. For example, in scientific collaboration networks, if one paper has more than two authors and another paper has two of the authors among these authors in the first paper, then such relationship can no longer be modeled by a graph. A hypergraph is a family of non-empty subsets (hyperedges) from a fixed set of vertices. It is a natural extension of a graph. It thus captures multivariate relationships in complex networks [1–5]. Examples include communities and collaboration teams in social networks, cliques in communication networks, and molecules in chemistry.

Trees are among the most essential, useful, and understandable objects in all of graph theory, which is also common sense in hypergraph theory. In this paper, we generalize the concept of trees for hypergraphs and prove bounds on the edge numbers of such hypergraphs.

The concept of a tree in graph theory has many equivalent descriptions which can reflect different properties of the tree from different perspectives [6]. The analogous situation for hypergraphs is unclear, because the descriptions of the corresponding concept of a tree in hypergraph theory is more complicated.

In 1989, Berge [7] proposed the concept of an arboreal hypergraph based on a property of hypergraphs and one of concepts of cycles in hypergraph theory. In 1999, Nieminen et al. [8] introduced two of generalizations of a tree into hypergraphs known as hypertrees and weak hypertrees, respectively. Recently, Katona et al. [9] discussed bounds on the edge number in another interesting generalization of trees also called hypertrees.

It is well-known from elementary graph theory that any two vertices in a tree are connected by a unique path. This description of the definition of a tree can easily reflect the edge-connectivity property of the tree and means that the tree would be disconnected if any edge is deleted from the tree. Owing to the definition of a tree in graph theory, it has exactly two components by deleting any edge e from the tree $T(V,\ E)$, i.e., $\forall e \in E(T), \omega(T-e)=2$. On the one hand, it is meaningful to define the similar structures in hypergraphs by this kind of method. On the other hand, the concept of the hypertree we usually consider (as the same in this paper) is a special case of the concept of the hypertree defined by Nieminen. A connected hypergraph $HT(V,\ \varepsilon)$ is a hypertree if the removal of any of its edges E results in a disconnected hypergraph $HT-E$ and the number of components of the hypergraph $HT-E$ is $|E|$. This concept of the generalization of a tree to hypergraphs has been introduced and studied in the literature, which can be regarded as a continuation of the classical generalization effort. For more details, see [8,10,11].

In the following, we further develop the concept proposed by Nieminen into the generalized hypertrees (GHT). Our main concern is that the number of components of $GHT-E$ is exactly 2 for any edge E of the GHT. After introducing the nomenclature and definitions in hypergraphs, we discuss lower and upper bounds on the edge number of $GHTs$.

Now, we want to clarify some basic terms and results in hypergraphs about our research object. The undefined concepts in this paper can be found in [7,12] for hypergraphs and [6] for graphs.

Nomenclature and Definitions

A hypergraph H is a pair $(V,\ \varepsilon)$ where V is the vertex set of H and ε is a collection of the nonempty subsets of V. An element in ε is a hyperedge, and it is also called an edge of H. Note that we do not allow a hypergraph to have isolated vertices which are not included in any edge. The order of the hypergraph $H=(V,\ \varepsilon)$ is the cardinality of V, i.e., $|V|=n$; its size is the cardinality of ε, i.e., $|\varepsilon|=m$.

The degree of a vertex v in H, denoted by $d_H(v)$ or $d(v)$, is the number of edges in H containing v. Let $\varepsilon=\{E_1,\ E_2,\ \cdots,\ E_m\}$, the cardinality of an edge $E_i \in \varepsilon$ is $|E_i|$. A hypergraph H is simple if $E_i \subseteq E_j$ implies that $i=j$ for any $i,\ j$ with $1 \le i,\ j \le m$. We consider only simple hypergraphs, unless specified otherwise. A hypergraph H is linear if it is simple and $|E_i \cap E_j| \le 1$ for any $i,\ j$ with $1 \le i,\ j \le m$. Two vertices in a hypergraph are adjacent if there is an edge which contains both vertices. Two edges in a hypergraph are adjacent if their intersection is not empty. A hypergraph H is r-uniform if $|E_i|=r$ for all i with $1 \le i \le m$. Thus, a simple graph is a simple 2-uniform hypergraph, and vice versa.

Let $H=(V,\ \varepsilon)$ be a hypergraph, and let $k \ge 2$ be an integer. An $(u,\ v)$-walk in H is a finite vertex-edge alternative sequence $W=(u=v_0,\ E_1,\ v_1,\ E_2,\ v_2,\cdots,\ E_k,\ v_k=v)$, such that $v_i,\ v_{i+1} \in E_{i+1}$, for any i with $0 \le i \le k-1$. The integer k is the length of the walk W. The shortest length of $(u,\ v)$-walk is the distance (also called path length) between u and v, which is denoted by $d(u,\ v)$. If $v_0, v_1, \cdots, v_k$ are distinct vertices, $E_1, E_2, \cdots, E_k$ are distinct edges, and $u=v_0=v_k=v$, the walk is called a cycle of length k in H. A hypergraph is connected if for any pair of vertices, there is a walk which connects both vertices; it is disconnected otherwise. It means that a connected hypergraph has exactly one component and a disconnected hypergraph has more than one component. We denote the number of components of the hypergraph H by $\omega(H)$. In Figure 1, we show an illustrative example of a hypergraph with seven vertices where the vertices are connected by four edges.

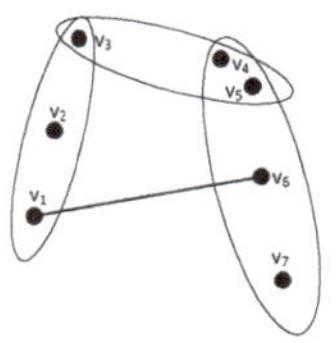

Figure 1. The above hypergraph H: $V = \{v_1, v_2, \cdots, v_7\}$, and $\varepsilon = \{\{v_1, v_2, v_3\}, \{v_3, v_4, v_5\}, \{v_4, v_5, v_6, v_7\}, \{v_1, v_6\}\}$. The degree of v_1 is 2. Hypergraph H is simple but not linear.

The notion of edge-connectivity $\lambda(G)$ of a graph G, defined as the minimum number of edges whose removal renders G disconnected, has attracted much attention [13,14]. The definition of edge-connectivity of graphs extends to hypergraphs in a natural way [15].

The following Theorem 1 is the basic conclusion of describing hypertrees from the perspective on edge-connectivity. By the definitions of a tree and a hypertree, it is also true for the tree in graph theory.

Theorem 1. *Let H be a connected hypergraph on n vertices. If H is a hypertree then* $\sum_{i=1}^{m} (|E_i| - 1) = n - 1$ *[10].*

By the concept of a hypertree HT, the number of components in the hypergraph $HT - E$ is $|E|$ after deleting any edge E from HT, i.e., $\forall E \in \varepsilon(HT)$, $\omega(HT - E) = |E|$, and vice versa (see Figure 2).

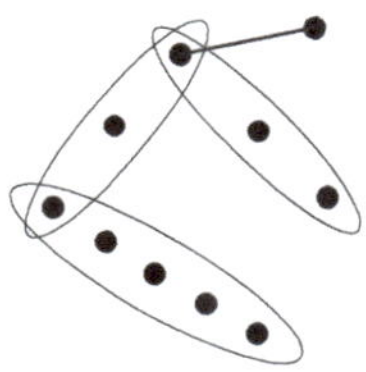

Figure 2. An example hypertree.

For the r-uniform hypergraph, Lai et al. obtained another related result in [16] as follows:

Theorem 2. *Let H be an r-uniform hypergraph on n vertices. If H is connected, then* $m(H) \geq (n-1)/(r-1)$. *Moreover, the equality holds if and only if for any edge* $E \in \varepsilon(H)$, $H - E$ *has r components.*

It gives the lower bound of the edge number of connected r-uniform hypergraphs and it describes the class of hypergraphs when the equality holds. These hypergraphs, commonly known as hypertrees, have results of deep studies because they have special structures [11]. However, it does not mean that this concept is the analogous situation of a tree in hypergraphs. In this paper, we call an analogous concept a generalized hypertree which attempts to give an in-depth description of the tree structure in hypergraphs.

Definition 1. (Generalized hypertree). *A connected hypergraph GHT is a generalized hypertree if the removal of its any edge E results in a disconnected hypergraph GHT − E and the number of components of GHT − E is exactly k* $(k \geq 2)$.

From a certain perspective, this definition is also a continuation of the common trees in hypergraphs. This definition is illustrated Figure 3.

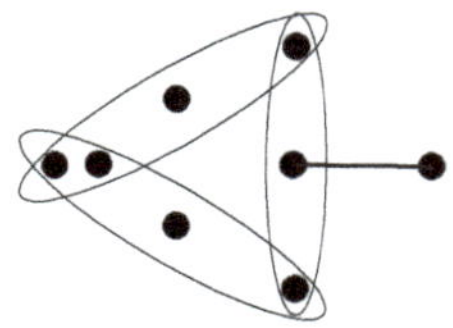

Figure 3. A generalized hypertree GHT: $\omega(GHT - E) = 2$ for all any edge E.

Our main results are the lower and upper bounds on the edge number of $GHTs$. Theorem 3 in Section 2 generalizes the well-known fact that a tree $T(V, E)$ on n vertices has $|E(T)| = n - 1$ edges. It is the most important result in this paper. It turns out that the lower bound in Theorem 3 can be improved for $GHT - E$ which has exactly k components for any edge $E \in \varepsilon(GHT)$ if $2 \leq k \leq r - 1$. The corresponding result is Theorem 4 whose discussion and proof are presented in Section 2. Based on the notion of generalized hypertree, Theorem 5 and Corollary 2 in Section 3, modify Theorems 1 and 2, respectively. When the lower bound holds in Corollary 2, we consider the relationship between generalized hypertrees and hypertrees, as shown in Section 4.

2. The Bounds of the Edge Number of *r*-Uniform *GHTs*

In this section, we mainly research the bounds (lower bound and upper bound) on edge number of r-uniform $GHTs$ and the number of components in hypergraph $GHT - E$ is exactly 2.

Theorem 3. *Let H be a connected r-uniform hypergraph on n vertices ($r \geq 3$, $n \geq 3$). If $H - E$ has exactly two components for any edge $E \in \varepsilon(H)$, then $2n/(r+1) \leq m(H) \leq n - r + 1$. Moreover, the lower and upper bounds are sharp.*

In order to prove Theorem 3, we need the following preliminaries.

A classical result for a graph $G = (V, E)$ with $V = \{v_1, v_2, \cdots, v_n\}$ and $E = \{e_1, e_2, \cdots, e_m\}$ states that $\sum\limits_{v \in V} d(v) = 2m$. Its parallel generalization in hypergraph theory is as follows.

Proposition 1. *Let $H = (V, \varepsilon)$ be a hypergraph, we have [12]*

$$\sum_{v \in V} d(v) = \sum_{E \in \varepsilon} |E|,$$

when H is an r-uniform hypergraph, the above equation can be simplified to

$$\sum_{v \in V} d(v) = rm.$$

The characterization of the connected r-uniform hypergraphic sequences was given by Tusyadej [7]. The following generalization of this result was proposed by Lai at al [16].

Proposition 2. *A n-tuple $d_1 \geq d_2 \geq \cdots \geq d_n$ is the degree sequence of a k-edge connected r-uniform hypergraph if and only if*

(1) $d_i \geq k$ *for* $i = 1, 2, \cdots, n$;

(2) $\sum\limits_{i=1}^{n} d_i \geq \frac{r(n-1)}{r-1}$ *if* $k = 1$.

Lemmas 1 and 2 show us some of the characteristics about the r-uniform GHT whose component number exactly equals 2 after deleting any edge.

Lemma 1. *Let H be a connected r-uniform hypergraph on n vertices ($r \geq 3$, $n \geq 3$). If $H - E$ has exactly two components for any edge $E \in \varepsilon(H)$, then there must exist an edge containing exactly one vertex of degree 1 in H.*

Proof. If the edge containing vertex with degree 1 does not exist in H, then the degree of all vertices of H is equal or greater than 2. By Proposition 2, H is at least 2-edge connected. Therefore, $H - E$ is connected for any edge $E \in \varepsilon(H)$. Therefore, there exists an edge E^* containing one vertex with degree 1 in H. Additionally, if E^* contains more than two vertices with degree 1, then the number of components is more than 2 after deleting E^*. □

Lemma 2. *Let H be a connected r-uniform hypergraph on n vertices ($r \geq 3$, $n \geq 3$). If $H - E$ has exactly two components for any edge $E \in \varepsilon(H)$, then $2 \leq u \leq m$, where u is the number of edges containing exactly one vertex with degree 1 (the same below).*

Proof. Obviously, $u \leq m$. The main effort is the proof for $u \geq 2$. By Lemma 1, we only need to prove $u \neq 1$. Let us suppose that $u = 1$. If $m = 1$, then we can conclude that the result is not correct, because the component number of $H - E$ is r ($r \geq 3$). If $m \geq 2$, let us assume E is the unique edge containing exact one vertex with degree 1. For $E^* \in \varepsilon(H) \backslash \{E\}$, the degree of vertex which contained in edge E^*, is not less than 2. By Proposition 2, we can deduce that $H - E^*$ is connected. It is contrary with the condition. □

Now we are ready to prove Theorem 3.

Proof of Theorem 3. In order to interpret clearly, we divide the bounds into lower bound and upper bound.

Case 1. The proof on lower bound of the edge number in Theorem 3.

By Proposition 1, Proposition 2 and Lemma 2, we conclude that

$$mr = u + \sum_{j=1}^{m-u} d_j.$$

For a fixed value r, we minimize the value on the right side of the above equation. All degrees of vertices bigger than 1 are taken as 2 for this purpose. That is, $d_j = 2$ for all j with $j = 1, 2, \cdots, m - u$. At the same time, we must maximize the number of vertices with degree 1, namely $u = m$. Additionally, we have

$$mr = m + 2(n - m).$$

It follows that

$$\frac{2n}{r+1}.$$

So far, we have completed the proof on lower bound of the edge number in Theorem 3.

Figure 4 is an r-uniform GHT example that the lower bound of edge number holds.

Case 2. The proof on upper bound of the edge number in Theorem 3.

If $r \geq 3$, and $m \geq 2$, then we argue by induction on n.

The base case of this induction is $n = r + 1$. Since the connected r-uniform hypergraph H has two components by deleting any edge, it is easy to see that the edge number of this hypergraph is 2.

Now we assume that the result is a true statement, for the vertex number is equal or less than $n - 1$, then we consider this case that the vertex number equals n. Let H be a connected r-uniform hypergraph on n vertices. If two components are H_1, H_2 after deleting an edge from H, and their edge numbers are m_1 and m_2, and their vertex numbers are n_1 and n_2, respectively, where $n_1 + n_2 = n$, then we discuss the following three cases. By the inductive hypothesis

(i) If $n_1,\ n_2 \geq r+1$, then we have

$$m = m_1 + m_2 + 1 = (n_1 - r + 1) + (n_2 - r + 1) + 1 = n - 2r + 3 < n - r + 1;$$

(ii) If $n_1 = n - r,\ n_2 = r$, then we have

$$m = m_1 + 1 + 1 = [(n-r) - r + 1] + 2 = n - 2r + 3 < n - r + 1;$$

(iii) If $n_1 = n - 1,\ n_2 = 1$, then we have

$$m = m_1 + 1 = [(n-1) - r + 1] + 1 = n - r + 1.$$

According to (i), (ii) and (iii), for H on n vertices, the result is also true. Based on the proof above, Theorem 3 holds. □

Figure 5 is an r-uniform GHT example that the upper bound of edge number holds.

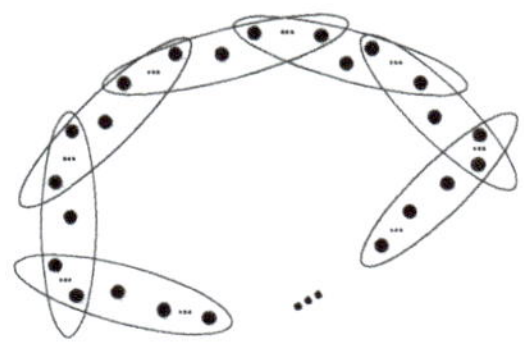

Figure 4. An r-uniform GHT with the minimum number of edges with $\omega = 2$ and $u = m$.

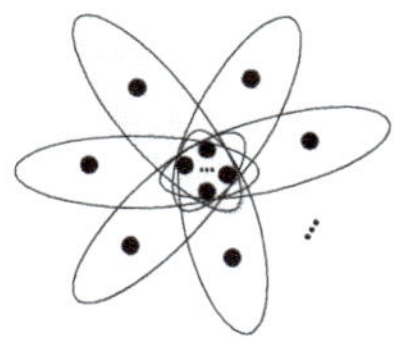

Figure 5. An r-uniform GHT with the maximum number of edges with $\omega = 2$ and $u = m$.

We call the r-uniform hypergraph shown in Figure 5 a sunflower which is denoted by SF_n.

In order to compare with the edge number of a tree in graph theory, we consider another structure in $GHTs$. If $u = 2$, then we have that the number of edges (i.e., edge E), which made none of the two components of $GHT - E$ containing an isolated vertex, is $m - 2$ by Lemma 1. To make m as small as possible, we need to make the degree of each vertex is not greater than 2. Hence $\sum_{i=1}^{n} d_i = 2n - 2$. Then, we have by Proposition 1,

$$m = \frac{2n-2}{r}.$$

For each r with $2 \leq r \leq n-1$,

$$\frac{2n-2}{r} - \frac{2n}{r+1} = \frac{2(n-r-1)}{r(r+1)} \geq 0.$$

Therefore, under the premise of $\omega(GHT - E) = 2$ for any edge E of the GHT, the edge number of this kind of $GHTs$ is not the smallest.

The Figure 6 is an r-uniform GHT example that the edge number is $m = (2n-2)/r$.

Figure 6. An r-uniform GHT with the minimum number of edges with $\omega = 2$ and $u = 2$.

In the previous part, we have discussed the bounds of edge number of a connected r-uniform GHT, whose component number exactly equals 2 by deleting any edge. Theorem 4 gives the bounds of edge number of connected r-uniform $GHTs$ except hypertrees. Theorem 4 below, generalizes Theorem 3.

Theorem 4. *Let H be a connected r-uniform hypergraph on n vertices ($r \geq 3$, $n \geq 3$). If $H - E$ has exactly k ($2 \leq k \leq r-1$) components for any edge $E \in \varepsilon(H)$, then $2n/(r+k-1) \leq m(H) \leq n-r+1$. Moreover, the lower and upper bounds are sharp, and the upper bound holds if and only if H is a SF_n.*

Proof. According to the definition of a connected r-uniform GHT, the maximum number of edges decreases with the increment of k. Therefore, the upper bound is true.

The proof of the lower bound in Theorem 4 is similar to the proof of the lower bound in Theorem 3. To obtain the minimum value of m, each edge of H contains $k-1$ vertices with degree 1 and $r-k+1$ vertices with degrees 2. It follows that

$$m(k-1) + 2[n - m(k-1)] = mr,$$

which implies

$$\frac{2n}{r+k-1}.$$

□

According to the result of Theorem 4, if $r = 2$ and $k = 1$, then we find that the hypergraph H is C_n, the cycle on n vertices.

When the lower bound holds, the hypergraph is not unique. One of such structures is shown in Figure 7, where each edge contains $k-1$ vertices with degree 1.

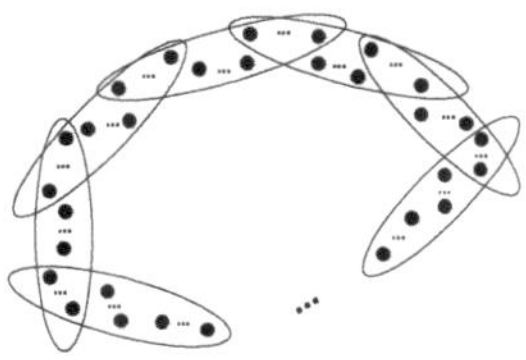

Figure 7. An r-uniform GHT with the minimum number of edges with $\omega = k$, and each. edge contains $k-1$ vertices with degree 1.

Since when $r = 2$, $n - r + 1$ gets the maximum value $n - 1$. Theorem 4 yields the following Corollary 1.

Corollary 1. *Let H be a connected r-uniform hypergraph on n vertices. If H is a GHT, then $m(H) \leq n-1$.*

3. The Edge Number of a Class of Connected r-Uniform Hypergraphs

Theorem 5. *Let H be a connected hypergraph on n vertices. If there exists an edge E leading to the number of components of $H - E$ less than $|E|$, then $\sum_{i=1}^{m} (|E_i| - 1) \geq n$.*

Proof. Since H is connected, and there exists an edge E leading to the number of components of $H - E$ less than $|E|$, hence H is not hypertree definitely. Based on Theorem 1, we have $\sum_{i=1}^{m} (|E_i| - 1) \geq n$. □

Let H be an r-uniform hypergraph, we can get the following result directly.

Corollary 2. *Let H be a connected r-uniform hypergraph on n vertices. If there exists an edge E leading to the number of components of $H - E$ less than r, then $m(H) \geq n/(r-1)$.*

An example hypergraph whose edge number is $n/(r-1)$ are described in Figure 8. In this hypergraph, any pair of adjacent edges share one common vertex.

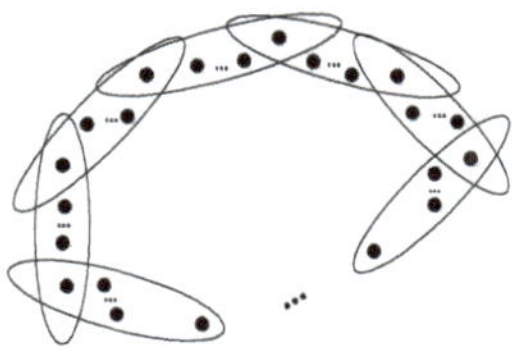

Figure 8. An r-uniform GHT with the minimum number of edges with $\omega = r-1$, and each hyperedge contains $r-2$ vertices with degree 1.

4. Characterizations of a Class of r-Uniform $GHTs$

Let H be a connected r-uniform hypergraph on n vertices. If H is a GHT and $m(H) = n/(r-1)$, then H is generated by identifying two vertices which are from two different edges of the GHT's corresponding hypertree.

If two vertices are contracted together from two nonadjacent edges, the generated hypergraph is a linear hypergraph with a cycle. An example is shown in Figure 9.

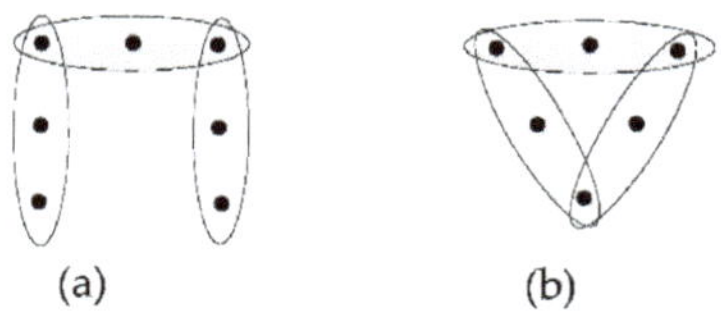

Figure 9. The linear variation (shown as (**b**)) is from the 3-uniform GHT (shown as (**a**)), and the number of vertices of (**b**) is 1 less than that of (**a**).

If two vertices contracted together are from the two adjacent edges, the generated hypergraph is nonlinear. An example is shown in Figure 10.

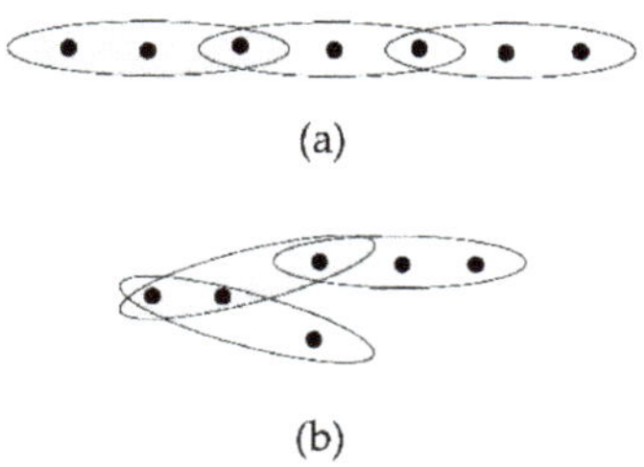

Figure 10. The non-linear variation (shown as (**b**)) is from the 3-uniform GHT (shown as (**a**)), and the number of vertices of (**b**) is 1 less than that of (**a**).

Let us specify the variations from another perspective, then we can obtain the following Theorem 6.

Theorem 6. *Let H be a connected GHT, if $m(H) = n/(r-1)$, then $H - E$ possesses $r-1$ or r (not all r) components for any edge $E \in \varepsilon(H)$.*

5. Conclusions

Many structures in the real world can be modeled with hypergraphs. For example, authors and papers are supposed to vertices and edges in the scientific collaboration hypernetworks based on the hypergraph theory. The hypergraph theory provides us with effective approach to depict relationships among various objects accurately.

A generalized hypertree is the generalization of the concept of a tree in graph theory. The main result of this paper is to obtain the bounds of the edge number in generalized hypertrees, inspiring further study in the direction of edge-minimal and edge-maximal generalizations of trees. Another class of hard questions involve various generalizations which seem to be closely related to hypernetwork reliability. We believe that the challenging open problems in the topic makes it worthy of future study.

The number of components of a tree $T(n)$ happens to be two after deleting an arbitrary edge, and the edge number of trees is $n-1$. However, the description of the corresponding concepts of a tree in hypergraph theory is much more complicated. The reason why we focus on the generalized hypertree whose component number is exactly 2 after deleting any edge, is that it has broad application prospects in real-life network design and network fault detection.

In graph theory, the star $S(n)$ and the path $P(n)$ are two classes of graphs with special properties. For example, $S(n)$ is the most reliable structure in $\mathcal{T}(n)$ (set of trees on n vertices), and $P(n)$ is the worst reliable structure in $\mathcal{T}(n)$ under the situation of vertex failure. For another example, $S(n)$ is the tree of the smallest average path length in $\mathcal{T}(n)$, and $P(n)$ is the largest one. In this paper, two extreme hypergraphs whose component number is exactly 2 after deleting an arbitrary edge, are not the correspondence hypergraphs of $S(n)$ and $P(n)$ in $\mathcal{T}(n)$. Thus, the corresponding structures and properties in graph theory are not necessarily established in hypergraph theory, even if it is a natural generalization of some basic concepts in graph theory. As mentioned in many references [1,12], the properties of the hypergraph cannot be simply transformed into corresponding graphs to study and analyze, such as the two-section graph, the associated graph and the line graph, etc. Therefore, we should have to pay more attention to study of the hypergraph theory itself.

Author Contributions: Conceptualization, K.Z.; Funding acquisition, H.Z.; Investigation, K.Z.; Methodology, H.Z.; Project administration, H.Z.; Resources, K.Z.; Software, K.Z., Z.Y., Y.Z. and L.W.; Supervision, H.Z.; Visualization, K.Z.; Writing—Original Draft, K.Z.; Writing—Review & Editing, Z.Y., Y.Z. and L.W. All authors read and approved the final version of the paper.

Funding: This research is supported by the National Natural Science Foundation of China (Grant Number: 11661069, 61663041 and 61763041), by the Science Found of Qinghai Province (Grant Number: 2018-ZJ-718 and 2015-ZJ-718) and by the Fundamental Research Funds for the Central Universities, China (Grant Number: 2017TS045).

Acknowledgments: The authors wish to express their gratitude to the anonymous reviewers and the associate editors for their rigorous comments and valuable suggestions improved this paper.

Conflicts of Interest: The authors declare no conflict of interest.

References

1. Estrada, E.; Rodriguez-Velazquez, J.A. Subgraph entrality and clustering in complex hyper-networks. *Phys. A* **2006**, *364*, 581–594. [CrossRef]
2. Johnson, J. Hypernetworks for reconstructing the dynamics of multilever systems. Presented at the European Conference on Complex Systerns, Said Business School, Oxford, UK, 25–29 September 2006.
3. Guo, J.L.; Suo, Q. Brand effect versus competitiveness in hypernetworks. *CHAOS* **2015**, *25*, 023102. [CrossRef] [PubMed]
4. Zhang, H.; Song, L.; Han, Z.; Zhang, Y. *Hypergraph Theory in Wireless Communication Networks*; Springer: New York, NY, USA, 2017.
5. Konstantinova, E.V.; Skorobogatov, V.A. Application of hypergraph theory in chemistry. *Discret. Math.* **2001**, *235*, 365–383. [CrossRef]
6. Bondy, J.A.; Murty, U.S.R. *Graph Theory*; Springer: New York, NY, USA, 2008.

7. Berge, C. *Hypergraphs: Combinatorics of Finite Sets*; Elsevier: Amsterdam, The Netherlands, 1989.
8. Nieminen, J.; Peltola, M. Hypertrees. *Appl. Math. Lett.* **1999**, *12*, 35–38. [CrossRef]
9. Katona, G.Y.; Szabó, P.G.N. Bounds on the number of edges in hupertrees. *Discret. Math.* **2016**, *339*, 1884–1991. [CrossRef]
10. Xu, X.M.; Sun, Y.G.; Yang, S. Hypergraph Theory with Applications. *Acta Electron. Sin.* **1994**, *22*, 65–71.
11. Goodall, A.; Mier, D.A. Spanning trees of 3-uniform hypergraphs. *Adv. Appl. Math.* **2011**, *47*, 840–868. [CrossRef]
12. Alain, B. *Hypergraph Theory: An Introduction*; Springer: New York, NY, USA, 2013.
13. Hellwig, A.; Volkmann, L. Maximally edge-connected graphs and digraphs-a survey. *Discret. Math.* **2008**, *308*, 3265–3296. [CrossRef]
14. Dankelmann, P.; Hellwig, A.; Volkmann, L. Inverse degree and edge-connectivity. *Discret. Math.* **2009**, *309*, 2943–2947. [CrossRef]
15. Dankelmann, P.; Meierling, D. Maximally edge-connected hypergraphs. *Discret. Math.* **2016**, *339*, 33–38. [CrossRef]
16. Gu, X.F.; Lai, H.J. Realizing degree sequences with k-edge-connected uniform hypergraphs. *Discret. Math.* **2013**, *313*, 1394–1400. [CrossRef]

mathematics

MDPI

Article

The A_α-Spectral Radii of Graphs with Given Connectivity

Chunxiang Wang [1,†] and Shaohui Wang [2,*,†]

1 School of Mathematics and Statistics and Hubei key Laboratory Mathematics Sciences, Central China Normal University, Wuhan 430079, China; wcxiang@mail.ccnu.edu.cn
2 Department of Mathematics, Savannah State University, Savannah, GA 31419, USA
* Correspondence: shaohuiwang@yahoo.com; Tel.: +1-352-665-3381
† These authors contributed equally to this work.

Received: 22 November 2018; Accepted: 24 December 2018; Published: 4 January 2019

Abstract: The A_α-matrix is $A_\alpha(G) = \alpha D(G) + (1-\alpha)A(G)$ with $\alpha \in [0,1]$, given by Nikiforov in 2017, where $A(G)$ is adjacent matrix, and $D(G)$ is its diagonal matrix of the degrees of a graph G. The maximal eigenvalue of $A_\alpha(G)$ is said to be the A_α-spectral radius of G. In this work, we determine the graphs with largest $A_\alpha(G)$-spectral radius with fixed vertex or edge connectivity. In addition, related extremal graphs are characterized and equations satisfying $A_\alpha(G)$-spectral radius are proposed.

Keywords: adjacent matrix; signless Laplacian; spectral radius; connectivity

1. Introduction

We consider simple finite connected graph G with the vertex set $V(G)$ and the edge set $E(G)$. The number of vertices $|V(G)| = n$ is the order of a graph, and the number of edges $|E(G)|$ is the size of a graph. Denote the neighborhood of $v \in V(G)$ by $N(v) = \{u \in V(G), vu \in E(G)\}$, and the degree of v by $d_G(v) = |N(v)|$ (or briefly d_v). For $L \subseteq V(G)$ and $R \subseteq E(G)$, let $w(G-L)$ or $w(G-R)$ be the number of components of $G-L$ or $G-R$. L(or R) be a vertex(edge) cut set if $w(G-L \text{ (or } R)) \geq 2$ and $E(w,L) = \{wu \in E(G), u \in L\}$. For $U \subseteq V(G)$, $G[U]$ denote the induced subgraph of G, that is, $V(G[U]) = U$ and $E(G[U]) = \{uv | uv \in E(G), u, v \in U\}$.

If $A(G)$ is adjacency matrix of a graph G, and $D(G)$ is its diagonal matrix of the degrees of G, then the signless Laplacian matrix of G is $D(G) + A(G)$. With the successful studies of these matrices, Nikiforov [1] proposed the A_α-matrix

$$A_\alpha(G) = \alpha D(G) + (1-\alpha)A(G)$$

with $\alpha \in [0,1]$. Obviously, $A_0(G)$ is the adjacent matrix and $A_{\frac{1}{2}}$ is the half of signless Laplacian matrix of G, respectively. For undefined terminologies and notations, we refer to [2].

The research of (adjacency, signless Laplacian) spectral radius is an intriguing topic during past decades [3–9]. For instances, Lovász and J. Pelikán studied the spectral radius of trees [10]. The minimal Laplacian spectral radius of trees with given matching number is given by Feng et al. [7]. The properties of spectra of graphs and their line graphs are studied by Chen [11]. The signless Laplacian spectra of graphs is explored by Cvetković et al. [12]. Zhou [13] found bounds of signless Laplacian spectral radius and its hamiltonicity. Graphs having none or one signless Laplacian eigenvalue larger than three are obtained by Lin and Zhou [14]. At the same time, the maximal adjacency or signless Laplacian spectral radius have attracted many interests among the mathematical literature including algebra and graph theory. Ye et al. [6] gave the maximal adjacency or signless Laplacian spectral radius of graphs subject to fixed connectivity.

Inspired by these outcomes, we determine the graphs with largest $A_\alpha(G)$-spectral radius with given vertex or edge connectivity. In addition, the corresponding extremal graphs are provided and the equations satisfying the $A_\alpha(G)$-spectral radius are obtained.

2. Preliminary

In this section, we provide some important concepts and lemmas that will be used in the main proofs.

Denote by G a graph such that $V(G) = \{v_1, v_2, \cdots, v_n\}$ is its vertex set and $E(G)$ is its edge set. The A_α-matrix of G has the (i,j)-entry of $A_\alpha(G)$ is $1-\alpha$ if $v_iv_j \in E(G)$; $\alpha d(v_i)$ if $i = j$, and otherwise 0. For $\alpha \in [0,1]$, let $\lambda_1(A_\alpha(G)) \geq \lambda_2(A_\alpha(G)) \geq \cdots \geq \lambda_n(A_\alpha(G))$ be the eigenvalues of $A_\alpha(G)$. The A_α-spectral radius of G is considered as the maximal eigenvalue $\rho := \lambda_1(A_\alpha(G))$. Let $X = (x_{v_1}, x_{v_2}, \cdots, x_{v_n})^T$ be a real vector of ρ.

By $A_\alpha(G) = \alpha D(G) + (1-\alpha)A(G)$, we have the quadratic formula of $X^TA_\alpha(G)X$ can be expressed that

$$X^TA_\alpha(G)X = \alpha \sum_{v_i \in V(G)} x_{v_i}^2 d_{v_i} + 2(1-\alpha) \sum_{v_iv_j \in E(G)} x_{v_i}x_{v_j}.$$

Because $A_\alpha(G)$ is a real symmetric matrix, and by Rayleigh principle, we have the formula

$$\rho(G) = max_{X \neq 0} \frac{X^TA_\alpha(G)X}{X^TX}.$$

As we know that once X is an eigenvector of $\rho(G)$ for a connected graph G, X should be unique and positive. The corresponding eigenequations for $A_\alpha(G)$ is rewritten as

$$\rho(G)x_{v_i} = \alpha d_{v_i} x_{v_i} + (1-\alpha) \sum_{v_iv_j \in E(G)} x_{v_j}. \tag{1}$$

As $A_1(G) = D(G)$, we study the A_α-matrix for $\alpha \in [0,1)$ below. Based on the definition of A_α-spectral radius, we have

Lemma 1. *[4,15] Let $A_\alpha(G)$ be the A_α-matrix of a connected graph G ($\alpha \in [0,1)$), $v, w \in V(G)$, $u \in T \subset V(G)$ such that $T \subset N(v) \setminus (N(w) \cup \{w\})$. Let G^* be a graph with vertex set $V(G)$ and edge set $E(G) \setminus \{uv, u \in T\} \cup \{uw, u \in T\}$, and X a unit eigenvector to $\rho(A_\alpha(G))$. If $x_w \geq x_v$ and $|T| \neq 0$, then $\rho(G^*) > \rho(G)$.*

If G is a connected graph, then $A_\alpha(G)$ is a nonnegative irreducible symmetric matrix. By the results of [1,16,17] and adding extra edges to a connected graph, then A_α-spectral radius will increase and the following lemma is straightforward.

Lemma 2. *(i) If G^* is any proper subgraph of connected graph G, and ρ is the A_α-spectral radius, then $\rho(G^*) < \rho(G)$.*

(ii) If X is a positive vector and r is a positive number such that $A_\alpha(G)X < rX$, then $\rho(G) < r$.

Recall that the vertex connectivity (respectively, edge connectivity) of a graph G is the smallest number of vertices (respectively, edges) such that if we remove them, the graph will be disconnected or be a single vertex. For convenience, let $\mathcal{F}_n$ be the set of all graphs of order n, and $\mathcal{F}_n^k$ (respectively, $\overline{\mathcal{F}}_n^k$) $(k \geq 0)$ be the set of such graphs with order n and vertex (resp., edge) connectivity k. Note that $\mathcal{F}_n^0 = \overline{\mathcal{F}}_n^0$ having some disconnected graphs of order n, and $\mathcal{F}_n^{n-1} = \overline{\mathcal{F}}_n^{n-1}$ consisting of the unique graph K_n. Obviously, $\mathcal{F}_n = \cup_k \mathcal{F}_n^k = \cup_k \overline{\mathcal{F}}_n^k$.

Recall the graph $K(p,q)(p \geq q \geq 0)$ obtained from K_p by attaching a vertex together with edges connecting this vertex to q vertices of K_p. $K(p,q)$ is was found by Brualdi and Solehid in terms of stepwise adjacency matrix, but it is Peter Rowlinson who gives the purely combinatorial definition of such graph. For the property of $K(p,q)$, we refer to [18–20]. Clearly, $K(p,0)$ is K_p with an additional isolated vertex. It's not hard to see that $K(p,q)$ is of vertex (resp., edge) connectivity q. Let δ, Δ be the smallest and largest degrees of vertices in the graph G, respectively.

Lemma 3. *The graph K_n is the graph in $\mathcal{F}_n$ having the largest A_α-spectral radius, and $K_{n-1} \cup K_1 = K(n-1,0)$ is the graph in $\mathcal{F}_n^0$ or $\overline{\mathcal{F}}_n^0$ having the smallest A_α-spectral radius.*

Proof. By Lemma 2, the first statement is clear. For the second one, let G be a graph which attains the maximum A_α-spectral radius in $\mathcal{F}_n^0$, then G only has two unique connected components: K_{n-1}, K_1; if not, any component of G will be a proper subgraph of K_{n-1}. Then $\rho(G) < \rho(K_{n-1}) = \rho(K_{n-1} \cup K_1)$, a contradiction. Then this lemma is proved. □

Lemma 4. *For $k \in [1, n-2]$, $K(n-1,k)$ is the graph having the largest A_α-spectral radius in $\mathcal{F}_n^k$.*

Proof. Denote by G a graph having the largest A_α-spectral radius in $\mathcal{F}_n^k$. x is a unit (positive) Perron vector of A_α. Let U be the vertex cut of G having k vertices, and these components of $G-U$ be $G_1, G_2, \cdots, G_s$, for $s \geq 2$. We declare that $s=2$; if not, adding all possible edges within the graph $G_1 \cup G_2 \cup \cdots \cup G_{s-1}$, we would get a graph belonging to $\mathcal{F}_n^k$ (because U is the smallest vertex cut set) and with a larger A_α-spectral radius. Similarly, induced subgraph $G[U]$, the subgraphs G_1 and G_2 are complete subgranph, and every vertex of U connects these vertices of G_1 and G_2. Next we prove that one of G_1, G_2 will be a singleton, which has a unique vertex. If not, suppose that G_1, G_2 have orders greater than one. Without loss of generality, denote by u a vertex of G_1 having a smallest value for x among vertices in $G_1 \cup G_2$. Deleting these edges of G_1 incident to u, and connecting all possible edges between $G_1 - u$ and G_2, we get a graph $\widetilde{G} = K(n-1,k)$ still in $\mathcal{F}_n^k$. By Lemma 1, $\rho(\widetilde{G}) > \rho(G)$, which yields a contradiction. So one of G_1, G_2 is a singleton, and G is the desired graph $K(n-1,k)$. □

Lemma 5. *For $k \in [1, n-2]$, $K(n-1,k)$ is the graph having maximum A_α-spectral radius in $\overline{\mathcal{F}}_n^k$.*

Proof. Denote by G a graph having the largest A_α-spectral radius in $\mathcal{F}_n^k$. x is a unit (positive) Perron vector of A_α. We know that each vertex of G has degree greater than or equal to k. Otherwise $G \notin \overline{\mathcal{F}}_n^k$. If there is a vertex u in G with degree k, then the edges adjacent to u are an edge cut such that $G-u$ is complete. The statement follows in this case. Then we will suppose that all vertices in G have degrees greater than k. Let E_c be an edge cut set of G having k edges. So $G - E_c$ consists of only two components G_1, G_2, respectively, of order n_1, n_2. Obviously G_1, G_2 are both complete. In addition, neither of G_1, G_2 is a singleton. Otherwise G would contain a vertex of degree k, which contradicted to the above assumption. So G_1, G_2 contain more than 1 vertex, i.e., $n_1 \geq 2$ and $n_2 \geq 2$.

Without loss of generality, suppose that G_1 contains a vertex w_1 having a minimal value given by x within all vertices of $G_1 \cup G_2$, and consists of vertices $w_1, w_2, \cdots, w_{n_1}$ such that $x(w_1) \leq x(w_2) \leq \cdots \leq x(w_{n_1})$. Assume that w_1 joins t vertices of G_2. Surely $t \leq min\{k, n_2\}$.

If $t = k$, there exist no edges joining $G_1 - w_1$ and G_2, and $n_2 \geq k+2$ otherwise G_2 contains a vertex of degree k. Denote by G' a new graph with vertex set $V(G)$ and edge set $E(G) \setminus E(w_1, N) \cup E(N, v')$, where $N = N(w_1) \cap V(G_1)$, and $v' \in V(G_2) - N(w_1) \cap V(G_2)$, by Lemma 1, we have $\rho(G') > \rho(G)$. Let G'' be another new graph with vertex set $V(G')$ and adding all possible edges between $G_1 - w_1$ and G_2. Note that $G'' = K(n-1,k)$, and G' is a proper subgraph of G''. By Lemma 2, we have $\rho(G'') > \rho(G')$. Thus, $\rho(G'') > \rho(G)$, a contradiction.

If $t < k$. Partition the set $V(G_1) - w_1$ as: $V_{11} = \{w_i : i = 2, 3, \cdots, n_1 - (k-t)\}$, $V_{12} = \{w_j : j = n_1 - (k-t) + 1, \cdots, n_1\}$. Thus, $|V_{11}| = n_1 - (k-t) - 1$; $|V_{12}| = k - t$.

Let $N = N(w_1) \cap V_{11}$, then $N \neq \emptyset$ since $d(w_1) > k$. Note there is vertex $v' \in V(G_2) - N(w_1) \cap V(G_2)$ since $n_2 \geq k+2$. Let G' be a new graph having vertex set $V(G)$ and edge set $E(G) \setminus E(w_1, N) \cup E(N, v')$, where $N = N(w_1) \cap V_{11}$, and $v' \in V(G_2) - N(w_1) \cap V(G_2)$, by Lemma 1, we have $\rho(G') > \rho(G)$. Let G'' be another new graph having vertex set $V(G')$ and adding all possible edges between $G_1 - w_1$ and G_2, adding all edges between w_1 and V_{12}. Note that $G'' = K(n-1, k)$, and G' is a proper subgraph of G''. Lemma 2 implies that $\rho(G'') > \rho(G')$. Thus, $\rho(G'') > \rho(G)$, a contradiction. The result follows. □

3. Main Results

In this section, we will determine maximizing A_α-spectral radius of of graphs with given connectivity. By Lemmas 4 and 5, we obtain the following Theorem:

Theorem 1. *The graph K_n is the graph in $\mathcal{F}_n$ with A_α-spectral radius , and $K_{n-1} \cup K_1 = K(n-1, 0)$ is the unique one in $\mathcal{F}_n^0$ or $\overline{\mathcal{F}}_n^0$ with A_α-spectral radius. For $k \in [1, n-2]$, $K(n-1, k)$ is the graph with maximum A_α-spectral radius in $\mathcal{F}_n^k$ or $\overline{\mathcal{F}}_n^k$.*

Proof. By the Lemmas 3–5, we obtain the results. □

Lemma 6. [20] *Given a partition $\{1, 2, \cdots, n\} = \Delta_1 \cup \Delta_2 \cup \cdots \cup \Delta_m$ with $|\Delta_i| = n_i > 0$, A be any matrix partitioned into blocks A_{ij}, where A_{ij} is an $n_i \times n_j$ block. Suppose that the block A_{ij} has constant row sums b_{ij} , and let $B = (b_{ij})$. Then the spectrum of B is contained in the spectrum of A (taking into account the multiplicities of the eigenvalues).*

Since $K(n-1, k)$ contains K_{n-1}, we can partition $K(n-1, k)$ into three different subsets: $\{u\}, T, S$, in which u is the vertex connecting a complete subgraph K_{n-1} with k edges, a subset S is in K_{n-1} connecting u, and $T = V(K_{n-1} \setminus S)$. Let x be a Perron vector of $K(n-1, k)$. $S = \{u_1, u_2, \cdots, u_k\}$ and $T = \{v_1, v_2, \cdots, v_t\}$. Note that $k + t + 1 = n$.

Theorem 2. *Label the vertices of $K(n-1, k)$ as $u, u_1, u_2, \cdots, u_k, v_1, v_2 \cdots, v_t$ with $k, t \geq 0$. The maximum eigenvalues of $A_\alpha(K(n-1, k))$ satisfy the equation: $f(\rho) = (\rho - k\alpha)(\rho - k\alpha - n + k + 2)(\rho - n\alpha + 1) - k(1-\alpha)(\rho - k\alpha - \alpha + 1)(\rho - n\alpha + \alpha + 1) + k(1-\alpha)^3(n-k-1) = 0$.*

Proof. Since the matrix $A_\alpha = \alpha D + (1-\alpha)A$, where D has on the diagonal the vector $(k, n-1, n-2)$ and A consists of the following three row-vectors, in the order: $(0, k, 0)$; $(1, k-1, n-k-1)$; $(0, k, n-k-2)$. Thus, by the Lemma 6, x is a constant value β_2 on the vertex set S, and constant value β_3 on the vertex set T. Defining $x(u) =: \beta_1, \rho(K(n-1, k)) =: \rho$, also by (1), we get

$$(\rho - \alpha k)\beta_1 = k(1-\alpha)\beta_2$$

$$(\rho - \alpha(n-1))\beta_2 = (1-\alpha)(\beta_1 + (k-1)\beta_2 + t\beta_3), \text{ and}$$

$$(\rho - \alpha(n-2))\beta_3 = (1-\alpha)(k\beta_2 + (t-1)\beta_3).$$

Then we get

$$(\rho - \alpha(n-1)) = \frac{k(1-\alpha)^2}{\rho - k\alpha} + \frac{kt(1-\alpha)^2}{\rho - k\alpha - t + 1} + (k-1)(1-\alpha).$$

Note that for $n = t + k + 1$, that is, $n - 1 = k + t$. Then we have:

$$(\rho - k\alpha) = \frac{k(1-\alpha)^2}{\rho - k\alpha} + \frac{kt(1-\alpha)^2}{\rho - k\alpha - t + 1} + (k-1)(1-\alpha) + t\alpha.$$

Then we obtain that

$$(\rho - k\alpha)(\rho - k\alpha - n + k + 2)(\rho - n\alpha + 1) - k(1-\alpha)(\rho - k\alpha \\ -\alpha + 1)(\rho - n\alpha + \alpha + 1) + k(1-\alpha)^3(n-k-1) = 0.$$

Thus, our proof is finished. □

Corollary 1. *Let G be a graph of order n having vertex/edge connectivity k, where* $1 \le k \le n-2$*, the maximum adjacency spectral radius is the largest root of the* $f(\lambda) = \lambda^3 - (n-3)\lambda^2 - (n+k-2)\lambda + k(n-k-2) = 0$.

Proof. By Theorem 2, let $\alpha = 0$, then $f(\lambda) = \lambda^3 - (n-3)\lambda^2 - (n+k-2)\lambda + k(n-k-2) = 0$. It is obvious since $A_0 = A(G)$. □

By letting the special values for α, we have the following corollary.

Corollary 2. *Let G be a graph of order n having vertex/edge connectivity k, where* $1 \le k \le n-2$*, the signless Laplacian spectral radius* $\lambda_1 = \frac{2n+k-4+\sqrt{(2n-k-4)^2+8k}}{2}$.

Proof. By Theorem 2, let $\alpha = \frac{1}{2}$, then $f(\lambda) = \lambda^3 - \frac{1}{2}(3n+k-6)\lambda^2 + (\frac{1}{4}(n-4)(2n+3k) + k + 2)\lambda - \frac{1}{4}k(n^2 - 5n + 6) = 0$. It is obvious since $2A_{\frac{1}{2}} = D + Q$. Thus,

$$\begin{aligned} 8f(\lambda) &= 8[\lambda^3 - \frac{1}{2}(3n+k-6)\lambda^2 + (\frac{1}{4}(n-4)(2n+3k) + k + 2)\lambda \\ &\quad -\frac{1}{4}k(n^2 - 5n + 6)] \\ &= (2\lambda)^3 - (3n+k-6)(2\lambda)^2 + ((n-4)(2n+3k) + 4k + 8)(2\lambda) \\ &\quad -2k(n^2 - 5n + 6) \\ &= (\lambda_1)^3 - (3n+k-6)(\lambda_1)^2 + ((n-4)(2n+3k) + 4k + 8)(\lambda_1) \\ &\quad -2k(n^2 - 5n + 6). \end{aligned}$$

Let $\lambda_1 = 2\lambda$ and

$$\begin{aligned} F(\lambda_1) &= (\lambda_1)^3 - (3n+k-6)(\lambda_1)^2 + ((n-4)(2n+3k) + 4k + 8)(\lambda_1) \\ &\quad -2k(n^2 - 5n + 6) = 0. \end{aligned}$$

Then we get:

$$\lambda_1 = \frac{2n+k-4+\sqrt{(2n-k-4)^2+8k}}{2}.$$

□

The above result is the same as [6].

Author Contributions: All authors have contributed equally to this work. Investigation and Methodology: C.W.; Methodology and Correction: S.W.

Funding: This work was partially supported by the National Natural Science Foundation of China under Grants 11771172 and 11571134.

Conflicts of Interest: The authors declare no conflict of interest.

References

1. Nikiforov, V. Merging the A- and Q-spectral theories. *Appl. Anal. Discret. Math.* **2017**, *11*, 81–107. [CrossRef]
2. Bollobás, B. *Modern Graph Theory*; Springer: New York, NY, USA, 1998.
3. Xing, R.; Zhou, B. On the least eigenvalue of cacti with pendant vertices. *Linear Algebra Appl.* **2013**, *438*, 2256–2273. [CrossRef]
4. Xue, J.; Lin, H.; Liu, S.; Shu, J. On the A_α-spectral radius of a graph. *Linear Algebra Appl.* **2018**, *550*, 105–120. [CrossRef]
5. Yu, A.; Lu, M.; Tian, F. On the spectral radius of graphs. *Linear Algebra Appl.* **2004**, *387*, 41–49. [CrossRef]
6. Ye, M.-L.; Fan, Y.-Z.; Wang, H.-F. Maximizing signless Laplacian or adjacency spectral radius of graphs subject to fixed connectivity. *Linear Algebra Appl.* **2010**, *433*, 1180–1186. [CrossRef]
7. Feng, L.; Li, Q.; Zhang, X.-D. Minimizing the Laplacian spectral radius of trees with given matching number. *Linear Multilinear Algebra* **2007**, *55*, 199–207. [CrossRef]
8. Li, S.; Zhang, M. On the signless Laplacian index of cacti with a given number of pendant vertices. *Linear Algebra Appl.* **2012**, *436*, 4400–4411. [CrossRef]
9. Wu, J.; Deng, H.; Jiang, Q. On the spectral radius of cacti with k-pendant vertices. *Linear Multilinear Algebra* **2010**, *58*, 391–398. [CrossRef]
10. Lovász, L.; Peliken, J. On the eigenvalues of trees. *Period. Math. Hungar.* **1973**, *3*, 175–182. [CrossRef]
11. Chen, Y. Properties of spectra of graphs and line graphs. *Appl. Math. J. Chin. Univ. Ser. B* **2002**, *17*, 371–376.
12. Cvetković, D.; Rowlinson, P.; Simić, S.K. Signless Laplacians of finite graphs. *Linear Algebra Appl.* **2007**, *423*, 155–171. [CrossRef]
13. Zhou, B. Signless Laplacian spectral radius and Hamiltonicity. *Linear Algebra Appl.* **2010**, *432*, 566–570. [CrossRef]
14. Lin, H.; Zhou, B. Graphs with at most one signless Laplacian eigenvalue exceeding three. *Linear Multilinear Algebra* **2015**, *63*, 377–383.
15. Nikiforov, V.; Pastén, G.; Rojo, O.; Soto, R.L. On the A_α-spectra of trees. *Linear Algebra Appl.* **2017**, *520*, 286–305. [CrossRef]
16. Berman, A.; Plemmons, R.J. *Nonnegative Matrices in the Mathematical Sciences*; SIAM: Philadelphia, PA, USA, 1994.
17. Collatz, L.; Sinogowitz, U. Spektrcn endlicher Graten. *Abh. Math. Scm. Univ. Hamburg* **1957**, *21*, 63–77.
18. Brualdi, R.A.; Solheid, E.S. On the spectral radius of connected graphs. *Publ. Inst. Math.* **1986**, *39*, 45–54.
19. Rowlinson, P. On the maximal index of graphs with a prescribed number of edges. *Linear Algebra Appl.* **1988**, *110*, 43–53. [CrossRef]
20. Cvetkovic, D.; Rowlinson, P.; Simic, S. *An Introduction to the Theory of Graph Spectra*; Cambridge University Press: Cambridge, UK, 2009.

mathematics

Article

Some Root Level Modifications in Interval Valued Fuzzy Graphs and Their Generalizations Including Neutrosophic Graphs

Naeem Jan [1], Kifayat Ullah [1], Tahir Mahmood [1], Harish Garg [2,*], Bijan Davvaz [3], Arsham Borumand Saeid [4] and Said Broumi [5]

1 Department of Mathematics and Statistics, International Islamic University Islamabad Pakistan, Islamabad Capital Territory 44000, Pakistan; naeem.phdma73@iiu.edu.pk (N.J.); kifayat.phdma72@iiu.edu.pk (K.U.); tahirbakhat@iiu.edu.pk (T.M.)
2 School of Mathematics, Thapar Institute of Engineering & Technology, Deemed University, Patiala 147004, Punjab, India
3 Department of Mathematics, Yazd University, Yazd 89195-741, Iran; davvaz@yazd.ac.ir
4 Department of Pure Mathematics, Faculty of Mathematics and Computer, Shahid Bahonar University of Kerman, Kerman 7616913439, Iran; a_b_saeid@yahoo.com
5 Laboratory of Information Processing, University Hassan II, Casablanca 20000, Morocco; broumisaid78@gmail.com
* Correspondence: harishg58iitr@gmail.com; Tel.: +91-86990-31147

Received: 22 November 2018; Accepted: 3 January 2019; Published: 10 January 2019

Abstract: Fuzzy graphs (FGs) and their generalizations have played an essential role in dealing with real-life problems involving uncertainties. The goal of this article is to show some serious flaws in the existing definitions of several root-level generalized FG structures with the help of some counterexamples. To achieve this, first, we aim to improve the existing definition for interval-valued FG, interval-valued intuitionistic FG and their complements, as these existing definitions are not well-defined; i.e., one can obtain some senseless intervals using the existing definitions. The limitations of the existing definitions and the validity of the new definitions are supported with some examples. It is also observed that the notion of a single-valued neutrosophic graph (SVNG) is not well-defined either. The consequences of the existing definition of SVNG are discussed with the help of examples. A new definition of SVNG is developed, and its improvement is demonstrated with some examples. The definition of an interval-valued neutrosophic graph is also modified due to the shortcomings in the current definition, and the validity of the new definition is proved. An application of proposed work is illustrated through a decision-making problem under the framework of SVNG, and its performance is compared with existing work.

Keywords: interval-valued fuzzy graph; intuitionistic fuzzy graph; interval-valued intuitionistic fuzzy graph; single-valued neutrosophic graph; interval-valued neutrosophic graph; complement

---

1. Introduction

The framework of the fuzzy set (FS) was introduced by Zadeh [1] in 1965 as a generalization of crisp sets, which describe the membership of an object to a set by assigning a membership grade from [0,1]. The concept of FS was further generalized to the interval-valued fuzzy set (IVFS) in [2] where the membership grade is defined by closed sub-intervals $[S^L, S^U] \subseteq [0,1]$. The idea of FS was further generalized to an enhanced framework of the intuitionistic fuzzy set (IFS) in [3] by considering the degree of non-membership of an object as well as its membership grade, such that the sum of the degrees is less than one. Later on, Atanassov and Gargov [4] extended the theory of IFS to an interval-valued IFS (IVIFS). To consider the neutral opinion or abstinence information in the analysis,

Smarandache introduced a framework of neutrosophic logic and consequently a neutrosophic set [5]. A neutrosophic set is characterized by three kinds of functions describing the degree of membership, abstinence and non-membership of an object. The concept of the single-valued neutrosophic set (SVNS) was developed by Wang et al. [6], characterized by three characteristic functions, named as truth, falsity and indeterminate, with a restriction that their sum is not greater than 3. The concept of SVNS was further generalized to an interval-valued neutrosophic set (IVNS) in Wang et al. [7] where the membership, abstinence and non-membership grades are defined in terms of closed sub-intervals of the unit interval. Under the above-mentioned sets, various researchers have addressed different kinds of algorithms to solve decision-making problems [8–25].

Further, apart from this, the concept of the fuzzy graph (FG), initially introduced by Kauffman [26], was thoroughly developed by Rosenfeld [27] in 1975. A generalization of FG known as the interval-valued FG (IVFG) was introduced by Akram and Dudek [28] where the nodes and edges are in the form of interval-valued fuzzy numbers (IVFNs). To deal with the study of FGs and IVFG, we refer the readers to the articles in [29–38]. After successful applications, Parvathi and Karunambigai [39] introduced the concept of intuitionistic fuzzy graphs (IFG), which is a generalization of the FG. The concept of interval-valued IFG (IVIFG) is proposed in [40] and some other terms and notions related to IVIFG can be found in [41,42]. Further, by following the notion of SVNS, the concept of the single-valued neutrosophic graph (SVNG) was introduced by Broumi et al. [43] and has been extensively used in several problems such as the shortest path problem [44,45], communication problem [46], and decision-making [47]. The concept of SVNG was further generalized to the interval-valued neutrosophic graph (IVNG) in [48], where the membership, abstinence and non-membership grades are described by closed subintervals of the unit interval. The framework of IVNG is the most sophisticated among all existing graph structures. For some other work on IVNG, one may refer to [49–51].

In the theory of fuzzy algebraic structures and from the above literature survey, we found some serious shortcomings which motivate us to develop some new definitions for the existing concepts. The problems in the existing literature and steps toward their solution are discussed item by item below.

(1) It is observed that the concept of IVFG [28] has some limitations which lead us to some undefined results. The lower and upper values of the edge interval in IVFG are defined as $\hat{S}^L\left(v_i, v_j\right) \leq \min\left(S^L(v_i), S^L\left(v_j\right)\right)$ and $\hat{S}^U\left(v_i, v_j\right) \leq \min\left(S^U(v_i), S^U\left(v_j\right)\right)$. According to this definition for two adjacent vertices (0.2, 0.4) and (0.2, 0.3), we obtained an undefined edge value of (0.2, 0.0) by the above approach. The same type of problem occurs in the framework of IVIFG [40] in membership as well as non-membership intervals. Further, the existing definition of IVIFG is not a generalization of IFG; therefore, one must develop a definition for IVIFG which clearly generalizes IFG. These types of flaws in the existing literature lead us to develop some new definitions for IVFG and IVIFG;

(2) The lower and upper values of edges in the complement of an IVFG are defined as $\left(\hat{S}^L\right)^c\left(v_i, v_j\right) = \min\left(S^L(v_i), S^L\left(v_j\right)\right) - \hat{S}^L\left(\left(v_i, v_j\right)\right)$ and $\left(\hat{S}^U\right)^c\left(v_i, v_j\right) = \min\left(S^U(v_i), S^U\left(v_j\right)\right) - \hat{S}^U\left(v_i, v_j\right)$. Let (0.2, 0.5) and (0.2, 0.3) be two vertices and (0.1, 0.3) their edge. Then, upon calculating the complement, we get the corresponding edge value (0.1, 0.0) which is an undefined interval. Similar types of flaws exist in the existing definitions of complements of IVIFGs, therefore leading us to develop some new definitions for the complement of IVFG as well as of IVIFG.;

(3) It is observed that the existing definition of SVNG [43] is not a generalization of IFG as claimed in the existing literature. Also, the existing notion of the SVNG and of its complement as defined in [43] is not well posed, as after taking the complement of a SVNG, the resulting graph obtained is not a SVNG. Further, the existing definition of a complement does not satisfy the property

$(G^c)^c = G$. These facts are described briefly in Section 5, and based on these defects, a new definition for SVNG and its complement have been developed and their validity is proved;

(4) Finally, we improve the definition of the IVNG as the existing definition contradicts the fact that IVNG is a generalization of IVIFG and IVFG. Further, the existing definition of the complement of IVNG is not well defined, which is described with the help of examples, and a new definition for complements is proposed and validated with examples.

The rest of the article is organized as follows: Section 1 provides a historical background of FGs and their generalized algebraic structures. The motivation and aim of our proposed work is also discussed in this section. In Section 2, some basic concepts of FGs and IFGs are discussed and supported with examples. Section 3 is based on a description of the limitations of IVFGs and IVIFGs and some improvements in their basic definitions. In Section 4, we discuss the drawbacks of the complements of IVFG and IVIFG and propose new definitions of complements along with some examples in their support. In Section 5, the drawbacks of the existing concepts of SVNGs, IVNGs and their complements are described. Further, some new improved definitions are proposed for SVNG, IVNGs and the complement of IVNG, supported by some examples. Section 6 is based on an application of SVNGs in decision-making, while Section 7 provides a summary of the article along with some future work.

2. Basic Concepts

In this section, some definitions of FG and its complement are defined and demonstrated with the help of an example. The primary definition of IFG is also discussed in this section, followed by an example for illustration.

Definition 1. [27] *An FG is a pair $G = (V, E)$ where V is the set of nodes and E is the collection of edges between these nodes such that*

(1) Every $v \in V$ is characterized by a function $S : V \to [0,1]$ denoting the degree of membership of $v \in V$;
(2) Every $e \in E$ is characterized by a function $\hat{S} : V \times V \to [0,1]$ denoting the degree of membership $e \in V \times V$ satisfying the condition $\hat{S}(v_i, v_j) \le \min(S(v_i), S(v_j))$.

Definition 2. [29] *The complement of FG $G = (V, E)$ is defined by $G^c = (V^c, E^c)$ where $V^c = V$ and the membership grade of $e \in E$ is defined by $(\hat{S})^c(v_i, v_j) = \min(S(v_i), S(v_j)) - \hat{S}((v_i, v_j))$.*

Example 1. *The following Figure 1 is an example of an FG while Figure 2 represents its complement.*

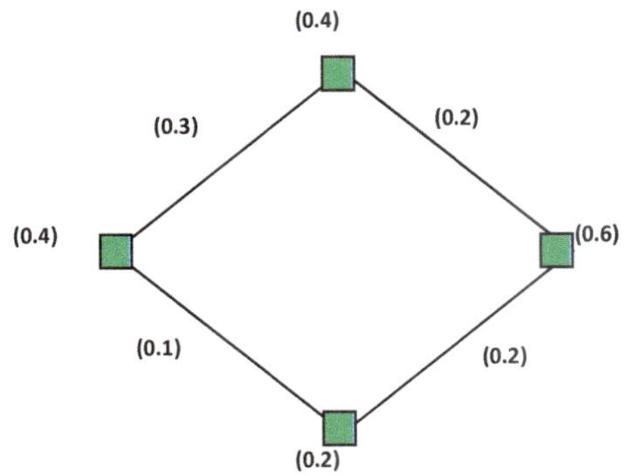

Figure 1. Fuzzy graph (FG) compatible with Definition 1.

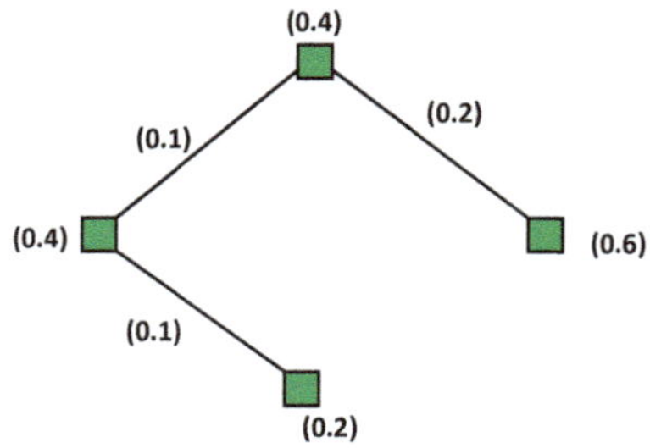

Figure 2. Complement of FG depicted in Figure 1.

Definition 3. [39] *An IFG is a duplet* $G = (V, E)$ *where the set of nodes is denoted by* V *and* E *is the collection of edges between these nodes such that*

(1) *Every* $v \in V$ *is characterized by two functions* $S : V \to [0,1]$ *and* $D : V \to [0, 1]$ *denoting the membership and non-membership degree of* $v \in V$ *which satisfies the condition that* $0 \le S + D \le 1$. *Moreover, the term* R *defined by* $R = 1 - S - D$, *denotes the hesitancy level of* $\in E$;

(2) *Every* $e \in E$ *is characterized by two functions* $\hat{S} : V \times V \to [0, 1]$ *and* $Đ : V \times V \to [0, 1]$ *denoting the membership and non-membership degree of* $e \in V \times V$ *satisfying the conditions:*

$$\hat{S}(v_i, v_j) \le \min(S(v_i), S(v_j))$$

$$Đ(v_i, v_j) \le \max(D(v_i), D(v_j))$$

with a condition that $0 \le \hat{S} + Đ \le 1$. *Moreover, the term* $\check{R}$ *denotes the hesitancy level of* $e \in E$ *such that* $\check{R} = 1 - \hat{S} - Đ$.

Example 2. *The following Figure 3 is an example of an IFG.*

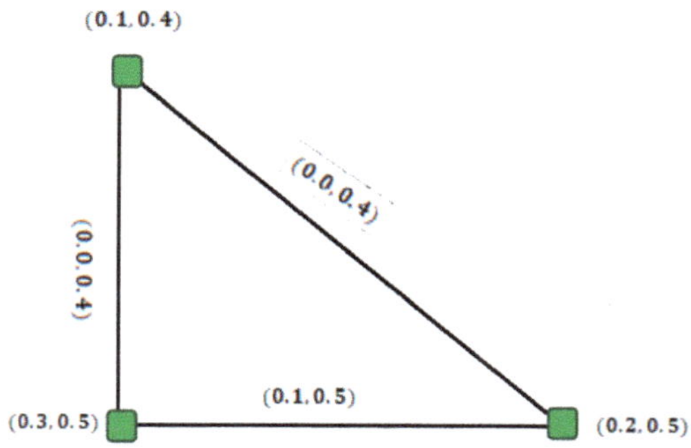

Figure 3. Intuitionistic fuzzy graph (IFG) compatible with Definition 2.

The rest of the basic notions are discussed in Sections 3–5, in which we present a comparative analysis with our proposed work.

3. Improvements in Interval-Valued Fuzzy Graphs and Interval-Valued Intuitionistic Fuzzy Graphs

In this section, first, the existing definitions of IVFG and IVIFG are reviewed and the shortcomings of these algebraic structures are pointed out with the help of examples. New definitions for IVFG and IVIFG are developed and their fitness is verified with the help of some examples.

First, we review the definition of IVFG defined in [28] and provide an example to show the limitation of the definitions.

Definition 4. [28] *An IVFG is a duplet $G = (V, E)$ where V denotes a collection of nodes and E denotes the collection of edges between these vertices such that*

(1) *Every $v \in V$ is characterized by a function S representing the membership degree of $v \in V$. Basically, $S = [S^L, S^U]$ is a closed subinterval [0, 1];*
(2) *Every $e \in E$ is characterized by a function $\hat{S}$ denoting the degree of membership $e \in V \times V$. Basically, $\hat{S} = [\hat{S}^L, \hat{S}^U]$ satisfying the conditions:*

$$\hat{S}^L(v_i, v_j) \leq \min(S^L(v_i), S^L(v_j))\hat{S}^U(v_i, v_j) \leq \min(S^U(v_i), S^U(v_j))$$

Example 3. *Consider the following Figure 4, where we observed that, while determining the edge values, we get undefined intervals as highlighted. This shows that the current notion of an IVFG is not satisfactory, which leads us to develop a new definition.*

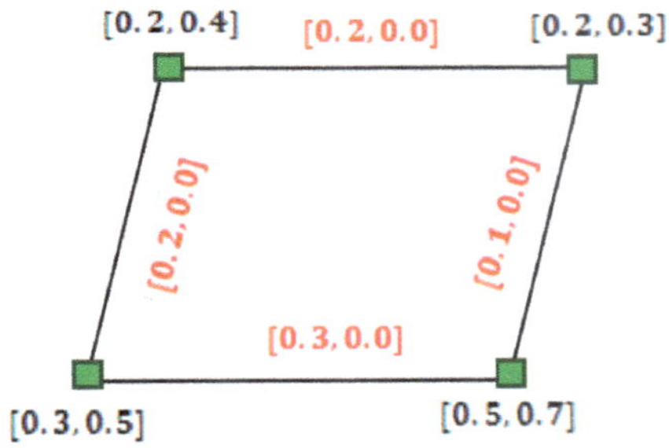

Figure 4. Interval-valued fuzzy graph containing undefined intervals.

Definition 5. *An IVFG is a duplet $G = (V, E)$ where V denotes a collection of nodes and E denotes the collection of edges between these vertices such that*

(1) *Every $v \in V$ is characterized by a function S representing the membership grade of $v \in V$. Basically, $S = [S^L, S^U]$ is a closed subinterval of the unit interval [0, 1];*
(2) *Every $e \in E$ is characterized by a function $\hat{S}$ denoting the membership grade of $e \in V \times V$. The closed subinterval $\hat{S} = [\hat{S}^L, \hat{S}^U]$ satisfying the conditions:*

$$\hat{S}^L(v_i, v_j) \leq \min(S^L(v_i), S^L(v_j)) \text{ and}$$
$$\hat{S}^U(v_i, v_j) \leq \min(S^U(v_i), S^U(v_j)) \text{ such that } \hat{S}^U(v_i, v_j) \geq \min(S^L(v_i), S^L(v_j)).$$

This definition is supported with the help of following example in which we show that none of the intervals is undefined and that Definition 5 works perfectly.

Example 4. *The following IVFG depicted in Figure 5 is in accordance with Definition 5.*

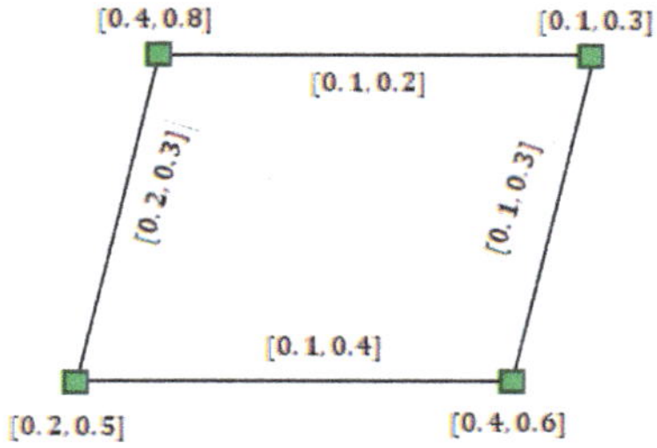

Figure 5. Interval-valued fuzzy graph compatible with Definition 5.

Next, we provide a critical study of IVIFGs with a definition which is based on IVIFG which yields undefined closed sub-intervals. First, we review the existing definition and the existing shortcomings with the help of examples. Then, the new definition is developed and supported by an example.

Definition 6. *[40] An IVIFG is a duplet $G = (V, E)$ where the set of nodes is denoted by V and E is the collection of edges between these nodes such that*

(1) Every $v \in V$ is characterized by two functions S and D denoting the membership and non-membership grades of $v \in V$. Basically, $S = [S^L, S^U]$ and $D = [D^L, D^U]$ are closed subintervals of the unit interval [0, 1] with a condition that $0 \le S^U + D^U \le 1$. Moreover, the term $R = [R^L, R^U]$ denote the hesitancy level of $v \in V$ such that $R^U = 1 - S^L - D^L$ and $R^L = 1 - S^U - D^U$.

(2) Every $e \in E$ is characterized by two functions $\hat{S}$ and $Đ$ denoting the membership and non-membership grades of $e \in V \times V$. Basically, $\hat{S} = [\hat{S}^L, \hat{S}^U]$ and $Đ = [Đ^L, Đ^U]$ satisfying the conditions:

$$\hat{S}^L(v_i, v_j) \le \min(S^L(v_i), S^L(v_j)) \quad \hat{S}^U(v_i, v_j) \le \min(S^U(v_i), S^U(v_j))$$
$$Đ^L(v_i, v_j) \ge \max(D^L(v_i), D^L(v_j)) \quad Đ^U(v_i, v_j) \ge \max(D^U(v_i), D^U(v_j))$$

provided that $0 \le \hat{S}^U + Đ^U \le 1$. Moreover, the term $\check{R} = [\check{R}^L, \check{R}^U]$ denote the hesitancy level of $e \in E$ such that $\check{R}^U = 1 - \hat{S}^L - Đ^L$ and $\check{R}^L = 1 - \hat{S}^U - Đ^U$.

Definition 6 seems to be weak in two ways. The first reason is that IVIFG is a generalization of IFG where the non-membership grade of edge is defined as $Đ(v_i, v_j) \le \max(D(v_i), D(v_j))$, which is not followed in defining IVIFG, raising a question regarding its perfection. Another shortcoming in Definition 6 is that we may sometimes have obtained undefined closed intervals, which should be avoided. For further illustration, consider the example as follows.

Example 5. *Consider the following IVIFG in Figure 6, where it is observed that we get some undefined intervals by applying Definition 6. The undefined closed intervals are highlighted.*

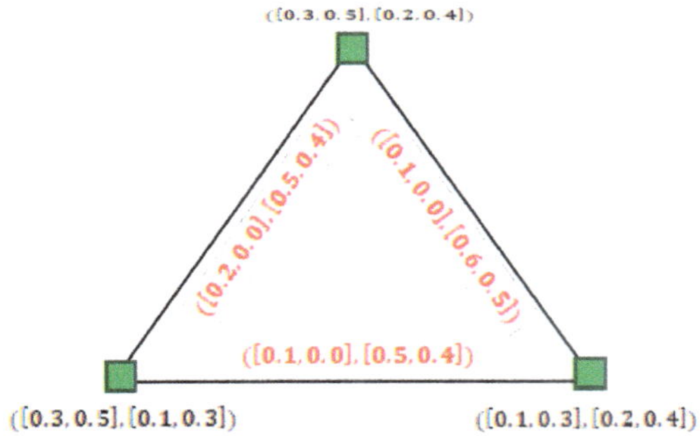

Figure 6. Interval-valued intuitionistic fuzzy graph (IVIFG) obtained using Definition 6 with undefined intervals.

Keeping in mind the weakness of Definition 6 as demonstrated by Example 5, we propose a new definition of IVIFG which is well defined and truly a generalization of IFG.

Definition 7. *An IVIFG is a pair $G = (V, E)$ where the set of nodes is denoted by V and E is the collection of edges between these nodes such that*

(1) *Every $v \in V$ is characterized by two functions S and D denoting the membership and non-membership grades of $v \in V$. Basically, $S = [S^L, S^U]$ and $D = [D^L, D^U]$ are closed subintervals of the unit interval [0, 1] with a condition that $0 \le S^U + D^U \le 1$. Moreover, the term $R = [R^L, R^U]$ denote the hesitancy level of $v \in V$ such that $R^U = 1 - S^L - D^L$ and $R^L = 1 - S^U - D^U$;*

(2) *Every $e \in E$ is characterized by two functions $\hat{S}$ and Đ denoting the membership and non-membership grades of $e \in V \times V$. Basically, $\hat{S} = [\hat{S}^L, \hat{S}^U]$ and $Đ = [Đ^L, Đ^U]$ are such that*

$$\hat{S}^L(v_i, v_j) \le \min(S^L(v_i), S^L(v_j)).$$
$$\hat{S}^U(v_i, v_j) \le \min(S^U(v_i), S^U(v_j)) \text{ such that } \hat{S}^U(v_i, v_j) \ge \min(S^L(v_i), S^L(v_j)).$$
$$Đ^L(v_i, v_j) \le \max(D^L(v_i), D^L(v_j)).$$
$$Đ^U(v_i, v_j) \le \max(D^U(v_i), D^U(v_j)) \text{ such that } Đ^U(v_i, v_j) \ge \max(D^L(v_i), D^L(v_j)).$$

provided that $0 \le \hat{S}^U + Đ^U \le 1$. Moreover, the term $\check{R} = [\check{R}^L, \check{R}^U]$ denote the hesitancy level of $e \in E$ such that $\check{R}^U = 1 - \hat{S}^L - Đ^L$ and $\check{R}^L = 1 - \hat{S}^U - Đ^U$.

The following example shows that Definition 7 improves the concept of IVIFG and that there is no chance of getting undefined intervals.

Example 6. *Consider the graph depicted in Figure 7 demonstrating the new definition of IVIFG.*

Figure 7. Interval-valued intuitionistic fuzzy graph.

The next section is based on the shortcomings of existing definitions of the complement of an IVFG and of an IVIFG together with the development of new definitions. The existing and new definitions are demonstrated with examples.

4. New Definitions for the Complement of the Interval-Valued Fuzzy Graph and Interval-Valued Intuitionistic Fuzzy Graph

It is observed that the definition of the complement of IVFG did not provide justifiable results on some occasions. The results obtained using the current definition of IVFG leads us to obtain some undefined intervals, and therefore in this section we propose a new definition for the complement of IVFG.

For the first time, the complement of IVFG was defined in [38]. First, we review the existing definition and, with the help of an example, we point out its shortcomings.

Definition 8. *[38] The complement of an IVFG $G = (V,\ E)$ is defined by $G^c = (V^c,\ E^c)$ where $V^c = V$ and the membership and non-membership grades of E satisfying the conditions:*

$$\left(\hat{S}^L\right)^c\left(v_i,\ v_j\right) = \min\left(S^L(v_i),\ S^L\left(v_j\right)\right) - \hat{S}^L\left(\left(v_i,\ v_j\right)\right)$$
$$\left(\hat{S}^U\right)^c\left(v_i,\ v_j\right) = \min\left(S^U(v_i),\ S^U\left(v_j\right)\right) - \hat{S}^U\left(v_i,\ v_j\right)$$

Example 7. *In the following, consider an IVFG depicted in Figure 8, then based on the existing definition 8, we compute its complement and it is given in Figure 9. It is observed from the Figure 9 that there are undefined intervals which are highlighted in red color.*

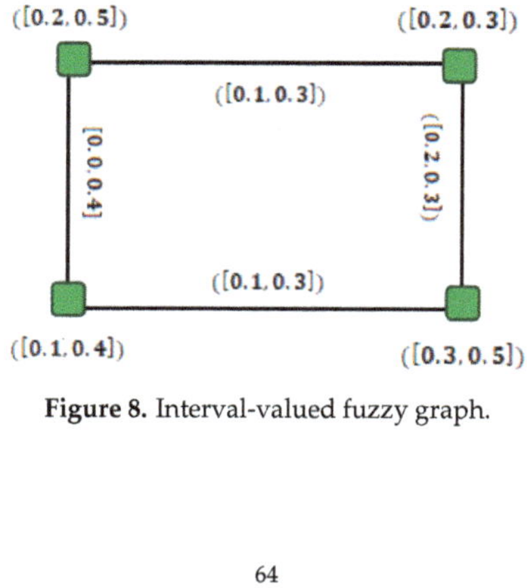

Figure 8. Interval-valued fuzzy graph.

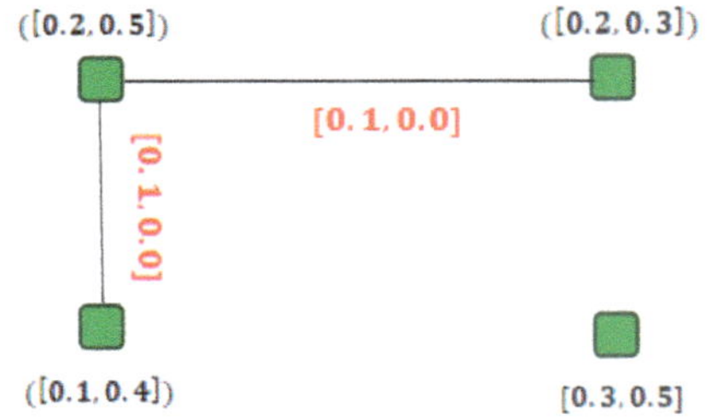

Figure 9. Complement of graph depicted in Figure 8 based on Definition 8, giving us undefined intervals.

Another definition of the complement of IVFG is defined in [11] which is described as follows:

Definition 9. *[28] The complement of an IVFG $G = (V, E)$ is defined by $G^c = (V^c, E^c)$ where $V^c = V$ and the membership functions of E^c satisfies the following:*

$$\left(\hat{S}^L\right)^c\left(v_i, v_j\right) = \left\{ \begin{array}{cc} 0 & if \;\; S^L\left(v_i, v_j\right) \geq 0 \\ \min\left(S^L(v_i), S^L\left(v_j\right)\right) & if \;\; S^L\left(v_i, v_j\right) = 0 \end{array} \right\}$$
$$\left(\hat{S}^U\right)^c\left(v_i, v_j\right) = \left\{ \begin{array}{cc} 0 & if \;\; S^U\left(v_i, v_j\right) \geq 0 \\ \min\left(S^U(v_i), S^U\left(v_j\right)\right) & if \;\; S^U\left(v_i, v_j\right) = 0 \end{array} \right\}$$

This definition is not well defined as it is not valid for all types of IVFGs. The problem in this definition of the complement is that it does not possesses the property $(G^c)^c = G$. This is demonstrated in the following example.

Example 8. *The complement of IVFG as exhibited in Figure 8 is depicted in Figure 10 below and it is observed that all the edges have disappeared surprisingly, which seems insignificant. Now, we determine the complement of Figure 10 again and put it on display in Figure 11, which clearly indicates that the basic result of the complement, i.e., $(G^c)^c = G$, does not hold true for Definition 9.*

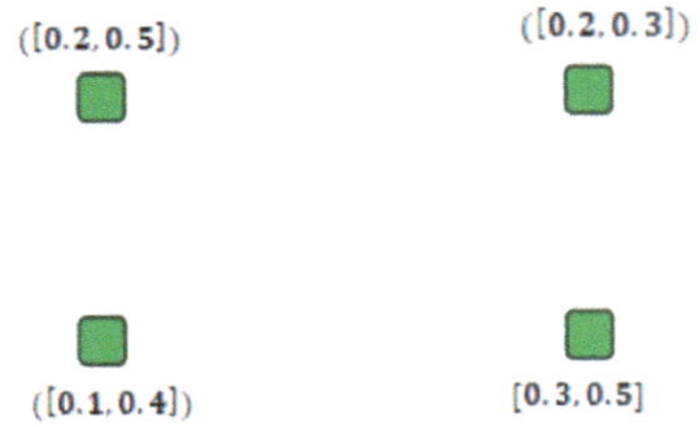

Figure 10. Complement of Figure 8.

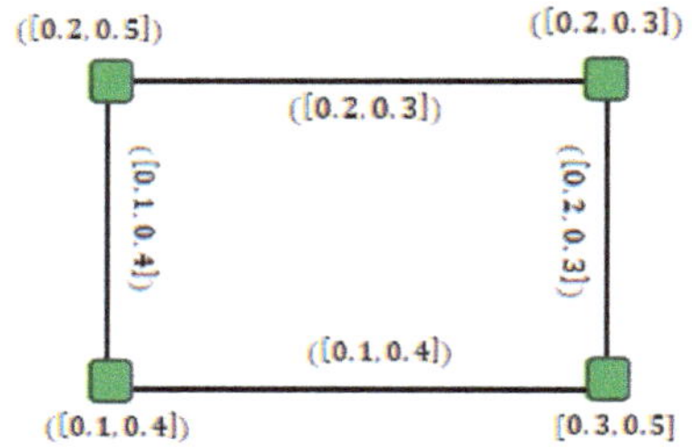

Figure 11. Complement of Figure 10 based on Definition 9.

Moreover, Definition 9 is used to find the complement of a complete IVFG in [28] by Akram and Dudek, where this definition possesses the property of a complement; i.e., $(G^c)^c = G$. However, we need a definition of a complement which can be used for the entire range of IVFG and not only complete IVFGs; therefore, we develop a new well-defined definition for the complement of IVFGs as follows.

Definition 10. *The complement of IVFG $G = (V, E)$ is defined by $G^c = (V^c, E^c)$ where $V^c = V$ and the membership grades of E satisfies the following properties:*

$$\left(\hat{S}^L\right)^c\left(v_i, v_j\right) = \min\left(S^L(v_i), S^L\left(v_j\right)\right) - \hat{S}^L\left(\left(v_i, v_j\right)\right)$$
$$\left(\hat{S}^U\right)^c\left(v_i, v_j\right) = \min\left(S^U(v_i), S^U\left(v_j\right)\right) - \hat{S}^U\left(v_i, v_j\right) + \min\left(S^L(v_i), S^L\left(v_j\right)\right).$$

In the following examples, we not only demonstrate the new definition but also verify that the basic property of the notion of the complement as presented in Definition 10 is available; i.e., $(G^c)^c = G$.

Example 9. *In what follows, Figure 12 represents the complement of* Figures 8 and 13 *represents the complement of Figure 12. Hence, as we claimed, the equality $(G^c)^c = G$ holds true for our proposed new definition.*

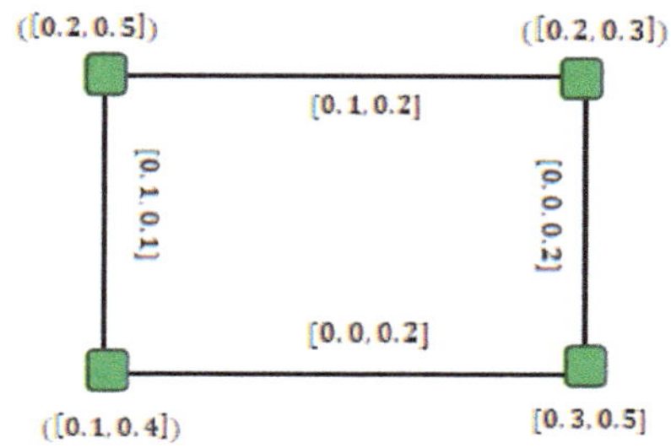

Figure 12. Complement of graph depicted in Figure 8.

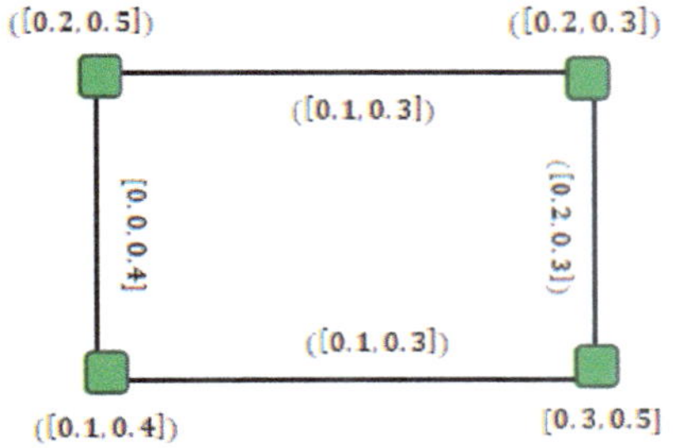

Figure 13. The complement of graph depicted in Figure 12.

Now, we develop the definition of the complement for IVIFGs because the existing definition does not make any sense in some cases, as we described for IVFGs.

Definition 11. *The complement of IVIFG $G = (V, E)$ is defined by $G^c = (V^c, E^c)$ where $V^c = V$ and the membership and non-membership grades of E satisfied the conditions:*

$$\left(\hat{S}^L\right)^c(v_i, v_j) = \min\left(S^L(v_i), S^L(v_j)\right) - \hat{S}^L\left((v_i, v_j)\right).$$
$$\left(\hat{S}^U\right)^c(v_i, v_j) = \min\left(S^U(v_i), S^U(v_j)\right) - \hat{S}^U(v_i, v_j) + \min\left(S^L(v_i), S^L(v_j)\right).$$
$$\left(Đ^L\right)^c(v_i, v_j) = \max\left(D^L(v_i), D^L(v_j)\right) - Đ^L\left((v_i, v_j)\right).$$
$$\left(Đ^U\right)^c(v_i, v_j) = \max\left(D^U(v_i), D^U(v_j)\right) - Đ^U(v_i, v_j) + \max\left(D^L(v_i), D^L(v_j)\right).$$

Example 10. *Consider the following IVIFGs whose graph is shown in Figure 14 while its complement graph is shown in Figure 15. Moreover, through some easy calculations, one can easily verify that $(G^c)^c = G$.*

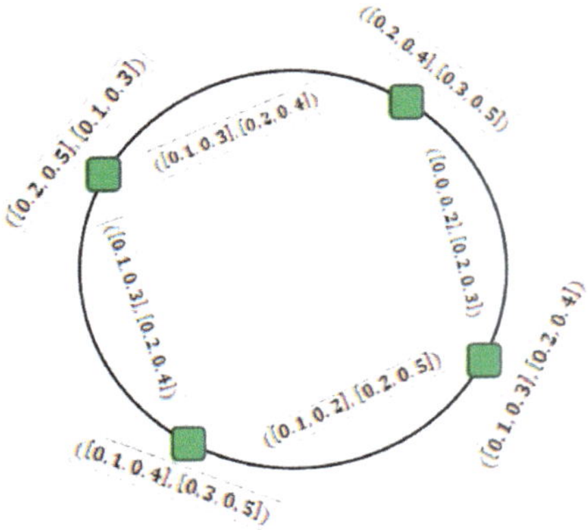

Figure 14. Interval-valued intuitionistic fuzzy graph.

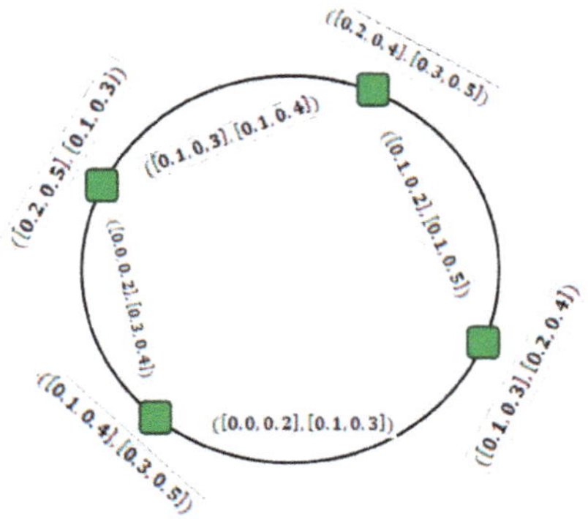

Figure 15. Complement of IVIFG depicted in Figure 14.

In the next section, we discuss the limitations of SVNGS, IVNGs, and their complements. In addition, we develop some new valid definitions.

5. Improvements in Single-Valued Neutrosophic Graphs and Interval-Valued Neutrosophic Graphs

In this section, we study the SVNGs proposed in [26] and IVNGs developed in [48] and we describe their shortcomings with some examples. Then, a new definition is proposed for both SVNGs and IVNGs, and their complement is discussed. In support of these new definitions, we present some examples.

Definition 12. *[43] An SVNG is a pair $G = (V,\ E)$ where V is the set of nodes and E is the collection of edges between these nodes such that*

(1) *Every $v \in V$ is characterized by three functions $S : V \to [0,\ 1]$, $I : V \to [0,\ 1]$ and $D : V \to [0,\ 1]$ denoting the membership, neutral and non-membership grades of $v \in V$ which satisfy $0 \le S + I + D \le 3$;*
(2) *Every $e \in E$ is characterized by three functions $\hat{S} : V \times V \to [0,\ 1]$, $\hat{I} : V \times V \to [0,\ 1]$ and $\text{Đ} : V \times V \to [0,\ 1]$ denoting the membership, neutral and non-membership grades of $e \in V \times V$ satisfies*

$$\hat{S}(v_i,\ v_j) \le \min(S(v_i),\ S(v_j))$$
$$\hat{I}(v_i,\ v_j) \ge \max(I(v_i),\ I(v_j))$$
$$\text{Đ}(v_i,\ v_j) \ge \max(D(v_i),\ D(v_j))$$

which satisfy the inequality $0 \le \hat{S} + \hat{I} + \text{Đ} \le 3$.

It is observed that this definition is not suitable and needs to be modified for two reasons. The first reason lies in the fact that as SVNG is a generalization of IFG, by exempting the indeterminacy value, Definition 12 should reduce to Definition 3. However, this does not happen, as the non-membership degree of an edge in IFG and in SVNG are defined in a different way; i.e., the non-membership degree for IFG is defined as $\text{Đ}(v_i,\ v_j) \le \max(D(v_i),\ D(v_j))$ while in case of SVNGs it is defined as $\text{Đ}(v_i,\ v_j) \ge \max(D(v_i),\ D(v_j))$.

The second reason is based on the definition of complement of SVNGs, as the complement of an SVNG needs to be an SVNG, which did not happen on some occasions. We present the current definition of the complement of an SVNG and with the help of an example describe that the complement of an SVNG is not well defined.

Definition 13. [43] *The complement of an SVNG $G = (V, E)$ is defined by $G^c = (V^c, E^c)$ where $V^c = V$ and the membership, neutral value and non-membership grades of E satisfy the conditions:*

$$(\hat{S})^c(v_i, v_j) = \min(S(v_i), S(v_j)) - \hat{S}((v_i, v_j)).$$
$$(\hat{I})^c(v_i, v_j) = \max(I(v_i), I(v_j)) - \hat{I}((v_i, v_j)).$$
$$(\text{Đ})^c(v_i, v_j) = \max(D(v_i), D(v_j)) - \text{Đ}((v_i, v_j)).$$

Keep in mind Definition 13 and consider the following example, where the complement of an SVNG does not remain an SVNG. All this leads to some modification in the basic definition of SVNG which not only generalizes an IFG but is also compatible to satisfy the basic properties of taking complements.

Example 11. *In the following figures, Figure 16 represents SVNGs and Figure 17 represents their complement. It is clear from Figure 17 that the complement of an SVNG is not an SVNG.*

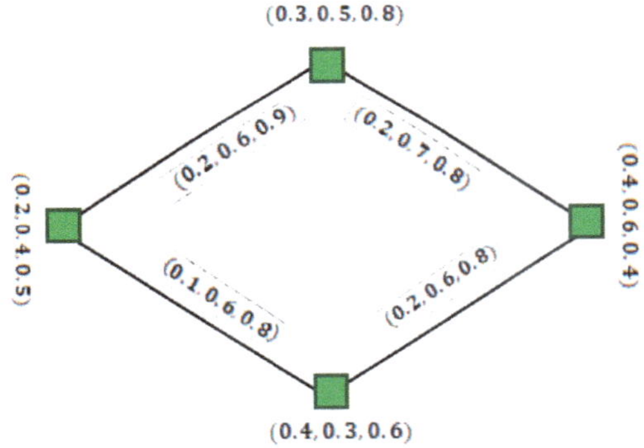

Figure 16. Single-valued neutrosophic graph.

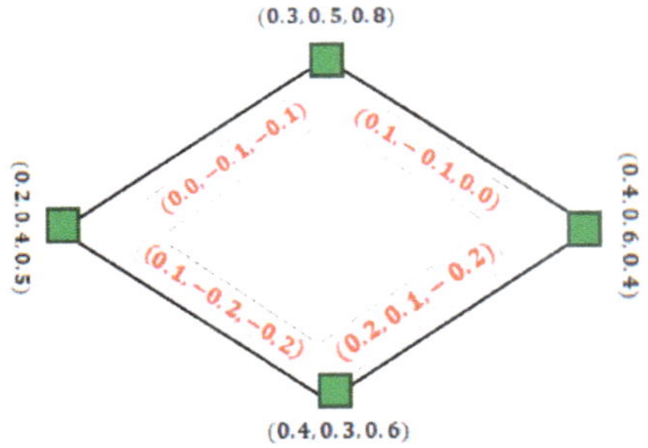

Figure 17. Complement of the SVNG depicted in Figure 16.

Now, we present a new definition for SVNG as a generalization of IFG which is well-defined.

Definition 14. *An SVNG is a pair $G = (V, E)$ where V is the set of nodes and E is the collection of edges between these nodes such that*

1. *Every $v \in V$ is characterized by three functions $S : V \to [0, 1]$, $I : V \to [0, 1]$ and $D : V \to [0, 1]$ denoting the membership, neutral and non-membership grades of $v \in V$ which satisfy the inequality $0 \leq S + I + D \leq 3.$;*

2. *Every* $e \in E$ *is characterized by three functions* $\hat{S}: V \times V \to [0, 1]$, $\hat{I}: V \times V \to [0, 1]$ *and* $Ð: V \times V \to [0, 1]$ *denoting the membership, neutral and non-membership grades of* $e \in V \times V$ *satisfying*

$$\hat{S}(v_i, v_j) \leq \min(S(v_i), S(v_j))$$
$$\hat{I}(v_i, v_j) \leq \max(I(v_i), I(v_j))$$
$$Ð(v_i, v_j) \leq \max(D(v_i), D(v_j))$$

provided that $0 \leq \hat{S} + \hat{I} + Ð \leq 3$.

Example 12. *In the following figures, Figure 18 represents an SVNG compatible with the new definition and Figure 19 represents the complement of the SVNG depicted in Figure 18.*

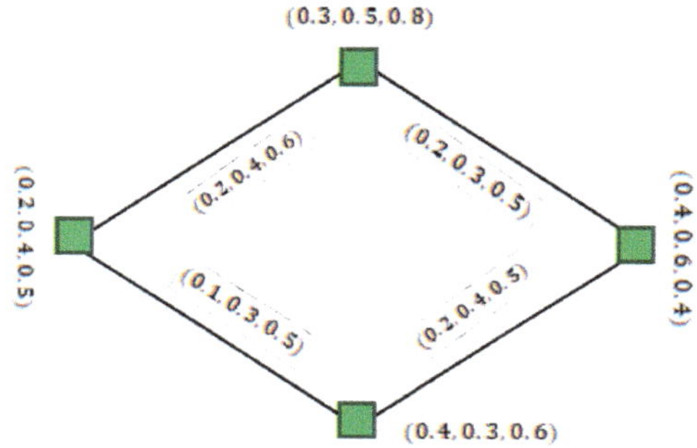

Figure 18. Single-valued neutrosophic graph.

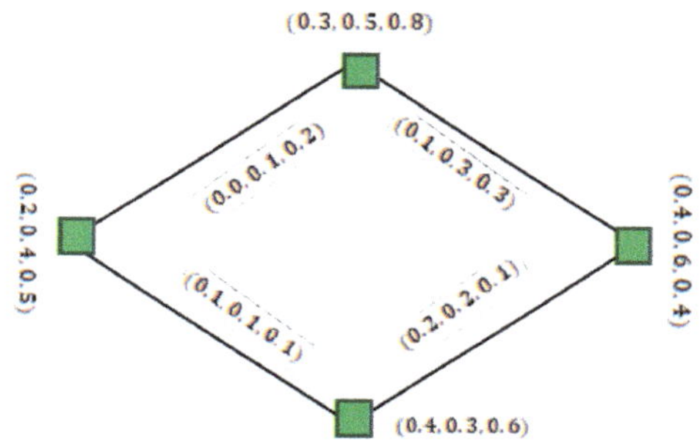

Figure 19. Complement of the graph depicted in Figure 18.

Likewise, in SVNGs the same kinds of flaw exist in the definition of IVNGs. First, we present the existing definition of IVNG and point out its limitation with the help of an example. Then, we develop the new definition for IVNG and support it with an example. After developing the new definition for IVNGs, we discuss the concept of complements for IVNGs by considering the defects in forming complements of IVFG and IVIFG.

Definition 15. [48] *An IVNG is a pair* $G = (V, E)$ *where V is the set of nodes and E is the collection of edges between these nodes such that*

1. *Every* $v \in V$ *is characterized by three functions S, I and D denoting the membership, neutral and non-membership grades of* $v \in V$. *Basically,* $S = [S^L, S^U]$, $I = [I^L, I^U]$ *and* $D = [D^L, D^U]$ *are closed subintervals of the unit interval [0, 1] which satisfy the inequality* $0 \leq S^U + I^U + D^U \leq 3$.;

2. *Every e ∈ E is characterized by three functions* $\hat{S}$, $\hat{I}$ *and* Đ *denoting the membership, neutral and non-membership grades of* $e \in V \times V$. *Basically,* $\hat{S} = \left[\hat{S}^L, \hat{S}^L\right]$, $\hat{I} = \left[\hat{I}^L, \hat{I}^U\right]$ *and* $\text{Đ} = \left[\text{Đ}^L, \text{Đ}^L\right]$ *are such that:*

$$\begin{array}{ll} \hat{S}^L(v_i, v_j) \le \min\left(S^L(v_i), S^L(v_j)\right) & \hat{S}^U(v_i, v_j) \le \min\left(S^U(v_i), S^U(v_j)\right) \\ \hat{I}^L(v_i, v_j) \ge \max\left(I^L(v_i), I^L(v_j)\right) & \hat{I}^U(v_i, v_j) \ge \max\left(I^U(v_i), I^U(v_j)\right) \\ \text{Đ}^L(v_i, v_j) \ge \max\left(D^L(v_i), D^L(v_j)\right) & \text{Đ}^U(v_i, v_j) \ge \max\left(D^U(v_i), D^U(v_j)\right) \end{array}$$

which satisfy the inequality $0 \le \hat{S}^U + \hat{I}^U + \text{Đ}^U \le 3$.

Likewise, IVFG and IVIFG are defined, and Definition 15 sometime leads to undefined closed sub-intervals as shown in the following example.

Example 13. *Consider the following example of an IVNG as exhibited in Figure 20 based on Definition 15, leading us to some undefined intervals.*

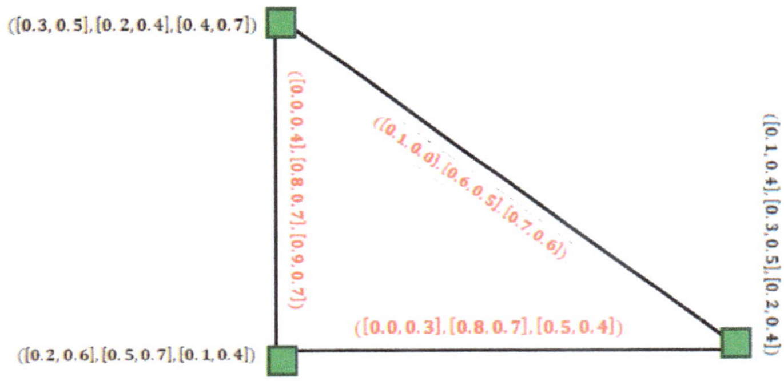

Figure 20. Interval-valued neutrosophic graph with undefined intervals.

Now, we present a new definition of IVNG as follows.

Definition 16. *An IVNG is a pair* $G = (V, E)$ *where V is the set of nodes and E is the collection of edges between these nodes such that*

(1) *Every* $v \in V$ *is characterized by three functions S, I and D denoting the membership, neutral and non-membership grades of* $v \in V$. *Basically,* $S = \left[S^L, S^U\right]$, $I = \left[I^L, I^U\right]$ *and* $D = \left[D^L, D^U\right]$ *are closed subintervals of the unit interval [0, 1] with a condition that* $0 \le S^U + I^U + D^U \le 3$.;

(2) *Every* $e \in E$ *is characterized by three functions* $\hat{S}$, $\hat{I}$ *and* Đ *denoting the membership, neutral and non-membership grades of* $e \in V \times V$. *Basically* $\hat{S} = \left[\hat{S}^L, \hat{S}^L\right]$, $\hat{I} = \left[\hat{I}^L, \hat{I}^U\right]$ *and* $\text{Đ} = \left[\text{Đ}^L, \text{Đ}^L\right]$ *are such that*

$$\begin{array}{ll} \hat{S}^L(v_i, v_j) \le \min\left(S^L(v_i), S^L(v_j)\right) & \hat{S}^U(v_i, v_j) \le \min\left(S^U(v_i), S^U(v_j)\right) \\ \hat{I}^L(v_i, v_j) \le \max\left(I^L(v_i), I^L(v_j)\right) & \hat{I}^U(v_i, v_j) \le \max\left(I^U(v_i), I^U(v_j)\right) \\ \text{Đ}^L(v_i, v_j) \le \max\left(D^L(v_i), D^L(v_j)\right) & \text{Đ}^U(v_i, v_j) \le \max\left(D^U(v_i), D^U(v_j)\right) \end{array}$$

provided that $\hat{S}^U(v_i, v_j) \le \hat{S}^L(v_i, v_j)$, $\hat{I}^L(v_i, v_j) \le \hat{I}^U(v_i, v_j)$, $\text{Đ}^L(v_i, v_j) \le \text{Đ}^U(v_i, v_j)$ *and* $0 \le \hat{S}^U + \hat{I}^U + \text{Đ}^U \le 3$.

Example 14. *Consider the graph depicted in Figure 21 demonstrating the new definition of IVNGs.*

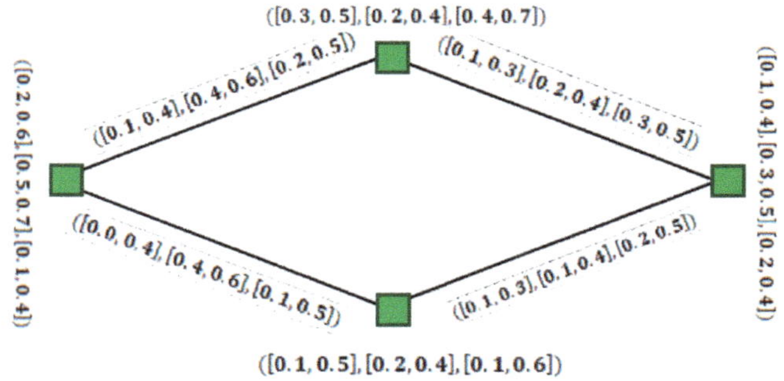

Figure 21. Interval-valued neutrosophic graph with undefined intervals.

Next, we discuss the existing definition for taking the complement of IVNG and show its limitations with the help of examples. Then, we propose a new definition for the complement of IVNG as the existing definition leads us to some undefined intervals.

Definition 17. [48] *The complement of IVNG $G = (V, E)$ is defined by $G^c = (V^c, E^c)$ where $V^c = V$ and the membership, abstinence and non-membership grades of E satisfy*

$$\begin{aligned}
\left(\hat{S}^L\right)^c\left(v_i, v_j\right) &= \min\left(S^L(v_i), S^L\left(v_j\right)\right) - \hat{S}^L\left(\left(v_i, v_j\right)\right).\\
\left(\hat{S}^U\right)^c\left(v_i, v_j\right) &= \min\left(S^U(v_i), S^U\left(v_j\right)\right) - \hat{S}^U\left(v_i, v_j\right).\\
\left(\hat{I}^L\right)^c\left(v_i, v_j\right) &= \max\left(I^L(v_i), I^L\left(v_j\right)\right) - \hat{I}^L\left(\left(v_i, v_j\right)\right).\\
\left(\hat{I}^U\right)^c\left(v_i, v_j\right) &= \max\left(I^U(v_i), I^U\left(v_j\right)\right) - \hat{I}^U\left(v_i, v_j\right).\\
\left(\text{Đ}^L\right)^c\left(v_i, v_j\right) &= \max\left(D^L(v_i), D^L\left(v_j\right)\right) - \text{Đ}^L\left(\left(v_i, v_j\right)\right).\\
\left(\text{Đ}^U\right)^c\left(v_i, v_j\right) &= \max\left(D^U(v_i), D^U\left(v_j\right)\right) - \text{Đ}^U\left(v_i, v_j\right).
\end{aligned}$$

Example 15. *The following IVNG depicted in Figure 22 shows that the Definition 17 for the complement of IVNG leads us to some undefined intervals.*

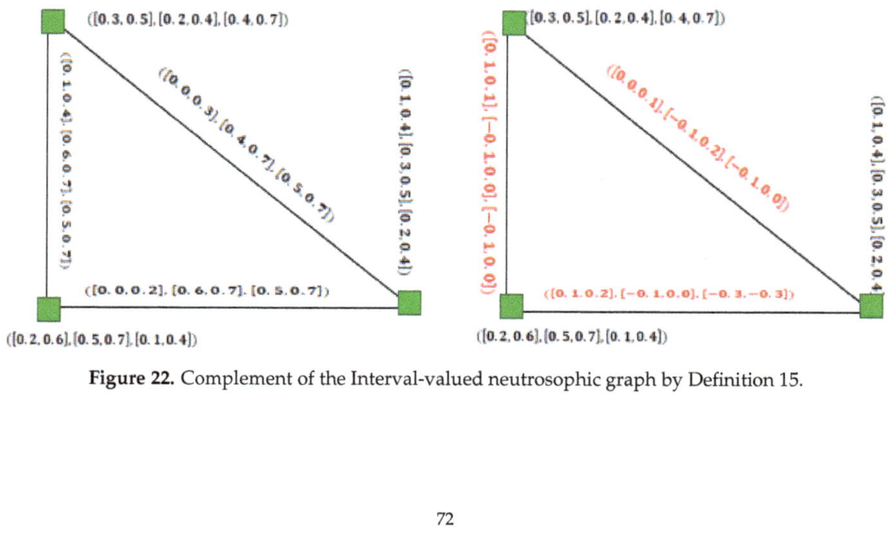

Figure 22. Complement of the Interval-valued neutrosophic graph by Definition 15.

As the IVNG is a generalization of IVFG and IVIFG, its complement defined in [31] is not well-defined; therefore, we propose a new definition for the complement of IVNG as follows.

Definition 18. *The complement of IVNG $G = (V, E)$ is defined by $G^c = (V^c, E^c)$ where $V^c = V$ and the membership, abstinence and non-membership grades of E are satisfying the conditions:*

$$\left(\hat{S}^L\right)^c\left(v_i, v_j\right) = \min\left(S^L(v_i), S^L\left(v_j\right)\right) - \hat{S}^L\left(\left(v_i, v_j\right)\right).$$
$$\left(\hat{S}^U\right)^c\left(v_i, v_j\right) = \min\left(S^U(v_i), S^U\left(v_j\right)\right) - \hat{S}^U\left(v_i, v_j\right) + \min\left(S^L(v_i), S^L\left(v_j\right)\right).$$
$$\left(\hat{I}^L\right)^c\left(v_i, v_j\right) = \max\left(I^L(v_i), I^L\left(v_j\right)\right) - \hat{I}^L\left(\left(v_i, v_j\right)\right).$$
$$\left(\hat{I}^U\right)^c\left(v_i, v_j\right) = \max\left(I^U(v_i), I^U\left(v_j\right)\right) - \hat{I}^U\left(v_i, v_j\right) + \max\left(I^L(v_i), I^L\left(v_j\right)\right).$$
$$\left(Đ^L\right)^c\left(v_i, v_j\right) = \max\left(D^L(v_i), D^L\left(v_j\right)\right) - Đ^L\left(\left(v_i, v_j\right)\right).$$
$$\left(Đ^U\right)^c\left(v_i, v_j\right) = \max\left(D^U(v_i), D^U\left(v_j\right)\right) - Đ^U\left(v_i, v_j\right) + \max\left(D^L(v_i), D^L\left(v_j\right)\right).$$

Example 16. *Consider the following IVNGs shown in Figure 23 and its complement graph is depicted in Figure 24. Through some easy calculation, one can easily verify that $(G^c)^c = G$.*

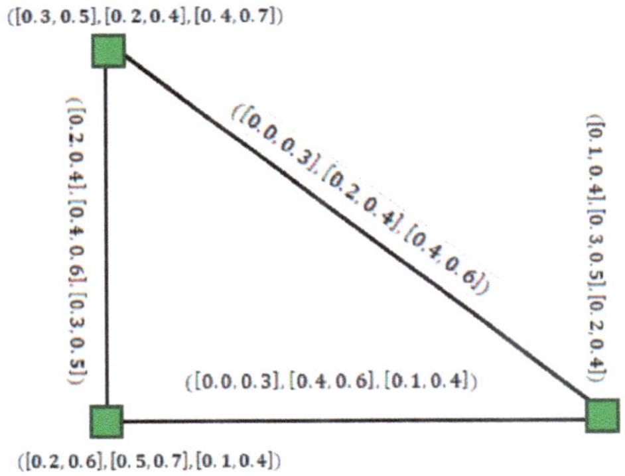

Figure 23. Interval-valued neutrosophic graph.

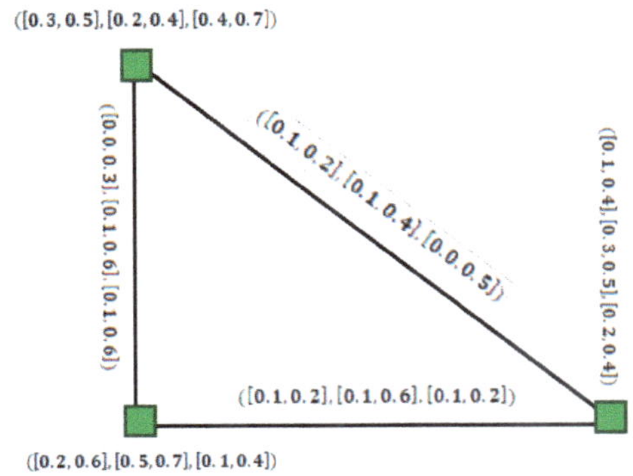

Figure 24. Complement of the graph depicted in Figure 23.

So far in our study, we have developed the theory of IVFGs and its generalizations, providing examples that explained the limitations in the existing definitions and the fitness of the new definitions.

Remark 1. *The work proposed is applicable whether the graphs are directed or undirected.*

Now, a real-life application is presented to show the feasibility of the proposed work. The process of ranking universities is explained based on SVNGs.

6. Illustrative Example

The higher education commission (HEC) of Pakistan needed to evaluate the Pakistani universities from research perspectives. HEC Pakistan planned to give away some financial support to those universities whose research was comparatively better. A university with a good research score will earn greater financial support. Here, we shall discuss the case of the federal capital Islamabad. Four universities, z_1, z_2, z_3 and z_4, from Islamabad are chosen, representing Quaid e Azam university, National University of Science and Technology, Comsats university and International Islamic University Islamabad. The criterion for the ranking of universities is "research productivity". The members of the quality assurance cell (QEC) of HEC provide their information in the form of single-valued neutrosophic preference relations (SVNPRs) as given in Figure 25. The weight vector corresponding to the them is given by $w = (0.3, 0.2, 0.3, 0.2)^T$. The detailed steps of algorithm for this process are explained followed by stepwise calculations of this problem.

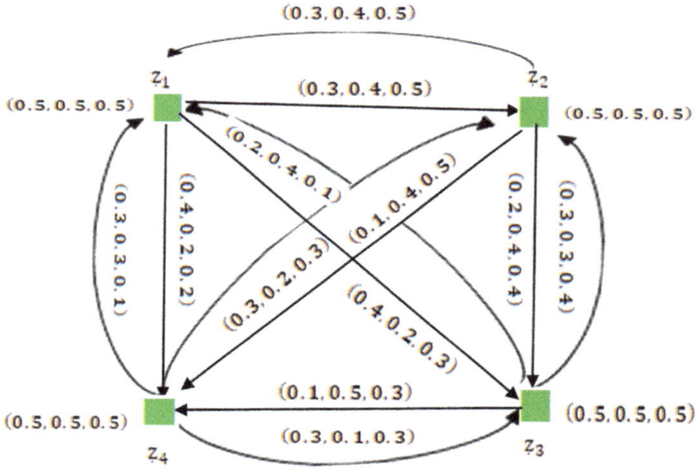

Figure 25. Directed network with single-valued neutrosophic information.

Steps of the algorithm:

Step 1: Establishing SVNPRs: This step involves the evaluation of objects by decision-makers in the form of SVNGs;

Step 2: The SVNPRs form a relational matrix known as the single-valued neutrosophic relational matrix (SVNRM) denoted by $\dot{R}M = \left(m\dot{r}_{ij}\right)_{n\times n}$ where $\dot{R}m_{ij} = \left(S_{ij}, I_{ij}, D_{ij}\right)$ such that $\dot{R}0 \leq S_{ij}, I_{ij}, D_{ij} \leq 1$ and $\dot{R}S_{ij} + I_{ij} + D_{ij} \leq 3$;

Step 3: Aggregating the different preferences by using the weighted averaging aggregation operator named as single-valued neutrosophic weighted averaging (SVNWA) operators [8] as follows:

$$m\dot{r}_i = SVNWA(m\dot{r}_{i1}, m\dot{r}_{i2}, \ldots, m\dot{r}_{in}) = \left(1 - \prod_{j=1}^{n}\left(1 - S_{ij}\right)^{w_j}, \left(\prod_{j=1}^{n} I_{ij}\right)^{w_j}, \left(\prod_{j=1}^{n} D_{ij}\right)^{w_j}\right)$$

where w_j is the weight vector for each preference such that $w_j > 0$ and $\sum_{j=1}^{n} w_j = 1$;

Step 4: Computing the score value of the aggregated number $m_i = (S_i, I_i, D_i)$ $(i = 1, 2, \ldots, n)$ by using the score function

$$S(m_i) = \frac{S_i + 1 - I_i + 1 - D_i}{3};$$

Step 5: Ranking all the alternatives based on the descending order of the score values.

Now, by applying the above-mentioned steps to the considered data, we summarized their computational steps as below:

Step 1: The information from QEC of HEC Pakistan related to each university is taken in the form of SVNGs and summarized in Figure 25.

Here, the importance of a new definition of an SVNG is clear. In decision-making problems, we usually have two types of attributes, known as attributes of cost type and of benefit types. In the case that the attributes are of cost types, the data is normalized using the definition of complement of SVNGs. Therefore, if we utilize the existing definition of SVNGs and their complement as we have done in Example 11, we obtain some undefined results. The case is similar with other structures. All this shows the importance of the new proposed definitions of several fuzzy graphs and their complements.

Step 2: From the SVNPRs provided in Figure 25, we have the following relational matrix.

$$M\dot{R} = \left(m\dot{r}_{ij}\right)_{4\times 4} = \begin{pmatrix} (0.5, 0.5, 0.5) & (0.3, 0.4, 0.5) & (0.4, 0.2, 0.3) & (0.4, 0.2, 0.2) \\ (0.3, 0.4, 0.5) & (0.5, 0.5, 0.5) & (0.2, 0.4, 0.4) & (0.1, 0.4, 0.5) \\ (0.2, 0.4, 0.1) & (0.3, 0.3, 0.4) & (0.5, 0.5, 0.5) & (0.1, 0.5, 0.3) \\ (0.3, 0.3, 0.1) & (0.3, 0.2, 0.3) & (0.3, 0.1, 0.3) & (0.5, 0.5, 0.5) \end{pmatrix}$$

Step 3: Using single-valued neutrosophic weighted averaging operators on the relational matrix to aggregate the data,

$$m\dot{r}_i = SVNWA(m\dot{r}_{i1}, m\dot{r}_{i2}, \ldots, m\dot{r}_{in}) = \left(1 - \prod_{j=1}^{n}\left(1 - S_{ij}\right)^{w_j}, \left(\prod_{j=1}^{n} I_{ij}\right)^{w_j}, \left(\prod_{j=1}^{n} D_{ij}\right)^{w_j}\right)$$

and the aggregated values are obtained as

$$\begin{aligned} m\dot{r}_1 &= (0.414150269,\ 0.3024252,\ 0.3571309); \\ \dot{r}m_2 &= (0.283693469,\ 0.4182558,\ 0.4676242); \\ \dot{r}m_3 &= (0.307394227,\ 0.4222088,\ 0.2663957); \\ \dot{r}m_4 &= (0.345556087,\ 0.2203618,\ 0.2389762). \end{aligned}$$

Step 4: The score values of the aggregated data are obtained as

$$S(m_1) = 0.584864723,\ \underset{\cdot}{s}S(m\dot{r}_2) = 0.465937823,\ S\underset{\cdot}{s}(m\dot{r}_3) = 0.539596576,\ \underset{\cdot}{s}S(m\dot{r}_4) = 0.62873936$$

Step 5: Since $S(m_4) > S(m_1) > S(m_3) > S(m_2)$, we obtained the ranking order of the given universities as $z_4 \succ z_1 \succ z_3 \succ z_2$, where $\succ$ refers to "preferred to". Thus, the ranking results indicate that International Islamic University Islamabad performs best in research productivity among the four universities of federal capital of Islamabad in the evaluation of the quality assurance cell of HEC.

7. Conclusions

In this manuscript, we successfully pointed out the shortcomings in the current definitions of IVFG, IVIFG with the help of examples and developed a new improved definition for these concepts. It is also discussed that the current definitions of complements for IVFG and IVIFG lead us to some undefined results; therefore, some new definitions for complements of IVFG and IVIFG are proposed and supported by examples. Further, it is observed that the definition of SVNG has some serious flaws, described by examples. Thus, a new modified definition for SVNG has been developed and supported by some examples. The concept of IVNG is also modified, and the validity of the modified definition is tested with examples. Throughout this article, we defined all those terms which were not previously defined well. In the near future, we shall investigate some other results of IVFGs, IVIFG, IVNG, SVNGs and extend them to solve decision-making problems under fuzzy or uncertain environments [52–58].

Author Contributions: Conceptualization, N,J; K.U; and T.M; ; Methodology, N,J; K.U; T.M. and B.D.; Validation, N.J.; K.U.; T.M.; H.G.; and A.B.S; Formal Analysis, N.J.; K.U.; T.M.; A.B.S; and S.B.; Investigation, N.J.; K.U.; T.M.; B.D..; Writing-Original Draft Preparation, N.J.; K.U.; T.M.; H.G..; Writing-Review & Editing, N.J.; K.U.; T.M.; H.G.; Visualization, N.J.; K.U.; T.M.; H.G.; S.B.

Funding: This research received no external funding.

Acknowledgments: The authors are thankful to the editor and anonymous reviewers for their useful suggestions regarding the improvement of this manuscript.

Conflicts of Interest: The authors declare no conflict of interest.

References

1. Zadeh, L.A. Fuzzy sets. *Inf. Control* **1965**, *8*, 338–353. [CrossRef]
2. Gorzalczany, M.B. A method of inference in approximate reasoning based on interval—Valued fuzzy sets. *Fuzzy Sets Syst.* **1987**, *21*, 1–17. [CrossRef]
3. Atanassov, K.T. Intuitionistic fuzzy sets. *Fuzzy Sets Syst.* **1986**, *20*, 87–96. [CrossRef]
4. Atanassov, K.; Gargov, G. Interval-valued intuitionistic fuzzy sets. *Fuzzy Sets Syst.* **1989**, *31*, 343–349. [CrossRef]
5. Smarandache, F. *Neutrosophy. Neutrosophic Probability, Set, and Logic, ProQuest Information & Learning*; American Research Press Rehoboth: Ann Arbor, MI, USA, 1998.
6. Wang, H.; Smarandache, F.; Zhang, Y.Q.; Sunderraman, R. Single valued neutrosophic sets. *Multisp. Multistruct.* **2010**, *4*, 410–413.
7. Wang, H.; Smarandache, F.; Zhang, Y.Q.; Smarandache, R. *Interval Neutrosophic Sets and Logic: Theory and Applications in Computing*; Hexis: Phoenix, AZ, USA, 2005.
8. Peng, J.J.; Wang, J.Q.; Wang, J.; Zhang, H.Y.; Chen, Z.H. Simplified neutrosophic sets and their applications in multi-criteria group decision-making problems. *Int. J. Syst. Sci.* **2016**, *47*, 2342–2358. [CrossRef]
9. Garg, H. New exponential operational laws and their aggregation operators for interval-valued Pythagorean fuzzy multicriteria decision—Making. *Int. J. Intell. Syst.* **2018**, *33*, 653–683. [CrossRef]
10. Peng, J.J.; Wang, J.Q.; Zhang, H.Y.; Chen, X.H. An outranking approach for multi-criteria decision-making problems with simplified neutrosophic sets. *Appl. Soft Comput.* **2014**, *25*, 336–346. [CrossRef]
11. Garg, H. Some robust improved geometric aggregation operators under interval-valued intuitionistic fuzzy environment for multi-criteria decision-making process. *J. Ind. Manag. Optim.* **2018**, *14*, 283–308. [CrossRef]
12. Garg, H. Novel intuitionistic fuzzy decision making method based on an improved operation laws and its application. *Eng. Appl. Artif. Intell.* **2017**, *60*, 164–174. [CrossRef]
13. Peng, X.; Dai, J. Algorithms for interval neutrosophic multiple attribute decision-making based on MABAC, similarity measure, and EDAS. *Int. J. Uncertain. Quantif.* **2017**, *7*, 395–421. [CrossRef]
14. Peng, X.D.; Dai, J.G. Approaches to single-valued neutrosophic madm based on mabac, topsis and new similarity measure with score function. *Neural Comput. Appl.* **2018**, *29*, 939–954. [CrossRef]
15. Garg, H. Some picture fuzzy aggregation operators and their applications to multicriteria decision-making. *Arab. J. Sci. Eng.* **2017**, *42*, 5275–5290. [CrossRef]

16. Garg, H.; Kumar, K. A novel exponential distance and its based TOPSIS method for interval-valued intuitionistic fuzzy sets using connection number of SPA theory. *Artif. Intell. Rev.* **2018**, 1–30. [CrossRef]
17. Smarandache, F. Subtraction and division of neutrosophic numbers. *Crit. Rev.* **2016**, *13*, 103–110.
18. Xu, Z.S. Intuitionistic fuzzy aggregation operators. *IEEE Trans. Fuzzy Syst.* **2007**, *15*, 1179–1187.
19. Garg, H.; Arora, R. Generalized intuitionistic fuzzy soft power aggregation operator based on t-norm and their application in multi criteria decision-making. *Int. J. Intell. Syst.* **2019**, *34*, 215–246. [CrossRef]
20. Garg, H.; Kumar, K. Distance measures for connection number sets based on set pair analysis and its applications to decision making process. *Appl. Intell.* **2018**, *48*, 3346–3359. [CrossRef]
21. Garg, H.; Nancy. Non-linear programming method for multi-criteria decision making problems under interval neutrosophic set environment. *Appl. Intell.* **2018**, *48*, 2199–2213. [CrossRef]
22. Wei, G.; Zhao, X. An approach to multiple attribute decision making with combined weight information in interval-valued intuitionistic fuzzy environmental. *Control Cybern.* **2012**, *41*, 97–112.
23. Wei, G.W. Picture 2-tuple linguistic bonferroni mean operators and their application to multiple attribute decision making. *Int. J. Fuzzy Syst.* **2017**, *19*, 997–1010. [CrossRef]
24. Garg, H.; Nancy. Some hybrid weighted aggregation operators under neutrosophic set environment and their applications to multicriteria decision-making. *Appl. Intell.* **2018**, *48*, 4871–4888. [CrossRef]
25. Garg, H.; Nancy. New logarithmic operational laws and their applications to multiattribute decision making for single-valued neutrosophic numbers. *Cogn. Syst. Res.* **2018**, *52*, 931–946. [CrossRef]
26. Kaufmann, A. *Introduction to Fuzzy Subsets*; Academic Press: New York, NY, USA, 1975; Volume 1.
27. Zadeh, L.A.; Fu, K.S.; Tanaka, K. (Eds.) Fuzzy sets and their applications to cognitive and decision processes. In Proceedings of the US–Japan Seminar on Fuzzy Sets and Their Applications, Berkeley, CA, USA, 1–4 July 1974.
28. Akram, M.; Dudek, W.A. Interval-valued fuzzy graphs. *Comput. Math. Appl.* **2011**, *61*, 289–299. [CrossRef]
29. Bhattacharya, P. Some remarks on fuzzy graphs. *Pattern Recognit. Lett.* **1987**, *6*, 297–302. [CrossRef]
30. Bhutani, K.R. On automorphisms of fuzzy graphs. *Pattern Recognit. Lett.* **1989**, *9*, 159–162. [CrossRef]
31. Bhutani, K.R.; Rosenfeld, A. Strong arcs in fuzzy graphs. *Inf. Sci.* **2003**, *152*, 319–322. [CrossRef]
32. Mordeson, J.N.; Chang-Shyh, P. Operations on fuzzy graphs. *Inf. Sci.* **1994**, *79*, 159–170. [CrossRef]
33. Rashmanlou, H.; Jun, Y.B. Complete interval-valued fuzzy graphs. *Ann. Fuzzy Math. Inform.* **2013**, *6*, 677–687.
34. Pal, M.; Rashmanlou, H. Irregular interval valued fuzzy graphs. *arXiv*, 2014; arXiv:1407.6190.
35. Akram, M. Interval-valued fuzzy line graphs. *Neural Comput. Appl.* **2012**, *21*, 145–150. [CrossRef]
36. Rashmanlou, H.; Pal, M. Balanced interval-valued fuzzy graphs. *J. Phys. Sci.* **2013**, *17*, 43–57.
37. Pramanik, T.; Samanta, S.; Pal, M. Interval-valued fuzzy planar graphs. *Int. J. Mach. Learn. Cybern.* **2016**, *7*, 653–664. [CrossRef]
38. Talebi, A.; Rashmanlou, H. Isomorphism on interval-valued fuzzy graphs. *Ann. Fuzzy Math. Inform.* **2013**, *6*, 47–58.
39. Parvathi, R.; Karunambigai, M. Intuitionistic fuzzy graphs. In *Computational Intelligence, Theory and Applications*; Springer: New York, NY, USA, 2006; pp. 139–150.
40. Mishra, S.; Pal, A. Product of interval valued intuitionistic fuzzy graph. *Ann. Pure Appl. Math.* **2013**, *5*, 37–46.
41. Mishra, S.; Pal, A. Regular interval-valued intuitionistic fuzzy graphs. *J. Inform. Math. Sci.* **2017**, *9*, 609–621.
42. Ismayil, A.M.; Ali, A.M. On strong interval-valued intuitionistic fuzzy graph. *Int. J. Fuzzy Math. Syst.* **2014**, *4*, 161–168.
43. Broumi, S.; Talea, M.; Bakali, A.; Smarandache, F. Single valued neutrosophic graphs. *Florentin Smarandache Surapati Pramanik* **2015**, 187.
44. Broumi, S.; Talea, M.; Bakali, A.; Smarandache, F.; Kumar, P.K. Shortest path problem on single valued neutrosophic graphs. In Proceedings of the 2017 International Symposium on Networks, Computers and Communications (ISNCC), Marrakech, Morocco, 16–18 May 2017; pp. 1–6.
45. Broumi, S.; Bakal, A.; Talea, M.; Smarandache, F.; Vladareanu, L. Applying dijkstra algorithm for solving neutrosophic shortest path problem. In Proceedings of the 2016 International Conference on Advanced Mechatronic Systems (ICAMechS), Melbourne, VIC, Australia, 30 November–3 December 2016; pp. 412–416.
46. Broumi, S.; Ullah, K.; Bakali, A.; Talea, M.; Singh, P.K.; Mahmood, T.; Smarandache, F.; Bahnasse, A.; Patro, S.K.; Oliveira, A. Novel System and Method for Telephone Network Planing Based on Neutrosophic Graph. *Glob. J. Comput. Sci. Technol.* **2018**, *18*, 1–10.
47. Akram, M.; Shahzadi, S. Neutrosophic soft graphs with application. *J. Intell. Fuzzy Syst.* **2017**, *32*, 841–858. [CrossRef]

48. Broumi, S.; Talea, M.; Bakali, A.; Smarandache, F. Interval valued neutrosophic graphs. *Crit. Rev. XII* **2016**, *2016*, 5–33.
49. Akram, M.; Nasir, M. Concepts of Interval-Valued Neutrosophic Graphs. *Infin. Study* **2017**, *6*, 22–41. [CrossRef]
50. Broumi, S.; Smarandache, F.; Talea, M.; Bakali, A. Operations on Interval Valued Neutrosophic Graphs; Florentin Smarandache Surapati Pramanik. *Infin. Study* **2016**.
51. Broumi, S.; Talea, M.; Bakali, A.; Smarandache, F. On strong interval valued neutrosophic graphs. *Crit. Rev.* **2016**, *12*, 49–71.
52. Garg, H.; Arora, R. Dual hesitant fuzzy soft aggregation operators and their application in decision making. *Cogn. Comput.* **2018**, *10*, 769–789. [CrossRef]
53. Garg, H.; Rani, D. Some generalized complex intuitionistic fuzzy aggregation operators and their application to multicriteria decision-making process. *Arab. J. Sci. Eng.* **2018**, *10*, 769–789. [CrossRef]
54. Rani, D.; Garg, H. Complex intuitionistic fuzzy power aggregation operators and their applications in multi-criteria decision-making. *Expert Syst.* **2018**, *35*, e12325. [CrossRef]
55. Kaur, G.; Garg, H. Multi—Attribute decision—Making based on bonferroni mean operators under cubic intuitionistic fuzzy set environment. *Entropy* **2018**, *20*, 65. [CrossRef]
56. Garg, H.; Nancy. Multi-criteria decision-making method based on prioritized muirhead mean aggregation operator under neutrosophic set environment. *Symmetry* **2018**, *10*, 280. [CrossRef]
57. Wei, G.; Garg, H.; Gao, H.; Wei, C. Interval-Valued Pythagorean Fuzzy Maclaurin Symmetric Mean Operators in Multiple Attribute Decision Making. *IEEE Access* **2018**, *6*, 67866–67884. [CrossRef]
58. Garg, H. Hesitant Pythagorean fuzzy Maclaurin symmetric mean operators and its applications to multiattribute decision making process. *Int. J. Intell. Syst.* **2018**. [CrossRef]

mathematics

Article

k-Rainbow Domination Number of $P_3 P_n$

Ying Wang [1,2], Xinling Wu [3], Nasrin Dehgardi [4], Jafar Amjadi [5], Rana Khoeilar [5] and Jia-Bao Liu [6,*]

1 Department of network technology, South China Institute of Software Engineering, Guangzhou 510990, China; wying@sise.com.cn
2 Institute of Computing Science and Technology, Guangzhou University, Guangzhou 510006, China
3 South China Business College, Guang Dong University of Foreign Studies, Guangzhou 510545, China; xinlingwu.guangzhou@gmail.com
4 Department of Mathematics and Computer Science, Sirjan University of Technology, Sirjan 7813733385, Iran; n.dehgardi@sirjantech.ac.ir
5 Department of Mathematics, Azarbaijan Shahid Madani University, Tabriz 5375171379, Iran; j-amjadi@azaruniv.ac.ir (J.A.); khoeilar@azaruniv.ac.ir (R.K.)
6 School of Mathematics and Physics, Anhui Jianzhu University, Hefei 230601, China
* Correspondence: liujiabaoad@163.com

Received: 12 January 2019; Accepted: 19 February 2019; Published: 21 February 2019

Abstract: Let k be a positive integer, and set $[k] := \{1, 2, \ldots, k\}$. For a graph G, a *k-rainbow dominating function* (or *kRDF*) of G is a mapping $f : V(G) \to 2^{[k]}$ in such a way that, for any vertex $v \in V(G)$ with the empty set under f, the condition $\bigcup_{u \in N_G(v)} f(u) = [k]$ always holds, where $N_G(v)$ is the open neighborhood of v. The weight of *kRDF* f of G is the summation of values of all vertices under f. The *k-rainbow domination number* of G, denoted by $\gamma_{rk}(G)$, is the minimum weight of a *kRDF* of G. In this paper, we obtain the *k-rainbow domination number* of grid $P_3 \square P_n$ for $k \in \{2, 3, 4\}$.

Keywords: *k-rainbow dominating function*; *k-rainbow domination number*; grids

1. Introduction

For a graph G, we denote by $V(G)$ and $E(G)$ the vertex set and the edge set of G, respectively. For a vertex $v \in V(G)$, the open neighborhood of v, denoted by $N_G(v)$, is the set $\{u \in V(G) : uv \in E(G)\}$ and the closed neighborhood of v, denoted by $N_G[v]$, is the set $N_G(v) \cup \{v\}$. The degree of a vertex $v \in V(G)$, denoted by $d_G(v)$, is defined by $d_G(v) = |N_G(v)|$. We let $\delta(G)$ and $\Delta(G)$ denote the minimum degree and maximum degree of a graph G, respectively.

Let k be a positive integer, and $[k] := \{1, 2, \ldots, k\}$. For a graph G, a *k-rainbow dominating function* (or *kRDF*) of G is a mapping $f : V(G) \to 2^{[k]}$ in such a way that for any vertex $v \in V(G)$ with the empty set under f, the condition $\bigcup_{u \in N_G(v)} f(u) = [k]$ always holds. The weight of a *kRDF* f of G is the value $\omega(f) := \sum_{v \in V(G)} |f(v)|$. The *k-rainbow domination number* of G, denoted by $\gamma_{rk}(G)$, is the minimum weight of a *kRDF* of G. A *kRDF* f of G is a *γ_{rk}-function* if $\omega(f) = \gamma_{rk}(G)$. The *k-rainbow domination number* was introduced by Brešar, Henning, and Rall [1] was studied by several authors (see, for example [2–15]).

For graphs F and G, we let $F \square G$ denote the Cartesian product of F and G. Vizing [16] conjectured that for arbitrary graphs F and G, $\gamma(F \square G) \geq \gamma(F)\gamma(G)$. This conjecture is still open, and the domination number or its related invariants of $F \square G$ are extensively studied with the motivation from Vizing's conjecture.

Concerning the *k-rainbow domination number* of $F \square G$, one problem naturally arises: Given two graphs F and G under some conditions, determine $\gamma_{rk}(F \square G)$ for all k. In [3], the authors determined $\gamma_{rk}(P_2 \square P_n)$ for $k = 3, 4, 5$.

In this paper, we examine grid graphs $P_3\square P_n$, and determine the value $\gamma_{rk}(P_3\square P_n)$ for $k \in \{2,3,4\}$ and all n, where P_m is the path of order m.

2. 2-Rainbow Domination Number of $P_3\square P_n$

We write $V(P_3\square P_n) = \{v_i, u_i, w_i \mid 0 \le i \le n-1\}$ and let $E(P_3\square P_n) = \{v_iu_i, u_iw_i \mid 0 \le i \le n-1\} \cup \{v_iv_{i+1}, u_iu_{i+1}, w_iw_{i+1} \mid 0 \le i \le n-1\}$ (see Figure 1). A 2RDF f is given in three lines, where in the first line there are values of the function f for vertices $\{v_0, v_1, \ldots, v_{n-1}\}$, in the second line of the vertices $\{u_0, u_1, \ldots, u_{n-1}\}$, and in the third line of the vertices $\{w_0, w_1, \ldots, w_{n-1}\}$ (see Figure 2). Furthermore, we use $0,1,2,3$ to encode the sets $\emptyset, \{1\}, \{2\}, \{1,2\}$.

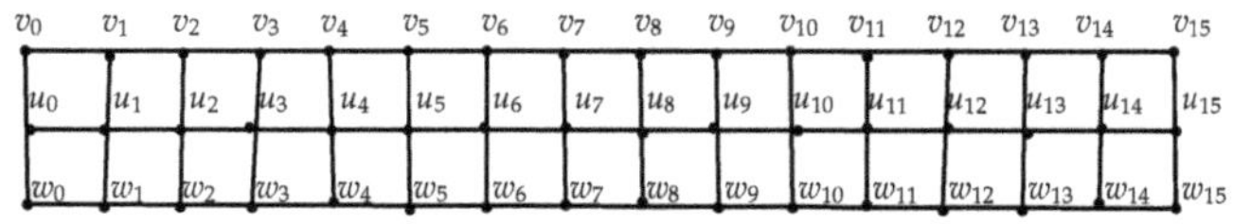

Figure 1. The grid graph $P_3\square P_{16}$.

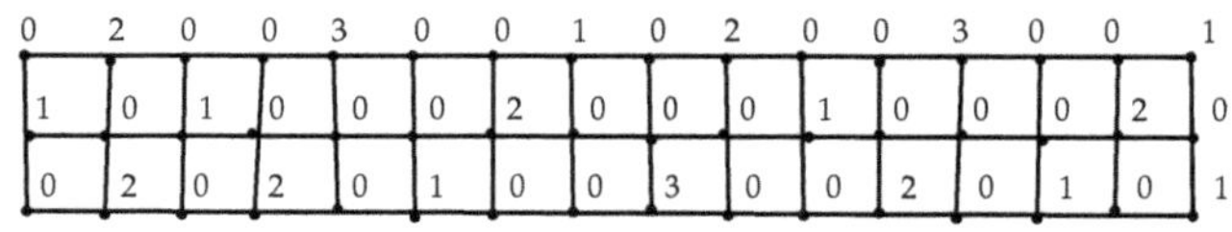

Figure 2. A 2RDF of $P_3\square P_n$.

To provide a complete answer, we need the following fact that can easily be proved as an exercise.

Fact 1. $\gamma_{r2}(P_3\square P_3) = 4$, $\gamma_{r2}(P_3\square P_4) = 6$, $\gamma_{r2}(P_3\square P_5) = 7$, $\gamma_{r2}(P_3\square P_6) = 8$, $\gamma_{r2}(P_3\square P_7) = 10$.

Theorem 1. *For* $n \ge 8$, $\gamma_{r2}(P_3\square P_n) = \left\lceil \frac{5n+3}{4} \right\rceil$.

Proof. First, we present constructions of a 2RDF of $P_3\square P_n$ of the desired weight.

1. $n \equiv 0 \pmod 8$:

   0200 30010200...30010200 3001

   1010 00200010...00200010 0020

   0202 01003002...01003002 0101
2. $n \equiv 1 \pmod 8$:

   0200 30010200...30010200 30010

   1010 00200010...00200010 00202

   0202 01003002...01003002 01010
3. $n \equiv 2 \pmod 8$:

   0200 30010200...30010200 300101

   1010 00200010...00200010 002020

   0202 01003002...01003002 010101
4. $n \equiv 3 \pmod 8$:

   0200 30010200...30010200 3001001

   1010 00200010...00200010 0020220

   0202 01003002...01003002 0101001

5. $n \equiv 4 \pmod 8$:

   0200 30010200...30010200 30010010

   1010 00200010...00200010 00202202

   0202 01003002...01003002 01010010
6. $n \equiv 5 \pmod 8$:

   0200 30010200...30010200 300102020

   1010 00200010...00200010 002000101

   0202 01003002...01003002 010030020
7. $n \equiv 6 \pmod 8$:

   0200 30010200...30010200 30

   1010 00200010...00200010 01

   0202 01003002...01003002 01
8. $n \equiv 7 \pmod 8$:

   0200 30010200...30010200 301

   1010 00200010...00200010 002

   0202 01003002...01003002 010

To show that these are also lower bounds, we prove there is a $\gamma_{r2}(P_3\Box P_n)$-function, f such that for every $0 \leq i \leq n-1$, $\omega(f_i) = |f(v_i)| + |f(u_i)| + |f(w_i)| \geq 1$. Let $n \geq 8$ and f be a $\gamma_{r2}(P_3\Box P_n)$-function such that the cardinality of $S = \{i \mid 0 \leq i \leq n-1 \text{ and } \omega(f_i) = 0\}$ is as small as possible. We claim that $|S| = 0$. Suppose, to the contrary, that $|S| \geq 1$ and let s be the smallest positive integer for which $\omega(f_s) = 0$. Then, $\omega(f_{s-1}) + \omega(f_{s+1}) \geq 6$. Then, we consider the following cases.

Case 1. $s = 1$ (the case $s = n-1$ is similar).
Then, we have $f(v_1) = f(u_1) = f(w_1) = \{1,2\}$ and the function g defined by $g(u_0) = \{1\}$, $g(v_1) = g(w_1) = \{2\}$, $g(u_2) = f(u_2) \cup \{1\}$, $g(v_0) = g(w_0) = g(u_1) = \emptyset$ and $g(x) = f(x)$ otherwise, is a 2RDF of $P_3\Box P_n$ of weight at most $\omega(f)$, which contradicts the choice of f.

Case 2. $s = 1$ ($s = n-2$ is similar).
Then, $\omega(f_0) + \omega(f_2) \geq 6$ and the function g defined by $g(u_0) = g(u_2) = \{1\}$, $g(v_1) = g(w_1) = \{2\}$, $g(v_3) = f(v_3) \cup \{2\}$, $g(w_3) = f(w_3) \cup \{2\}$, $g(v_0) = g(w_0) = g(u_1) = g(v_2) = g(w_2) = \emptyset$ and $g(x) = f(x)$ otherwise, is an 2RDF of $P_3\Box P_n$ of weight at most $\omega(f)$, which contradicts the choice of f.

Case 3. $2 \leq s \leq n-3$.
Since $\omega(f_{s-2}) \geq 1$, then $|f(v_{s-2})| + |f(u_{s-2})| + |f(w_{s-2})| \geq 1$. First, let $|f(u_{s-2})| \geq 1$. We may assume that $\{1\} \subseteq f(u_{s-2})$. It is easy to see that the function g defined by $g(v_{s-1}) = g(v_{s+1}) = g(w_{s-1}) = g(w_{s+1}) = \{2\}$, $g(u_s) = \{1\}$, $g(u_{s+2}) = f(u_{s+2}) \cup \{1\}$, $g(u_{s-1}) = g(v_s) = g(w_s) = g(u_{s+1}) = \emptyset$ and $g(x) = f(x)$ otherwise, is an 2RDF of $P_3\Box P_n$ of weight at most $\omega(f)$, which contradicts the choice of f. Now, let $|f(w_{s-2})| \geq 1$ ($|f(v_{s-2})| \geq 1$ is similar). We may assume that $\{1\} \subseteq f(w_{s-2})$. Hence, the function g defined by $g(v_{s-2}) = f(v_{s-2}) \cup \{1\}$, $g(v_{s+1}) = g(u_{s-1}) = g(w_{s+1}) = \{2\}$, $g(u_s) = \{1\}$, $g(u_{s+2}) = f(u_{s+2}) \cup \{1\}$, $g(u_{s-1}) = g(v_s) = g(w_s) = g(w_{s-1}) = g(u_{s+1}) = \emptyset$ and $g(x) = f(x)$ otherwise, is an 2RDF of $P_3\Box P_n$ of weight $\omega(f)$, which is contradicting the choice of f. Therefore, $|S| = 0$.

We can see that for every $0 \leq i \leq n-2$, if $\omega(f_i) = \omega(f_{i+1}) = \omega(f_{i+2}) = 1$, then $\omega(f_{i-1}), \omega(f_{i+3}) > 1$. In addition, there is the function f such that, if $\omega(f_0) = 1$ ($\omega(f_{n-1}) = 1$ is similar), then $\omega(f_1) > 1$ and $\omega(f_1) + \omega(f_2) + \omega(f_3) + \omega(f_4) \geq 6$ and if $\omega(f_0) = 2$ ($\omega(f_{n-1}) = 2$ is similar), then $\omega(f_0) + \omega(f_1) + \omega(f_2) + \omega(f_3) \geq 6$.

If $\omega(f_0) = 1$ and $\omega(f_{n-1}) = 1$, then

$$\begin{aligned} 4\omega(f) &= 4 \sum_{0 \le i \le n-1} \omega(f_i) \\ &= [3\omega(f_0) + 2\omega(f_1) + \omega(f_2)] + [3\omega(f_{n-1}) + 2\omega(f_{n-2}) + \omega(f_{n-3})] \\ &\quad + \sum_{i \in \{0,\ldots,n-4\}-\{1,n-5\}} (\omega(f_i) + \omega(f_{i+1}) + \omega(f_{i+2}) + \omega(f_{i+3})) \\ &\quad + [\omega(f_1) + \omega(f_2) + \omega(f_3) + \omega(f_4)] + [\omega(f_{n-5}) + \omega(f_{n-4}) + \omega(f_{n-3}) + \omega(f_{n-2})] \\ &\ge 8 + 8 + 5(n-5) + 12 \\ &= 5(n-3) + 18. \end{aligned}$$

If $\omega(f_0) = 1$ and $\omega(f_{n-1}) = 2$, then

$$\begin{aligned} 4\omega(f) &= 4 \sum_{0 \le i \le n-1} \omega(f_i) \\ &= [3\omega(f_0) + 2\omega(f_1) + \omega(f_2)] + [3\omega(f_{n-1}) + 2\omega(f_{n-2}) + \omega(f_{n-3})] \\ &\quad + \sum_{i \in \{0,\ldots,n-4\}-\{1\}} (\omega(f_i) + \omega(f_{i+1}) + \omega(f_{i+2}) + \omega(f_{i+3})) \\ &\quad + [\omega(f_1) + \omega(f_2) + \omega(f_3) + \omega(f_4)] \\ &\ge 8 + 9 + 5(n-4) + 6 \\ &= 5(n-3) + 18. \end{aligned}$$

If $\omega(f_0) = 2$ and $\omega(f_{n-1}) = 2$, then

$$\begin{aligned} 4\omega(f) &= 4 \sum_{0 \le i \le n-1} \omega(f_i) \\ &= [3\omega(f_0) + 2\omega(f_1) + \omega(f_2)] + [3\omega(f_{n-1}) + 2\omega(f_{n-2}) + \omega(f_{n-3})] \\ &\quad + \sum_{i \in \{0,\ldots,n-4\}-\{1\}} [\omega(f_i) + \omega(f_{i+1}) + \omega(f_{i+2}) + \omega(f_{i+3})] \\ &\quad + [\omega(f_1) + \omega(f_2) + \omega(f_3) + \omega(f_4)] \\ &\ge 9 + 9 + 5(n-3) \\ &= 5(n-3) + 18. \end{aligned}$$

Thus, $\omega(f) = \lceil \frac{5n+3}{4} \rceil$. □

3. 3-Rainbow Domination Number of $P_3 \Box P_n$

As in the previous section, a 3RDF is given in three lines and we use $0, 1, 2, 3$ to encode the sets $\emptyset, \{1\}, \{2\}, \{3\}$.

To provide a complete answer, we need the following fact.

Fact 2. $\gamma_{r3}(P_3 \Box P_3) = 5$, $\gamma_{r3}(P_3 \Box P_4) = 8$.

Theorem 2. *For $n \ge 5$,*

$$\gamma_{r3}(P_3 \Box P_n) = \begin{cases} (3n+1)/2 & \text{if} \quad n \equiv 1 \text{ (mode 2)}, \\ (3n+2)/2 & \text{if} \quad n \equiv 0 \text{ (mode 2)}, \end{cases}$$

Proof. First, we present constructions of a 3RDF of $P_3\Box P_n$ of the desired weight.

1. $n \equiv 0 \pmod 4$:

   2010 ... 2010 2201

   0303 ... 0303 0030

   1020 ... 1020 1102
2. $n \equiv 1 \pmod 4$:

   2010 ... 2010 2

   0303 ... 0303 0

   1020 ... 1020 1
3. $n \equiv 2 \pmod 4$:

   2010 ... 2010 201201

   0303 ... 0303 030030

   1020 ... 1020 102102
4. $n \equiv 3 \pmod 4$:

   2010 ... 2010 201

   0303 ... 0303 030

   1020 ... 1020 102

To show that these are also lower bounds, we prove there is a $\gamma_{r3}(P_3\Box P_n)$-function, f that satisfies the following conditions:

1. For every $0 \le i \le n-1$, $\omega(f_i) = |f(v_i)| + |f(u_i)| + |f(w_i)| \ge 1$,
2. For every $1 \le i \le n-2$, if $\omega(f_i) = 1$, then $\omega(f_{i-1}) + \omega(f_{i+1}) \ge 4$. In particular, if $\omega(f_i) = 1$, then $(\omega(f_{i-1}) + \omega(f_i)) + (\omega(f_i) + \omega(f_{i+1})) \ge 6$,
3. $\omega(f_0) \ge 2$ and $\omega(f_{n-1}) \ge 2$.

First, we show that for every $\gamma_{r3}(P_3\Box P_n)$-function f, $\omega(f_i) = |f(v_i)| + |f(u_i)| + |f(w_i)| \ge 1$ when $0 \le i \le n-1$. Let $n \ge 5$ and f be a $\gamma_{r3}(P_3\Box P_n)$-function and $S = \{i \mid 0 \le i \le n-1 \text{ and } \omega(f_i) = 0\}$. We claim that $|S| = 0$. Assume to the contrary that $|S| \ge 1$. Then, we consider the following cases

Case 1. $0 \in S$ (the case $n-1 \in S$ is similar).
Then, we have $f(v_1) = f(u_1) = f(w_1) = \{1,2,3\}$ and it is easy to see that the function g defined by $g(v_0) = \{1\}$, $g(u_1) = \{3\}$, $g(w_0) = \{2\}$, $g(v_2) = f(v_2) \cup \{2\}$, $g(w_2) = f(w_2) \cup \{1\}$, $g(u_0) = g(v_1) = g(w_1) = \emptyset$ and $g(x) = f(x)$ otherwise, is an 3RDF of $P_3\Box P_n$ of weight less than $\omega(f)$, which is a contradiction.

Let s be the smallest positive integer for which $\omega(f_s) = 0$. Then, $s \ge 1$ and $\omega(f_{s-1}) + \omega(f_{s+1}) \ge 9$.

Case 2. $s = 1$ ($s = n-2$ is similar).
Then, the function g defined by $g(v_0) = g(u_0) = g(w_0) = \{1\}$, $g(v_1) = \{2\}$, $g(w_1) = \{1\}$, $g(u_2) = \{3\}$, $g(v_3) = f(v_3) \cup \{1\}$, $g(w_3) = f(w_3) \cup \{2\}$, $g(u_1) = g(v_2) = g(w_2) = \emptyset$ and $g(x) = f(x)$ otherwise, is an 3RDF of $P_3\Box P_n$ of weight less than $\omega(f)$, which is a contradiction.

Case 3. $2 \le s \le n-3$.
The function g defined by $g(u_{s-1}) = g(u_{s+1}) = \{3\}$, $g(v_s) = \{2\}$, $g(w_s) = \{1\}$, $g(v_{s-2}) = f(v_{s-2}) \cup \{1\}$, $g(w_{s-2}) = f(w_{s-2}) \cup \{2\}$, $g(v_{s+2}) = f(v_{s+2}) \cup \{1\}$, $g(w_{s+2}) = f(w_{s+2}) \cup \{2\}$, $g(v_{s-1}) = g(v_{s+1}) = g(u_s) = g(w_{s-1}) = g(w_{s+1}) = \emptyset$ and $g(x) = f(x)$ otherwise, is an 3RDF of $P_3\Box P_n$ of weight less than $\omega(f)$, which is a contradiction. Therefore, $|S| = 0$.

Now, let f be a $\gamma_{r3}(P_3\Box P_n)$-function. It is easy to see that, if $\omega(f_i)=1$, then $\omega(f_{i-1})+\omega(f_{i+1})\geq 4$ when $1\leq i\leq n-2$.

Finally, we show that there is $\gamma_{r3}(P_3\Box P_n)$-function f such that $\omega(f_0)\geq 2$ ($\omega(f_{n-1})\geq 2$ is similar). Let f be a $\gamma_{r3}(P_3\Box P_n)$-function such that $\omega(f_0)=1$. If $|f(v_0)|=1$ ($|f(w_0)|=1$ is similar), then $|f(w_0)|=|f(u_0)|=0$, $|f(u_1)|\geq 2$ and $|f(w_1)|=3$. We may assume that $\{1,2\}\subseteq f(u_1)$. It is easy to see that the function g defined by $g(w_0)=\{3\}$, $g(w_2)=\{3\}$, $g(w_1)=\emptyset$ and $g(x)=f(x)$ otherwise, is an 3RDF of $P_3\Box P_n$ of weight less than $\omega(f)$, which is a contradiction. Now, let $|f(u_0)|=1$. Then, $|f(w_0)|=|f(v_0)|=0$, $|f(v_1)|\geq 2$ and $|f(w_1)|\geq 2$. It is easy to see that the function g defined by $g(w_0)=\{1\}$, $g(w_2)=\{2\}$, $g(u_1)=\{3\}$, $g(v_2)=f(v_2)\cup\{1\}$, $g(w_2)=f(w_2)\cup\{2\}$, $g(u_1)=g(u_2)=\emptyset$ and $g(x)=f(x)$ otherwise, is an 3RDF of $P_3\Box P_n$ of weight $\omega(f)$.

Hence, there is a $\gamma_{r3}(P_3\Box P_n)$-function, f that satisfies the following conditions:

1. For every $0\leq i\leq n-1$, $\omega(f_i)\geq 1$;
2. For every $1\leq i\leq n-2$, if $\omega(f_i)=1$, then $\omega(f_{i-1})+\omega(f_{i+1})\geq 4$; and
3. $\omega(f_0)\geq 2$ and $\omega(f_{n-1})\geq 2$.

If n is odd, then

$$\begin{aligned}2\omega(f) &= 2\sum_{0\leq i\leq n-1}\omega(f_i)\\ &= \omega(f_0)+\omega(f_{n-1})+\sum_{0\leq i\leq n-2}(\omega(f_i)+\omega(f_{i+1}))\\ &\geq 4+3(n-1).\end{aligned}$$

Then, $\omega(f)=\frac{3n+1}{2}$ when n is odd. Now, let n is even. Then, there is $s\neq n-1$ such that $\omega(f_s)+\omega(f_{s+1})\geq 4$. Hence,

$$\begin{aligned}2\omega(f) &= 2\sum_{0\leq i\leq n-1}\omega(f_i)\\ &= \omega(f_s)+\omega(f_{s+1})+\omega(f_0)+\omega(f_{n-1})+\sum_{0\leq i\leq n-2, i\neq s}(\omega(f_i)+\omega(f_{i+1}))\\ &\geq 8+3(n-2).\end{aligned}$$

Therefore, $\omega(f)=\frac{3n+2}{2}$ when n is even. □

4. 4-Rainbow Domination Number of $P_3\Box P_n$

As above, a 4RDF is given in three lines and we use 0, 1, 2, 5 to encode the sets $\emptyset,\{1\},\{2\},\{3,4\}$. To provide a complete answer, we need the following fact.

Fact 3. $\gamma_{r4}(P_3\Box P_3)=6$, $\gamma_{r4}(P_3\Box P_4)=9$.

Theorem 3. *For $n\geq 5$, $\gamma_{r4}(P_3\Box P_n)=2n$.*

Proof. First, we show that $\gamma_{r4}(P_3\Box P_n)\leq 2n$. To do this, we present constructions of a 4RDF of $P_3\Box P_n$ of the desired weight.

1. $n\equiv 0 \pmod 4$:

   2010 . . . 2010 2201

   0505 . . . 0505 0050

   1020 . . . 1020 1102

2. $n \equiv 1 \pmod 4$:

   2010...2010 2

   0505...0505 0

   1020...1020 1
3. $n \equiv 2 \pmod 4$:

   2010...2010 201201

   0505...0505 050050

   1020...1020 102102
4. $n \equiv 3 \pmod 4$:

   2010...2010 201

   0505...0505 050

   1020...1020 102

To prove the inverse inequality, we show that every $\gamma_{r4}(P_3\Box P_n)$-function f satisfies the following conditions:

1. For every $0 \le i \le n-1$, $\omega(f_i) = |f(v_i)| + |f(u_i)| + |f(w_i)| \ge 1$;
2. For every $1 \le i \le n-2$, if $\omega(f_i) = 1$, then $\omega(f_{i-1}) + \omega(f_{i+1}) \ge 6$; and
3. $\omega(f_0) \ge 2$ and $\omega(f_{n-1}) \ge 2$.

First, we show that for every $\gamma_{r4}(P_3\Box P_n)$-function f, $\omega(f_i) = |f(v_i)| + |f(u_i)| + |f(w_i)| \ge 1$ when $0 \le i \le n-1$. Let $n \ge 5$ and f be a $\gamma_{r4}(P_3\Box P_n)$-function and $S = \{i \mid 0 \le i \le n-1 \text{ and } \omega(f_i) = 0\}$. We claim that $|S| = 0$. Assume to the contrary that $|S| \ge 1$. Then, we consider the following cases.

Case 1. $0 \in S$ (the case $n-1 \in S$ is similar).
Then, we have $f(v_1) = f(u_1) = f(w_1) = \{1,2,3,4\}$ and the function g defined by $g(v_C) = \{1\}$, $g(u_1) = \{3,4\}$, $g(w_0) = \{2\}$, $g(v_2) = f(v_2) \cup \{2\}$, $g(w_2) = f(w_2) \cup \{1\}$, $g(u_0) = g(v_1) = g(w_1) = \emptyset$ and $g(x) = f(x)$ otherwise, is an 4RDF of $P_3\Box P_n$ of weight less than $\omega(f)$, which is a contradiction.

Let $\omega(f_s) = 0$. Then, $s \ge 1$ and $\omega(f_{s-1}) + \omega(f_{s+1}) \ge 12$.

Case 2. $s = 1$ ($s = n-2$ is similar).
The function g defined by $g(v_0) = g(u_0) = g(w_0) = \{1\}$, $g(v_1) = \{2\}$, $g(w_1) = \{1\}$, $g(u_2) = \{3,4\}$, $g(v_3) = f(v_3) \cup \{1\}$, $g(w_3) = f(w_3) \cup \{2\}$, $g(u_1) = g(v_2) = g(w_2) = \emptyset$ and $g(x) = f(x)$ otherwise, is an 4RDF of $P_3\Box P_n$ of weight less than $\omega(f)$, which is a contradiction.

Case 3. $2 \le s \le n-3$.
Then, it is easy to see that the function g defined by $g(u_{s-1}) = g(u_{s+1}) = \{3,4\}$, $g(v_s) = \{2\}$, $g(w_s) = \{1\}$, $g(v_{s-2}) = f(v_{s-2}) \cup \{1\}$, $g(w_{s-2}) = f(w_{s-2}) \cup \{2\}$, $g(v_{s+2}) = f(v_{s+2}) \cup \{1\}$, $g(w_{s+2}) = f(w_{s+2}) \cup \{2\}$, $g(v_{s-1}) = g(v_{s+1}) = g(u_s) = g(w_{s-1}) = g(w_{s+1}) = \emptyset$ and $g(x) = f(x)$ otherwise, is an 4RDF of $P_3\Box P_n$ of weight less than $\omega(f)$, which is a contradiction.
Therefore, $|S| = 0$.

Now, let f be a $\gamma_{r4}(P_3\Box P_n)$-function. It is easy to see that, if $\omega(f_i) = 1$, then $\omega(f_{i-1}) + \omega(f_{i+1}) \ge 6$ when $1 \le i \le n-2$.

We show that for every $\gamma_{r4}(P_3\Box P_n)$-function f $\omega(f_0) \ge 2$ ($\omega(f_{n-1}) \ge 2$ is similar). Let f be a $\gamma_{r4}(P_3\Box P_n)$-function such that $\omega(f_0) = 1$. If $|f(v_0)| = 1$ ($|f(w_0)| = 1$ is similar), then $|f(w_0)| = |f(u_0)| = 0$, $|f(u_1)| \ge 3$ and $|f(w_1)| = 4$. We may assume that $\{1,2,3\} \subseteq f(u_1)$. The function g defined by $g(w_0) = \{4\}$, $g(w_2) = \{4\}$, $g(w_1) = \emptyset$ and $g(x) = f(x)$ otherwise, is an 4RDF of $P_3\Box P_n$ of weight less than $\omega(f)$, which is a contradiction. Now, let $|f(u_0)| = 1$. Then, $|f(w_0)| = |f(v_0)| = 0$,

$|f(v_1)| \geq 3$ and $|f(w_1)| \geq 3$. The function g defined by $g(w_0) = \{1\}$, $g(w_2) = \{2\}$, $g(u_1) = \{3,4\}$, $g(v_2) = f(v_2) \cup \{1\}$, $g(w_2) = f(w_2) \cup \{2\}$, $g(u_1) = g(u_2) = \emptyset$ and $g(x) = f(x)$ otherwise, is an 4RDF of $P_3 \square P_n$ of weight less than $\omega(f)$, which is a contradiction.

Hence, every $\gamma_{r4}(P_3 \square P_n)$-function f satisfies the following conditions:

1. For every $0 \leq i \leq n-1$, $\omega(f_i) \geq 1$;
2. For every $1 \leq i \leq n-2$, if $\omega(f_i) = 1$, then $\omega(f_{i-1}) + \omega(f_{i+1}) \geq 6$. In particular $(\omega(f_{i-1}) + \omega(f_i)) + (\omega(f_i) + \omega(f_{i+1})) \geq 8$; and
3. $\omega(f_0) \geq 2$ and $\omega(f_{n-1}) \geq 2$.

Hence,

$$\begin{aligned} 2\omega(f) &= 2 \sum_{0 \leq i \leq n-1} \omega(f_i) \\ &= \sum_{0 \leq i \leq n-2} (\omega(f_i) + \omega(f_{i+1})) + \omega(f_0) + \omega(f_{n-1}) \\ &\geq 4(n-1) + 4. \end{aligned}$$

Hence, $\omega(f) = 2n$. □

Author Contributions: R.K. contributes for supervision, methodology, validation, project administration and formal analysing. N.D., J.A., Y.W., J.-B.L. contribute for resources, some computations and wrote the initial draft of the paper which were investigated and approved by Y.W., X.W., J.-B.L., and J.A. wrote the final draft.

Funding: This research was funded by the National Natural Science Foundation of China (Grant No. 11701118), Guangdong Provincial Engineering and Technology Research Center ([2015]1487), Guangdong Provincial Key Platform and Major Scientific Research Projects (Grant No. 2016KQNCX238), Key Supported Disciplines of Guizhou Province - Computer Application Technology (Grant No. QianXueWeiHeZi ZDXK [2016]20), and the Specialized Fund for Science and Technology Platform and Talent Team Project of Guizhou Province (Grant No. QianKeHePingTaiRenCai [2016]5609), the China Postdoctoral Science Foundation under Grant 2017M621579; the Postdoctoral Science Foundation of Jiangsu Province under Grant 1701081B; Project of Anhui Jianzhu University under Grant no. 2016QD116 and 2017dc03.

Conflicts of Interest: The author declares no conflict of interest.

References

1. Brešar, B.; Henning, M.A.; Rall, D.F. Rainbow domination in graphs. *Taiwan. J. Math.* **2008**, *12*, 213–225. [CrossRef]
2. Amjadi, J.; Asgharshrghi, L.; Dehgardi, N.; Furuyai, M.; Sheikholeslami, S.M.; Volkmann, L. The *k*-rainbow reinforcement numbers in graphs. *Discrete Appl. Math.* **2017**, *217*, 394–404. [CrossRef]
3. Amjadi, J.; Dehgardi, N.; Furuya, M.; Sheikholeslami, S.M. A sufficient condition for large rainbow domination number. *Int. J. Comp. Math. Comp. Syst. Theory* **2017**, *2*, 53–65. [CrossRef]
4. Brešar, B.; Šumenjak, T.K. On the 2-rainbow domination in graphs. *Discrete Appl. Math.* **2007**, *155*, 2394–2400. [CrossRef]
5. Chang, G.J.; Wu, J.; Zhu, X. Rainbow domination on trees. *Discrete Appl. Math.* **2010**, *158*, 8–12. [CrossRef]
6. Chunling, T.; Xiaohui, L.; Yuansheng, Y.; Meiqin, L. 2-rainbow domination of generalized Petersen graphs $P(n,2)$. *Discrete Appl. Math.* **2009**, *157*, 1932–1937.
7. Dehgardi, N.; Sheikholeslami, S.M.; Volkmann, L. The rainbow domination subdivision number of a graph. *Mat. Vesnik* **2015**, *67*, 102–114. [CrossRef]
8. Dehgardi, N.; Sheikholeslami, S.M.; Volkmann, L. The *k*-rainbow bondage number of a graph. *Discrete Appl. Math.* **2014**, *174*, 133–139. [CrossRef]
9. Meierling, D.; Sheikholeslami, S.M.; Volkmann, L. Nordhaus-Gaddum bounds on the *k*-rainbow domatic number of a graph. *Appl. Math. Lett.* **2011**, *24*, 1758–1761. [CrossRef]

10. Shao, Z.; Jiang, H.; Wu, P.; Wang, S.; Žerovnik, J.; Zhang, X.; Liu, J.B. On 2-rainbow domination of generalized Petersen graphs. *Discrete Appl. Math.* **2018**. [CrossRef]
11. Shao, Z.; Liang, M.; Yin, C.; Xu, X.; Pavlič, P.; Žerovnik, J. On rainbow domination numbers of graphs. *Inform. Sci.* **2014**, *254*, 225–234. [CrossRef]
12. Shao, Z.; Sheikholeslami, S.M.; Wang, B.; Wu, P.; Zhang, X. Trees with equal total domination and 2-rainbow domination numbers. *Filomat* **2018**, *32*, 599–607. [CrossRef]
13. Sheikholeslami, S.M.; Volkmann, L. The k-rainbow domatic number of a graph. *Discuss. Math. Graph Theory* **2012**, *32*, 129–140. [CrossRef]
14. Wu, Y.; Jafari Rad, N. Bounds on the 2-rainbow domination number of graphs. *Graphs Combin.* **2013**, *29*, 1125–1133. [CrossRef]
15. Xu, G. 2-rainbow domination of generalized Petersen graphs $P(n,3)$. *Discrete Appl. Math.* **2009**, *157*, 2570–2573. [CrossRef]
16. Vizing, V.G. Some unsolved problems in graph theory. *Uspehi Mater. Nauk* **1968**, *23*, 117–134. [CrossRef]

mathematics

MDPI

Article

More Results on the Domination Number of Cartesian Product of Two Directed Cycles

Ansheng Ye [1,2], Fang Miao [1,2], Zehui Shao [3], Jia-Bao Liu [4,*], Janez Žerovnik [5,6] and Polona Repolusk [6,7,8]

1 School of Geophysics, Chengdu University of Technology, Chengdu 610059, China; yas@cdu.edu.cn (A.Y.); mf@cdut.edu.cn (F.M.)
2 School of Information Science and Engineering, Chengdu University, Chengdu 610106, China
3 Institute of Computing Science and Technology, Guangzhou University, Guangzhou 510006, China; zshao@gzhu.edu.cn
4 School of Mathematics and Physics, Anhui Jianzhu University, Hefei 230601, China
5 Faculty of Mechanical Engineering, University of Ljubljana, SI-1000 Ljubljana, Slovenia; janez.zerovnik@fs.uni-lj.si
6 Institute of Mathematics, Physics and Mechanics, SI-1000 Ljubljana, Slovenia; polona.repolusk@um.si
7 Faculty of Natural Sciences and Mathematics, University of Maribor, SI-2000 Maribor, Slovenia
8 Faculty of Information Studies, P.O. Box 603, 8000 Novo Mesto, Slovenia
* Correspondence: liujiabaoad@163.com

Received: 12 December 2018; Accepted: 21 February 2019; Published: 24 February 2019

Abstract: Let $\gamma(D)$ denote the domination number of a digraph D and let $C_m \Box C_n$ denote the Cartesian product of C_m and C_n, the directed cycles of length $n \geq m \geq 3$. Liu et al. obtained the exact values of $\gamma(C_m \Box C_n)$ for m up to 6 [Domination number of Cartesian products of directed cycles, Inform. Process. Lett. 111 (2010) 36–39]. Shao et al. determined the exact values of $\gamma(C_m \Box C_n)$ for $m = 6, 7$ [On the domination number of Cartesian product of two directed cycles, Journal of Applied Mathematics, Volume 2013, Article ID 619695]. Mollard obtained the exact values of $\gamma(C_m \Box C_n)$ for $m = 3k + 2$ [M. Mollard, On domination of Cartesian product of directed cycles: Results for certain equivalence classes of lengths, Discuss. Math. Graph Theory 33(2) (2013) 387–394.]. In this paper, we extend the current known results on $C_m \Box C_n$ with m up to 21. Moreover, the exact values of $\gamma(C_n \Box C_n)$ with n up to 31 are determined.

Keywords: domination number; Cartesian product; directed cycle

1. Introduction

In this paper, we only consider digraphs without multiple edges or loops. For a digraph $D = (V, A)$ (or $D = (V(G), A(G))$) with vertex set V, arc set A, and a vertex $v \in V$, $N_D^+(v)$ and $N_D^-(v)$ denote the set of out-neighbors and in-neighbors of v, $d_D^+(v) = |N_D^+(v)|$ and $d_D^-(v) = |N_D^-(v)|$ denote the out-degree and in-degree of v in D, respectively. For two vertices u and v in D, we say u dominates v if $u = v$ or $uv \in A$. Let $N_D^+[v] = N_D^+(v) \cup \{v\}$. A vertex v dominates all vertices in $N_D^+[v]$. A set $S \subseteq V$ is a dominating set of D if S dominates $V(D)$. The domination number of D, denoted by $\gamma(D)$, is the minimum cardinality of a dominating set of D. A dominating set S is called a γ-set of D if $|S|$ is the minimum cardinality over all dominating sets of D.

The *Cartesian product* of graphs D_1 and D_2 is the graph $D_1 \Box D_2$ with the vertex set $V(D_1) \times V(D_2)$, and $(a, b)(a', b') \in A(D_1 \Box D_2)$ if either $aa' \in A(D_1)$ and $b = b'$, or $bb' \in A(D_2)$ and $a = a'$. For more information on the Cartesian product of graphs [1]. It can be seen that if D_1 and D_2 are directed graphs, then $D_1 \Box D_2$ is also a directed graph. We denote by C_n and P_n the directed cycle and directed path with n vertices, respectively. Then we study the domination number of directed graphs $C_m \Box C_n$ for $n \geq m \geq 3$.

Graph domination and associated concepts have been studied for many years and there are more than 200 papers to study on the subject [2–4]. Among them, many authors study the domination number of products of graphs [5,6], especially for cylinders [7], torus [8,9] and grids [10]. Liu et al. initiated the study of the domination number of two directed cycles, and they [11,12] determined the exact values of $\gamma(C_m \Box C_n)$ for m up to 6 and showed that

Theorem 1. *$\gamma(C_m \Box C_n) = \frac{mn}{3}$ if $m \equiv 0$ (mod 3) and $n \equiv 0$ (mod 3).*

Shao et al. [13] extended the exact values of $\gamma(C_m \Box C_n)$ for $m \in \{7, 10\}$. Mollard [14] determined exact values of $\gamma(C_m \Box C_n)$ for $m = 3k + 2$.

In this paper, we again apply the "*dynamic algorithm*", used in [15], to extend the exact values of $\gamma(C_m \Box C_n)$ for m up to 21. These results go much further from the previous known results. Moreover, the exact values of $\gamma(C_n \Box C_n)$ with n up to 31 are determined.

2. The Approach—General Outline

The "*dynamic algorithm*" was proposed to compute invariants of fasciagraphs and rotagraphs [16], and thus it can be used to compute the chromatic number, independence number, domination number etc. of products of graphs. For example, Shao et al. [13] apply the "*dynamic algorithm*" to compute the domination number of Cartesian product of two cycles with one cycle of length 7 or 10 and in [15] authors derived some formulas for the Roman domination number of products of paths and cycles.

In [17], the approach was used to derive closed expressions for domination numbers $\gamma(P_n \Box C_k)$ (for $k \leq 11, n \in \mathbb{N}$) and domination numbers $\gamma(C_n \Box P_k)$ and $\gamma(C_n \Box C_k)$ (for $k \leq 7, n \in \mathbb{N}$). We recall a brief formal description from [17].

Let $D_1, \ldots, D_n$ be arbitrary mutually disjoint (di)graphs, here called *monographs*, and $X_1, \ldots, X_n$ a sequence of sets of arcs. An arc of X_i joins a vertex of $V(D_i)$ with a vertex of $V(D_{i+1})$ ($X_i \subseteq V(D_i) \times V(D_{i+1})$ for $i = 1, \ldots, n$). Furthermore, for convenience we set $D_{n+1} = D_1$. A *polygraph* $\Omega_n = \Omega_n(D_1, \ldots D_n; X_1, \ldots X_n)$ over monographs $D_1, \ldots, D_n$ is defined in the following way:

$$V(\Omega_n) = V(D_1) \cup \ldots \cup V(D_n),$$

$$A(\Omega_n) = A(D_1) \cup X_1 \cup \ldots \cup A(D_n) \cup X_n.$$

For a polygraph Ω_n and for $i = 1, \ldots, n$ we also define

$$L_i = \{u \in V(D_i) \mid \exists v \in D_{i+1} : uv \in X_i\},$$

$$R_i = \{u \in V(D_{i+1}) \mid \exists v \in D_i : uv \in X_i\}.$$

In general, $R_i \cap L_{i+1}$ does not have to be empty. Special types of polygraphs are rotagraphs and fasciagraphs. If all (di)graphs D_i are isomorphic to a fixed (di)graph D and all sets X_i are equal to a fixed set X, we call such a graph *rotagraph*, $\omega_n(D; X)$. More precisely, in a rotagraph (1) there are isomorphisms $\varphi_i : V(D_i) \longrightarrow V(D)$ for $i = 1, \ldots, n+1$, and $\varphi_{n+1} = \varphi_1$ and (2) all sets X_i are equal to a fixed set $X \subseteq V(D) \times V(D)$ ($(u, v) \in X \iff \left(\varphi_i^{-1}(u), \varphi_{i+1}^{-1}(v)\right) \in X_i$ for all i). Losely speaking, a *fasciagraph*, $\psi_n(D; X)$ is a rotagraph without edges between the last and the first copy of D. Formally, in a fasciagraph we have $X_n = \emptyset$ and $X_1 = X, \ldots, X_{n-1} = X$. Note that in a rotagraph, all sets L_i and R_i are equal to fixed sets L and R (i.e., $L_i = \varphi_i^{-1}(L)$ and $R_i = \varphi_{i+1}^{-1}(R)$). This holds for fasciagraphs with one exception, namely $L_n = \emptyset$ and $R_n = \emptyset$. Clearly, the Cartesian products of paths $P_n \Box P_k$ are examples of fasciagraphs and Cartesian products of cycles $C_n \Box C_k$ are examples of rotagraphs. Products of a path and a cycle can be seen either as rotagraphs or as fasciagraphs. In previous work [15,17,18], this fact has been used in studies of undirected graphs to obtain a number of results regarding the products of paths and cycles.

We now recall the definition of a *semiring* $\mathcal{P} = (P, \oplus, \circ, e^{\oplus}, e^{\circ})$. It is a set P together with two binary operations, $\oplus$ and $\circ$ such that the following hold:

1. $(P, \oplus)$ is a commutative monoid with $e^{\oplus}$ as a unit;
2. $(P, \circ)$ is a monoid with e° as a unit;
3. $\circ$ is left– and right–distributive over $\oplus$;
4. $\forall x \in P, x \circ e^{\oplus} = e^{\oplus} = e^{\oplus} \circ x$.

A *path algebra* is an idempotent semiring. It is well known that a semiring is a path algebra exactly when $e^{\circ} \oplus e^{\circ} = e^{\circ}$ holds for e°, the unit of the monoid $(P, \circ)$. Let $\mathbb{N}_0$ denote the set of nonnegative integers and $\mathbb{N}$ the set of positive integers. An example of a path algebra that is important here is $\mathcal{P}_1 = (\mathbb{N}_0 \cup \{\infty\}, \min, +, \infty, 0)$.

Let $\mathcal{P} = (P, \oplus, \circ, e^{\oplus}, e^{\circ})$ be a path algebra and let $\mathcal{M}_n(\mathcal{P})$ be the set of all $n \times n$ matrices over P. Let $M, N \in \mathcal{M}_n(\mathcal{P})$ and define operations $\oplus$ and $\circ$ in the usual way:

$$
\begin{aligned}
(M \oplus N)_{ij} &= M_{ij} \oplus N_{ij}, \\
(M \circ N)_{ij} &= \bigoplus_{k=1}^{n} M_{ik} \circ N_{kj}.
\end{aligned}
$$

Note that in the case of the path algebra $\mathcal{P}_1 = (\mathbb{N}_0 \cup \{\infty\}, \min, +, \infty, 0)$, the above definitions can be rewritten as:

$$
\begin{aligned}
(M \oplus N)_{ij} &= \min\{M_{ij}, N_{ij}\}, \\
(M \circ N)_{ij} &= \min_{k \in \{1,\dots,n\}} \{M_{ik} + N_{kj}\}.
\end{aligned}
$$

$\mathcal{M}_n(\mathcal{P})$ equipped with above operations is a path algebra where the units of semiring are the zero matrix and the unit matrix. In our example, $\mathcal{P}_1 = (\mathbb{N}_0 \cup \{\infty\}, \min, +, \infty, 0)$, all elements of the zero matrix are ∞, the unit of the monoid $(P, \min)$, and the unit matrix is a diagonal matrix with diagonal elements equal to $e^{\circ} = 0$ and all other elements equal to $e^{\oplus} = \infty$.

Let D be a labeled digraph with a labeling function ℓ which assigns to every arc of D an element of a path algebra P. Let $V(D) = \{v_1, v_2, \dots, v_n\}$. The labeling ℓ of D can be extended to paths in the following way: For a path $Q = (v_{i_0}, v_{i_1})(v_{i_1}, v_{i_2}) \dots (v_{i_{k-1}}, v_{i_k})$ of D let

$$
\ell(Q) = \ell\left(v_{i_0}, v_{i_1}\right) \circ \ell\left(v_{i_1}, v_{i_2}\right) \circ \dots \circ \ell\left(v_{i_{k-1}}, v_{i_k}\right).
$$

Let S_{ij}^k be the set of all paths of order k from v_i to v_j in D and let $M(D)$ be the matrix defined by:

$$
M(D)_{ij} = \begin{cases} \ell\left(v_i, v_j\right); & \text{if } (v_i, v_j) \text{ is an arc of } D \\ e^{\oplus}; & \text{otherwise} \end{cases}
$$

It is well-known [19] that

$$
\left(M(D)^k\right)_{ij} = \bigoplus_{Q \in S_{ij}^k} \ell(Q).
$$

Let $\omega_n(D; X)$ be a rotagraph and $\psi_n(D; X)$ a fasciagraph. Set $U_i = L_i \cup R_i$. Note that in the case of fasciagraphs or rotagraphs, all sets L_i and R_i are equal to fixed sets L and R, respectively. Therefore we can write $U = L \cup R$, keeping in mind that this is a disjoint union of sets L and R, where $L \subseteq D_i$ and $R \subseteq D_{i+1}$. Denote $N = 2^{|U|}$. The labeled digraph $\mathcal{D} = \mathcal{D}(D; X)$ is defined as follows: (1) elements of vertex set of $\mathcal{D}$ are the subsets of U, denoted V_i; (2) the label (or, weight) of the arc that joins a subset V_i with a subset V_j is the contribution of the monograph D_i to the solution, assuming the sets V_i and V_j are part of the solution. In general however, for example when independent domination is considered, some of the pairs of sets V_i and V_j do not allow any feasible solution. In such case either a very large label is given (∞), or the arc V_iV_j is deleted from $\mathcal{D} = \mathcal{D}(D;X)$, and consequently $\mathcal{D}$ need not be a complete graph.

Consider for a moment $\psi_3(D;X)$ and let $V_i \subseteq L_1 \cup R_1$ and $V_j \subseteq L_2 \cup R_2$ (of course $R_1 = R_2 = R$ and $L_1 = L_2 = L$ (see Figure 1)). Let $\gamma_{i,j}(D;X)$ stand for the size of minimum dominating set of $D_2 \setminus \left((V_i \cap R_1) \cup (L_2 \cap V_j)\right)$. Then we define a labeling of $\mathcal{D}$, $\ell : A(\mathcal{D}) \longrightarrow \mathbb{N}_0 \cup \{\infty\}$, in the following way:

$$\ell(V_i, V_j) = |V_i \cap R| + \gamma_{i,j}(D;X) + |L \cap V_j| - |V_i \cap R \cap L \cap V_j|. \tag{1}$$

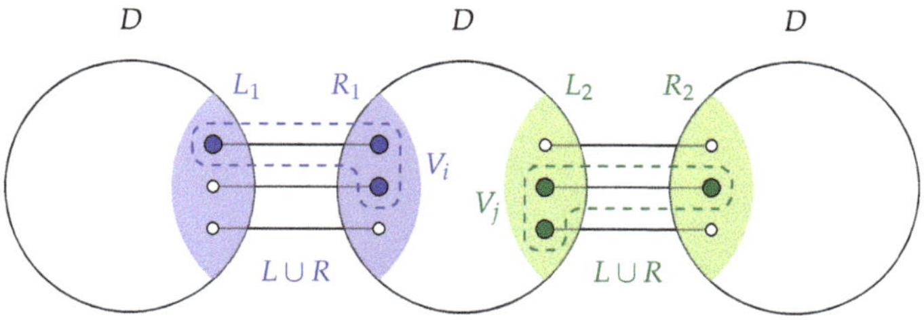

Figure 1. An example of a fasciagraf $\psi_3(D;X)$ with corresponding sets, defined above.

We now recall Algorithm 1, first proposed in [16], that computes the domination number of a fasciagraph or a rotagraph in $O(\log n)$ time:

Algorithm 1 The domination number.

Step 1: Let $\mathcal{P}_1 = (\mathbb{N}_0 \cup \{\infty\}, \min, +, \infty, 0)$ be a path algebra.

Step 2: Label $\mathcal{D}(D;X)$ with the labeling, defined in (1).

Step 3: In $\mathcal{M}(\mathcal{P}_1)$ calculate $M(\mathcal{D})^n$.

Step 4: Let $\gamma(\psi_n(D;X)) = (M(\mathcal{D})^n)_{00}$ and $\gamma(\omega_n(D;X)) = \min_i (M(\mathcal{D})^n)_{ii}$.

This algorithm can be considerably improved because in some cases, computing the powers of $M(\mathcal{D})^n = M_n$ in $O(C)$ time is possible. For example, it is well-known that improvements using special structure of the matrices give rise to constant time algorithm for computation of the domination numbers:

Lemma 1. *([20]) Let $k = |V(\mathcal{D}(D;X))|$ and $K = |V(D)|$. Then there is an index $q \leq (2K+2)^{k^2}$ such that $M_q = M_p + C$ for some index $p < q$ and some constant matrix C. Let $P = q - p$. Then for every $r \geq p$ and every $s \geq 0$ we have*

$$M_{r+sP} = M_r + sC.$$

Note that this phenomena is not restricted to domination type problems. For example, an analogous lemma for distance based invariants was elaborated already in [21].

Hence, if we assume that the size of the monograph is a given constant (and n is a variable), the algorithm will run in constant time. But it is important to emphasize that the algorithm is useful for practical purposes only if the number of vertices of the monograph is relatively small. Namely, the matrices $\mathcal{M}(\mathcal{D})^k$ are of size $|V(\mathcal{D})| \times |V(\mathcal{D})|$, and manipulating such matrices implies that the time complexity is in general exponential in the number of vertices of the monograph. Therefore, a straightforward implementation of the algorithm may have prohibitively long running times.

A method that is particulary useful for fasciagraphs is the following. Observe that one may consider each row of the matrices separately. Thus, instead of using $O(|V(\mathcal{D})|^2)$ space, only $O(|V(\mathcal{D})|)$ is required. Furthermore, depending on the problem, it may not be necessary to compute all the rows of $\mathcal{M}(\mathcal{D})^n$. Instead of calculating the whole matrices $\mathcal{M}(\mathcal{D})^n$, the correct answer is obtained if only those rows which are important for the result are calculated and the difference of the new row against the previously stored rows is checked until a constant difference is detected. This idea was used in [15] on Roman domination problem and in [22] for pentomino exclusion problem. Formally, it is based on the following lemma

Lemma 2. *[15] Assume that the j-th row of M^{n+P} and M^n differ by a constant, $M_{ji}^{n+P} = M_{ji}^n + C$ for all i. Then $\min_i M_{ji}^{n+P} = \min_i M_{ji}^n + C$.*

The idea may be even easier to explain directly in terms of graphs, avoiding the matrix notation. As we study products of cycles in this paper, we restrict attention to rotagraph type problems. Observe that a dominating set in $C_m \square C_n$ corresponds to a cycle of length n in the auxiliary graph $\mathcal{D}$. This leads to the problem of finding a cycle of length n with minimal weight. Above considerations translate into the ideas that if there is a subset of vertices of $\mathcal{D}$ that must be in any solution, then we can start the search for cycles on these vertices only.

In this paper we make use of another idea [13]. Namely, if there are vertices of $\mathcal{D}$ that can not be on any minimal cycle, then we can ignore them.

3. The Approach: Details

Below, we define the labelled auxiliary digraph $\mathcal{D}$ that can be used to solve our domination problem. Then we provide a procedure for reducing the number of vertices of the auxiliary digraph $\mathcal{D}$, similarly as in [13]. The dynamic algorithm is then applied on the reduced auxiliary graph.

Before we continue, let us denote $V(C_n) = \{0, 1, \ldots, n-1\}$ and C_m^i a C_m-layer of $C_m \square C_n$, which corresponds to vertex $i \in V(C_n)$. Similarly as in [13], the vertices v of the auxiliary digraph are represented with a "pattern", i.e., a dominating set on two consecutive C_m-layers. A pattern here is a sequence of 0 s and 1 s, where 1 stands for a vertex in a dominating set of $C_m^i \cup C_m^{i+1}$ and 0 for a vertex that needs to be dominated from a neighboring vertex. The weight of a vertex v, $w(v)$, is defined to be the number of 1 s in a given pattern, i.e., $w(v) = |S \cap (C_m^i \cup C_m^{i+1})|$, where S is a dominating set of $C_m \square C_n$. Clearly, for a cycle C of length n we have $w(C) = 2|S|$, since every vertex in S appears in two consecutive patterns.

The following procedure is used to provide a subset of patterns that will be the vertices of $\mathcal{D}$.

Procedure *ReduceNumberOfVertices*:

Step 1: for a fixed m, find an upper bound of $\gamma(C_m \square C_n)$ in the form of $an + b$ for some $a, b \in \mathbb{Q}$;
Step 2: establish a lower bound $\ell_b > 0$ for the number of vertices in a minimum dominating set of $C_m^i \cup C_m^{i+1}$ for any $i \in \{0, \ldots, n-2\}$.
Step 3: remove each vertex v in $\mathcal{D}$ such that $w(v) - \ell_b > 2na + 2b - n\ell_b$.

Remark 1. ***Procedure*** ReduceNumberOfVertices *can reduce the order of $\mathcal{D}$ for finding the domination number of $C_m \square C_n$.*

Let S be a minimum dominating set of $C_m \square C_n$. Steps 1 and 2 of **Procedure** ReduceNumberOfVertices provide $a, b \in \mathbb{Q}$, and $\ell_b > 0$, such that

$$\gamma(C_m \square C_n) \leq an + b, \tag{2}$$

and

$$|S \cap (C_m^i \cup C_m^{i+1})| \geq \ell_b \text{ for any } i \in \{0, 1, \cdots, n-2\}. \tag{3}$$

Assume that S is such a dominating set, that for some i there is a corresponding vertex v with $w(v) - \ell_b > 2na + 2b - n\ell_b$ and recall that $w(v) = |S \cap (C_m^i \cup C_m^{i+1})|$.
Let $\xi_j = |S \cap (C_m^j \cup C_m^{j+1})| - \ell_b$ for each $j \in \{0, 1, \ldots, n-2\}$. It is clear $\xi_j \geq 0$. Let

$$\xi = \sum_{j=0}^{n-1} \xi_j = 2|S| - n\ell_b. \tag{4}$$

Since $\xi_j \geq 0$ for each j, we have $\xi = 2|S| - n\ell_b \geq \xi_j = |S \cap (C_m^j \cup C_m^{j+1})| - \ell_b$. Since $|S \cap (C_m^j \cup C_m^{j+1})| - \ell_b > 2na + 2b - n\ell_b$, we have $|S| > na + b$, contradicting Equation (2) (and the minimality of S). Therefore, there is no vertex v of $\mathcal{D}$ with $w(v) - \ell_b > 2na + 2b - n\ell_b$ and those vertices can be removed from the auxiliary digraph.

In practice, the procedure *ReduceNumberOfVertices* reduce the vertices of auxiliary graph dramatically for almost all cases, and we give the following examples:

Example 1. *For $m = 10$, let $\ell_b = \frac{7n}{2}$. We label the graph $P_2 \square C_n$ and the constructed auxiliary digraph has 328362 vertices. By removing unnecessary vertices, the auxiliary digraph can be reduced to 16,575 vertices by letting $(w_1, w_2) \in \{(3,3), (3,4), (4,3), (4,4), (3,5), (5,3), (4,5), (5,4)\}$, where w_1, w_2 are the weight restricted to the first and second column of $P_2 \square C_{10}$, respectively.*

Example 2. *For $m = 11$, let $\ell_b = 4n$. We label the graph $P_2 \square C_{11}$ and the constructed auxiliary digraph has 1,169,558 vertices. By removing unnecessary vertices, the auxiliary digraph can be reduced to 2442 vertices by letting $(w_1, w_2) \in \{(4,4), (3,5), (5,3), (3,4), (4,3), (2,4), (4,2)\}$, where w_1, w_2 are the weight restricted to the first and second column of $P_2 \square C_{11}$, respectively.*

We apply *Algorithm 1* with the auxiliary graph $\mathcal{D}$ reduced by procedure *ReduceNumberOfVertices*, and it is carried out on a 2.6-GHz Intel(R) Core(TM) i7-5600U CPU with a memory of 20G. As a result we are able to extend the formulae for $\gamma(C_m \square C_n)$ with m up to 20.

4. Results

The formulae for $\gamma(C_m \square C_n)$ are listed below for $m \leq 20$ and arbitrary $n \geq m$. For $m \leq 6$ the formulae were proved by Liu, Zhang, Chen, and Meng [11,12]. Shao et al. proved the formulae for $m = 7, 10$. All these formulae were confirmed by the algorithm described in this paper and furthermore formulae for $m \leq 21$ were determined. In particular, the domination numbers of the square torus graphs $C_n \square C_n$ with n up to 31 ($n \not\equiv 0 \pmod 3$) are presented in Table 1, where a dominating set with 290 vertices in $C_{29} \square C_{29}$ is presented in Figure 2 and a dominating set with 331 vertices in $C_{31} \square C_{31}$ is presented in Figure 3, the sets of black vertices stand for the corresponding dominating sets, and the direction of the cycle is top to bottom and left to right.

Table 1. Exact values of some square torus $C_n \square C_n$.

n	19	20	22	23	25	26	28	29	31
$\gamma(C_n \square C_n)$	127	140	169	184	217	234	271	290	331

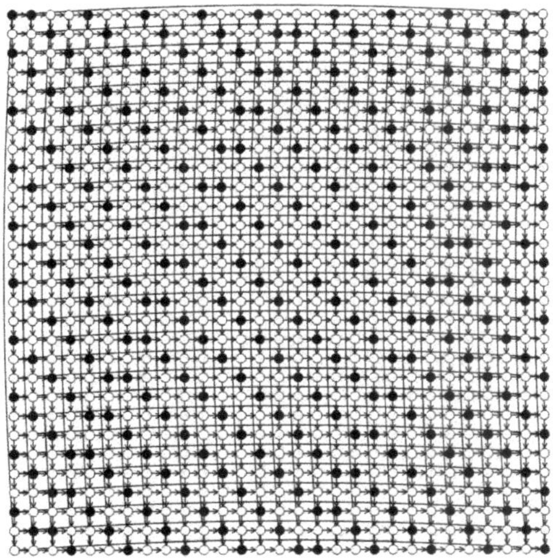

Figure 2. A dominating set with 290 vertices in $C_{29}\square C_{29}$.

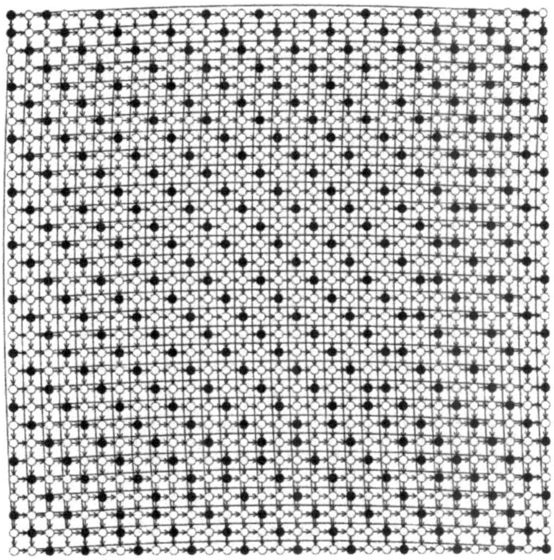

Figure 3. A dominating set with 331 vertices in $C_{31}\square C_{31}$.

- $\gamma(C_3\square C_n) = \begin{cases} n, & n \equiv 0 \pmod 3; \\ n+1, & \text{otherwise.} \end{cases}$
- $\gamma(C_4\square C_n) = \begin{cases} \frac{3n}{2}, & n \equiv 0 \pmod 8; \\ n + \lceil \frac{n+1}{2} \rceil, & \text{otherwise.} \end{cases}$
- $\gamma(C_5\square C_n) = 2n.$
- $\gamma(C_6\square C_n) = \begin{cases} 2n, & n \equiv 0 \pmod 3; \\ 2n+2, & \text{otherwise.} \end{cases}$
- $\gamma(C_7\square C_n) = \begin{cases} \lfloor \frac{5n}{2} \rfloor, & n \equiv 0 \pmod{14}; \\ \lfloor \frac{5n}{2} \rfloor + 1, & n \equiv 3,4,6,8,10,11 \pmod{14}; \\ \lfloor \frac{5n}{2} \rfloor + 2, & n \equiv 1,2,5,7,9,12,13 \pmod{14}. \end{cases}$
- $\gamma(C_8\square C_n) = 3n.$

- $\gamma(C_9\Box C_n) = \begin{cases} 3n, & n \equiv 0 \pmod 3; \\ 3n+3, & \text{otherwise.} \end{cases}$
- $\gamma(C_{10}\Box C_n) = \begin{cases} \lfloor \frac{7n}{2} \rfloor, & n \equiv 0 \pmod{20}; \\ \lfloor \frac{7n}{2} \rfloor + 1, & n \equiv 3,6,14,17 \pmod{20}; \\ \lfloor \frac{7n}{2} \rfloor + 2, & n \equiv 4,7,8,9,10,11,12,13,16 \pmod{20}; \\ \lfloor \frac{7n}{2} \rfloor + 3, & n \equiv 1,2,5,15,18,19 \pmod{20}. \end{cases}$
- $\gamma(C_{11}\Box C_n) = 4n.$
- $\gamma(C_{12}\Box C_n) = \begin{cases} 4n, & n \equiv 0 \pmod 3; \\ 4n+4, & \text{otherwise.} \end{cases}$
- $\gamma(C_{13}\Box C_n) = \begin{cases} \lfloor \frac{9n}{2} \rfloor, & n \equiv 0 \pmod{26}; \\ \lfloor \frac{9n}{2} \rfloor + 1, & n \equiv 3,6,20,23 \pmod{26}; \\ \lfloor \frac{9n}{2} \rfloor + 2, & n \equiv 9,10,12,14,16,17 \pmod{26}; \\ \lfloor \frac{9n}{2} \rfloor + 3, & n \equiv 4,7,8,11,13,15,18,19,22 \pmod{26}; \\ \lfloor \frac{9n}{2} \rfloor + 4, & n \equiv 1,2,5,21,24,25 \pmod{26}. \end{cases}$
- $\gamma(C_{14}\Box C_n) = 5n.$
- $\gamma(C_{15}\Box C_n) = \begin{cases} 5n, & n \equiv 0 \pmod 3; \\ 5n+5, & \text{otherwise.} \end{cases}$
- $\gamma(C_{16}\Box C_n) = \begin{cases} \lfloor \frac{11n}{2} \rfloor, & n \equiv 0 \pmod{32}; \\ \lfloor \frac{11n}{2} \rfloor + 1, & n \equiv 3,6,26,29 \pmod{32}; \\ \lfloor \frac{11n}{2} \rfloor + 2, & n \equiv 9,12,20,23 \pmod{32}; \\ \lfloor \frac{11n}{2} \rfloor + 3, & n \equiv 10,13,14,15,16,17,18,19,22 \pmod{32}; \\ \lfloor \frac{11n}{2} \rfloor + 4, & n \equiv 4,7,8,11,21,24,25,28 \pmod{32}; \\ \lfloor \frac{11n}{2} \rfloor + 5, & n \equiv 1,2,5,27,30,31 \pmod{32}. \end{cases}$
- $\gamma(C_{17}\Box C_n) = 6n.$
- $\gamma(C_{18}\Box C_n) = \begin{cases} 6n, & n \equiv 0 \pmod 3; \\ 6n+6, & \text{otherwise.} \end{cases}$
- $\gamma(C_{19}\Box C_n) = \begin{cases} \lfloor \frac{11n}{2} \rfloor, & n \equiv 0 \pmod{38}; \\ \lfloor \frac{11n}{2} \rfloor + 1, & n \equiv 3,6,32,35 \pmod{38}; \\ \lfloor \frac{11n}{2} \rfloor + 2, & n \equiv 9,12,26,29 \pmod{38}; \\ \lfloor \frac{11n}{2} \rfloor + 3, & n \equiv 15,16,18,20,22,23 \pmod{38}; \\ \lfloor \frac{11n}{2} \rfloor + 4, & n \equiv 10,13,14,17,19,21,24,25,28 \pmod{38}; \\ \lfloor \frac{11n}{2} \rfloor + 5, & n \equiv 4,8,11,34 \pmod{38}; \\ \lfloor \frac{11n}{2} \rfloor + 6, & n \equiv 1,2,5,33,36,37 \pmod{38}. \end{cases}$
- $\gamma(C_{20}\Box C_n) = 7n.$
- $\gamma(C_{21}\Box C_n) = \begin{cases} 7n, & n \equiv 0 \pmod 3; \\ 7n+7, & \text{otherwise.} \end{cases}$

Author Contributions: Z.S. and J.Ž. contribute for supervision, methodology, validation, project administration and formal analysing. A.Y., F.M., J.-B.L. and P.R. contribute for resources, some computations and wrote the initial draft of the paper which were investigated and approved by Z.S., J.Ž. and J.-B.L., the final draft was written by A.Y. and Z.S.

Funding: This work is supported by The National Key Research and Development Program under grant 2016YFB0800600, Natural Science Foundation of Guangdong Province under grant 2018A0303130115, the China Postdoctoral Science Foundation under Grant 2017M621579; the Postdoctoral Science Foundation of Jiangsu Province under Grant 1701081B; Project of Anhui Jianzhu University under Grant No. 2016QD116 and 2017dc03, Research of J. Žerovnik and P. Repolusk was supported in part by Slovenian Research Agency under grants P2-0248, J1-7051, N1-0071, J1-8155, P1-0383 and J1-9109.

Conflicts of Interest: The authors declare no conflict of interest.

References

1. Hammack, R.; Imrich, W.; Klavžar, S. *Handbook of Product Graphs*, 2nd ed.; CRC Press: Boca Raton, FL, USA, 2011.
2. Fu, X.; Yang, Y.; Jiang, B. On the domination number of generalized Petersen graphs $P(n,2)$. *Discret. Math.* **2009**, *309*, 2445–2451. [CrossRef]
3. Klavžar, S.; Ma, M. The domination number of exchanged hypercubes. *Inf. Process. Lett.* **2014**, *114*, 159–162. [CrossRef]
4. Shao, Z.; Wu, P.; Jiang, H.; Li, Z.; Žerovnik, J.; Zhang, X. Discharging approach for double Roman domination in graphs. *IEEE Access* **2018**, *6*, 63345–63351. [CrossRef]
5. Crevals, S.; Östergard, P.R.J. On the domination number of 2-dimensional torus graphs. *Utilitas Math.* **2018**, *106*, 289–300.
6. El-Zahar, M.; Pareek, C.M. Domination number of products of graphs. *Ars Comb.* **1991**, *31*, 223–227.
7. Crevals, S.; Wang, H.; Kim, H.K.; Baek, H. Domination number of the directed cylinder. *Aust. J. Comb.* **2015**, *61*, 192–209.
8. Shao, Z.; Klavžar, S.; Li, Z.; Wu, P.; Xu, J. On the signed Roman k-domination: Complexity and thin torus graphs. *Discret. Appl. Math.* **2017**, *233*, 175–186. [CrossRef]
9. Shao, Z.; Xu, J.; Sheikholeslami, S.M.; Wang, S. The domination complexity and related extremal values of large 3D torus. *Complexity* **2018**, *2018*, 3041426. [CrossRef]
10. Alanko, S.; Crevals, S.; Isopoussu, A.; Östergard, P.R.J.; Pettersson, V. Computing the domination number of grid graphs. *Electron. J. Comb.* **2011**, *18*, P141.
11. Liu, J.; Zhang, X.; Chen, X.; Meng, J. On domination number of Cartesian product of directed cycles. *Inf. Process. Lett.* **2010**, *110*, 171–173. [CrossRef]
12. Zhang, X.; Liu, J.; Chen, X.; Meng, J. Domination number of Cartesian products of directed cycles. *Inf. Process. Lett.* **2010**, *111*, 36–39. [CrossRef]
13. Shao, Z.; Zhu, E.; Lang, F. On the domination number of Cartesian product of two directed cycles. *J. Appl. Math.* **2013**, *2013*, 619695. [CrossRef]
14. Mollard, M. On the domination of Cartesian product of directed cycles: Results for certain equivalence classes of lengths. *Discuss. Math. Graph Theory* **2013**, *33*, 387–394. [CrossRef]
15. Pavlič, P.; Žerovnik, J. Roman domination number of the Cartesian products of paths and cycles. *Electron. J. Comb.* **2012**, *19*, P19.
16. Klavžar, S.; Žerovnik, J. Algebraic approach to fasciagraphs and rotagraphs. *Discret. Appl. Math.* **1996**, *68*, 93–100. [CrossRef]
17. Pavlič, P.; Žerovnik, J. A note on the domination number of the Cartesian products of paths and cycles. *Kragujev. J. Math.* **2013**, *37*, 275–285.
18. Repolusk, P.; Žerovnik, J. Formulas for various domination numbers of products of paths and cycles. *Ars Comb.* **2018**, *137*, 177–202.
19. Carré, B. *Graphs and Networks*; Clarendon Press: Oxford, UK, 1979.
20. Žerovnik, J. Deriving formulas for domination numbers of fasciagraphs and rotagraphs. *Lect. Notes Comput. Sci.* **1999**, *1684*, 559–568.
21. Juvan, M.; Mohar, B.; Žerovnik, J. Distance-related Invariants on Polygraphs. *Discret. Appl. Math.* **1997**, *80*, 57–71. [CrossRef]
22. Žerovnik, J. New formulas for the pentomino exclusion problem. *Aust. J. Comb.* **2006**, *36*, 197–212.

mathematics

MDPI

Article

Distance Degree Index of Some Derived Graphs

Jianzhong Xu [1], Jia-Bao Liu [2,*], Ahsan Bilal [3], Uzma Ahmad [3], Hafiz Muhammad Afzal Siddiqui [4,*], Bahadur Ali [3] and Muhammad Reza Farahani [5]

1 Department of Electronics and Information Engineering, Bozhou University, Bozhou 236800, China; xujianzhongok@163.com
2 School of Mathematics and Physics, Anhui Jianzhu University, Hefei 230601, China
3 Department of Mathematics, University of the Punjab, Lahore 54590, Pakistan; ahsanbilal06@gmail.com (A.B.); uzma.math@pu.edu.pk (U.A.); bahadurali65math@gmail.com (B.A.)
4 Department of Mathematics, COMSATS University Islamabad, Lahore Campus, Lahore 54590, Pakistan
5 Department of Applied Mathematics, Iran University of Science and Technology, Narmak, Tehran 16844, Iran; Mrfarahani88@gmail.com
* Correspondence: liujiabaoad@163.com (J.-B.L.); hmasiddiqui@gmail.com (H.M.A.S.)

Received: 21 January 2019; Accepted: 14 March 2019; Published: 19 March 2019

Abstract: Topological indices are numerical values associated with a graph (structure) that can predict many physical, chemical, and pharmacological properties of organic molecules and chemical compounds. The distance degree (DD) index was introduced by Dobrynin and Kochetova in 1994 for characterizing alkanes by an integer. In this paper, we have determined expressions for a DD index of some derived graphs in terms of the parameters of the parent graph. Specifically, we establish expressions for the DD index of a line graph, subdivision graph, vertex-semitotal graph, edge-semitotal graph, total graph, and paraline graph.

Keywords: DD index; Wiener index; Edge Wiener; degree of a vertex; distance between two vertices

MSC: 05C10; 05C12; 05C76

1. Introduction

Graph theory is becoming interestingly significant as it is being actively applied in biochemistry, nanotechnology, electrical engineering, computer science, and operations research [1–4]. The powerful combinatorial method found in graph theory has also been used to prove the results of pure mathematics. A topological index, also known as a graph descriptor, is a real number associated with a graph. Topological indices are helpful for predicting certain physical, chemical, and pharmacological properties of organic molecules and chemical compounds.

A molecular graph is a graphical description of the structural formula of a chemical compound with the help of graph theory. In the construction of a chemical graph, atoms are represented by nodes or vertices and bonds between the atoms are represented by lines or edges [5]. In all major fields of chemistry, chemical graphs are used for many different purposes [6]. As has already been proved in [7], the said indices are useful for characterizing alkanes by an integer. Although the computed expression does not have direct application, we believe that it can be used to compute the said indices for chemical and molecular graphs, which are useful for characterizing alkanes by an integer.

All graphs considered in this paper are simple and connected. Let $G = G(V, E)$ be a simple and connected graph with vertex set $V = V(G)$ and edge set $E = E(G)$. The degree of vertex i in G is denoted by $\delta_G(i)$ and is defined as the number of edges incident with vertex i. Also, the degree of an edge $e = ij$ in G is denoted by $\delta_G(e)$ and is defined as the number of edges incident to both its end vertices, i and j. Mathematically, $\delta_G(e) = \delta_G(i) + \delta_G(j) - 2$. A complete graph on n vertices is denoted by K_n and defined as a graph in which every vertex is adjacent to all other vertices. Let H and K be two subgraphs of G, such that $V(H) \cap V(K) = \phi$. Let $i \in V(H)$ and $j \in V(K)$ be the vertices such that

$$d_G(i,j) = \min\{d_G(u,v) : u \in V(H), v \in V(K)\};$$

then i and j are called the terminal vertices of subgraphs H and K in G, as shown in the following figure.

In Figure 1, H and K are subgraphs of graph G. In graph G, the distance between vertex c of H and the vertex f of K is 1, which is the minimum of all vertices of H and K. So, c and f are known as terminal vertices of H and K in G.

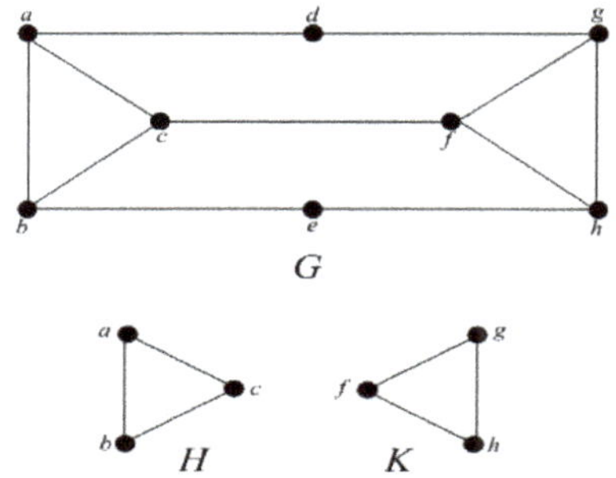

Figure 1. Graph G with its subgraphs H and K.

The first and second Zagreb indices were defined by Gutman and Trinajastić in 1972 [8] and very useful in QSPR and QSAR [9–11]. The first and second Zagreb index are defined as:

$$M_1 = M_1(G) = \sum_{ij \in E(G)} [\delta_G(i) + \delta_G(j)] = \sum_{i \in V(G)} \delta_G(i)^2$$

and

$$M_2 = M_2(G) = \sum_{ij \in E(G)} \delta_G(i)\delta_G(j).$$

Another vertex-degree based topological index was found to be useful in the earliest work on Zagreb indices [8], but later it was totally ignored. Quite recently, some interest was shown in it [12,13]. It is called the Forgotten index or simply F-index and defined as:

$$F = F(G) = \sum_{i \in V(G)} \delta_G(i)^3 = \sum_{ij \in E(G)} [\delta_G(i)^2 + \delta_G(j)^2].$$

In 2008, Došlić introduced the first Zagreb coindex [14], which is defined as:

$$\overline{M_1} = \overline{M_1}(G) = \sum_{ij \notin E(G)} [\delta_G(i) + \delta_G(j)].$$

In 1947, the chemist Harold Wiener introduced the Wiener index [15], which correlates to the boiling point and structure of the molecule of paraffins. The Wiener index is the oldest topological index and is defined as:

$$W(G) = \sum_{\{i,j\} \subseteq V(G)} d_G(i,j),$$

where $d_G(i,j)$ is the distance in G between the vertices i and j. The Wiener index attracts many chemists and mathematicians and has a long history in the literature. For details, see [16–21]. For the applications of the molecular structures, see [22].

The edge version of the Wiener index was introduced in 2010 [23], and is defined as:

$$W_e(G) = \sum_{\{e,f\}\subseteq E(G)} [d_G(e,f)+1],$$

where $d_G(e,f)$ is the distance between the edges $e = xy$ and $f = ij$ in G and defined as:

$$d_G(e,f) = \min\{(d_G(x,i), d_G(x,j), d_G(y,i), d_G(y,j)\}.$$

The degree distance index (DD index) was introduced in [7] by Dobrynin and Kochetova and is defined as:

$$DD(G) = \sum_{\{i,j\}\subseteq V(G)} [\delta_G(i)+\delta_G(j)]d_G(i,j).$$

For a brief look at the different results of DD index, see [24–29]. We define an edge version of DD index as,

$$DD_e(G) = \sum_{\{e,f\}\subseteq (E(G))} [\delta_G(e)+\delta_G(f)][d_G(e,f)+1].$$

The distance between the edge $e = xy$ and a vertex i is defined as

$$d_G(e,i) = \min\{d_G(i,x), d_G(i,y)\}.$$

For basic terminology, see [30].

1.1. Some Derived Graphs

Line Graph: Line graph of a graph G is denoted by $L = L(G)$ such that $V(L) = E(G)$ and there is an edge between two vertices of L if and only if corresponding edges are incident in G. Clearly, $|V(L)| = m$ and by hand shaking-lemma one can easily see that $|E(L)| = \frac{M_1}{2} - m$.
Subdivision Graph: Subdivision graph of a graph G is obtained by inserting a vertex of degree 2 in each edge. Therefore, $|V(S)| = |V(G)| + |E(G)| = n + m$ and $|E(S)| = 2|E(G)| = 2m$.
Vertex-Semitotal Graph: Vertex-Semitotal graph of a graph G is denoted as $T_1 = T_1(G)$ and is obtained by adding a new vertex to each edge of G and then joining a new vertex to the end vertices of the corresponding edge. Thus, $|V(T_1)| = |V(G)| + |E(G)| = n + m$ and

$$|E(T_1)| = |E(S)| + |E(G)| = 2m + m = 3m.$$

Edge-Semitotal Graph: The edge-semitotal graph of a graph G is denoted as $T_2 = T_2(G)$ and is obtained by inserting a new vertex ateach edge of G, joining those new vertices by edges whose corresponding edges are incident in G. We have $|V(T_2)| = |E(G)| + |V(G)| = n + m$ and

$$|E(T_2)| = |E(S)| + |E(L)| = m + \frac{M_1}{2}.$$

Total Graph: The union of edge-semitotal graph and vertex-semitotal graph is called total graph of a graph G. It is denoted by $T = T(G)$. Also,

$$|V(T)| = |V(G)| + |E(G)| = n + m \text{ and}$$
$$|E(T)| = |E(G)| + |E(S)| + |E(L)| = 2m + \tfrac{M_1}{2}.$$

Paraline Graph: The paraline graph $PL = PL(G)$ is the line graph of subdivision graph denoted by $PL = PL(G) = L(S(G))$. Also $|V(PL)| = |E(S)| = 2m$ and

$$|E(PL)| = \frac{M_1(S)}{2} - 2m.$$

In [20], one sees that $M_1(S) = M_1 + 4m$. Thus,

$$|E(PL)| = \frac{M_1 + 4m}{2} - 2m = \frac{M_1}{2}.$$

In Figure 2, self-explanatory examples of these derived graphs are given for a particular graph. In every derived graph of G (except the paraline graph $PL(G)$), the vertices corresponding to the vertices of G are denoted by circles and the vertices corresponding to the edges of G are denoted by squares.

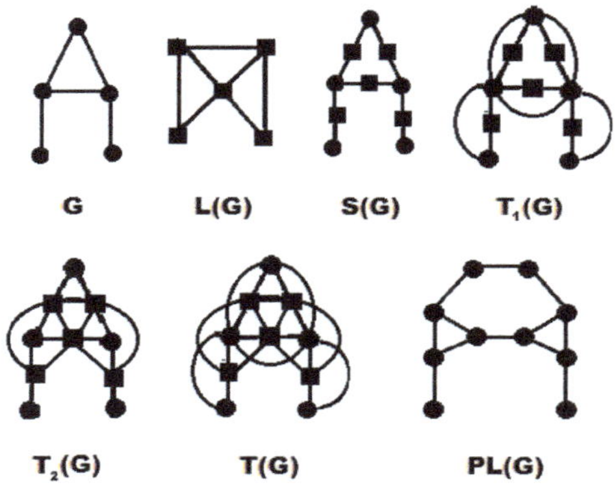

Figure 2. Different graphs derived from G.

The following is a known result of Zagreb indices.

Lemma 1. [31] *For a graph G with n vertices and m edges, we have*

$$\overline{M_1}(G) = 2m(n-1) - M_1(G).$$

1.2. DD Index of Some Derived Graphs

In [31–33] the authors studied the expressions for Zagreb indices and multiplicative Zagreb indices of aforementioned derived graphs. In the past authors dealt only with degree-based indices to derive expressions for the derived graphs. The following proposition encourages us to deal with degree distance-based index. In this section, we study degree distance-based index, DD index for these derived graphs.

Proposition 1. *If $L = L(G)$ is line graph of a graph G, then*

$$DD(L) = DD_e(G).$$

For a subdivision graph S, a vertex-semitotal graph T_1, an edge-semitotal graph T_2, and a total graph T, we can categorize the vertex set into two types of vertices. The first is the set of vertices of G and the second is the set of vertices corresponding to the edges of G. We name them α-type and β-type vertices, respectively. On the basis of this division, there are three types of edges in these graphs:

1. $\alpha\alpha$-edge, an edge between two α-type vertices,
2. $\beta\beta$-edge, an edge between two β-type vertices,
3. $\alpha\beta$-edge, an edge between an α-type vertex and a β-type vertex.

Theorem 1. *If $S = S(G)$ is a subdivision graph of graph G, then*

$$DD(S) = 2DD(G) + 8W_e(G) + 2m(m+n) + 2 \sum_{i_\alpha \in V(G), j_\beta \in E(G)} [\delta_G(i_\alpha) + 2] d_G(i_\alpha, j_\beta).$$

Proof. We can see that for any α-type vertex i_α of S

$$\delta_S(i_\alpha) = d_G(i_\alpha)$$

and for any β type vertex j_β of S

$$\delta_S(j_\beta) = 2.$$

Also,

$$\begin{array}{ll} d_S(i_\alpha, j_\alpha) = 2d_G(i_\alpha, j_\alpha); & i_\alpha, j_\alpha \in V(G) \\ d_S(i_\beta, j_\beta) = 2[d_G(i_\beta, j_\beta) + 1]; & i_\beta, j_\beta \in E(G) \\ d_S(i_\alpha, j_\beta) = 2d_G(i_\alpha, j_\beta) + 1; & i_\alpha \in V(G), j_\beta \in E(G). \end{array}$$

By definition of the DD index, we have

$$\begin{aligned} DD(S) &= \sum_{\{i,j\} \subseteq V(S)} [\delta_S(i) + \delta_S(j)] d_S(i,j) \\ &= \sum_{\{i_\alpha, j_\alpha\} \subseteq V(S)} [\delta_S(i_\alpha) + \delta_S(j_\alpha)] d_S(i_\alpha, j_\alpha) + \sum_{\{i_\beta, j_\beta\} \subseteq V(S)} [\delta_S(i_\beta) + \delta_S(j_\beta)] d_S(i_\beta, j_\beta) \\ &+ \sum_{\{i_\alpha, j_\beta\} \subseteq V(S)} [\delta_S(i_\alpha) + \delta_S(j_\beta)] d_S(i_\alpha, j_\beta) \\ &= \sum_{\{i_\alpha, j_\alpha\} \subseteq V(G)} [\delta_G(i_\alpha) + \delta_G(j_\alpha)] 2d_G(i_\alpha, j_\alpha) + \sum_{\{i_\beta, j_\beta\} \subseteq E(G)} [2+2] 2[d_G(i_\beta, j_\beta) + 1] \\ &+ \sum_{i_\alpha \in V(G), j_\beta \in E(G)} [\delta_G(i_\alpha) + 2][2d_G(i_\alpha, j_\beta) + 1] \\ &= 2DD(G) + 8W_e(G) + 2m(m+n) + 2 \sum_{i_\alpha \in V(G), j_\beta \in E(G)} [\delta_G(i_\alpha) + 2] d_G(i_\alpha, j_\beta). \end{aligned}$$

□

Theorem 2. *If T_1 is the vertex-semitotal graph of graph G, then*

$$2DD(T_1) = DD(G) + 4W_e(G) + 2m(2m - 1 + n) + 2 \sum_{i_\alpha \in V(G), j_\beta \in E(G)} [\delta_G(i_\alpha) + 1] d_G(i_\alpha, j_\beta).$$

Proof. First note that for any α-type vertex i_α of T_1

$$d_{T_1}(i_\alpha) = 2d_G(i_\alpha)$$

and for any β-type vertex j_β of T_1

$$d_{T_1}(j_\beta) = 2.$$

Also,

$$\begin{array}{ll} d_{T_1}(i_\alpha, j_\alpha) = d_G(i_\alpha, j_\alpha); & i_\alpha, j_\alpha \in V(G) \\ d_{T_1}(i_\beta, j_\beta) = d_G(i_\beta, j_\beta) + 2; & i_\beta, j_\beta \in E(G) \\ d_{T_1}(i_\alpha, j_\beta) = d_G(i_\alpha, j_\beta) + 1; & i_\alpha \in V(G), j_\beta \in E(G). \end{array}$$

We have

$$
\begin{aligned}
DD(T_1) &= \sum_{\{i,j\}\subseteq V(T_1)} [\delta_{T_1}(i)+\delta_{T_1}(j)]d_{T_1}(i,j) \\
&= \sum_{\{i_\alpha,j_\alpha\}\subseteq V(T_1)} [\delta_{T_1}(i_\alpha)+\delta_{T_1}(j_\alpha)]d_{T_1}(i_\alpha,j_\alpha) \\
&+ \sum_{\{i_\beta,j_\beta\}\subseteq V(T_1)} [\delta_{T_1}(i_\beta)+\delta_{T_1}(j_\beta)]d_{T_1}(i_\beta,j_\beta) \\
&+ \sum_{\{i_\alpha,j_\beta\}\subseteq V(T_1)} [\delta_{T_1}(i_\alpha)+\delta_{T_1}(j_\beta)]d_{T_1}(i_\alpha,j_\beta) \\
&= \sum_{\{i_\alpha,j_\alpha\}\subseteq V(G)} [2\delta_G(i_\alpha)+2\delta_G(j_\alpha)]d_G(i_\alpha,j_\alpha) + \sum_{\{i_\beta,j_\beta\}\subseteq E(G)} [2+2][d_G(i_\beta,j_\beta)+2] \\
&+ \sum_{i_\alpha\in V(G),j_\beta\in E(G)} [2\delta_G(i_\alpha)+2][d_G(i_\alpha,j_\beta)+1]. \\
&= 2\sum_{\{i_\alpha,j_\alpha\}\subseteq V(G)} [\delta_G(i_\alpha)+\delta_G(j_\alpha)]d_G(i_\alpha,j_\alpha) + 4\sum_{\{i_\beta,j_\beta\}\subseteq E(G)} [d_G(i_\beta,j_\beta)+1] \\
&+4\sum_{\{i_\beta,j_\beta\}\subseteq E(G)} 1 + \sum_{i_\alpha\in V(G),j_\beta\in E(G)} [2\delta_G(i_\alpha)+2]d_G(i_\alpha,j_\beta) \\
&+ \sum_{i_\alpha\in V(G),j_\beta\in E(G)} [2\delta_G(i_\alpha)+2]. \\
&= 2DD(G)+4W_e(G)+4\sum_{\{i_\beta,j_\beta\}\subseteq E(G)} 1 + \sum_{i_\alpha\in V(G),j_\beta\in E(G)} 2\delta_G(i_\alpha) + \sum_{i_\alpha\in V(G),j_\beta\in E(G)} 2 + 2\sum_{i_\alpha\in V(G),j_\beta\in E(G)} [\delta_G(i_\alpha)+1]d_G(i_\alpha,j_\beta). \\
&= 2DD(G)+4W_e(G)+4\sum_{\{i_\beta,j_\beta\}\subseteq E(G)} 1 + 2m(2m)+2mn+2\sum_{i_\alpha\in V(G),j_\beta\in E(G)} [\delta_G(i_\alpha)+1]d_G(i_\alpha,j_\beta).
\end{aligned}
$$

Since

$$\sum_{\{i_\beta,j_\beta\}\subseteq E(G)} 1 = \frac{m(m-1)}{2}.$$

Therefore,

$$DD(T_1) = 2DD(G)+4W_e(G)+2m(3m-1+n)+2\sum_{i_\alpha\in V(G),j_\beta\in E(G)} [\delta_G(i_\alpha)+1]d_G(i_\alpha,j_\beta).$$

□

Theorem 3. *If T_2 is an edge-semitotal graph of graph G, then*

$$
\begin{aligned}
DD(T_2) &= DD(G)+DD_e(G)+4W_e(G)+nM_1(G) \\
&+ \sum_{i_\alpha\in V(G),j_\beta\in E(G)} [\delta_G(i_\alpha)+\delta_G(j_\beta)+2]d_{(G)}(i_\alpha,j_\beta).
\end{aligned}
$$

Proof. To prove this theorem, we proceed in a similar way.

First note that for any α-type vertex i_α of T_2

$$d_{T_2}(i_\alpha) = d_G(i_\alpha)$$

and for any β-type vertex j_β of T_2

$$
\begin{aligned}
& d_{T_2}(j_\beta) = d_G(j_\beta)+2. \\
\text{Also,}\quad & d_{T_2}(i_\alpha,j_\alpha) = d_G(i_\alpha,j_\alpha)+1; \quad i_\alpha,j_\alpha\in V(G) \\
& d_{T_2}(i_\beta,j_\beta) = d_G(i_\beta,j_\beta)+1; \quad i_\beta,j_\beta\in E(G) \\
& d_{T_2}(i_\alpha,j_\beta) = d_G(i_\alpha,j_\beta)+1; \quad i_\alpha\in V(G),j_\beta\in E(G).
\end{aligned}
$$

So,

$$
\begin{aligned}
DD(T_2) &= \sum_{\{i,j\}\subseteq V(T_2)} [\delta_{T_2}(i)+\delta_{T_2}(j)]d_{T_2}(i,j) \\
&= \sum_{\{i_\alpha,j_\alpha\}\subseteq V(T_2)} [\delta_{T_2}(i_\alpha)+\delta_{T_2}(j_\alpha)]d_{T_2}(i_\alpha,j_\alpha) \\
&+ \sum_{\{i_\beta,j_\beta\}\subseteq V(T_2)} [\delta_{T_2}(i_\beta)+\delta_{T_2}(j_\beta)]d_{T_2}(i_\beta,j_\beta) \\
&+ \sum_{\{i_\alpha,j_\beta\}\subseteq V(T_2)} [\delta_{T_2}(i_\alpha)+\delta_{T_2}(j_\beta)]d_{T_2}(i_\alpha,j_\beta) \\
&= \sum_{\{i_\alpha,j_\alpha\}\subseteq V(G)} [\delta_G(i_\alpha)+\delta_G(j_\alpha)][d_G(i_\alpha,j_\alpha)+1] \\
&+ \sum_{\{i_\beta,j_\beta\}\subseteq E(G)} [\delta_G(i_\beta)+\delta_G(j_\beta)+4][d_G(i_\beta,j_\beta)+1] \\
&+ \sum_{i_\alpha\in V(G),j_\beta\in E(G)} [\delta_G(i_\alpha)+\delta_G(j_\beta)+2][d_G(i_\alpha,j_\beta)+1]. \\
&= \sum_{\{i_\alpha,j_\alpha\}\subseteq V(G)} [\delta_G(i_\alpha)+\delta_G(j_\alpha)]d_G(i_\alpha,j_\alpha) + \sum_{\{i_\alpha,j_\alpha\}\subseteq V(G)} [\delta_G(i_\alpha)+\delta_G(j_\alpha)] \\
&+ \sum_{\{i_\beta,j_\beta\}\subseteq E(G)} [\delta_G(i_\beta)+\delta_G(j_\beta)][d_G(i_\beta,j_\beta)+1] + 4\sum_{\{i_\beta,j_\beta\}\subseteq E(G)} [d_G(i_\beta,j_\beta)+1] \\
&+ \sum_{i_\alpha\in V(G),j_\beta\in E(G)} [\delta_G(i_\alpha)+\delta_G(j_\beta)+2][d_G(i_\alpha,j_\beta)] \\
&+ \sum_{i_\alpha\in V(G),j_\beta\in E(G)} [\delta_G(i_\alpha)+\delta_G(j_\beta)+2] \\
&= \sum_{\{i_\alpha,j_\alpha\}\subseteq V(G)} [\delta_G(i_\alpha)+\delta_G(j_\alpha)]d_G(i_\alpha,j_\alpha) + \sum_{i_\alpha j_\alpha\in E(G)} [\delta_G(i_\alpha)+\delta_G(j_\alpha)] \\
&+ \sum_{i_\alpha j_\alpha\notin E(G)} [\delta_G(i_\alpha)+\delta_G(j_\alpha)] + \sum_{\{i_\beta,j_\beta\}\subseteq E(G)} [\delta_G(i_\beta)+\delta_G(j_\beta)][d_G(i_\beta,j_\beta)+1] \\
&+4\sum_{\{i_\beta,j_\beta\}\subseteq E(G)} [d_G(i_\beta,j_\beta)+1] + \sum_{i_\alpha\in V(G),j_\beta\in E(G)} [\delta_G(i_\alpha)+\delta_G(j_\beta)+2][d_G(i_\alpha,j_\beta)] \\
&+ \sum_{i_\alpha\in V(G),j_\beta\in E(G)} [\delta_G(i_\alpha)+\delta_G(j_\beta)+2]
\end{aligned}
$$

$$
DD(T_2) = DD(G)+M_1(G)+\overline{M_1}(G)+DD_e(G)+4W_e(G)+\sum_{i_\alpha\in V(G),j_\beta\in E(G)}\delta_G(i_\alpha)+\sum_{i_\alpha\in V(G),j_\beta\in E(G)}\delta_G(j_\beta)+\sum_{i_\alpha\in V(G),j_\beta\in E(G)}2+
$$
$$
+\sum_{i_\alpha\in V(G),j_\beta\in E(G)} [\delta_G(i_\alpha)+\delta_G(j_\beta)+2]d_{(G)}(i_\alpha,j_\beta).
$$

Using the Lemma 1 and relation

$$
\sum_{e\in E(G)} \delta_G(e) = \sum_{i\in V(L)} \delta_l(i) = 2|E(L)| = M_1 - 2m.
$$

So,

$$
\begin{aligned}
&DD(T_2) = DD(G)+M_1(G)+\overline{M_1}(G)+DD_e(G)+4W_e(G)+m(2m)+n(M_1(G)-2m)+2(mn) \\
&+ \sum_{i_\alpha\in V(G),j_\beta\in E(G)} [\delta_G(i_\alpha)+\delta_G(j_\beta)+2]d_{(G)}(i_\alpha,j_\beta). \\
&DD(T_2) = DD(G)+DD_e(G)+4W_e(G)+nM_1(G) \\
&+2m[m+n-1] + \sum_{i_\alpha\in V(G),j_\beta\in E(G)} [\delta_G(i_\alpha)+\delta_G(j_\beta)+2]d_{(G)}(i_\alpha,j_\beta).
\end{aligned}
$$

□

Theorem 4. *If T is the total graph of graph G, then*

$$
\begin{aligned}
DD(T) &= 2DD(G)+DD_e(G)+4W_e(G)+4m^2+nM_1(G) \\
&+ \sum_{i_\alpha\in V(G),j_\beta\in E(G)} [2\delta_G(i_\alpha)+\delta_G(j_\beta)+2]d_G(i_\alpha,j_\beta)
\end{aligned}
$$

Proof. First note that for any α-type vertex i_α of T

$$
d_T(i_\alpha) = 2d_G(i_\alpha)
$$

and for any β-type vertex j_β of T

$$
d_T(j_\beta) = d_G(j_\beta)+2.
$$

Also,

$$\begin{aligned}
&d_T(i_\alpha, j_\alpha) = d_G(i_\alpha, j_\alpha); \qquad i_\alpha, j_\alpha \in V(G)\\
&d_T(i_\beta, j_\beta) = d_G(i_\beta, j_\beta) + 1; \qquad i_\beta, j_\beta \in E(G)\\
&d_T(i_\alpha, j_\beta) = d_G(i_\alpha, j_\beta) + 1; \qquad i_\alpha \in V(G), j_\beta \in E(G).\\
&DD(T) = \sum_{\{i,j\}\subseteq V(T)} [\delta_T(i) + \delta_T(j)]d_T(i,j)\\
&= \sum_{\{i_\alpha,j_\alpha\}\subseteq V(T)} [\delta_T(i_\alpha) + \delta_T(j_\alpha)]d_T(i_\alpha, j_\alpha)\\
&+ \sum_{\{i_\beta,j_\beta\}\subseteq V(T)} [\delta_T(i_\beta) + \delta_T(j_\beta)]d_T(i_\beta, j_\beta)\\
&+ \sum_{\{i_\alpha,j_\beta\}\subseteq V(T)} [\delta_T(i_\alpha) + \delta_T(j_\beta)]d_T(i_\alpha, j_\beta)\\
&= \sum_{\{i_\alpha,j_\alpha\}\subseteq V(G)} [2\delta_G(i_\alpha) + 2\delta_G(j_\alpha)]d_G(i_\alpha, j_\alpha)\\
&+ \sum_{\{i_\beta,j_\beta\}\subseteq E(G)} [\delta_G(i_\beta) + \delta_G(j_\beta) + 4][d_G(i_\beta, j_\beta) + 1]\\
&+ \sum_{i_\alpha\in V(G),j_\beta\in E(G)} [2\delta_G(i_\alpha) + \delta_G(j_\beta) + 2][d_G(i_\alpha, j_\beta) + 1].\\
&= \sum_{\{i_\alpha,j_\alpha\}\subseteq V(G)} [2\delta_G(i_\alpha) + 2\delta_G(j_\alpha)]d_G(i_\alpha, j_\alpha)\\
&+ \sum_{\{i_\beta,j_\beta\}\subseteq E(G)} [\delta_G(i_\beta) + \delta_G(j_\beta)][d_G(i_\beta, j_\beta) + 1] + 4\sum_{\{i_\beta,j_\beta\}\subseteq E(G)} [d_G(i_\beta, j_\beta) + 1]\\
&+ \sum_{i_\alpha\in V(G),j_\beta\in E(G)} [2\delta_G(i_\alpha) + \delta_G(j_\beta) + 2]\\
&+ \sum_{i_\alpha\in V(G),j_\beta\in E(G)} [2\delta_G(i_\alpha) + \delta_G(j_\beta) + 2]d_G(i_\alpha, j_\beta).\\
&= 2DD(G) + DD_e(G) + 4W_e(G) + 4m^2 + n(M_1 - 2m) + 2mn\\
&+ \sum_{i_\alpha\in V(G),j_\beta\in E(G)} [2\delta_G(i_\alpha) + \delta_G(j_\beta) + 2]d_G(i_\alpha, j_\beta)\\
&DD(T) = 2DD(G) + DD_e(G) + 4W_e(G) + 4m^2 + nM_1(G)\\
&+ \sum_{i_\alpha\in V(G),j_\beta\in E(G)} [2\delta_G(i_\alpha) + \delta_G(j_\beta) + 2]d_G(i_\alpha, j_\beta).
\end{aligned}$$

□

Lemma 2. *For graph G the following equality holds.*

$$\sum_{\{i,j\}\subseteq V(G)} [\delta_G(i)\delta_G(j) - \delta_G(i) - \delta_G(j)][\delta_G(i) + \delta_G(j)] = M_1(G)(2m - n) - 4m^2 + 2M_1(G) - F(G).$$

Proof. First we calculate

$$\begin{aligned}
&\sum_{i\in V(G)}\sum_{j\in V(G)} [\delta_G(i)\delta_G(j) - \delta_G(i) - \delta_G(j)][\delta_G(i) + \delta_G(j)]\\
&= \sum_{j\in V(G)} [M_1(G)\delta_G(j) + 2m\delta_G(j)^2 - M_1(G) - 4m\delta_G(j) - n\delta_G(j)^2]\\
&= 2M_1(G)(2m - n) - 8m^2.
\end{aligned}$$

Now the above calculated expression can be written as follows:

$$\begin{aligned}
&\sum_{i\in V(G)}\sum_{j\in V(G)} [\delta_G(i)\delta_G(j) - \delta_G(i) - \delta_G(j)][\delta_G(i) + \delta_G(j)]\\
&= 2\sum_{\{i,j\}\subseteq V(G)} [\delta_G(i)\delta_G(j) - \delta_G(i) - \delta_G(j)][\delta_G(i) + \delta_G(j)]\\
&+ \sum_{i=j\in V(G)} [\delta_G(i)\delta_G(j) - \delta_G(i) - \delta_G(j)][\delta_G(i) + \delta_G(j)],
\end{aligned}$$

which gives us the equation

$$M_1(G)(2m-n)-4m^2 = \sum_{\{i,j\}\subseteq V(G)} [\delta_G(i)\delta_G(j)-\delta_G(i)-\delta_G(j)][\delta_G(i)+\delta_G(j)]$$
$$+F(G)-2M_1(G),$$

which completes the proof. □

Theorem 5. *Let G be a graph having no cycle of even length. If $PL = PL(G)$ is a paraline graph of G, then*

$$DD(PL) = M_1(G)(2m-n+1)-4m^2+2\sum_{\{i,j\}\subseteq V(G)} [\delta_G(i)+\delta_G(j)]\delta_G(i)\delta_G(j)d_G(i,j).$$

Proof. It can be noted that for any vertex $i \in V(G)$ there are $\delta_G(i)$ vertices in PL having the same degree as the vertex i such that all the $\delta_G(i)$ vertices are connected with each other. In fact, $PL(G)$ can be obtained from G by replacing every vertex i by $K_{\delta_G(i)}$.

By definition of the DD index,

$$DD(PL) = \sum_{\{u,v\}\subseteq V(PL)} [\delta_{PL}(u)+\delta_{PL}(v)]d_{PL}(u,v).$$

Now, either u or v belongs to the same $K_{\delta_G(i)}$ or two different $K_{\delta_G(i)}$ values, where $i \in V(G)$. So,

$$DD(PL) = \sum_{u,v\in K_{d_G(i)}, i\in V(G)} [\delta_{PL}(u)+\delta_{PL}(v)]d_{PL}(u,v)$$
$$+ \sum_{u\in K_{d_G(i)}, v\in K_{d_G(j)}, \{i,j\}\subseteq V(G)} [\delta_{PL}(u)+\delta_{PL}(v)]d_{PL}(u,v).$$

Here we discuss the two terms one by one.

I. In the first term, for any vertex $i \in V(G)$ there are a total of $\frac{\delta_G(i)(\delta_G(i)-1)}{2}$ pairs of vertices in $PL(G)$ of degrees $\delta_G(i)$ and having a distance of 1 between them in the corresponding $K_{\delta_G(i)}$.

II. Before discussing the second term, it is important to see that there is exactly one shortest path in G between any two different vertices, because G has no cycle of even length. Consequently, there is exactly one shortest path between their corresponding complete graphs in $PL(G)$.

Now, in the second term, for any pair of vertices i and j in G, we have the following cases in corresponding pair of complete graphs $K_{\delta_G(i)}$ and $K_{\delta_G(j)}$ in $PL(G)$:

***Case-1*:** There is exactly one pair of terminal vertices in $K_{\delta_G(i)}$ and $K_{\delta_G(j)}$, having distance $[2d_G(i,j)-1]$ between them.

***Case-2*:** There are $[\delta_G(i)-1][\delta_G(j)-1]$ pairs of non-terminal vertices in $K_{\delta_G(i)}$ and $K_{\delta_G(j)}$, having distance $[2d_G(i,j)+1]$ between them.

***Case-3*:** There are $(\delta_G(i)-1)$ pairs of vertices having a non-terminal vertex in $K_{\delta_G(i)}$ and a terminal vertex in $K_{\delta_G(j)}$, and having distance $2d_G(i,j)$ between them. Similarly, there are $(\delta_G(j)-1)$ pairs of vertices having a terminal vertex in $K_{\delta_G(i)}$ and a non-terminal vertex in $K_{\delta_G(j)}$, and having distance $2d_G(i,j)$ between them. So we have a total of $[\delta_G(i)+\delta_G(j)-2]$ such pairs.

Thus, in light of the above discussion, we have

$$
\begin{aligned}
DD(PL) &= \sum_{i\in V(G)} \frac{\delta_G(i)(\delta_G(i)-1)}{2}[\delta_G(i)+\delta_G(i)](1) \\
&+ \sum_{\{i,j\}\subseteq V(G)} (1)[\delta_G(i)+\delta_G(j)][2d_G(i,j)-1] \\
&+ \sum_{\{i,j\}\subseteq V(G)} [\delta_G(i)-1][\delta_G(j)-1][\delta_G(i)+\delta_G(j)][2d_G(i,j)+1] \\
&+ \sum_{\{i,j\}\subseteq V(G)} [\delta_G(i)+\delta_G(j)-2][\delta_G(i)+\delta_G(j)][2d_G(i,j)] \\
&= \sum_{i\in V(G)} [\delta_G(i)^3-\delta_G(i)^2] + \sum_{\{i,j\}\subseteq V(G)} [\delta_G(i)+\delta_G(j)][-1+2d_G(i,j) \\
&+\delta_G(i)\delta_G(j)-\delta_G(i)-\delta_G(j)+1+2\delta_G(i)\delta_G(j)d_G(i,j)-2\delta_G(i)d_G(i,j) \\
&-2\delta_G(j)d_G(i,j)+2d_G(i,j)+2\delta_G(i)d_G(i,j)+2\delta_G(j)d_G(i,j)-4d_G(i,j)] \\
&= \sum_{i\in V(G)} [\delta_G(i)^3-\delta_G(i)^2] \\
&+ \sum_{\{i,j\}\subseteq V(G)} [\delta_G(i)+\delta_G(j)][-\delta_G(i)-\delta_G(j)+\delta_G(i)\delta_G(j)] \\
&+2 \sum_{\{i,j\}\subseteq V(G)} [\delta_G(i)+\delta_G(j)]\delta_G(i)\delta_G(j)d_G(i,j)
\end{aligned}
$$

By using Lemma 2, we have

$$
\begin{aligned}
DD(PL) &= F - M_1(G) + M_1(G)(2m-n) - 4m^2 - F + 2M_1(G) \\
&+2 \sum_{\{i,j\}\subseteq V(G)} [\delta_G(i)+\delta_G(j)]\delta_G(i)\delta_G(j)d_G(i,j) \\
&= M_1(G)(2m-n+1) - 4m^2 + 2 \sum_{\{i,j\}\subseteq V(G)} [\delta_G(i)+\delta_G(j)]\delta_G(i)\delta_G(j)d_G(i,j).
\end{aligned}
$$

□

2. Conclusions

In this paper, we studied the *DD* index of derived graphs, which involves distance and degrees. We computed the expressions to find the *DD* index of the derived graph by using the parent graph. More specifically, we found the expressions of the *DD* index using some topological indices of the parent graph. As has already been proved, the said indices are useful for characterizing alkanes by an integer. Although the computed expression does not have direct application, we believe that it can be used to compute the said indices for chemical and molecular graphs, which are useful for characterizing alkanes by an integer.

Author Contributions: These authors contributed equally to this work.

Funding: This work was supported by the Teaching Groups in Anhui Province (2016jytd080, 2015jytd048); the Natural Science Foundation of the Education Department of Anhui Province (KJ2017A704), the Key Program of the Excellent Young Talents Support of Higher Education in Anhui Province (gxyq2018116); the Natural Science Foundation of Bozhou University (BYZ2018B03).

Acknowledgments: Authors are thankful to the referees for their valuable comments to improve this paper.

Conflicts of Interest: The authors declare no conflict of interest.

References

1. Tadic, B.; Ziukovic, J. The graph theory frame work for odeling nano scale systems. *Nanoscale Syst. MMTA* **2013**, *2*, 30–48.
2. Balaban, A.T. Applications of graph theory in chemistry. *J. Chem. Inf. Comput. Sci* **1985**, *25*, 334–343. [CrossRef]

3. Abdullahi, S. Anapplication of graph theory to the electrical circuit using matrix method. *J. Math.* **2014**, 164–166. [CrossRef]
4. Shirinivas, S.G.; Vetrivel, S.; Elango, N.M. Applications of graph theory in computer science an overview. *J. Eng. Technol.* **2010**, 2, 4610–4621.
5. Gold, V.; Loening, K.L.; McNaught, A.D.; Shemi, P. *Compendium of Chemical Terminology*; Blackwell Science Oxford: Oxford, UK, 1997; Volume 1669.
6. Bonchev, D.; Rouvray, D.H. *Chemical Graph Theory Introductions and Fundamentals*; CRC Press: Boca Raton, FL, USA, 1991; Volume 1.
7. Dobrynin, A.A.; Kochetova, A.A. Degree distance of a graph: A degree analogue of the Wiener index. *J. Chem. Inf. Comput. Sci.* **1994**, *34*, 1082–1086. [CrossRef]
8. Gutman, I.; Trinajstić, N. Graph theory and molecular orbitals III. *Total Electron Energy Altern. Hydrocarb. Chem. Phys. Lett.* **1972**, *17*, 535–538.
9. Balaban, A.T. *From Chemical Topology to Three-Dimensional Geometry*; Plenum Press: New York, NY, USA, 1997.
10. Devillers, J.; Balaban, A.T. (Eds.) *Topological Indices and Related Descriptors in QSAR and QSPR*; Gordon and Breach: Amsterdam, The Netherlands, 1999.
11. Todeschini, R.; Consonni, V. *Handbook of Molecular Descriptors*; Wiley–VCH: Weinheim, Germany, 2000
12. Furtula, B.; Gutman, I. A forgotten topological index. *J. Math. Chem.* **2015**, *53*, 1184–1190. [CrossRef]
13. De, N.; Nayeem, S.M.A.; Pal, A. The F-coindex of some graph operations. *Springer Plus* **2016**, *5*, 221. [CrossRef] [PubMed]
14. Došlic, T. Vertex-weighted Wiener polynomials for composite graphs. *Ars Math. Contemp.* **2008**, *1*, 66–80. [CrossRef]
15. Wiener, H. Structural determination of Paraffin boiling points. *J. Am. Chem. Soc.* **1947**, *69*, 17–20. [CrossRef]
16. Ashrafi, A.R.; Yousafi, S. An exact expression for the Wiener index of a polyhex nanotorus. *MATCH Commun. Math. Comput. Chem.* **2006**, *56*, 169–178.
17. Ashrafi, A.R.; Yousafi, S. A new algorithm for computing distance matrix and Wiener index of zig-zag polyhex nano-tubes, nano-scale. *Res. Lett.* **2007**, *2*, 202–206.
18. Chepoi, V.; Kalvzar, S. The Wiener index and the Szeged index of benzoid systems in linear time. *J. Chem. Inf. Comput. Sci.* **1997**, *37*, 752–755. [CrossRef]
19. Cai, X.; Zhou, B. Reverse Wiener index of connected graphs, MATCH Commun. Math. *Comput. Chem.* **2008**, *60*, 95–105.
20. Dobrynin, A.A.; Entringer, R.; Gutman, I. Wiener index of trees, Theory and applications. *Acta App. Math.* **2011**, *66*, 211–249. [CrossRef]
21. Dobrynin, A.A.; Gutman, I.; Klavzar, S.; Zigerl, P. Wiener index of hexagonal systems. *Accta. Appl. Math.* **2001**, *72*, 247–294. [CrossRef]
22. Sandberg, T.O.; Weinberger, C.; Smått, J.H. Molecular dynamics on wood-derived lignans analyzed by intermolecular network theory. *Molecules* **2018**, *23*, 1990. [CrossRef] [PubMed]
23. Iranmanesh, A.; Gutman, I.; Khormali, O.; Mahmiani, A. The edge versions of Wiener index. *Match Commun. Math. Comput. Chem.* **2009**, *61*, 663–672.
24. Ali, P.; Mukwembi, S.; Munyria, S. Degree distance and vertex connectivity. *Disc. Appl. Math.* **2013**, *161*, 2802–2811. [CrossRef]
25. Du, Z.; Zhou, B. Degree distance of a graph. *Filomat* **2010**, *24*, 95–120. [CrossRef]
26. Ilic, A.; Stevanovic, D.; Feng, L.; Yu, G.; Danklemann, P. Degree distance of unicyclic graphs. *Discr. Appl. Math.* **2011**, *159*, 779–788. [CrossRef]
27. Tomescu, I. Unicyclic and bicyclic graphs having minmum degree distance. *Discr. Appl. Math.* **2008**, *156*, 125–130. [CrossRef]
28. Tomescu, I. Properties of connected graphs having minimum degree distance. *Discr. Appl. Math.* **2008**, *309*, 2745–2748. [CrossRef]
29. Tomsecu, I. Ordering connected graphs having small degree distance. *Discr. Appl. Math.* **2010**, *158*, 1714–1717. [CrossRef]
30. Bondy, J.A.; Murty, U.S.R. *Graph Theory*; Springer: Berlin, Germany, 2008.
31. Gutman, I.; Furtula, B.; Kovijanic, Z.; Vukicevic; Popivoda, G. On Zagreb indices and coindices. *Match Commun. Math. Comput. Chem.* **2015**, *74*, 5–16.

32. Basavanagoud, B.; Gutman, I.; Gali, C.S. On second Zagreb index and coindex of some derived graphs. *Kragujev. J. Sci.* **2015**, *37*, 113–121.
33. Basavanagoud, B.; Patil, S. Multiplicative Zagreb indices and coindices of some derived graphs. *Opusc. Math.* **2016**, *36*, 287–299. [CrossRef]

mathematics

MDPI

Article

On the Normalized Laplacian and the Number of Spanning Trees of Linear Heptagonal Networks

Jia-Bao Liu [1,2], Jing Zhao [1], Zhongxun Zhu [3] and Jinde Cao [2,*]

1 School of Mathematical Sciences, Anhui Jianzhu University, Hefei 230601, China; liujiabaoad@163.com (J.-B.L.); zhaojing94823@163.com (J.Z.)

2 School of Mathematics, Southeast University, Nanjing 210096, China

3 Department of Mathematics and Statistics, South Central University for Nationalities, Wuhan 430074, China; zzxun73@mail.scuec.edu.cn

* Correspondence: jdcao@seu.edu.cn

Received: 22 February 2019; Accepted: 25 March 2019; Published: 28 March 2019

Abstract: The normalized Laplacian plays an important role on studying the structure properties of non-regular networks. In fact, it focuses on the interplay between the structure properties and the eigenvalues of networks. Let H_n be the linear heptagonal networks. It is interesting to deduce the degree-Kirchhoff index and the number of spanning trees of H_n due to its complicated structures. In this article, we aimed to first determine the normalized Laplacian spectrum of H_n by decomposition theorem and elementary operations which were not stated in previous results. We then derived the explicit formulas for degree-Kirchhoff index and the number of spanning trees with respect to H_n.

Keywords: normalized Laplacian; resistance distance; degree-Kirchhoff index; spanning tree

1. Introduction

A network sometimes can be viewed as a graph. In this context, we just consider connected, simple and undirected graphs. It means the graphs without the multiple edges or loops. A given graph $G = (V_G, E_G)$ can be presented by a set $V_G = \{v_1, v_2, \ldots, v_n\}$ of vertices and a set $E_G = \{e_1, e_2, \ldots, e_m\}$ of edges. For other terminology and notation not stated here, refer to Reference [1,2].

The *adjacency matrix* $A(G)$ of G is a square matrix with order n, which elements α_{ij} are 1 or 0, depending on whether there is an edge or not between vertices i and j. Denoted by $D(G) = diag(d_1, d_2, \ldots, d_n)$, the degree diagonal matrix of G, where $d_1, d_2, \ldots, d_n$ are the degree of vertices $v_1, v_2, \ldots, v_n$. Combining the adjacency and degree matrix, we get the *Laplacian matrix*, for which expressions can be written as $L(G) = D(G) - A(G)$, respectively. One may deeply understand them with the help of Reference [3–6]. The *normalized Laplacian* [7] is given by:

$$(\mathcal{L}(G))_{ij} = \begin{cases} 1, & i = j; \\ -\frac{1}{\sqrt{d_i d_j}}, & i \neq j, \text{ and } v_i \text{ is adjacent to } v_j; \\ 0, & \text{otherwise.} \end{cases} \tag{1}$$

Obviously, $\mathcal{L}(G) = D(G)^{-\frac{1}{2}} L(G) D(G)^{-\frac{1}{2}} = I - D(G)^{-\frac{1}{2}}(D^{-1}A)D(G)^{-\frac{1}{2}}$.

As we know, the normalized Laplacian plays an important role in studying the structure properties of non-regular graphs. In fact, the spectral graph theory focuses on the interplay between the structure properties and the eigenvalues of a graph.

In mathematical chemistry, topological descriptors are always used to make a prediction for the physico-chemical, biological, and structural properties of some molecule graphs. For instance, one of the most studied topological indices the Harary index. The larger the Harary index, the larger the compactness of the molecule. Other topological indices were also used to study the structure properties of graphs. One well-known valency-based topological descriptor was the Wiener index [8,9], which is defined as:

$$W(G) = \sum_{i<j} d_{ij},$$

where d_{ij} is the distance between vertices i and j, that is, the length of the shortest path between vertices i and j.

As a weighted version of Wiener index, Gutman has introduced the *Gutman index* in Reference [10] as:

$$Gut(G) = \sum_{i<j} d_i d_j d_{ij}.$$

According to the theory of electrical networks N, Klein and Randić proposed a new distance function, named resistance distance, in 1993 [11]. He denoted r_{ij} as the resistance distance for any pair of vertices between i and j of a graph, which means the effective resistance between the vertices i and j in N with replacing each edge of a graph by a unit resistor. The Kirchhoff index [12,13] was defined as:

$$Kf(G) = \sum_{i<j} r_{ij}.$$

It is equal to the sum of resistance distances between all pairs of vertices of G. For analog considerations to the Gutman index, Chen and Zhang [14] defined the weighted version of the Kirchhoff index, which was called the degree-Kirchhoff index and given by:

$$Kf^*(G) = \sum_{i<j} d_i d_j r_{ij}.$$

Recently, more and more researchers have concentrated on the Kirchhoff index and degree-Kirchhoff index, due to its wide applications. Despite all that, it seems hard to deal with the Kirchhoff index and degree-Kirchhoff index of complex graphs. Thus, some researchers have tried to find some techniques to solve the problems about the Kirchhoff and degree-Kirchhoff index, such as general star-mesh transformation [15], combinatorial formulas, and others (see Reference [16] for details). In Reference [17,18], Klein and Lovász proved independently that:

$$Kf(G) = n \sum_{k=2}^{n} \frac{1}{\mu_k}. \tag{2}$$

Based on the definition of the Laplacian matrix, one finds the sum of each rows of the Laplacian matrix is zero. Futher, $0 = \mu_1 < \mu_2 \leq \cdots \leq \mu_n$ are the eigenvalues of $L(G)$.

In Reference [14], Chen and Zhang also proposed a method to obtain the formula of degree-Kirchhoff index, which is associated with the eigenvalues of the normalized Laplacian, namely,

$$Kf^*(G) = 2m \sum_{k=1}^{n} \frac{1}{\lambda_k}, \tag{3}$$

where $\lambda_1 \leq \lambda_2 \leq \cdots \leq \lambda_n$ are the eigenvalues of $\mathcal{L}(G)$.

The number of spanning trees [7] of a given graph G, also known as the complexity of G, is the number of subgraphs which contains all the vertices of G. Further, all those subgraphs must be trees.

In 1985, Yang et al. [19] proposed a general solution in theoretically treating linear viscoelasticity by using the knowledge in graph theory. On the basis of this solution, many problems in linear systems have been solved. For instance, based on the decomposition theorem of Laplacian matrix, Y. Yang et al. [20] obtained the Kirchhoff index of linear hexagonal chains in 2007. Y. Pan et al. constructed a crossed hexagonal by adding two pairs of crossed edges in linear hexagonal chains, and they got the Kirchhoff and degree-Kirchhoff index [21], respectively (see Reference [5,22,23] for details.)

Let H_n be the heptagonal networks as illustrated in Figure 1, of which two heptagons have two common edges. That is, two heptagons can be seen as adding two P_2 and attaching them. Notice that there are some applications of the heptagons derived from different aspects, which are listed as follows. The 20p and 50p coins of the United Kingdom have two heptagons. Although there are less heptagonal floor plans, a remarkable example is the Mausoleum of Prince Ernst. Meanwhile, cacti are the most common plants with heptagons in natural structures.

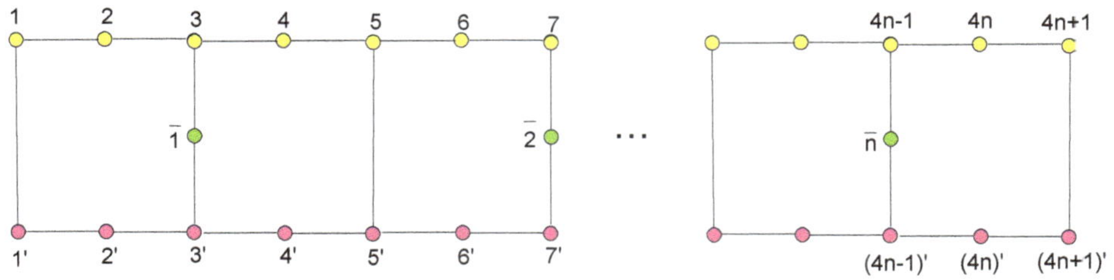

Figure 1. Linear Heptagonal Networks.

For the exited results of linear networks [20–23], there were only two types of vertices considered in these paper. The structure of linear heptagonal networks is much more complex, of which have three types of vertices, shown in Figure 1. It was challenging and meaningful for us to study the structural properties of linear heptagonal networks. On the other hand, the degree-Kirchhoff index and the number of spanning trees are the important parameters to predict the structure properties of graphs. Therefore, computing the degree-Kirchhoff index and the number of spanning trees of linear heptagonal networks are interesting works. In the rest of the context, we recall some necessary notations and methods used in the proofs of the main results in Section 2. We aimed to first determine the normalized Laplacian spectrum of H_n by the decomposition theorem for the normalized Laplacian matrix of H_n. We then derived the explicit formulas for the degree-Kirchhoff index and the number of spanning trees of H_n through using the relationships between the roots and coefficients in Section 3. The discussion of the previous results is given in Section 4. A conclusion is summarized in Section 5.

2. Materials and Methods

In this section, we first recall some basic notations and introduce a classical method, which laid the foundations for this paper. All equations that we introduce below can be found in Reference [1,19].

Fix a square matrix M with order n, denote the submatrix of M the $M[i_1, \ldots, i_k]$, and yield by the deletion of the i_1-th, $\ldots$, i_k-th rows and corresponding columns. Assume that $P_M(x) = \det(xI - M)$ is the characteristic polynomial of the $n \times n$ matrix M.

Let H_n be the graph, as illustrated in Figure 1, with $V_0 = \{\bar{1}, \bar{2}, \ldots, \bar{n}\}$,$V_1 = \{1, 2, \ldots, 4n+1\}$ and $V_2 = \{1', 2', \ldots, (4n+1)'\}$. It is straightforward to verify that $\pi = (\bar{1})(\bar{2}) \cdots (\bar{n})(1, 1')(2, 2') \cdots (4n+1, (4n+1)')$ is an automorphism. Then the normalized laplacian matrix of H_n can be given as the following block matrices.

$$\mathcal{L}(H_n) = \begin{pmatrix} \mathcal{L}_{V_0V_0} & \mathcal{L}_{V_0V_1} & \mathcal{L}_{V_0V_2} \\ \mathcal{L}_{V_1V_0} & \mathcal{L}_{V_1V_1} & \mathcal{L}_{V_1V_2} \\ \mathcal{L}_{V_2V_0} & \mathcal{L}_{V_2V_1} & \mathcal{L}_{V_2V_2} \end{pmatrix}. \tag{4}$$

Notice that, $\mathcal{L}_{V_0V_1} = \mathcal{L}_{V_0V_2}, \mathcal{L}_{V_1V_0} = \mathcal{L}_{V_2V_0}, \mathcal{L}_{V_1V_1} = \mathcal{L}_{V_2V_2}$ and $\mathcal{L}_{V_1V_2} = \mathcal{L}_{V_2V_1}$.

Given:

$$T = \begin{pmatrix} I_n & 0 & 0 \\ 0 & \frac{1}{\sqrt{2}}I_n & \frac{1}{\sqrt{2}}I_n \\ 0 & \frac{1}{\sqrt{2}}I_n & -\frac{1}{\sqrt{2}}I_n \end{pmatrix},$$

then

$$T\mathcal{L}(H_n)T' = \begin{pmatrix} \mathcal{L}_A & 0 \\ 0 & \mathcal{L}_S \end{pmatrix}, \tag{5}$$

where T' is the transposition of T.

$$\mathcal{L}_A = \begin{pmatrix} I_n & \sqrt{2}\mathcal{L}_{V_0V_1} \\ \sqrt{2}\mathcal{L}_{V_1V_0} & \mathcal{L}_{V_1V_1} + \mathcal{L}_{V_1V_2} \end{pmatrix}, \ \mathcal{L}_s = \mathcal{L}_{V_1V_1} - \mathcal{L}_{V_1V_2}. \tag{6}$$

At this place, the matrices introduced above are given in the following. In view of Equation (1) and Figure 1, one has $\mathcal{L}_{V_0V_0} = I_n$.

$$\begin{aligned}\mathcal{L}_{V_0V_1} &= (l_{ij})_{n\times(4n+1)} \\ &= \begin{pmatrix} 0 & 0 & -\frac{1}{\sqrt{6}} & 0 & 0 & 0 & 0 & \cdots & 0 & 0 & 0 \\ 0 & 0 & 0 & 0 & 0 & 0 & -\frac{1}{\sqrt{6}} & \cdots & 0 & 0 & 0 \\ 0 & 0 & 0 & 0 & 0 & 0 & 0 & \cdots & 0 & 0 & 0 \\ \vdots & \vdots & \vdots & \vdots & \vdots & \vdots & \vdots & \ddots & \vdots & \vdots & \vdots \\ 0 & 0 & 0 & 0 & 0 & 0 & 0 & \cdots & 0 & 0 & 0 \\ 0 & 0 & 0 & 0 & 0 & 0 & 0 & \cdots & -\frac{1}{\sqrt{6}} & 0 & 0 \end{pmatrix}_{n\times(4n+1)},\end{aligned}$$

where $l_{ij} = -\frac{1}{\sqrt{6}}$ and $l_{ij} = 0$, if $j = 4i-1$ and otherwise, respectively.

$$\mathcal{L}_{V_1V_1} = \begin{pmatrix} 1 & -\frac{1}{2} & 0 & 0 & \cdots & 0 & 0 & 0 \\ -\frac{1}{2} & 1 & \frac{-1}{\sqrt{6}} & 0 & \cdots & 0 & 0 & 0 \\ 0 & \frac{-1}{\sqrt{6}} & 1 & \frac{-1}{\sqrt{6}} & \cdots & 0 & 0 & 0 \\ 0 & 0 & \frac{-1}{\sqrt{6}} & 1 & \cdots & 0 & 0 & 0 \\ \vdots & \vdots & \vdots & \vdots & \ddots & \vdots & \vdots & \vdots \\ 0 & 0 & 0 & 0 & \cdots & 1 & \frac{-1}{\sqrt{6}} & 0 \\ 0 & 0 & 0 & 0 & \cdots & \frac{-1}{\sqrt{6}} & 1 & -\frac{1}{2} \\ 0 & 0 & 0 & 0 & \cdots & 0 & -\frac{1}{2} & 1 \end{pmatrix}_{(4n+1)\times(4n+1)}.$$

On the other hand, one gets:

$$\mathcal{L}_{V_1V_2} = (l_{ij})_{(4n+1)\times(4n+1)} = diag(-\frac{1}{2}, 0, 0, 0, -\frac{1}{3}, \ldots, -\frac{1}{3}, 0, 0, 0, -\frac{1}{2}),$$

where $l_{11} = l_{4n+1,4n+1} = -\frac{1}{2}$, $l_{4k+1,4k+1} = -\frac{1}{3}$ for $k \in \{1, 2, \ldots, n-1\}$ and 0 otherwise.

According to Equation (5), one has:

$$\mathcal{L}_{V_1V_1}+\mathcal{L}_{V_1V_2} = (l_{ij})_{(4n+1)\times(4n+1)}$$
$$= \begin{pmatrix} \frac{1}{2} & -\frac{1}{2} & 0 & 0 & 0 & \cdots & 0 & 0 & 0 \\ -\frac{1}{2} & 1 & \frac{-1}{\sqrt{6}} & 0 & 0 & \cdots & 0 & 0 & 0 \\ 0 & \frac{-1}{\sqrt{6}} & 1 & \frac{-1}{\sqrt{6}} & 0 & \cdots & 0 & 0 & 0 \\ 0 & 0 & \frac{-1}{\sqrt{6}} & 1 & \frac{-1}{\sqrt{6}} & \cdots & 0 & 0 & 0 \\ 0 & 0 & 0 & \frac{-1}{\sqrt{6}} & \frac{2}{3} & \cdots & 0 & 0 & 0 \\ \vdots & \vdots & \vdots & \vdots & \vdots & \ddots & \vdots & \vdots & \vdots \\ 0 & 0 & 0 & 0 & 0 & \cdots & 1 & \frac{-1}{\sqrt{6}} & 0 \\ 0 & 0 & 0 & 0 & 0 & \cdots & \frac{-1}{\sqrt{6}} & 1 & -\frac{1}{2} \\ 0 & 0 & 0 & 0 & 0 & \cdots & 0 & -\frac{1}{2} & \frac{1}{2} \end{pmatrix}_{(4n+1)\times(4n+1)},$$

where $l_{11} = l_{4n+1,4n+1} = \frac{1}{2}$, $l_{ii} = 1$ for $i \in \{2,3,\ldots,4n\}$ and $i \neq 4k+1$, $l_{4k+1,4k+1} = \frac{2}{3}$ for $k \in \{1,2,\ldots,n-1\}$. Also, $l_{12} = l_{21} = l_{4n,4n+1} = l_{4n+1,4n} = -\frac{1}{2}$, $l_{i,i+1} = l_{i+1,i} = -\frac{1}{\sqrt{6}}$ for $i \in \{2,3,\ldots,$4n-1$\}$ and $l_{ij} = 0$ for $|i-j| > 1$.

And:

$$\mathcal{L}_S = (l_{ij})_{(4n+1)\times(4n+1)}$$
$$= \begin{pmatrix} \frac{3}{2} & -\frac{1}{2} & 0 & 0 & 0 & \cdots & 0 & 0 & 0 \\ -\frac{1}{2} & 1 & \frac{-1}{\sqrt{6}} & 0 & 0 & \cdots & 0 & 0 & 0 \\ 0 & \frac{-1}{\sqrt{6}} & 1 & \frac{-1}{\sqrt{6}} & 0 & \cdots & 0 & 0 & 0 \\ 0 & 0 & \frac{-1}{\sqrt{6}} & 1 & \frac{-1}{\sqrt{6}} & \cdots & 0 & 0 & 0 \\ 0 & 0 & 0 & \frac{-1}{\sqrt{6}} & \frac{4}{3} & \cdots & 0 & 0 & 0 \\ \vdots & \vdots & \vdots & \vdots & \vdots & \ddots & \vdots & \vdots & \vdots \\ 0 & 0 & 0 & 0 & 0 & \cdots & 1 & \frac{-1}{\sqrt{6}} & 0 \\ 0 & 0 & 0 & 0 & 0 & \cdots & \frac{-1}{\sqrt{6}} & 1 & -\frac{1}{2} \\ 0 & 0 & 0 & 0 & 0 & \cdots & 0 & -\frac{1}{2} & \frac{3}{2} \end{pmatrix}_{(4n+1)\times(4n+1)},$$

where $l_{11} = l_{4n+1,4n+1} = \frac{3}{2}$, $l_{ii} = 1$ for $i \in \{2,3,\ldots,4n\}$ and $i \neq 4k+1$, $l_{4k+1,4k+1} = \frac{4}{3}$ for $k \in \{1,2,\ldots,n-1\}$. Also, $l_{12} = l_{21} = l_{4n,4n+1} = l_{4n+1,4n} = -\frac{1}{2}$, $l_{i,i+1} = l_{i+1,i} = -\frac{1}{\sqrt{6}}$ for $i \in \{2,3,\ldots,4n-1\}$ and $l_{ij} = 0$ for $|i-j| > 1$.

In what follows, the lemmas that we present will be used throughout the main results.

Lemma 1. *Assume that $\mathcal{L}_A, \mathcal{L}_s$ are defined as mentioned above [1]. Then:*

$$P_{\mathcal{L}(L_n)}(x) = P_{\mathcal{L}_A}(x)P_{\mathcal{L}_s}(x).$$

According to the relation between the number of spanning trees and its normalized laplacian eigenvalues, one arrives the following lemma.

Lemma 2. *Assume that G is a connected graph with order n and size m [7], then*

$$\prod_{i=1}^{n} d_G(v_i) \prod_{i=2}^{n} \lambda_i = 2m\tau(G),$$

where $\tau(G)$ is the number of spanning trees of G.

Lemma 3. *Let M_1, M_2, M_3 and M_4 be the $p \times p$, $p \times q$, $q \times p$ and $q \times q$ matrices, respectively [24]. Then:*

$$(a).\ det\begin{pmatrix} M_1 & M_2 \\ M_3 & M_4 \end{pmatrix} = det(M_4) \cdot det(M_1 - M_2M_4^{-1}M_3),$$

$$(b).\ det\begin{pmatrix} M_1 & M_2 \\ M_3 & M_4 \end{pmatrix} = det(M_1) \cdot det(M_4 - M_3M_1^{-1}M_2),$$

where M_1 and M_4 are invertible, and $M_1 - M_2M_4^{-1}M_3$ and $M_4 - M_3M_1^{-1}M_2$ are called the Schur complements of M_4 and M_1, respectively.

In what follows, a flowchart (Figure 2) is given on the basis of the steps we have processed, and it will facilitate the understanding of the proposed approach. The explanations of these notations that appear in the flowchart are presented in Section 3.

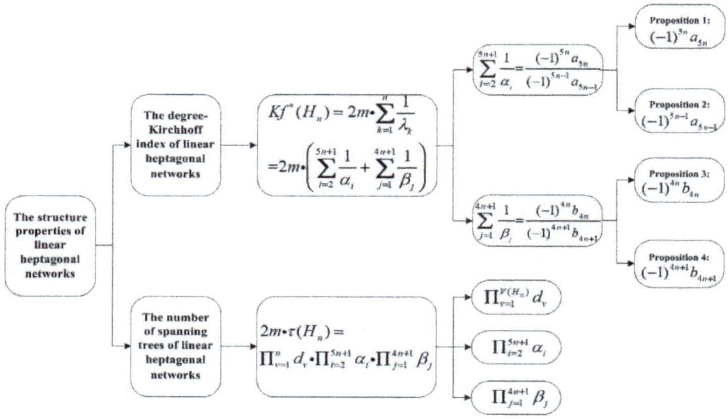

Figure 2. The steps processed in this paper.

3. Main Results

In this section, we first figure out the explicit formula for the degree-Kirchhoff index of H_n. The steps of computing the degree-Kirchhoff index follow the flowchart in Figure 2. Combining the eigenvalues of $\mathcal{L}_A$ and $\mathcal{L}_S$, it is easy to get the normalized Laplacian spectrum of H_n by Lemma 1. On the other hand, one can find that the number or spanning trees of H_n consists of the products of the degree of all vertices, the eigenvalues of $\mathcal{L}_A$ and $\mathcal{L}_S$, respectively. Suppose $0 = \alpha_1 < \alpha_2 \leq \cdots \leq \alpha_{5n+1}$ and $0 < \beta_1 \leq \beta_2 \leq \cdots \leq \beta_{4n+1}$ are the roots of $P_{\mathcal{L}_A}(x) = 0$ and $P_{L_S}(x) = 0$, respectively. By Equation (3), one has:

Lemma 4. *Suppose that H_n are the linear heptagonal networks. Then:*

$$Kf^*(H_n) \quad = \quad 2 \cdot (11n+1) \cdot (\sum_{i=2}^{5n+1} \frac{1}{\alpha_i} + \sum_{j=1}^{4n+1} \frac{1}{\beta_j}). \tag{7}$$

Evidently, one just needs to calculate the eigenvalues of the $\mathcal{L}_A$ and $\mathcal{L}_S$. Hence, the formulas of $\sum_{i=2}^{5n+1} \frac{1}{\alpha_i}$ and $\sum_{j=1}^{4n+1} \frac{1}{\beta_j}$ are given in the following lemmas.

Lemma 5. *Suppose that* $0 = \alpha_1 < \alpha_2 \leq \cdots \leq \alpha_{5n+1}$ *are the eigenvalues of* $\mathcal{L}_A$. *One has:*

$$\sum_{i=2}^{5n+1} \frac{1}{\alpha_i} = \frac{401n^3 - 170n^2 + 54n}{11n+1}.$$

Proof. It is straightforward to verify that:

$$P_{\mathcal{L}_A}(x) = x^{5n+1} + a_1 x^{5n} + \cdots + a_{5n-1}x^2 + a_{5n}x = x(x^{5n} + a_1 x^{5n-1} + \cdots + a_{5n-1}x + a_{5n}),$$

with $a_{5n} \neq 0$. Then, one knows that $\frac{1}{\alpha_2}, \frac{1}{\alpha_3}, \ldots, \frac{1}{\alpha_{5n+1}}$ are the roots of the following equation:

$$a_{5n} \cdot x^{5n} + a_{5n-1} \cdot x^{5n-1} + \cdots + a_1 \cdot x + 1 = 0.$$

According to the relationships between the roots and the coefficients and Vieta theorem, one arrives at:

$$\sum_{i=2}^{5n+1} \frac{1}{\alpha_i} = \frac{(-1)^{5n-1} \cdot a_{5n-1}}{(-1)^{5n} \cdot a_{5n}}.$$

□

Proposition 1. $(-1)^{5n} \cdot a_{5n} = \frac{11n+1}{8}(\frac{1}{6})^{2n-1}$.

Proof. Since the number $(-1)^{5n}a_{5n}$ is the sum of all those principal minors of $\mathcal{L}_A$ which have $5n$ rows and columns, we get:

$$(-1)^{5n} \cdot a_{5n} = \sum_{i=1}^{5n+1} \det \mathcal{L}_A[i]. \tag{8}$$

In view of Equation (6), one may find that i can be selected in the identity matrix I_n or $\mathcal{L}_{V_1V_1} + \mathcal{L}_{V_1V_2}$. Thus, all these cases are listed as below.

Case 1. One first considers the case $1 \leq i \leq n$. That is to say, $\mathcal{L}_A[i]$ is obtained by deleting the i row of the identity matrix I_n and $\sqrt{2}\mathcal{L}_{V_0V_1}$, and also the corresponding column of the I_n and $\sqrt{2}\mathcal{L}_{V_1V_0}$. Denoted by I_{n-1}, $R_{(n-1)\times(4n+1)}$, $R'_{(n-1)\times(4n+1)}$ and $U_{(4n+1)\times(4n+1)}$ are the resulting blocks, respectively. According to the Lemma 3, and applying elementary operations (see also in Reference[25], P8), one knows:

$$\det \mathcal{L}_A[i] = \begin{vmatrix} I_{n-1} & R_{(n-1)\times(4n+1)} \\ R'_{(n-1)\times(4n+1)} & U_{(4n+1)\times(4n+1)} \end{vmatrix} = \begin{vmatrix} I_{n-1} & 0 \\ 0 & M_1 \end{vmatrix},$$

where $M_1 = U_{(4n+1)\times(4n+1)} - R'_{(n-1)\times(4n+1)}R_{(n-1)\times(4n+1)} = (m_{pq})$, of which $m_{11} = m_{4n+1,4n+1} = \frac{1}{2}$, $m_{pp} = 1$ for $p \in \{2, 4, 6, \ldots, 4n\}$, $m_{pp} = \frac{2}{3}$ for $p \in \{3, 5, 7, 4n-1\}$ but $p \neq 4i-1$, $m_{4i-1,4i-1} = 1$. Also, $m_{12} = m_{21} = m_{4n,4n+1} = m_{4n+1,4n} = -\frac{1}{2}$, $m_{p,p+1} = m_{p+1,p} = -\frac{1}{\sqrt{6}}$ for $p \in \{2, 3, \ldots, 4n-1\}$ and $m_{pq} = 0$ for $|p-q| > 1$. With a explicit calculation, one gets:

$$\det \mathcal{L}_A[i] = \det M_1 = \frac{1}{8} \times \left(\frac{1}{6}\right)^{2n-1}. \tag{9}$$

Case 2. One now takes into account the case $n+1 \leq i \leq 5n+1$. Namely, $\mathcal{L}_A[i]$ is obtained by deleting the i row of the $\mathcal{L}_{V_1V_1} + \mathcal{L}_{V_1V_2}$ and $\sqrt{2}\mathcal{L}_{V_1V_0}$, also the corresponding column of the $\mathcal{L}_{V_1V_1} + \mathcal{L}_{V_1V_2}$ and $\sqrt{2}\mathcal{L}_{V_0V_1}$. Denote by I_n, $R_{n\times 4n}$, $R'_{n\times 4n}$ and $U_{4n\times 4n}$ the resulting blocks, respectively. According to Lemma 3 and applying elementary operations, we arrive at:

$$\det \mathcal{L}_A[i] = \begin{vmatrix} I_n & R_{n\times 4n} \\ R'_{n\times 4n} & U_{4n\times 4n} \end{vmatrix} = \begin{vmatrix} I_n & 0 \\ 0 & M_2 \end{vmatrix},$$

where $M_2 = U_{4n\times 4n} - R'_{n\times 4n}R_{n\times 4n}$. Set $M_{2'} = (m_{pq})$ with order $4n+1$, of which $m_{11} = m_{4n+1,4n+1} = \frac{1}{2}, m_{pp} = 1$ for $p \in \{2,4,6,\ldots,4n\}$, $m_{pp} = \frac{2}{3}$ for $p \in \{3,5,7,4n-1\}$. Also, $m_{12} = m_{21} = m_{4n,4n+1} = m_{4n+1,4n} = -\frac{1}{2}$, $m_{p,p+1} = m_{p+1,p} = -\frac{1}{\sqrt{6}}$ for $p \in \{2,3,\ldots,4n-1\}$ and $m_{pq} = 0$ for $|p-q| > 1$. Evidently, $M_2 = M_{2'}[i-n]$. With an explicit calculation, one gets:

$$\det \mathcal{L}_A[i] = \det M_{2'}[i-n] = \begin{cases} \frac{1}{4}\times\left(\frac{1}{6}\right)^{2n-1}, & if\ i = n+1\ or\ i = 5n+1; \\ \frac{1}{4}\times\left(\frac{1}{6}\right)^{2n-1}, & if\ n+2 \le i \le 5n,\ i\ is\ even; \\ \frac{1}{16}\times\left(\frac{1}{6}\right)^{2n-2}, & if\ n+2 \le i \le 5n,\ i\ is\ odd. \end{cases} \tag{10}$$

Combining Equations (8)–(10), we have:

$$\begin{aligned} (-1)^{5n}\cdot a_{5n} &= \sum_{i=1}^{5n+1} \det \mathcal{L}_A[i] = \sum_{i=1}^{n} \det \mathcal{L}_A[i] + \sum_{i=n+1}^{5n+1} \det \mathcal{L}_A[i] \\ &= n\cdot\frac{1}{8}\times\left(\frac{1}{6}\right)^{2n-1} + 2n\cdot\frac{1}{4}\times\left(\frac{1}{6}\right)^{2n-1} + (2n-1)\cdot\frac{1}{16}\times\left(\frac{1}{6}\right)^{2n-2} + 2\times\frac{1}{4}\times\left(\frac{1}{6}\right)^{2n-1} \\ &= \frac{11n+1}{8}\cdot\left(\frac{1}{6}\right)^{2n-1}. \end{aligned}$$

This completes the proof of the Proposition 1. □

In what follows, we will focus on the calculation of the $(-1)^{5n-1}a_{5n-1}$.

Proposition 2. $(-1)^{5n-1}a_{5n-1} = \frac{1}{48}(401n^3 - 170n^2 + 54n)\left(\frac{1}{6}\right)^{2n-2}$.

Proof. Since the number $(-1)^{5n-1}a_{5n-1}$ is the sum of all those principal minors of $\mathcal{L}_A$ which have $5n-1$ rows and columns, it is straightforward to obtain that:

$$(-1)^{5n-1}\cdot a_{5n-1} = \sum_{i=1}^{5n+1} \det \mathcal{L}_A[i,j]. \tag{11}$$

Similar to consideration of Proposition 1, all these cases are listed as follows.

Case 1. One first considers the case $1 \le i < j \le n$. That is to say, $\mathcal{L}_A[i,j]$ is obtained by deleting the i and j rows of the identity matrix I_n and $\sqrt{2}\mathcal{L}_{V_0V_1}$, also the corresponding columns of the I_n and $\sqrt{2}\mathcal{L}_{V_1V_0}$. Denote by I_{n-2}, $R_{(n-2)\times(4n+1)}$, $R'_{(n-2)\times(4n+1)}$ and $U_{(4n+1)\times(4n+1)}$ the resulting blocks, respectively. According to Lemma 3 and applying elementary operations, one has:

$$\det \mathcal{L}_A[i,j] = \begin{vmatrix} I_{n-2} & R_{(n-2)\times(4n+1)} \\ R'_{(n-2)\times(4n+1)} & U_{(4n+1)\times(4n+1)} \end{vmatrix} = \begin{vmatrix} I_{n-2} & 0 \\ 0 & M_3 \end{vmatrix},$$

where $M_3 = U_{(4n+1)\times(4n+1)} - R'_{(n-2)\times(4n+1)}R_{(n-2)\times(4n+1)} = (m_{pq})$, of which $m_{11} = m_{4n+1,4n+1} = \frac{1}{2}, m_{pp} = 1$ for $p \in \{2,4,6,\ldots,4n\}$, $m_{pp} = \frac{2}{3}$ for $p \in \{3,5,7,4n-1\}$ but $p \neq 4i-1$ and $p \neq 4j-1$, $m_{4i-1,4i-1} = m_{4j-1,4j-1} = 1$. Also, $m_{12} = m_{21} = m_{4n,4n+1} = m_{4n+1,4n} = -\frac{1}{2}$, $m_{p,p+1} = m_{p+1,p} = -\frac{1}{\sqrt{6}}$ for $p \in \{2,3,\ldots,4n-1\}$ and $m_{pq} = 0$ for $|p-q| > 1$. With an explicit calculation, one gets:

$$\det \mathcal{L}_A[i,j] = \det M_3 = \frac{1}{8} \cdot (j-i+2) \cdot \left(\frac{1}{6}\right)^{2n-1}. \tag{12}$$

Case 2. One now takes into account the case $n+1 \le i < j \le 5n+1$. Namely, $\mathcal{L}_A[i,j]$ is obtained by deleting the $(i-n)$-th and $(j-n)$-th rows of the $\mathcal{L}_{V_1V_1} + \mathcal{L}_{V_1V_2}$ and $\sqrt{2}\mathcal{L}_{V_1V_0}$, also the corresponding columns of the $\mathcal{L}_{V_1V_1} + \mathcal{L}_{V_1V_2}$ and $\sqrt{2}\mathcal{L}_{V_0V_1}$. Denote by I_n, $R_{n\times(4n-1)}$, $R'_{n\times(4n-1)}$ and $U_{(4n-1)\times(4n-1)}$ the resulting blocks, respectively. According to Lemma 3 and applying elementary operations, one arrives at:

$$\det \mathcal{L}_A[i,j] = \begin{vmatrix} I_n & R_{n\times(4n-1)} \\ R'_{n\times(4n-1)} & U_{(4n-1)\times(4n-1)} \end{vmatrix} = \begin{vmatrix} I_n & 0 \\ 0 & M_4 \end{vmatrix},$$

where $M_4 = U_{(4n-1)\times(4n-1)} - R'_{n\times(4n-1)}R_{n\times(4n-1)} = M_{2'}[i-n, j-n]$. With an explicit calculation, one gets:

$$\det \mathcal{L}_A[i,j] = \begin{cases} \frac{1}{8}(j-i)\left(\frac{1}{6}\right)^{2n-2}, & if\ n+2 \le i = n+l < j = n+k \le 5n,\ l\ is\ even,\ k\ is\ odd; \\ \frac{1}{12}(j-i)\left(\frac{1}{6}\right)^{2n-2}, & if\ n+2 \le i = n+l < j = n+k \le 5n,\ l\ is\ even,\ k\ is\ even; \\ \frac{1}{8}(j-i)\left(\frac{1}{6}\right)^{2n-2}, & if\ n+2 \le i = n+l < j = n+k \le 5n,\ l\ is\ odd,\ k\ is\ even; \\ \frac{3}{16}(j-i)\left(\frac{1}{6}\right)^{2n-2}, & if\ n+2 \le i = n+l < j = n+k \le 5n,\ l\ is\ odd,\ k\ is\ odd; \\ \frac{1}{8}(j-i)\left(\frac{1}{6}\right)^{2n-2}, & if\ i = n+1,\ j = n+k,\ k\ is\ odd; \\ \frac{1}{6}(j-i)\left(\frac{1}{6}\right)^{2n-2}, & if\ i = n+1,\ j = n+k,\ k\ is\ even; \\ \frac{1}{8}(j-i)\left(\frac{1}{6}\right)^{2n-2}, & if\ i = n+k,\ j = 5n+1,\ k\ is\ odd; \\ \frac{1}{12}(j-i)\left(\frac{1}{6}\right)^{2n-2}, & if\ i = n+k,\ j = 5n+1,\ k\ is\ even; \\ 2n\left(\frac{1}{6}\right)^{2n-1}, & if\ i = n+1,\ j = 5n+1. \end{cases} \tag{13}$$

Case 3. One devotes to consideration of the case $1 \le i \le n$, $n+1 \le j \le 5n+1$. That is, $\mathcal{L}_A[i,j]$ is obtained by deleting the i rows of the identity matrix I_n and $\sqrt{2}\mathcal{L}_{V_0V_1}$, $(j-n)$-th rows of the $\sqrt{2}\mathcal{L}_{V_1V_0}$ and $\mathcal{L}_{V_1V_1} + \mathcal{L}_{V_1V_2}$, respectively. In addition, the corresponding columns of the I_n, $\sqrt{2}\mathcal{L}_{V_1V_0}$, $\mathcal{L}_{V_1V_1} + \mathcal{L}_{V_1V_2}$ and $\sqrt{2}\mathcal{L}_{V_0V_1}$. Denote by I_{n-1}, $R_{(n-1)\times 4n}$, $R'_{(n-1)\times 4n}$ and $U_{4n\times 4n}$ the resulting blocks, respectively. According to Lemma 3 and applying elementary operations, one has:

$$\det \mathcal{L}_A[i,j] = \begin{vmatrix} I_{n-1} & R_{(n-1)\times 4n} \\ R'_{(n-1)\times 4n} & U_{4n\times 4n} \end{vmatrix} = \begin{vmatrix} I_n & 0 \\ 0 & M_5 \end{vmatrix},$$

where $M_5 = U_{4n\times 4n} - R'_{(n-1)\times 4n}R_{(n-1)\times 4n} = M_1[j-n]$. By an explicit calculation, we get:

$$\det \mathcal{L}_A[i,j] = \begin{cases} (\frac{1}{2}+i)\left(\frac{1}{6}\right)^{2n-1}, & if\ j = n+1; \\ \frac{1}{24}(j-n-2i)\left(\frac{1}{6}\right)^{2n-2}, & if\ n+2 \le j = n+k \le 5n,\ j-n-2i > 0,\ k\ is\ even; \\ \frac{1}{12}\left(\frac{1}{6}\right)^{2n-2}, & if\ n+2 \le j = n+k \le 5n,\ j-n-2i = 0,\ k\ is\ even; \\ \frac{1}{12}(|\frac{j-n-2i}{2}|+1)\left(\frac{1}{6}\right)^{2n-2}, & if\ n+2 \le j = n+k \le 5n,\ j-n-2i < 0,\ k\ is\ even; \\ \frac{1}{16}(j-n-2i)\left(\frac{1}{6}\right)^{2n-2}, & if\ n+2 \le j = n+k \le 5n,\ j-n-2i \ge 0,\ k\ is\ odd; \\ \frac{1}{16}(|j-n-2i|+2)\left(\frac{1}{6}\right)^{2n-2}, & if\ n+2 \le j = n+k \le 5n,\ j-n-2i < 0,\ k\ is\ odd; \\ \frac{1}{4}(4n-2i+1)\left(\frac{1}{6}\right)^{2n-1}, & if\ j = 5n+1. \end{cases} \quad (14)$$

Together with Equations (11)–(14), one gets:

$$\begin{aligned} (-1)^{5n-1}a_{5n-1} &= \sum_{1\le i\le j}^{5n+1} \det \mathcal{L}_A[i,j] \\ &= \sum_{1\le i\le j}^{n} \det \mathcal{L}_A[i,j] + \sum_{n+1\le i\le j}^{5n+1} \det \mathcal{L}_A[i,j] + \sum_{1\le i\le n, n+1\le j\le 5n+1}^{5n+1} \det \mathcal{L}_A[i,j] \\ &= \frac{n^3+12n^2+11n}{48}\left(\frac{1}{6}\right)^{2n-1} + \frac{290n^3-144n^2+25n}{36}\left(\frac{1}{6}\right)^{2n-2} \\ &\quad + \frac{40n^3+30n^2+29n}{144}\left(\frac{1}{6}\right)^{2n-2} \\ &= \frac{1}{48}(401n^3-170n^2+54n)\left(\frac{1}{6}\right)^{2n-2}. \end{aligned}$$

The proof of the Proposition 2 completed. □

By Propositions 1 and 2 , we have the desired result of Lemma 5.

Lemma 6. *Suppose that $\beta_1, \beta_2, \ldots, \beta_{4n+1}$ are the eigenvalues of $\mathcal{L}_S$. One has:*

$$\sum_{j=1}^{4n+1} \frac{1}{\beta_j} = \frac{\delta(n)+\zeta(n)}{441\left[(-24+2183\sqrt{2})(11-6\sqrt{2})^n - (24+2183\sqrt{2})(11+6\sqrt{2})^n\right]},$$

where $\delta(n) = 4\left(461133 - 44560\sqrt{2} - 6n(249938+99525\sqrt{2})\right)(11+6\sqrt{2})^n$,
$\zeta(n) = 4\left(461133 + 44560\sqrt{2} + 6n(-249938+99525\sqrt{2})\right)(11-6\sqrt{2})^n$.

Proof. It is straightforward to verify that:

$$P_{\mathcal{L}_S}(x) = x^{4n+1} + b_1 x^{4n} + \cdots + b_{4n-1}x^2 + b_{4n}x + b_{4n+1},$$

with $b_{4n+1} \neq 0$. Then, one knows that $\frac{1}{\beta_1}, \frac{1}{\beta_2}, \ldots, \frac{1}{\beta_{4n+1}}$ are the roots of the following equation:

$$b_{4n+1} \cdot x^{4n+1} + b_{4n} \cdot x^{4n} + \cdots + b_1 \cdot x + 1 = 0.$$

In the line with the relationships between the roots and the coefficients of $P_{\mathcal{L}_s}(x)$, one arrives at:

$$\sum_{j=1}^{4n+1} \frac{1}{\beta_j} = \frac{(-1)^{4n} b_{4n}}{(-1)^{4n+1} b_{4n+1}} = \frac{(-1)^{4n} b_{4n}}{\det \mathcal{L}_s}.$$

On the one hand, we first consider l-th order principal submatrix, D_l, and yield by the first l rows and columns of $\mathcal{L}_s, l = 1, 2, \ldots, 4n$. Set $d_l = \det D_l$. Then $d_1 = \frac{3}{2}, d_2 = \frac{5}{4}, d_3 = 1, d_4 = \frac{19}{24}, d_5 = \frac{8}{9}, d_6 = \frac{109}{144}, d_7 = \frac{263}{432}, d_8 = \frac{139}{288}$ and:

$$\begin{cases} d_{4i} = d_{4i-1} - \frac{1}{6}d_{4i-2}, \\ d_{4i+1} = \frac{4}{3}d_{4i} - \frac{1}{6}d_{4i-1}, \\ d_{4i+2} = d_{4i+1} - \frac{1}{6}d_{4i}, \\ d_{4i+3} = d_{4i+2} - \frac{1}{6}d_{4i+1}. \end{cases} \tag{15}$$

With a explicit calculation, we get:

$$\begin{cases} d_{4i} = \frac{460+3\sqrt{2}}{196\sqrt{2}}\left(\frac{11+6\sqrt{2}}{36}\right)^i + \frac{-460+3\sqrt{2}}{196\sqrt{2}}\left(\frac{11-6\sqrt{2}}{36}\right)^i, \\ d_{4i+1} = \frac{31+18\sqrt{2}}{24\sqrt{2}}\left(\frac{11+6\sqrt{2}}{36}\right)^i + \frac{-31+18\sqrt{2}}{24\sqrt{2}}\left(\frac{11-6\sqrt{2}}{36}\right)^i, \\ d_{4i+2} = \frac{27+15\sqrt{2}}{24\sqrt{2}}\left(\frac{11+6\sqrt{2}}{36}\right)^i + \frac{-27+15\sqrt{2}}{24\sqrt{2}}\left(\frac{11-6\sqrt{2}}{36}\right)^i, \\ d_{4i+3} = \frac{131+72\sqrt{2}}{144\sqrt{2}}\left(\frac{11+6\sqrt{2}}{36}\right)^i + \frac{-131+72\sqrt{2}}{144\sqrt{2}}\left(\frac{11-6\sqrt{2}}{36}\right)^i. \end{cases} \tag{16}$$

□

Proposition 3.

$$\det \mathcal{L}_s = \frac{540+493\sqrt{2}}{1152}\left(\frac{11+6\sqrt{2}}{36}\right)^{n-1} + \frac{540-493\sqrt{2}}{1152}\left(\frac{11-6\sqrt{2}}{36}\right)^{n-1}.$$

Proof. By expanding $\det \mathcal{L}_s$ with regards to the last row, one gets:

$$\begin{aligned} \det \mathcal{L}_s &= \frac{3}{2}d_{4n} - \frac{1}{4}d_{4n-1} = \frac{3}{2}\left[\frac{460+3\sqrt{2}}{196\sqrt{2}}\left(\frac{11+6\sqrt{2}}{36}\right)^n + \frac{-460+3\sqrt{2}}{196\sqrt{2}}\left(\frac{11-6\sqrt{2}}{36}\right)^n\right] \\ &\quad -\frac{1}{4}\left[\frac{131+72\sqrt{2}}{144\sqrt{2}}\left(\frac{11+6\sqrt{2}}{36}\right)^{n-1} + \frac{-131+72\sqrt{2}}{144\sqrt{2}}\left(\frac{11-6\sqrt{2}}{36}\right)^{n-1}\right] \\ &= \frac{540+493\sqrt{2}}{1152}\left(\frac{11+6\sqrt{2}}{36}\right)^{n-1} + \frac{540-493\sqrt{2}}{1152}\left(\frac{11-6\sqrt{2}}{36}\right)^{n-1}. \end{aligned}$$

This has completed the proof of the Proposition 3. □

On the other hand, we will take into account the k-th order principal submatrix, R_k, and yield by the last k rows and columns of $\mathcal{L}_s, k = 1, 2, \ldots, 4n$. Set $r_k = \det R_k$. Then $r_1 = \frac{3}{2}, r_2 = \frac{5}{4}, r_3 = 1, r_4 = \frac{19}{24}, r_5 = \frac{8}{9}, r_6 = \frac{109}{144}, r_7 = \frac{263}{432}, r_8 = \frac{139}{288}$ and:

$$\begin{cases} r_{4i} = r_{4i-1} - \frac{1}{6}r_{4i-2}, \\ r_{4i+1} = \frac{4}{3}r_{4i} - \frac{1}{6}r_{4i-1}, \\ r_{4i+2} = r_{4i+1} - \frac{1}{6}r_{4i}, \\ r_{4i+3} = r_{4i+2} - \frac{1}{6}r_{4i+1}. \end{cases} \tag{17}$$

Obviously, one finds that $d_{4i} = r_{4i}, d_{4i+1} = r_{4i+1}, d_{4i+2} = r_{4i+2}, d_{4i+3} = r_{4i+3}$.

Proposition 4.

$$
\begin{aligned}
(-1)^{4n} b_{4n} &= \frac{-92,607-46,462\sqrt{2}+6n(80,482+52,947\sqrt{2})}{127,008}\Big(\frac{11+6\sqrt{2}}{36}\Big)^{n-1} \\
&+\frac{-92,607+46,462\sqrt{2}-6n(-80,482+52,947\sqrt{2})}{127,008}\Big(\frac{11-6\sqrt{2}}{36}\Big)^{n-1}.
\end{aligned}
$$

Proof. Denoted l_{ii} the diagonal entries of $\mathcal{L}_S$. Since the number $(-1)^{4n}b_{4n}$ is the sum of all those principal minors of $\mathcal{L}_S$ which have $4n$ rows and columns, one arrives at:

$$
(-1)^{4n} b_{4n} = \sum_{i=1}^{4n+1} \det \mathcal{L}_S[i] = \sum_{i=1}^{4n+1} \det \begin{pmatrix} D_{i-1} & 0 \\ 0 & R_{4n+1-i} \end{pmatrix} = \sum_{i=1}^{4n+1} \det D_{i-1} \cdot \det R_{4n+1-i}, \quad (18)
$$

where

$$
R_{4n+1-i} = \begin{pmatrix} l_{i+1,i+1} & \cdots & 0 & 0 \\ \vdots & \ddots & \vdots & \vdots \\ 0 & \cdots & l_{4n,4n} & -\frac{1}{\sqrt{6}} \\ 0 & \cdots & -\frac{1}{\sqrt{6}} & l_{4n+1,4n+1} \end{pmatrix}.
$$

Let $d_0 := \det D_0 = 1$ and $\det R = 1$, if $i = 4n+1$. In line with Equation (18), one gets:

$$
\begin{aligned}
(-1)^{4n} b_{4n} &= \sum_{i=1}^{4n+1} \det \mathcal{L}_S[i] = \sum_{i=1}^{4n} \det \mathcal{L}_S[i] + d_{4n} = \sum_{k=1}^{n} \det \mathcal{L}_S[4k] \\
&+ \sum_{k=0}^{n-1} \det \mathcal{L}_S[4k+1] + \sum_{k=0}^{n-1} \det \mathcal{L}_S[4k+2] + \sum_{k=0}^{n-1} \det \mathcal{L}_S[4k+3] + d_{4n} \\
&= \sum_{k=1}^{n} d_{4(k-1)+3} d_{4(n-k)+1} + \sum_{k=1}^{n-1} d_{4k} d_{4(n-k)} + \sum_{k=0}^{n-1} d_{4k+1} d_{4(n-k-1)+3} \\
&+ \sum_{k=0}^{n-1} d_{4k+2} d_{4(n-k-1)+2} + 2d_{4n}. \quad (19)
\end{aligned}
$$

By a straightforward calculation, we can obtain the following expressions.

$$
\begin{aligned}
\sum_{k=1}^{n} d_{4(k-1)+3} d_{4(n-k)+1} &= n \cdot \frac{6653+4590\sqrt{2}}{6912 \cdot 36^{n-1}} (11+6\sqrt{2})^{n-1} + n \cdot \frac{6653-4590\sqrt{2}}{6912 \cdot 36^{n-1}} (11-6\sqrt{2})^{n-1} \\
&+ \frac{1}{12} \cdot \frac{-1469\sqrt{2}}{6912 \cdot 36^{n-1}} (11+6\sqrt{2})^{n} + \frac{1}{12} \cdot \frac{1469\sqrt{2}}{6912 \cdot 36^{n-1}} (11-6\sqrt{2})^{n}. \quad (20)
\end{aligned}
$$

$$
\begin{aligned}
\sum_{k=1}^{n-1} d_{4k} d_{4(n-k)} &= (n-1) \cdot \frac{105809+1380\sqrt{2}}{38416 \cdot 36^{n}} (11+6\sqrt{2})^{n} + (n-1) \cdot \frac{105809-1380\sqrt{2}}{38416 \cdot 36^{n}} (11-6\sqrt{2})^{n} \\
&+ \frac{1}{12} \cdot \frac{-105791\sqrt{2}}{38416 \cdot 36^{n}} (11+6\sqrt{2})^{n-1} + \frac{1}{12} \cdot \frac{105791\sqrt{2}}{38416 \cdot 36^{n}} (11-6\sqrt{2})^{n-1}. \quad (21)
\end{aligned}
$$

$$
\begin{aligned}
\sum_{k=0}^{n-1} d_{4k+1} d_{4(n-k-1)+3} &= n \cdot \frac{6653+4590\sqrt{2}}{6912 \cdot 36^{n-1}} (11+6\sqrt{2})^{n-1} + n \cdot \frac{6653-4590\sqrt{2}}{6912 \cdot 36^{n-1}} (11-6\sqrt{2})^{n-1} \\
&+ \frac{1}{12} \cdot \frac{-1469\sqrt{2}}{6912 \cdot 36^{n-1}} (11+6\sqrt{2})^{n} + \frac{1}{12} \cdot \frac{1469\sqrt{2}}{6912 \cdot 36^{n-1}} (11-6\sqrt{2})^{n}. \quad (22)
\end{aligned}
$$

$$\begin{aligned}\sum_{k=0}^{n-1} d_{4k+2}d_{4(n-k-1)+2} &= n\cdot\frac{131+90\sqrt{2}}{128\cdot 36^{n-1}}(11+6\sqrt{2})^{n-1}+n\cdot\frac{131-90\sqrt{2}}{128\cdot 36^{n-1}}(11-6\sqrt{2})^{n-1}\\ &+\frac{1}{12}\cdot\frac{-31\sqrt{2}}{128\cdot 36^{n-1}}(11+6\sqrt{2})^{n}+\frac{1}{12}\cdot\frac{31\sqrt{2}}{128\cdot 36^{n-1}}(11-6\sqrt{2})^{n}. \quad (23)\end{aligned}$$

$$2d_{4n} = \frac{6+460\sqrt{2}}{196\cdot 36^{n}}(11+6\sqrt{2})^{n}+\frac{6-460\sqrt{2}}{196\cdot 36^{n}}(11-6\sqrt{2})^{n}. \quad (24)$$

Substituting Equations (20)–(24) to (19), we can get the desired result. □

Together with the Propositions 3 and 4, one gets Lemma 6 immediately.
By Lemmas 5 and 6, we have the following theorem.

Theorem 1. *For graph H_n,*

$$Kf^*(H_n) = (22n+2)\cdot\left(\frac{401n^3-170n^2+54n}{11n+1}+\frac{(-1)^{4n}\cdot b_{4n}}{\det\mathcal{L}_s}\right),$$

where

$$\begin{aligned}(-1)^{4n}b_{4n} &= \frac{-92607-46462+6n(80482+52947\sqrt{2})}{127008}\left(\frac{11+6\sqrt{2}}{36}\right)^{n-1}\\ &+\frac{-92607+46462-6n(-80482+52947\sqrt{2})}{127008}\left(\frac{11-6\sqrt{2}}{36}\right)^{n-1}.\end{aligned}$$

$$\det\mathcal{L}_s = \frac{540+493\sqrt{2}}{1152}\left(\frac{11+6\sqrt{2}}{36}\right)^{n-1}+\frac{540-493\sqrt{2}}{1152}\left(\frac{11-6\sqrt{2}}{36}\right)^{n-1}.$$

Theorem 2. *For graph H_n,*

$$\tau(H_n) = 2^{3n+1}\cdot 3^{2n-1}\cdot\left[\frac{540+493\sqrt{2}}{1152}\left(\frac{11+6\sqrt{2}}{36}\right)^{n-1}+\frac{540-493\sqrt{2}}{1152}\left(\frac{11-6\sqrt{2}}{36}\right)^{n-1}\right].$$

Proof. By Lemma 2, one has $\prod_{i=1}^{|V(H_n)|} d_i\prod_{i=2}^{5n+1}\alpha_i\prod_{j=1}^{4n+1}\beta_j = 2|E(H_n)|\tau(H_n)$. Notice that:

$$\begin{aligned}\prod_{i=1}^{|V(H_n)|} d_i &= 2^{5n+4}\cdot 3^{4n-2},\\ \prod_{i=2}^{5n+1}\alpha_i &= (-1)^{5n}a_{5n} = \frac{11n+1}{8}\left(\frac{1}{6}\right)^{2n-1},\\ \prod_{j=1}^{4n+1}\beta_j &= \det\mathcal{L}_s = \frac{540+493\sqrt{2}}{1152}\left(\frac{11+6\sqrt{2}}{36}\right)^{n-1}+\frac{540-493\sqrt{2}}{1152}\left(\frac{11-6\sqrt{2}}{36}\right)^{n-1}.\end{aligned}$$

Evidently,

$$\tau(H_n) = 2^{3n+1}\cdot 3^{2n-1}\cdot\left[\frac{540+493\sqrt{2}}{1152}\left(\frac{11+6\sqrt{2}}{36}\right)^{n-1}+\frac{540-493\sqrt{2}}{1152}\left(\frac{11-6\sqrt{2}}{36}\right)^{n-1}\right].$$

This completes the proof of the theorem. □

4. Discussion

In recent decades, the resistance distance has attracted some attentions due to its practical applications. The spectral graph theory focuses on the interplay between the structure properties and eigenvalues of a graph. In Reference [17,18], Klein and Lovász independently found that the sum of resistance distance, namely, the Kirchhoff index, could be determined by the Laplacian eigenvalues of the graph. In later years, Chen and Zhang [14] defined the degree-Kirchhoff index. Meanwhile, they proved that the degree-Kirchhoff index could be given by the normalized Laplacian eigenvalues of the graph. Since the relationships between Kirchhoff (and degree-Kirchhoff, respectively) index and the Laplacian (normalized Laplacian, respectively) eigenvalues of the graph, the Kirchhoff and degree-Kirchhoff index are highly concerned. Y. Yang et al. [20] determined the Laplacian spectrum and Kirchhoff index of linear hexagonal chains by the decomposition theorem of the Laplacian polynomial in 2007. More surprising, they found the Wiener index of linear hexagonal chains is almost twice that of its Kirchhoff index. Reference [22] explored the normalized Laplacian spectrum and degree-Kirchhoff index of linear hexagonal chains by the decomposition theorem of normalized Laplacian polynomial. They also found the Gutman index of linear hexagonal chains is almost twice that of its degree-Kirchhoff index. Y. Pan et al. [21] constructed a crossed hexagonal by adding two pairs of crossed edges in linear hexagonal chains, and the Kirchhoff and degree-Kirchhoff indices are derived, respectively. Besides, they presented the Wiener (Gutman, respectively) index of linear crossed hexagonal chains is almost four times that of its Kirchhoff (degree-Kirchhoff, respectively) index. Applying similar methods, X. Ma et al. [26] determined the degree-Kirchhoff index and spanning trees of linear hexagonal Möbius graphs. For the results of linear phenylenes, see Reference [5,23]. Considering a more complex graph and different methods, the degree-Kirchhoff index and number of spanning trees of liner heptagonal networks were given in this paper.

5. Conclusions

In this paper, we were committed to computing the degree-Kirchhoff index and the number of spanning trees of liner heptagonal networks. Though we also used the decomposition theorem of the normalized Laplacian polynomial that appeared in the previous results, the methods of calculating the expression $\sum_{i=2}^{5n+1} \frac{1}{\alpha_i}$ were much different than previous results. The main techniques we used were Lemma 3 and elementary operations. Indeed, one can use the *Doolittle decomposition theorem* to solve the normalized Laplacian matrix of liner heptagonal networks, if one hopes to check the correction of the proofs. Additionally, it is interesting to deduce the Kirchhoff index, degree-Kirchhoff index, and the number of spanning trees of linear crossed heptagonal networks or linear heptagonal Möbius networks, due to its much more complex structures than liner heptagonal networks, as in Refs. [21,26].

Author Contributions: The authors made equal contributions to this paper.

Funding: The work was partly supported by China Postdoctoral Science Foundation (No. 2017M621579), Postdoctoral Science Foundation of Jiangsu Province (No. 1701081B) and Project of Anhui Jianzhu University (No. 2016QD116, 2017dc03 and 2017QD20).

Acknowledgments: The authors would like to express their sincere gratitude to the editor and referees for a very careful reading of the paper and for all their insightful comments and valuable suggestions, which led to a number of improvements in this paper.

Conflicts of Interest: The authors declare no conflict of interest.

References

1. He, C.; Li, S.C.; Luo, W.; Sun, L. Calculating the normalized Laplacian spectrum and the number of spanning trees of linear pentagonal chains. *J. Comput. Appl. Math.* **2018**, *344*, 381–393. [CrossRef]
2. Huang, J.; Li, S.C. The normalized Laplacians on both k-triangle graph and k-quadrilateral graph with their applications, *Appl. Math. Comput.* **2018**, *320*, 213–225.
3. Mohar, B. The Laplacian spectrum of graphs. In *Graph Theory, Combinatorics, and Applications*; Alavi, Y., Chartrand, G., Oellermann, O.R., Schwenk, A.J., Eds.; Wiley-Interscience: Hoboken, NJ, USA, 1991; Volume 2, pp. 871–898.
4. Liu, J.-B.; Pan, X.F.; Hu, F.T. The Laplacian polynomial of graphs derived from regular graphs and applications. *Ars Combin.* **2016**, *126*, 289–300.

5. Peng, Y.J.; Li, S.C. On the Kirchhoff index and the number of spanning trees of linear phenylenes. *MATCH Commun. Math. Comput. Chem.* **2017**, *77*, 765–780.
6. Liu, J.-B.; Pan, X.F.; Hu, F.T. The {1}-inverse of the Laplacian of subdivision-vertex and subdivision-edge coronae with applications. *Linear Multilinear Algebra* **2017**, *65*, 178–191. [CrossRef]
7. Chung, F.R.K. *Spectral Graph Theory*; American Mathematical Society: Providence, RI, USA, 1997.
8. Wiener, H. Structural determination of paraffin boiling points. *J. Am. Chem. Soc.* **1947**, *69*, 17–20. [CrossRef]
9. Dobrynin, A. Branchings in trees and the calculation of the Wiener index of a tree. *MATCH Commun. Math. Comput. Chem.* **2000**, *41*, 119–134.
10. Gutman, I. Selected properties of the Schultz molecular topological index. *J. Chem. Inf. Comput. Sci.* **1994**, *34*, 1087–1089. [CrossRef]
11. Klein, D.J.; Randić, M. Resistance distances. *J. Math. Chem.* **1993**, *12*, 81–95. [CrossRef]
12. Klein, D.J.; Ivanciuc, O. Graph cyclicity, excess conductance, and resistance deficit. *J. Math. Chem.* **2001**, *30*, 271–287. [CrossRef]
13. Klein, D.J. Resistance-distance sum rules. *Croat. Chem. Acta* **2002**, *75*, 633–649.
14. Chen, H.Y.; Zhang, F.J. Resistance distance and the normalized Laplacian spectrum. *Discrete Appl. Math.* **2007**, *155*, 654–661. [CrossRef]
15. Bedrosian, S. Converse of the star-mesh transformation. *IRE Trans. Circuit Theory* **1961**, *8*, 491-493. [CrossRef]
16. Huang, S.; Zhou, J.; Bu, C. Some results on Kirchhoff index and degree-Kirchhoff index. *MATCH Commun. Math. Comput. Chem.* **2016**, *75*, 207–222.
17. Gutman, I.; Mohar, B. The quasi-Wiener and the Kirchhoff indices coincide. *J. Chem. Inf. Comput. Sci.* **1996**, *36*, 982–985. [CrossRef]
18. Zhu, H.Y.; Klein, D.J.; Lukovits, I. Extensions of the Wiener number. *J. Chem. Inf. Comput. Sci.* **1996**, *36*, 420–428. [CrossRef]
19. Yang, Y.L.; Yu, T.Y. Graph theory of viscoelasticities for polymers with starshaped, multiple-ring and cyclic multiple -ring molecules. *Makromol. Chem.* **1985**, *186*, 609. [CrossRef]
20. Yang, Y.J.; Zhang, H.P. Kirchhoff Index of Linear Hexagonal Chains. *Int. J. Quantum Chem.* **2008**, *108*, 503–512. [CrossRef]
21. Pan, Y.; Li, J. Kirchhoff index, multiplicative degree-Kirchhoff index and spanning trees of the linear crossed hexagonal chains. *Int. J. Quantum Chem.* **2018**, *118*, e25787. [CrossRef]
22. Huang, J.; Li, S.C.; Sun, L. The normalized Laplacians degree-Kirchhoff index and the spanning trees of linear hexagonal chains. *Discrete Appl. Math.* **2016**, *207*, 67–79. [CrossRef]
23. Zhu, Z.X.; Liu, J.-B. The normalized Laplacian, degree-Kirchhoff index and the spanning tree numbers of generalized phenylenes. *Discrete Appl. Math.* **2019**, *254*, 256–267. [CrossRef]
24. Zhang, F. (Ed.) *The Schur Complement and Its Applications*; Springer: NewYork, NY, USA, 2005.
25. Gutman, I.; Cyvin, S.J. *Introduction to the Theory of Benzenoid Hydrocarbons*; Springer: Belin/Heidelberg, Germany, 1989.
26. Ma, X.; Bian, H. The normalized Laplacians, degree-Kirchhoff index and the spanning trees of hexagonal Möbius graphs. *Appl. Math. Comput.* **2019**, *355*, 33–46.

mathematics

MDPI

Article

The Bounds of Vertex Padmakar–Ivan Index on k-Trees

Shaohui Wang [1], Zehui Shao [2,*], Jia-Bao Liu [3] and Bing Wei [4]

[1] Department of Mathematics and Physics, Texas A&M International University, Laredo, TX 78041, USA; shaohuiwang@yahoo.com
[2] Institute of Computing Science and Technology, Guangzhou University, Guangzhou 510006, China
[3] School of Mathematics and Physics, Anhui Jianzhu University, Hefei 230601, China; liujiabaoad@163.com
[4] Department of Mathematics, The University of Mississippi, University, MS 38677, USA; bwei@olemiss.edu
* Correspondence: zshao@gzhu.edu.cn

Received: 26 January 2019 ; Accepted: 11 March 2019; Published: 1 April 2019

Abstract: The Padmakar–Ivan (PI) index is a distance-based topological index and a molecular structure descriptor, which is the sum of the number of vertices over all edges uv of a graph such that these vertices are not equidistant from u and v. In this paper, we explore the results of PI-indices from trees to recursively clustered trees, the k-trees. Exact sharp upper bounds of PI indices on k-trees are obtained by the recursive relationships, and the corresponding extremal graphs are given. In addition, we determine the PI-values on some classes of k-trees and compare them, and our results extend and enrich some known conclusions.

Keywords: extremal values; PI index; k-trees; distance

---

1. Introduction

Let G be a simple connected non-oriented graph with vertex set $V(G)$ and edge set $E(G)$. The distance $d(x,y)$ between the vertices $x,y \in V(G)$ is the minimum length of the paths between x and y in G. The oldest and most thoroughly examined molecular descriptor is Wiener index or path number [1], which was first considered in trees by Wiener in 1947 as follows: $W(G) = \sum_{\{x,y\}\subset V(G)} d(x,y)$. Compared to Wiener index, Szeged index was proposed by Gutman [2] in 1994 that, given $xy \in E(G)$, let $n_{xy}(x)$ be the number of vertices $w \in V(G)$ such that $d(x,w) < d(y,w)$, $Sz(G) = \sum_{xy\in E(G)} n_{xy}(x)n_{xy}(y)$. Based on the considerable success of Wiener index and Sz index, Khadikar proposed a new distance-based index [3] to be used in the field of nano-technology, that is edge Padmakar–Ivan (PI_e) index, $PI_e(G) = \sum_{xy\in E(G)}[n_e(x) + n_e(y)]$, where $n_e(x)$ denotes the number of edges which are closer to the vertex x than to the vertex y, and $n_e(y)$ denotes the number of edges which are closer to the vertex y than to the vertex x, respectively.

It is easy to see that the above concept does not count edges equidistant from both ends of the edge $e = xy$. Based on this idea, Khalifeh et al. [4] introduced a new PI index of vertex version that $PI(G) = PI_v(G) = \sum_{xy\in E(G)}[n_{xy}(x) + n_{xy}(y)]$. Note that, in order to obtain a good recursive formulas, we do not consider the vertices x,y for $n_{xy}(x)$ and $n_{xy}(y)$. Thus, $n_{xy}(x) + n_{xy}(y) \le n-2$.

Nowadays, Padmakar–Ivan indices are widely used in Quantitative Structure–Activity Relationship (QSAR) and Quantitative Structure–Property Relationship (QSPR) [5,6], and there are many interesting results [5,7–26] between graph theory and chemistry. For instances, Klavžar [27] provided PI-partitions and arbitrary Cartesian product. Pattabiraman and Paulraja [28] presented the formulas for vertex PI indices of the strong product of a graph and the complete multipartite graph. Ilić and Milosavljević [29] established basic properties of weighted vertex PI index and some lower and upper bounds on special graphs. Wang and Wei [30] studied vertex PI index on an extention of trees (cacti). In [31], Das and Gutman obtained a lower bound on the vertex PI index of a connected graph in terms of numbers of vertices, edges, pendent vertices, and clique number. Hoji et al. [32] provided exact formulas for the vertex PI indices of Kronecker product of a connected graph G and a complete graph. Since the tree is a basic class of graphs in mathematics and chemistry, and these results indicate that either the stars or the paths attain the maximal or minimal bounds for particular chemical indices, then a natural question is how about the situations for vertex Padmakar–Ivan index?

Because PI index is a distance-based index and not very easy to calculate, we first consider the bipartite graph G with n vertices. Note that the tree is a subclass of bipartite graphs which have no odd cycles. By the definition of $PI(G)$ and the assumption that we do not consider the vertices x, y for $n_{xy}(x)$ and $n_{xy}(y)$, one can obtain that every edge of G has the PI-value as $n-2$. Thus, the following observation is obtained.

Obervation 1. *For a bipartite graph G with n vertices and m edges, $PI(G) = (n-2)m$. In particular, if G is a tree, then $PI(G) = (n-1)(n-2)$.*

Next, we will consider the graphs with odd cycles. In particular, the general tree, k-tree, contains a lot of odd cycles. Then, we are going to consider the PI indices of k-trees and figure out whether or not a k-star or a k-path attains the maximal or minimal bound for PI-indices of k-trees. Our main results are as follows: Theorems 1 and 2 give the exact PI-values of k-stars, k-paths and k-spirals (see Definitions 1–5 below).

Theorem 1. *For any k-star S_n^k and k-path P_n^k with $n = kp + s$ vertices, where $p \geq 0$ is an integer and $s \in [2, k+1]$, we have*

$$\begin{aligned}(i) PI(S_n^k) &= k(n-k)(n-k-1),\\ (ii) PI(P_n^k) &= \tfrac{k(k+1)(p-1)(3kp+6s-2k-4)}{6} + \tfrac{(s-1)s(3k-s+2)}{3}.\end{aligned}$$

Theorem 2. *For any k-spiral $T_{n,c}^{k*}$ with $n \geq k$ vertices, where $c \in [1, k-1]$, we have*

$$PI(T_{n,c}^{k*}) = \begin{cases} \frac{(n-k)(n-k-1)(4k-n+2)}{3} & \text{if } n \in [k, 2k-c], \\ \frac{3c(n-2k+c-1)(n-2k+c)+(k-c)(2c^2+3nc-4kc+3kn-4k^2-6k+3n-2)}{3} & \text{if } n \geq 2k-c+1. \end{cases}$$

Theorem 3 proves that k-stars achieve the maximal values of PI-values for k-trees, and Theorem 4 shows that k-paths do not arrive the minimal values and certain PI-values of k-spirals are less than that of k-paths.

Theorem 3. *For any k-tree T_n^k with $n \geq k \geq 1$, we have $PI(T_n^k) \leq PI(S_n^k)$.*

Theorem 4. *For any k-spiral $T_{n,c}^{k*}$ with $n \geq k \geq 1$, we have*

$$\begin{aligned}(i) \quad & PI(P_n^k) \geq PI(T_{n,c}^{k*}) \text{ if } c \in [1, \tfrac{k+1}{2}),\\ (ii) \quad & PI(P_n^k) \leq PI(T_{n,c}^{k*}) \text{ if } c \in [\tfrac{k+1}{2}, k-1].\end{aligned}$$

2. Preliminary

In this section, we first give some notations and lemmas that are crucial in the following sections. As usual, $G = (V, E)$ is a connected finite simple undirected graph with vertex set $V = V(G)$ and edge set $E = E(G)$. Let $|G|$ or $|V|$ be the cardinality of V. For any $S \subseteq V(G)$ and $F \subseteq E(G)$, we use $G[S]$ to denote the subgraph of G induced by S, $G - S$ to denote the subgraph induced by $V(G) - S$ and $G - F$

to denote the subgraph of G obtained by deleting F. $w(G-S)$ is the number of components of $G-S$ and S is a cut set if $w(G-S)\geq 2$. For any $u,v\in V(G)$, P_{uv} is a path connecting u and v, $d(u,v)$ is the distance between u and v, $N(v)=N_G(v)=\{w\in V(G), vw\in E(G)\}$ is the neighborhood of v and $N[v]=N(v)\cup\{v\}$. For any integers a,b with $a\leq b$, the interval $[a,b]$ is the set of all integers between a and b including a,b. In addition, let $[a,b)=[a,b]-\{b\}$ and $(a,b]=[a,b]-\{a\}$. In particular, $[a,b]=\phi$ for $a>b$. $f'(x)$ is a derivative of any differentiable function $f(x)$, where x is the variable. $\lfloor x\rfloor$ is the largest integer that is less than or equal to x; $\lceil x\rceil$ is the smallest integer that is greater than or equal to x. It is clear that d is from 0 to the diameter of graphs. Other undefined notations are referred to [33].

It is commonly known that a chordal graph G with at least three vertices is a triangulated graph and contains a simplicial vertex, whose neighborhood induces a clique. During recent decades, there are many interesting studies related to chordal graphs. In 1969, Beineke and Pippert [7] gave the definition of k-trees, which is a significant subclass of chordal graphs. Now, we just give some definitions about k-trees below.

Definition 1. *For positive integers n,k with $n\geq k$, the k-tree, denoted by T_n^k, is defined recursively as follows: The smallest k-tree is the k-clique K_k. If G is a k-tree with $n\geq k$ vertices and a new vertex v of degree k is added and joined to the vertices of a k-clique in G, then the obtained graph is a k-tree with $n+1$ vertices.*

Definition 2. *For positive integers n,k with $n\geq k$, the k-path, denoted by P_n^k, is defined as follows: starting with a k-clique $G[\{v_1,v_2\dots v_k\}]$. For $i\in[k+1,n]$, the vertex v_i is adjacent to vertices $\{v_{i-1},v_{i-2}\dots v_{i-k}\}$ only.*

Definition 3. *For positive integers n,k with $n\geq k$, the k-star, denoted by S_n^k, is defined as follows: Starting with a k-clique $G[\{v_1,v_2\dots v_k\}]$ and an independent set S with $|S|=n-k$. For $i\in[k+1,n]$, the vertex v_i is adjacent to vertices $\{v_1,v_2\dots v_k\}$ only.*

Definition 4. *For positive integers n,k,c with $n\geq k$ and $c\in[1,k-1]$, let $v_1,v_2,\dots,v_{n-c}$ be the simplicial ordering of P_{n-c}^{k-c}. The k-spiral, denoted by $T_{n,c}^{k*}$, is defined as $P_{n-c}^{k-c}+K_c$, which is, $V(T_{n,c}^{k*})=\{v_1,v_2,\dots,v_n\}$ and $E(T_{n,c}^{k*})=E(P_{n-c}^{k-c})\cup E(K_c)\cup\{v_1v_l,v_2v_l,\dots,v_{n-c}v_l\}$, for $l\in[n-c+1,n]$.*

Definition 5. *Let $v\in V(T_n^k)$ be a vertex of degree k whose neighbors form a k-clique of T_n^k, then v is called a k-simplicial vertex. Let $S_1(T_n^k)$ be the set of all k-simplicial vertices of T_n^k, for $n\geq k+2$, and set $S_1(K_k)=\phi, S_1(K_{k+1})=\{v\}$, where v is any vertex of K_{k+1}. Let $G_0=G, G_i=G_{i-1}-v_i$, where v_i is a k-simplicial vertex of G_{i-1}, then $\{v_1,v_2\dots v_n\}$ is called a simplicial elimination ordering of the n-vertex graph G.*

In order to consider the PI-value of any k-tree G, let $G'=G\cup\{u\}$ be a k-tree obtained by adding a new vertex u to G. For any $v_1,v_2\in V(G)$, let $d(v_1,v_2)$ be the distance between v_1 and v_2 in G, $d'(v_1,v_2)$ be the distance between v_1 and v_2 in G'. Now, we define a function that measures the difference of PI-values of any edge relating a vertex from G to G' as follows: $f:\{w\in V(G'), xy\in E(G)\}$ to $\{1,0\}$ as follows:

$$f(w,xy)=\begin{cases}0, & \text{if} \quad w=u \text{ and } d'(x,w)=d'(y,w),\\ 0, & \text{if} \quad w\in V(G) \text{ and } d(x,w)-d'(x,w)=d(y,w)-d'(y,w),\\ 1, & \text{if} \quad \textit{otherwise}.\end{cases}$$

Using the construction of k-trees, we can derive the following lemmas. Note that $PI(xy)=n_{xy}(x)+n_{xy}(y)$ and $PI(xy)\leq n-2$.

Lemma 1. *Let xy be any edge of a k-tree G with at least $n\geq k+1$ vertices, then $PI(xy)\leq n-k-1$.*

Proof. Since every vertex of any k-tree G with at least $k+1$ vertices must be in some $(k+1)$-cliques, which is, $|N(x) \cap N(y)| \geq k-1$ for any $xy \in E(G)$, we have $PI(xy) \leq n-(k-1)-2 = n-k-1$. □

Lemma 2. *Let xy be any edge of a k-tree G with n vertices and $G' = G \cup \{u\}$ be a k-tree obtained by adding u to G. If $w \in V(G)$, then $f(w,xy) = 0$.*

Proof. By adding u to G, since G' is a k-tree, we can get that the distance of any pair of vertices of G will increase at most 1, then $f(w,xy) \leq 1$. If $w \in V(G)$, then there exists a shortest path P_{xw} or P_{yw} such that $u \notin V(P_{xw})$ or $V(P_{yw})$, that is, $f(w,xy) = 0$. □

Lemma 3. *For any k-path G with n vertices, where $n \geq k+2$, let $S_1(G) = \{v_1, v_n\}$ and $\{v_1, v_2, \ldots, v_n\}$ be the simplicial elimination ordering of G, then $d(v_i, v_j) = \lceil \frac{j-i}{k} \rceil$, for $i < j$ and $i, j \in [1,n]$. Furthermore, if $n = kp+s$ with $p \geq 1, s \in [2, k+1]$, then*

$$d(v, v_{kp+s}) = \begin{cases} p+1 & \text{if} \quad v \in \{v_1, v_2, \ldots, v_{s-1}\}, \\ p-i & \text{if} \quad v \in \{v_{ki+s}, v_{ki+s+1}, \ldots, v_{k(i+1)+s-1}\}, i \in [0, p-1]. \end{cases}$$

Proof. If $j-i \leq k$, then v_i, v_j must be in the same $(k+1)$-clique of G, and we have $d(v_i, v_j) = 1$; if $j-i \geq k+1$, then $P_{v_iv_j} = v_iv_{i+k}v_{i+2k}\cdots v_{i+(\lfloor \frac{j-i}{k} \rfloor -1)k}v_{i+\lfloor \frac{j-i}{k} \rfloor k}v_j$ is one of the shortest paths between v_i and v_j. Thus, $d(v_i, v_j) = \lceil \frac{j-i}{k} \rceil$ and Lemma 3 is proved. □

Lemma 4. *For any k-spiral $T^{k*}_{n,c}$ with n vertices and $v_i, v_j \in V(T^{k*}_{n,c})$ for $i < j$,*

$$d(v_i, v_j) = \begin{cases} 1, & \text{if} \quad j-i \leq k-c, i, j \in [1, n-c], \\ 1, & \text{if} \quad i \text{ or } j \in [n-c+1, n], \\ 2, & \text{if} \quad j-i \geq k-c+1, i, j \in [1, n-c]. \end{cases}$$

Proof. If $j-i \leq k-c$ with $i, j \in [1, n-c]$, by Definition 4, we can get that v_i, v_j must be in the same $(k+1)$-clique of G and $d(v_i, v_j) = 1$; If i or $j \in [n-c+1, n]$, without loss of generality, say v_i such that $i \in [n-c+1, n]$, then $N[v_i] = V(T^{k*}_{n,c})$, that is, $d(v_i, v_j) = 1$; If $j-i \geq k-c+1$ with $i, j \in [1, n-c]$, then $v_i \notin N(v_j)$ and $P_{v_iv_j} = v_iv_nv_j$ is one of the shortest paths between v_i and v_j, that is, $d(v_i, v_j) = 2$. Thus, Lemma 4 is proved. □

3. Main Proofs

In this section, we give the proofs of main results by inductions. For a k-tree T^k_n, if $n = k$ or $k+1$, then T^k_n is a k or (k+1)-clique, that is, $PI(T^k_n) = 0$. Thus, all of the theorems are true and we will only consider the case when $n \geq k+2$ below.

Proof of Theorem 1. For (i), let $V(S^k_n) = \{u_1, u_2, \ldots, u_n\}$, $G[\{u_1, \ldots, u_k\}]$ be a k-clique and $N(u_{l_0}) = \{u_1, u_2, \ldots, u_k\}$ for $l_0 \geq k+1$. Just by Definition 3, we can get that for $i, j \in [1,k]$, $N[u_i] = N[u_j] = V(S^k_n)$, then $PI(u_iu_j) = n_{u_iu_j}(u_i) + n_{u_iu_j}(u_j) = 0$; for $i \in [1,k]$ and $l_0 \in [k+1, n]$, $|N[u_i] - N[u_{l_0}]| = n-k-1$, then $PI(u_iu_l) = n-k-1$. Thus, we can get $PI(S^k_n) = \sum_{i,j \in [1,k]} PI(u_iu_j) + \sum_{i \in [1,k], l_0 \in [k+1,n]} PI(u_iu_{l_0}) = k(n-k)(n-k-1)$.

For (ii), we will proceed it by induction on $|P^k_n| = n \geq k+2$. If $n = k+2$, let $\{v_1, v_2, \ldots, v_{k+2}\}$ be the simplicial elimination ordering of P^k_{k+2}. By Lemma 3, we can get that $PI(v_1v_i) = 1, PI(v_iv_{i'}) = 0$ and $PI(v_iv_{k+2}) = 1$ for $i, i' \in [2, k+1]$. Thus, $PI(P^k_{k+2}) = \sum_{i=2}^{k+1} PI(v_1v_i) + \sum_{i=2}^{k+1} PI(v_iv_{k+2}) = 2k$. Assume that Theorem 1 is true for a k-path with at most $kp+s-1$ vertices, where $p \geq 1, 2 \leq s \leq k+1$. Let P^k_n be a k-path such that $|P^k_n| = kp+s$, $V(P^k_n) = \{v_1, v_2, \ldots, v_{kp+s}\}$ and $\{v_1, v_2, \ldots, v_{kp+s}\}$ be the simplicial elimination ordering of P_n^k. Set $P_{n-1}^k = P_n^k - \{v_{kp+s}\}$, then $\{v_1, v_2, \ldots, v_{kp+s-1}\}$ is the simplicial elimination ordering of P_{n-1}^k and for any edge $v_iv_j \in E(P_n^k)$, $d(v_i, v_j)$ or $d'(v_i, v_j)$ is the distance of v_i and v_j in P_{n-1}^k or P_n^k, respectively.

Let $\alpha = [\frac{k(k+1)(p-1)(3kp+6s-2k-4)}{6} + \frac{(s-1)s(3k-s+2)}{3}] - [\frac{k(k+1)(p-1)(3kp+6s-2k-10)}{6} + \frac{(s-2)(s-1)(3k-s+3)}{3}] = pk^2 + pk - k^2 - 3k + 2ks - s^2 + 3s - 2$. If we can show that by adding v_{kp+s} to P_{n-1}^k, $PI(P_n^k) = PI(P_{n-1}^k) + \alpha$, then Theorem 1 is true.

Set $w = v_{kp+s}$, $A_1 = \{v_1v_s, v_1v_{s+1}, \ldots, v_1v_{k+1}\}, A_2 = \{v_2v_s, \ldots, v_2v_{k+2}\}, \ldots, A_{s-1} = \{v_{s-1}v_s, \ldots, v_{s-1}v_{k+s-1}\}$ and $B_1 = \{v_1v_2, v_1v_3, \ldots, v_1v_{s-1}\}, B_2 = \{v_2v_3, \ldots, v_2v_{s-1}\}, \ldots, B_{s-2} = \{v_{s-2}v_{s-1}\}, B_{s-1} = \phi$. By Definition 2 and Lemma 3, we have $d'(v_1, v_{kp+s}) = p+1, d'(v_s, v_{kp+s}) = p$ and $d'(v_1, v_{kp+s}) = p+1, d'(v_2, v_{kp+s}) = p+1$, that is, $d'(v_1, v_{kp+s}) \neq d'(v_s, v_{kp+s})$ and $d'(v_1, v_{kp+s}) = d'(v_2, v_{kp+s})$. Thus, $f(w, v_1v_s) = 1$ and $f(w, v_1v_2) = 0$. Similarly, for any edge $v_{h_1}v_{h_2} \in \cup_{i=1}^{s-1} A_i$ with $h_1 < h_2$, we have $d'(v_{h_1}, v_{kp+s}) \neq d'(v_{h_2}, v_{kp+s})$, that is, $f(w, v_{h_1}v_{h_2}) = 1$; For $v_{h_1}v_{h_2} \in \cup_{i=1}^{s-1} B_i$ with $h_1 < h_2$, we have $d'(v_{h_1}, v_{kp+s}) = d'(v_{h_2}, v_{kp+s})$, that is, $f(w, v_{h_1}v_{h_2}) = 0$. Thus, we can get that

$$f(v_{kp+s}, xy) = \begin{cases} 1, & \text{if} \quad xy \in \cup_{i=1}^{s-1} A_i, \\ 0, & \text{if} \quad xy \in \cup_{i=1}^{s-1} B_i. \end{cases}$$

For $t \in [0, p-2]$, set $A_{kt+s} = \{v_{kt+s}v_{k(t+1)+s}\}$, $A_{kt+s+1} = \{v_{kt+s+1}v_{k(t+1)+s}, v_{kt+s+1}v_{k(t+1)+s+1}\}$, $\ldots, A_{k(t+1)+s-1} = \{v_{k(t+1)+s-1}v_{k(t+1)+s}, v_{k(t+1)+s-1}v_{k(t+1)+s+1}, \ldots, v_{k(t+1)+s-1}v_{k(t+2)+s-1}\}$, and $B_{kt+s} = \{v_{kt+s}v_{kt+s+1}, \ldots, v_{kt+s}v_{k(t+1)+s-1}\}, B_{kt+s+1} = \{v_{kt+s+1}v_{kt+s+2}, \ldots, v_{kt+s+1}v_{k(t+1)+s-1}\}, \ldots, B_{k(t+1)+s-2} = \{v_{k(t+1)+s-2}v_{k(t+1)+s-1}\}, B_{k(t+1)+s-1} = \phi$. For $t = 0$ and by Lemma 3, we have $d'(v_s, v_{kp+s}) = p$, $d'(v_{k+s}, v_{kp+s}) = p-1$ and $d'(v_s, v_{kp+s}) = p, d'(v_{s+1}, v_{kp+s}) = p$, that is, $d'(v_s, v_{kp+s}) \neq d'(v_{k+s}, v_{kp+s})$ and $d'(v_s, v_{kp+s}) = d'(v_{s+1}, v_{kp+s})$. Thus, $f(w, v_sv_{k+s}) = 1$ and $f(w, v_sv_{s+1}) = 0$. similarly, for any edge $v_{h_1}v_{h_2} \in \cup_{i=kt+s}^{k(t+1)+s-1} A_i$ with $h_1 < h_2$, we have $d'(v_{h_1}, v_{kp+s}) \neq d'(v_{h_2}, v_{kp+s})$, that is, $f(w, v_{h_1}v_{h_2}) = 1$; for $v_{h_1}v_{h_2} \in \cup_{i=kt+s}^{k(t+1)+s-1} B_i$ with $h_1 < h_2$, we have $d'(v_{h_1}, v_{kp+s}) = d'(v_{h_2}, v_{kp+s})$, that is, $f(w, v_{h_1}v_{h_2}) = 0$. Thus, we can get that

$$f(v_{kp+s}, xy) = \begin{cases} 1, & \text{if} \quad xy \in \cup_{i=kt+s}^{k(t+1)+s-1} A_i, \\ 0, & \text{if} \quad xy \in \cup_{i=kt+s}^{k(t+1)+s-1} B_i. \end{cases}$$

Next, we consider the edges in the $(k+1)$-clique $P_n^k[N[v_{kp+s}]]$. For any edge $v_{h_1}v_{h_2}$ with $h_1, h_2 \in [k(p-1)+s, kp+s-1]$, we have $d'(v_{h_1}, v_{kp+s}) = d'(v_{h_2}, v_{kp+s}) = 1$, that is, $f(w, v_{h_1}v_{h_2}) = 0$. For any edge v_hv_{kp+s} with $h \in [k(p-1)+s, kp]$, by Lemma 3, we can obtain that $d'(v_1, v_h) = p, d'(v_1, v_{kp+s}) = p+1$, $d'(v_{h-k}, v_h) = 1, d'(v_{h-k}, v_{kp+s}) = 2$ and when $h \neq k(p-1)+s$, $d'(v_{k(p-1)+s}, v_h) = 1, d'(v_{k(p-1)+s}, v_{kp+s}) = 1$, that is, $d'(v_1, v_h) \neq d'(v_1, v_{kp+s})$, $d'(v_{h-k}, v_h) \neq d'(v_{h-k}, v_{kp+s})$ and $d'(v_{k(p-1)+s}, v_h) = d'(v_{k(p-1)+s}, v_{kp+s})$. Similarly, we get that for $j \in [1, p-1]$, $j' \in [1, p]$ and $l \neq h$,

$$\begin{cases} d'(v_l, v_h) \neq d'(v_l, v_{kp+s}) & \text{if} \quad l \in [1, s-1] \cup [h-jk, k(p-j)+s-1], \\ d'(v_l, v_h) = d'(v_l, v_{kp+s}) & \text{if} \quad l \in [k(p-j')+s, h-j'k+k-1] \cup [h+1, kp+s-1]. \end{cases}$$

Thus, if $v_h = v_{k(p-1)+s}$, then $d'(v_l, v_{k(p-1)+s}) \neq d'(v_l, v_{kp+s})$ with $l \in [1, s-1] \cup \{\cup_{j=1}^{p-1}[k(p-1)+s-jk, (p-j)k+s-1]\} = [1, (p-1)k+s-1]$ and $d'(v_l, v_{k(p-1)+s}) = d'(v_l, v_{kp+s})$ with $l \in [(p-1)k+s+1, kp+s]$, that is, $PI(v_{k(p-1)+s}v_{kp+s}) = (p-1)k+s-1$; similarly, we can obtain that $PI(v_{k(p-1)+s+1}v_{kp+s}) = (p-1)(k-1)+s-1; PI(v_{k(p-1)+s+2}v_{kp+s}) = (p-1)(k-2)+s-1; \ldots; PI(v_{kp}v_{kp+s}) = (p-1)s+s-1$.

For any edge v_hv_{kp+s} with $h \in [kp+1, kp+s-1]$, by Lemma 3, we can obtain that $d'(v_{h-k}, v_h) = 1, d'(v_{h-k}, v_{kp+s}) = 2$ and $d'(v_{k(p-1)+s}, v_h) = 1, d'(v_{k(p-1)+s}, v_{kp+s}) = 1$, that is, $d'(v_{h-k}, v_h) \neq$

$d'(v_{h-k}, v_{kp+s})$ and $d'(v_{k(p-1)+s}, v_h) = d'(v_{k(p-1)+s}, v_{kp+s})$. Similarly, we get that for $j'' \in [1,p]$ and $l \neq h$,

$$\begin{cases} d'(v_l, v_h) \neq d'(v_l, v_{kp+s}) & \text{if} \quad l \in [h - j''k, k(p-j'') + s - 1], \\ d'(v_l, v_h) = d'(v_l, v_{kp+s}) & \text{if} \quad l \in [k(p-j'') + s, h - j''k + k - 1] \cup [h+1, kp+s-1]. \end{cases}$$

Thus, if $v_h = v_{kp+1}$, then $d'(v_l, v_{kp+1}) \neq d'(v_l, v_{kp+s})$ for $l \in \cup_{j''=1}^{p}[kp+1-j''k, k(p-j'')+s-1]$ and $d'(v_l, v_{kp+1}) = d'(v_l, v_{kp+s})$ with $l \in \{\cup_{j''=1}^{p}[k(p-j'')+s, k(p+1-j'')]\} \cup [h+1, kp+s-1]$, that is, $PI(v_{kp+1}v_{kp+s}) = (s-1)p$; similarly, we have $PI(v_{kp+1}v_{kp+s}) = (s-2)p; \ldots; PI(v_{kp+s-2}v_{kp+s}) = 2p; PI(v_{kp+s-1}v_{kp+s}) = p$.

Set $w \in V(P_{n-1}^k)$, if $xy \in E(P_n^k)$ with x or $y \neq v_{kp+s}$, by Lemma 2, we have $f(w, xy) = 0$. Thus,

$$\begin{aligned} PI(P_n^k) - PI(P_{n-1}^k) &= \textstyle\sum_{xy \in \cup_{i=1}^{k(p-1)+s-1}(A_i \cup B_i)} f(w, xy) + PI(v_{k(p-1)+s}v_{kp+s}) \\ &\quad + PI(v_{k(p-1)+s+1}v_{kp+s}) + \cdots + PI(v_{kp+s-1}v_{kp+s}) \\ &= [(k+2-s) + (k+3-s) + \cdots + k] + (1+2+\cdots+k)(p-1) \\ &\quad + [k(p-1)+s-1] + [(k-1)(p-1)+s-1] + [(k-2)(p-1)+s \\ &\quad -1] + \cdots + [s(p-1)+s-1] + [(s-1)p + (s-2)p + \cdots + 2p + p] \\ &= pk^2 + pk - k^2 - 3k + 2ks - s^2 + 3s - 2 \\ &= \alpha. \end{aligned}$$

Thus, $PI(P_n^k) = \frac{k(k+1)(p-1)(3kp+6s-2k-4)}{6} + \frac{(s-1)s(3k-s+2)}{3}$, for $|P_n^k| = kp+s$ and Theorem 1 is proved. □

Proof of Theorem 2. We will proceed with it by induction on $n \geq k+2$. If $n = k+2$, by Definition 4, we have $T_{n,c}^{k*}$ is also a k-path, that is, $PI(T_{n,c}^{k*}) = 2k$. If $n \geq k+3$, assume that Theorem 2 is true for the k-spiral with at most $n-1$ vertices, we will consider $T_{n,c}^{k*}$ with n vertices. Let $T_{n,c}^{k*}$ be a k-spiral with $V(T_{n,c}^{k*}) = V(T_{n-1,c}^{k*}) \cup \{v\}$ and $E(T_{n,c}^{k*}) = E(T_{n-1,c}^{k*}) \cup \{vv_{n-1}, vv_{n-2}, \ldots, vv_{n-k}\}$ such that $v_1, v_2, \ldots, v_{n-c-1}$ is the simplicial ordering of P_{n-c-1}^{k-c}, where $T_{n-1,c}^{k*} = P_{n-c-1}^{k-c} + K_c$ with $V(T_{n-1,c}^{k*}) = \{v_1, v_2, \ldots, v_{n-1}\}$ and $E(T_{n-1,c}^{k*}) = E(P_{n-c-1}^{k-c}) \cup E(K_c) \cup \{v_1v_l, v_2v_l, \ldots, v_{n-c-1}v_l\}$ for $l \in [n-c, n-1]$. For any edge $v_iv_j \in E(T_{n,c}^{k*})$, $d(v_i, v_j)$ or $d'(v_i, v_j)$ is the distance of v_i and v_j in $T_{n-1,c}^{k*}$ or $T_{n,c}^{k*}$, respectively.

For $k+2 \leq n \leq 2k-c$, let $\gamma = \frac{(n-k)(n-k-1)(4k-n+2)}{3} - \frac{(n-k-1)(n-k-2)(4k-n+3)}{3} = (n-k-1)(3k-n+2)$. If we can show that by adding v to $T_{n-1,c}^{k*}$, $PI(T_{n,c}^{k*}) = PI(T_{n-1,c}^{k*}) + \gamma$, then Theorem 2 is true.

Set $w = v$ and let $l \in [n-c, n-1]$, by Lemma 4, we have $d'(v_l, v) = 1$ and $d'(v_i, v) = 2$ for $i \in [1, n-k-1]$, that is, $f(w, v_lv_i) = 1$; $d'(v_l, v) = d'(v_i, v) = 1$ for $i \in [n-k, n-1]$, that is, $f(w, v_lv_i) = 0$. Set $C_1 = \{v_1v_2, v_1v_3, \ldots, v_1v_{n-k-1}\}, C_2 = \{v_2v_3, v_2v_4, \ldots, v_2v_{n-k-1}\}, \ldots, C_{n-k-2} = \{v_{n-k-2}v_{n-k-1}\}, C_{n-k-1} = \phi$, $D_1 = \{v_1v_{n-k}, v_1v_{n-k+1}, \ldots, v_1v_{k-c+1}\}, D_2 = \{v_2v_{n-k}, v_2v_{n-k+1}, \ldots, v_2v_{k-c+2}\}, \ldots, D_{n-k-1} = \{v_{n-k-1}v_{n-k}, v_{n-k-1}v_{n-k+1}, \ldots, v_{n-k-1}v_{n-c-1}\}$. By Lemma 4, we have $d'(v_1, v) = d'(v_2, v) = 2$ and $d'(v_{n-k}, v) = 1$, that is, $f(w, v_1v_2) = 0$ and $f(w, v_1v_{n-k}) = 1$. Similarly, for $v_{h_1}v_{h_2} \in \cup_{i=1}^{n-k-1}C_i$ with $h_1 < h_2$, we have $d'(v_{h_1}, v) = d'(v_{h_2}, v) = 2$, that is, $f(w, v_{h_1}v_{h_2}) = 0$; for $v_{h_1}v_{h_2} \in \cup_{i=1}^{n-k-1}D_i$ with $h_1 < h_2$, we have $d'(v_{h_1}, v) = 2$ and $d'(v_{h_2}, v) = 1$, that is, $f(w, v_{h_1}v_{h_2}) = 1$. Set $C_{n-k} = \{v_{n-k}v_{n-k+1}, v_{n-k}v_{n-k+2}, \ldots, v_{n-k}v_{n-c-1}\}, C_{n-k+1} = \{v_{n-k+1}v_{n-k+2}, v_{n-k+1}v_{n-k+3}, \ldots, v_{n-k+1}v_{n-c-1}\}, \ldots, C_{n-c-2} = \{v_{n-c-2}v_{n-c-1}\}$. By Lemma 4, we have $d'(v_{n-k}, v) = d'(v_{n-k-1}, v) = 1$, that is, $f(w, v_{n-k}v_{n-k-1}) = 0$. Similarly, for $v_{h_1}v_{h_2} \in \cup_{i=n-k}^{n-c-2}C_i$ with $h_1 < h_2$, we have $d'(v_{h_1}, v) = d'(v_{h_2}, v) = 1$, that is, $f(w, v_{h_1}v_{h_2}) = 0$.

Set $E_1 = \{vv_i, i \in [n-k, n-c-1]\}$, by Lemma 4, we have $d'(v_i, v) = 2, d'(v_i, v_{n-k}) = 1$ for $i \in [1, n-k-1]$ and $d'(v_j, v) = d'(v_j, v_{n-k}) = 1$ for $i \in [n-k+1, n]$. Thus, $PI(v_{n-k}v) = n-k-1$. Similarly, $PI(v_{n-k+1}v) = PI(v_{n-k+2}v) = \cdots = PI(v_{k-c+1}v) = n-k-1$. In addition, by Lemma 4, we have $d'(v_i, v) = 2, d'(v_i, v_{k-c+2}) = 1$ for $i \in [2, n-k-1]$, $d'(v_1, v) = d'(v_1, v_{k-c+2}) = 2$ and $d'(v_j, v) = d(v_j, v_{k-c+2}) = 1$ for $j \in [n-k, n]$. Thus, $PI(v_{k-c+2}v) = n-k-2$. Similarly, we have $PI(v_{k-c+3}v) = n-k-3, PI(v_{k-c+4}v) = n-k-4, \ldots, PI(v_{n-c-1}v) = 1$. Set $E_2 = \{vv_l, l \in$

$[n-c, n-1]\}$, since $N[v_l] - N[v] = n-k-1$, we have $PI(vv_l) = n-k-1$. Set $E_3 = \{v_i v_l, i \in [1, n-c-1], l \in [n-c, n-1]\}$, by Lemma 4, we have $d'(v_i, v) = 2$ for $i \in [1, n-k-1]$, $d'(v_i, v) = 1$ for $i \in [n-k, n-c-1]$, $d'(v_l, v) = 1$ for $l \in [n-c, n-1]$. Thus, $f(w, v_i v_l) = 1$ for $i \in [1, n-k-1]$ and $f(w, v_i v_l) = 0$ for $i \in [n-k, n-c-1]$.

Set $w \in V(T_n^{k*}) - \{v\}$, if $xy \in E(T_{n,c}^{k*})$ with x or $y \neq v$, by Lemma 2, we have $f(w, xy) = 0$. Thus,

$$\begin{aligned} PI(T_n^{k*}) - PI(T_{n-1}^{k*}) &= \textstyle\sum_{xy \in \cup_{i=1}^{n-c-2} C_i} f(w, xy) + \sum_{xy \in \cup_{i=1}^{n-k-1} D_i} f(w, xy) + \sum_{xy \in E_1 \cup E_2} PI(xy) + \\ &\quad \textstyle\sum_{xy \in E_3} f(w, xy) \\ &= 0 + [(2k-n-c+2) + (2k-n-c+3) + \cdots + (k-c)] \\ &\quad + [1 + 2 + \cdots + (n-k-2) + (n-k-1)(2k-n-c+2)] \\ &\quad + c(n-k-1) + c(n-k-1) \\ &= (n-k-1)(3k-n+2) \\ &= \gamma, \end{aligned}$$

and Theorem 2 is proved.

For $n \geq 2k-c+1$, let $\sigma = \frac{3c(n-2k+c-1)(n-2k+c)+(k-c)(2c^2+3nc-4kc+3kn-4k^2-6k+3n-2)}{3} - \frac{3c(n-2k+c-2)(n-1-2k+c)+(k-c)(2c^2+3(n-1)c-4kc+3k(n-1)-4k^2-6k+3n-2)}{3} = k^2 - 4kc + c^2 + 2nc - 3c + k$. If we can show that by adding v to $T_{n-1,c}^{k*}$, $PI(T_{n,c}^{k*}) = PI(T_{n-1,c}^{k*}) + \sigma$, then Theorem 2 is proved.

Set $w = v$, by Lemma 4, we have $d'(v_l, v) = 1$ for $l \in [n-c, n-1]$, $d'(v_i, v) = 2$ for $i \in [1, n-k-1]$ and $d'(v_j, v) = 1$ for $j \in [n-k, n-c-1]$. Thus, $f(w, v_l v_i) = 1$ and $f(w, v_l v_j) = 0$. Set $C_1 = \{v_1v_2, v_1v_3, \ldots, v_1v_{k-c+1}\}, C_2 = \{v_2v_3, v_2v_4, \ldots, v_2v_{k-c+2}\}, \ldots, C_{n-2k+c-1} = \{v_{n-2k+c-1}v_{n-2k+c}, v_{n-2k+c-1}v_{n-2k+c+1}, \ldots, v_{n-2k+c-1}v_{n-k-1}\}, C_{n-2k+c} = \{v_{n-2k+c}v_{n-2k+s+1}, v_{n-2k+c}v_{n-2k+s+2}, \ldots, v_{n-2k+c}v_{n-k-1}\}, C_{n-2k+c+1} = \{v_{n-2k+c+1}v_{n-2k+c+2}, v_{n-2k+c+1}v_{n-2k+c+3}, \ldots, v_{n-2k+c+1}v_{n-k-1}\}, \ldots, C_{n-k-1} = \phi, D_{n-2k+c} = \{v_{n-2k+c}v_{n-k}\}, D_{n-2k+c+1} = \{v_{n-2k+c+1}v_{n-k}, v_{n-2k+c+1}v_{n-k+1}\}, \ldots, D_{n-k-1} = \{v_{n-k-1}v_{n-k}, v_{n-k-1}v_{n-k+1}, \ldots, v_{n-k-1}v_{n-c-1}\}$.

By Lemma 4, we can get that $d'(v_1, v) = d'(v_2, v) = 2$ and $d'(v_{n-k}, v) = 1$, that is, $f(w, v_1v_2) = 0$ and $f(w, v_1v_{n-k}) = 1$. Similarly, for $v_{h_1}v_{h_2} \in \cup_{i=1}^{n-k-1} C_i$ with $h_1 < h_2$, we have $d'(v_{h_1}, v) = d'(v_{h_2}, v) = 2$, that is, $f(w, v_{h_1}v_{h_2}) = 0$; for $v_{h_1}v_{h_2} \in \cup_{i=n-2k+c}^{n-k-1} D_i$ with $h_1 < h_2$, we have $d'(v_{h_1}, v) = 2$ and $d'(v_{h_2}, v) = 1$, that is, $f(w, v_{h_1}v_{h_2}) = 1$. Set $C_{n-k} = \{v_{n-k}v_{n-k+1}, v_{n-k}v_{n-k+2}, \ldots, v_{n-k}v_{n-c-1}\}, C_{n-k+1} = \{v_{n-k+1}v_{n-k+2}, v_{n-k+1}v_{n-k+3}, \ldots, v_{n-k+1}v_{n-c-1}\}, \ldots, C_{n-c+2} = \{v_{n-c-2}v_{n-c-1}\}$. By Lemma 4, we can get that $d'(v_{n-k}, v) = d'(v_{n-k+1}, v) = 1$, that is, $f(w, v_{n-k}v_{n-k+1}) = 0$. Similarly, for $v_{h_1}v_{h_2} \in \cup_{i=n-k}^{n-c-2} C_i$ with $h_1 < h_2$, we have $d'(v_{h_1}, v) = d'(v_{h_2}, v) = 1$, that is, $f(w, v_{h_1}v_{h_2}) = 0$.

Set $E_1 = \{vv_i, i \in [n-k, n-c-1]\}$, by Lemma 4, we have $d'(v, v_{n-k-1}) = 2, d'(v_{n-c-1}, v_{n-k-1}) = 1$, $d'(v, v_i) = d'(v_{n-c-1}, v_i) = 1$ for $i \in [n-k, n-c-2] \cup [n-c, n-1]$ and $d'(v, v_j) = d(v_{n-c-1}, v_j) = 2$ for $j \in [1, n-k-2]$. Thus, $PI(vv_{n-c-1}) = 1$. Similarly, we have $PI(vv_{n-c-2}) = 2, PI(vv_{n-c-3}) = 3, \ldots, PI(vv_{n-k}) = k-c$. Set $E_2 = \{vv_l, l \in [n-c, n-1]\}$, since $N[v_l] - N[v] = n-k-1$, we have $PI(vv_l) = n-k-1$. Set $E_3 = \{v_iv_l, i \in [1, n-c-1], l \in [n-c, n-1]\}$, by Lemma 4, we have $d'(v, v_i) = 2, d'(v, v_l) = 1$ for $i \in [1, n-k-1]$ and $d'(v, v_i) = d'(v, v_l) = 1$ for $i \in [n-k, n-c-1]$. Thus, $f(w, v_iv_l) = 1$ for $i \in [1, n-k-1]$ and $f(w, v_iv_l) = 0$ for $i \in [n-k, n-c-1]$.

Set $w \in V(T_n^{k*}) - \{v\}$, if $xy \in E(T_{n,c}^{k*})$ with x or $y \neq v$, by Lemma 2, we have $f(w, xy) = 0$. Thus,

$$\begin{aligned} PI(T_n^{k*}) - PI(T_{n-1}^{k*}) &= \textstyle\sum_{xy \in \cup_{i=1}^{n-c-2} C_i} f(w, xy) + \sum_{xy \in \cup_{i=n-2k+c}^{n-k-1} D_i} f(w, xy) + \sum_{xy \in E_1 \cup E_2} PI(xy) \\ &\quad \textstyle + \sum_{xy \in E_3} f(w, xy) \\ &= 0 + [1 + 2 + 3 + \cdots + (k-c)] + [1 + 2 + 3 + \cdots + (k-c)] \\ &\quad + c(n-k-1) + c(n-k-1) \\ &= k^2 - 4kc + c^2 + 2nc - 3c + k \\ &= \sigma, \end{aligned}$$

and Theorem 2 is proved. □

Proof of Theorem 3. For $n \geq k+2$, we will proceed it by introduction on $|T_n^k| = n$. If $n = k+2$, T_n^k is also a k-path, that is, $PI(T_n^k) = 2k$. If $n \geq k+3$, assume that Theorem 3 is true for the k-tree with at most $n-1$ vertices, let $v \in S_1(T_n^k)$ and $T_{n-1}^k = T_n^k - v$, by the induction hypothesis, we have $PI(T_{n-1}^k) \leq PI(S_{n-1}^k) = k(n-k-1)(n-k-2)$. By adding back v, let $N(v) = \{x_1, x_2, \ldots, x_k\}$ and $w = v$. Since $T_n^k[v, x_1, x_2, \ldots, x_k]$ is a $(k+1)$-clique, we have $f(w, x_i x_j) = 0$ for $i, j \in [1, k]$. By Lemmas 1 and 2, we can obtain that $PI(vx_i) \leq n-k-1$ with $i \in [1, k]$ and $f(w, xy) \leq 1$ for any edge $xy \in E(T_n^k) - E(T_n^k[v, x_1, x_2, \ldots, x_k])$. Next, set $w \in V(T_n^k) - \{v\}$, by Lemma 2, if $xy \in E(T_n^k)$ with x or $y \neq v$, we have $f(w, xy) = 0$. Since $|E(T_n^k) - E(T_n^k[v, x_1, x_2, \ldots, x_k])| = k(n-k-1)$, we have

$$\begin{aligned}
PI(T_n^k) &= PI(T_{n-1}^k) + \textstyle\sum_{xy \in E(T_n^k - \{vx_i, i \in [1,k]\})} f(w, xy) + \sum_{i=1}^k PI(vx_i) \\
&\leq PI(S_{n-k}^k) + k(n-k-1) + k(n-k-1) \\
&= k(n-k-1)(n-k-2) + k(n-k-1) + k(n-k-1) \\
&= k(n-k)(n-k-1) \\
&= PI(S_n^k).
\end{aligned}$$

Thus, this finishes the proof of Theorem 3. □

Proof of Theorem 4. For $k = 1$ and by Fact 1, we can get that every tree with the same vertices has the same PI-value; then, Theorem 4 is obvious; for $k \geq 2$, if $k+2 \leq n \leq 2k-c$, let $n = kp + s$ with $p = 1$ and $s = n-k$, by Theorem 1, we have $PI(P_n^k) - PI(T_{n,c}^{k*}) = \frac{(s-1)s(3k-s+2)}{3} - \frac{(n-k)(n-k-1)(4k-n+2)}{3} = \frac{(n-k-1)(n-k)[3k-(n-k)+2]}{3} - \frac{(n-k)(n-k-1)(4k-n+2)}{3} = 0$, and Theorem 4 is true. If $n \geq 2k-c+1$, $p = \frac{n-s}{k}$ and by Theorems 1 and 2, define the new functions as follows: for $z \geq 2k-c+1$, $1 \leq c \leq k-1$ and $2 \leq s \leq k+1$,

$$\begin{aligned}
g(z) &= \tfrac{(k+1)(z-s-k)(3z+3s-2k-4)}{6} + \tfrac{s(s-1)(3k-s+2)}{3}, \\
h(z,c) &= c(z-2k+c-1)(z-2k+c) + \tfrac{(k-c)(2c^2+3zc-4kc+3kz-4k^2-6k+3z-2)}{3}, \\
l(z,c) &= g(z) - h(z,c) \\
&= (\tfrac{k}{2} + \tfrac{1}{2} - c)z^2 + (-c^2 + 2c + 4kc - \tfrac{11k^2}{6} - \tfrac{5k}{2} - \tfrac{2}{3})z \\
&+ \tfrac{ks^2}{2} - \tfrac{k^2 s}{6} - \tfrac{ks}{2} + \tfrac{5k^3}{3} + \tfrac{5k^2}{3} + \tfrac{s^2}{2} + \tfrac{4k}{3} - \tfrac{s^3}{3} - 6k^2 c + 3kc^2 - \tfrac{c^3}{3} - 4kc + c^2 - \tfrac{2c}{3}, \\
l_z(z) &= l_z(z,c) \\
&= (k+1-2c)z - c^2 + 2c + 4kc - \tfrac{11k^2}{6} - \tfrac{5k}{2} - \tfrac{2}{3}.
\end{aligned}$$

Then, it is enough to determine whether or not $l(z,c) \geq 0$ is true. By some calculations, we can obtain the following claim:

Claim 1. *$z_1 = 2k-c+1$, $z_2 = 2k-c+2$ are the two roots of $l(z,c) = 0$ with $c \neq \frac{k+1}{2}$.*

Proof. For any $c \in [1, k-1]$, let $z_1 = 2k-c+1$, $z_2 = 2k-c+2$, and we have $l(2k-c+1, c) = 0$, $l(2k-c+2, c) = 0$. If $c \neq \frac{k+1}{2}$, then Claim is true. □

For fixed $c \in [1, \frac{k+1}{2})$, that is, $\frac{k}{2} + \frac{1}{2} - c > 0$, then the function of $l(z,c)$ about z is open up. Since z is an integer and by Fact 2, we have $l(z,c) \geq 0$ for $z \geq 2k-c+1$ and Theorem 4 is true; if $c = \frac{k+1}{2}$ and $k \geq 1$, we have $l_z(z) = \frac{1-k^2}{12} \leq 0$, that is, $l(z, \frac{k+1}{2})$ is decreasing about z. By the proof of Fact 2, we have $l(2k-c+1, c) = 0$. For $z \geq 2k-c+1$, we can get that $l(z, \frac{k+1}{2}) \leq l(2k-c+1, \frac{k+1}{2}) = 0$ and Theorem 4 is true; for fixed $c \in (\frac{k+1}{2}, k-1]$, that is, $\frac{k}{2} + \frac{1}{2} - c < 0$, then the function of $l(z,c)$ about z is open down. Since z is an integer and by Claim, we can obtain that $l(z,c) \leq 0$ for $z \geq 2k-c+1$ and this finishes the proof of Theorem 4. □

4. Conclusions

We can see that the k-stars attain the maximal values of PI-values for k-trees. One of the guesses is that the k-paths attain the minimal values. Actually, it is not the case and some PI-values of k-spirals is even smaller than that of k-paths. Meanwhile, not all PI-values of k-spirals are less than the values of all other k-trees. This fact indicates an interesting problem—which type of k-trees will achieve the minimum PI-value?

Author Contributions: S.W. contributes for supervision, methodology, validation, project administration and formal analysing. S.W., Z.S., J.-B.L., B.W. contribute for resources, investigation some computations and wrcte the initial draft of the paper which were investigated and approved by S.W. and B.W. wrote the final draft.

Funding: This research was funded by Natural Science Foundation of Guangdong Province under grant 2018A0303130115.

Conflicts of Interest: The authors declare no conflict of interest.

References

1. Wiener, H. Structural Determination of Paraffin Boiling Points. *J. Am. Chem. Soc.* **1947**, *69*, 17–20. [CrossRef]
2. Gutman, I. A formula for the Wiener number of trees and its extension to graphs containing cycles. *Graph Theory Notes* **1994**, *27*, 9–15.
3. Khadikar, P.V. On a Novel Structural Descriptor PI. *Nat. Acad. Sci. Lett.* **2000**, *23*, 113–118.
4. Khalifeh, M.H.; Yousefi-Azari, H.; Ashrafi, A.R. Vertex and edge PI indices of Cartesian product graphs. *Discret. Appl. Math.* **2008**, *156*, 1780–1789. [CrossRef]
5. Das, K.C.; Gutman, I. Some properties of the second Zagreb index. *MATCH Commun. Math. Comput. Chem.* **2004**, *52*, 103–112.
6. Song, L.; Staton, W.; Wei, B. Independence polynomials of k-tree related graphs. *Discret. Appl. Math.* **2010**, *158*, 943–950. [CrossRef]
7. Beineke, L.W.; Pippert, R.E. The number labeled k-dimentional trees. *J. Comb. Theory* **1969**, *6*, 200–205. [CrossRef]
8. de Caen, D. An upper bound on the sum of squares of degrees in a graph. *Discret. Math.* **1998**, *185*, 245–248. [CrossRef]
9. Das, K.C. Maximizing the sum of the squares of the degrees of a graph. *Discret. Math.* **2004**, *285*, 57–66. [CrossRef]
10. Gutman, I. Multiplicative Zagreb indices of trees. *Bull. Soc. Math. Banja Luka* **2011**, *18*, 17–23.
11. Li, X.; Shi, Y. A survey on the randic index. *MATCH Commun. Math. Comput. Chem.* **2008**, *59*, 127–156.
12. Li, X.; Li, Y. The asymptotic behavior of the Estrada index for trees. *Bull. Malays. Math. Sci. Soc.* **2013**, *36*, 97–106.
13. Wang, S.; Wang, C.; Liu, J. On extremal multiplicative Zagreb indices of trees with given domination number. *Appl. Math. Comput.* **2018**, *332*, 338–350. [CrossRef]
14. Lang, R.; Deng, X.; Lu, H. Bipartite graphs with the maximal value of the second Zagreb index. *Bull. Malays. Math. Sci. Soc.* **2013**, *36*, 1–6.
15. Ma, G.; Bian, Q.; Wang, J. Bounds on the PI index of unicyclic and bicyclic graphs with given girth. *Discret. Appl. Math.* **2017**, *230*, 156–161. [CrossRef]
16. Shi, Y. Note on two generalizations of the randic index. *Appl. Math. Comput.* **2015**, *265*, 1019–1025. [CrossRef]
17. Shao, Z.; Wu, P.; Gao, Y.; Gutman, I.; Zhang, X. On the maximum ABC index of graphs without pendent vertices. *Appl. Math. Comput.* **2017**, *315*, 298–312. [CrossRef]
18. Shao, Z.; Wu, P.; Zhang, X.; Dimitrov, D.; Liu, J.B. On the maximum ABC index of graphs with prescribed size and without pendent vertices. *IEEE Access* **2018**, *6*, 27604–27616. [CrossRef]
19. Estes, J.; Wei, B. Sharp bounds of the Zagreb indices of k-trees. *J. Comb. Optim.* **2014**, *27*, 271–291. [CrossRef]
20. Gutman, I.; Trinajstić, N. Graph theory and molecular orbitals. Total π-electron energy of alternant hydrocarbons. *Chem. Phys. Lett.* **1972**, *17*, 535–538. [CrossRef]
21. Hosoya, H. Topological Index. A New Proposed Quantity Characterizing Topological Nature of Structural Isomers of Saturated Hydrocarbons. *Bull. Chem. Soc. Jpn.* **1972**, *45*, 2332–2339.
22. Iranmanesh, A.; Hosseinzadeh, M.A.; Gutman, I. On multiplicative Zagreb indices of graphs. *Iran. J. Math. Chem.* **2012**, *7*, 145–154.
23. Ahmad, Y.; Ali, U.; bilal, M.; Zafar, S.; Zahid, Z. Some new standard graphs labeled by 3-total edge product cordial labeling. *Appl. Math. Nonlinear Sci.* **2017**, *2*, 61–72. [CrossRef]

24. Sudhakar, S.; Francis, S.; Balaji, V. Odd mean labeling for two star graph. *Appl. Math. Nonlinear Sci.* **2017**, *2*, 195–200. [CrossRef]
25. Basavanagoud, B.; Desai, V.R.; Patil, S. (β, α)-Connectivity Index of Graphs. *Appl. Math. Nonlinear Sci.* **2017**, *2*, 21–30. [CrossRef]
26. Zhou, S.; Xu, L.; Xu, Y. A sufficient condition for the existence of a k-factor excluding a given r-factor. *Appl. Math. Nonlinear Sci.* **2017**, *2*, 13–20. [CrossRef]
27. Klavžar, S. On the PI index: PI-partitions and Cartesian product graphs. *MATCH Commun. Math. Comput. Chem.* **2007**, *57*, 573–586.
28. Pattabiraman, K.; Paulraja, P. Wiener and vertex PI indices of the strong product of graphs. *Discuss. Math. Graph Theory* **2012**, *32*, 749–769. [CrossRef]
29. Ilić, A.; Milosavljević, N. The weighted vertex PI index. *Math. Comput. Model.* **2013**, *57*, 623–631.
30. Wang, C.; Wang, S.; Wei, B. Cacti with Extremal PI Index. *Trans. Comb.* **2016**, *5*, 1–8.
31. Das, K.C.; Gutman, I. Bound for vertex PI index in terms of simple graph parameters. *Filomat* **2013**, *27*, 1583–1587. [CrossRef]
32. Hoji, M.; Luo, Z.; Vumar, E. Wiener and vertex PI indices of Kronecker products of graphs. *Discret. Appl. Math.* **2010**, *158*, 1848–1855. [CrossRef]
33. Bondy, J.A.; Murty, J.A. *Graph Theory*; Springer: New York, NY, USA, 2008.

mathematics

MDPI

Article

Reformulated Zagreb Indices of Some Derived Graphs

Jia-Bao Liu [1,*], Bahadur Ali [2], Muhammad Aslam Malik [2], Hafiz Muhammad Afzal Siddiqui [3,*] and Muhammad Imran [4,†]

1 School of Mathematics and Physics, Anhui Jianzhu University, Hefei 230601, China
2 Department of Mathematics, University of the Punjab, Lahore 54000, Pakistan; bahadurali65math@gmail.com (B.A.); malikpu@yahoo.com (M.A.M.)
3 Department of Mathematics, COMSATS University Islamabad, Lahore Campus, Lahore 54000, Pakistan
4 School of Natural Sciences, National University of Sciences and Technology, Islamabad H-12, Pakistan; imrandhab@gmail.com
* Correspondence: liujiabaoad@163.com (J.-B.L.); hmasiddiqui@gmail.com (H.M.A.S.); Tel.: +92-42-111-001-007 (H.M.A.S.)
† Current address: Department of Mathematical Sciences, United Arab Emirates University, Al Ain P.O. Box 15551, UAE.

Received: 15 March 2019; Accepted: 15 April 2019; Published: 22 April 2019

Abstract: A topological index is a numeric quantity that is closely related to the chemical constitution to establish the correlation of its chemical structure with chemical reactivity or physical properties. Miličević reformulated the original Zagreb indices in 2004, replacing vertex degrees by edge degrees. In this paper, we established the expressions for the reformulated Zagreb indices of some derived graphs such as a complement, line graph, subdivision graph, edge-semitotal graph, vertex-semitotal graph, total graph, and paraline graph of a graph.

Keywords: Zagreb indices; reformulated Zagreb indices; degree of vertex; degree of edge

1. Introduction

In the fields of *mathematical chemistry* and *chemical graph theory*, a topological index is a numerical parameter that is measured based on the molecular graph of a chemical constitution. Topological indices are extensively used in the study of quantitative structure-activity relationships (QSARs) to establish the correlation between different properties of molecules and/or the biological activity with their structure. Topological indices have also been applied in spectral graph theory to measure the resilience and robustness of complex networks [1].

Suppose that Ω denotes the set of all graphs. Then, a function $T : \Omega \to \mathbb{R}^+$ is called a *topological index* if for any pair of isomorphic graphs G and H, we have $T(G) = T(H)$. There are many topological indices that are useful in chemistry, biochemistry, and nanotechnology.

Throughout this paper, we are concerned only with the simple and connected graphs. Let $G = (V, E)$ be such a graph. We use the notations $V = V(G)$ and $E = E(G)$ for the vertex set and edge set of G, respectively. We use the notation $d_G(i)$ for the *degree* of a vertex i in G. The hand shaking lemma says that the sum of the degrees of all the vertices is equal to double the number of edges. Mathematically,

$$\sum_{i \in V(G)} d_G(i) = 2|E(G)|$$

For basic definitions and notations, see the book [2].

The first and second Zagreb indices are defined as:

$$M_1 = M_1(G) = \sum_{i \in V(G)} d_G(i)^2$$
$$M_2 = M_2(G) = \sum_{ij \in E(G)} d_G(i) d_G(j).$$

Another expression for the first Zagreb index is:

$$M_1 = M_1(G) = \sum_{ij \in E(G)} [d_G(i) + d_G(j)].$$

These two Zagreb indices defined by Gutman and Trinajstic in 1972 [3] are among the oldest topological indices, having many applications in mathematical chemistry. Zagreb indices are known to be very useful in quantitative structure-property relationships (QSPR) and QSAR [4–6]. In [7] and [8], the authors have derived the expressions for Zagreb indices of some derived graphs.

Zagreb indices were reformulated by Miličević [9] as:

$$EM_1 = EM_1(G) = \sum_{e \in E(G)} d_G(e)^2,$$
$$EM_2 = EM_2(G) = \sum_{e \sim f \in E(G)} d_G(e) d_G(f),$$

where $d_G(e)$ denotes the degree of an edge $e = ij$ in G and defined as the total number of edges incident with e. Mathematically, $d_G(e) = d_G(i) + d_G(j) - 2$. Here, $e \sim f$ indicates that the edges e and f are incident. Another expression for the first reformulated Zagreb index is:

$$EM_1 = EM_1(G) = \sum_{e \sim f \in E(G)} [d_G(e) + d_G(f)].$$

The first reformulated Zagreb index is closely related to Laplacian eigenvalues [10]. Much interest has been shown by researches and scientists in the reformulated Zagreb indices [11–15].

Another vertex-degree-based topological index was found to be useful in the earliest work on Zagreb indices [3,16], but later was totally ignored. Quite recently, some interest has been shown in it [17,18]. It is called the forgotten index or simply the F-index and is defined as:

$$F = F(G) = \sum_{i \in V(G)} d_G(i)^3 = \sum_{ij \in E(G)} [d_G(i)^2 + d_G(j)^2].$$

In general, for any real number "α", the generalized version of first Zagreb index is defined as:

$${}^{\alpha}M_1 = {}^{\alpha}M_1(G) = \sum_{i \in V(G)} d_G(i)^{\alpha} = \sum_{ij \in E(G)} [d_G(i)^{\alpha-1} + d_G(j)^{\alpha-1}].$$

Clearly, the first Zagreb index and F-index are special cases of ${}^{\alpha}M_1$ for $\alpha = 2$ and $\alpha = 3$, respectively.

Bearing in mind the reformulation of Zagreb indices, here we reformulate the F-index as:

$$\begin{aligned} EF = EF(G) &= \sum_{e \in E(G)} d_G(e)^3 \\ &= \sum_{e \sim f \in E(G)} [d_G(e)^2 + d_G(f)^2]. \end{aligned}$$

We prefer to call it the reformulated forgotten index or simply the reformulated F-index.

Some Derived Graphs

Let, as before, G be a simple and connected graph with vertex set $V(G)$ and edge set $E(G)$. Let $|V(G)| = n$ and $|E(G)| = m$. Now, we consider the following graphs derived from G:

- *Complement*: The complement $\overline{G}$ of G is the graph with the same set of vertices as G, but there is an edge between two vertices of $\overline{G}$ if and only if there is no edge between these vertices in G. Clearly, $|V(\overline{G})| = n$ and $\overline{m} = |E(\overline{G})| = \frac{n(n-1)}{2} - m$.
- *Line graph*: The line graph $L = L(G)$ of G is the graph in which the vertex set is the edge set of G, and there is an edge between two vertices of L if and only if their corresponding edges are incident in G. Thus, $|V(L)| = m$, and by hand shaking lemma,

$$\begin{aligned} |E(L)| &= \frac{1}{2} \sum_{i \in V(L)} d_L(i) = \frac{1}{2} \sum_{e \in E(G)} d_G(e) \\ &= \frac{1}{2} \sum_{ij \in E(G)} [d_G(i) + d_G(j) - 2] \\ &= \frac{1}{2}(M_1 - 2m) = \frac{M_1}{2} - m. \end{aligned}$$

- *Subdivision graph*: A subdivision graph of a graph G can be constructed by inserting a vertex on each edge of G, which will change that edge into a path of length two. This graph is denoted as $S = S(G)$.
  So, $|V(S)| = |V(G)| + |E(G)| = n + m$ and $|E(S)| = 2|E(G)| = 2m$.
- *Vertex-semitotal graph*: A vertex-semitotal graph $T_1 = T_1(G)$ is constructed from G by inserting a new vertex on each edge of G and then by joining every newly-inserted vertex to the end vertices of the corresponding edge. Thus, $|V(T_1)| = |V(G)| + |E(G)| = n + m$ and $|E(T_1)| = |E(S)| + |E(G)| = 2m + m = 3m$.
- *Edge-semitotal graph*: An edge-semitotal graph $T_2 = T_2(G)$ is made by putting a new vertex in each edge of G and then joining with edges those new vertices whose corresponding edges are incident in G. Thus, $|V(T_2)| = |V(G)| + |E(G)| = n + m$ and $|E(T_2)| = |E(S)| + |E(L)| = 2m + \frac{M_1}{2} - m = m + \frac{M_1}{2}$.
- *Total graph*: The total graph $T = T(G)$ is the union of the vertex-semitotal graph and the edge-semitotal graph. Thus, $|V(T)| = |V(G)| + |E(G)| = n + m$ and $|E(T)| = |E(G)| + |E(S)| + |E(L)| = m + 2m + \frac{M_1}{2} - m = 2m + \frac{M_1}{2}$.
- *Paraline graph*: The paraline graph $PL = PL(G)$ is the line graph of the subdivision graph denoted by $PL = PL(G) = L(S(G))$. Furthermore, $|V(PL)| = |E(S)| = 2m$ and:

$$|E(PL)| = \frac{M_1(S)}{2} - 2m.$$

In [7], one can easily see that $M_1(S) = M_1 + 4m$. Thus:

$$|E(PL)| = \frac{M_1 + 4m}{2} - 2m = \frac{M_1}{2}.$$

In Figure 1, self-explanatory examples of these derived graphs are given for a particular graph. In every derived graph of G (except the paraline graph $PL(G)$), the vertices corresponding to the vertices of G are denoted by circles and the vertices corresponding to the edges of G are denoted by squares.

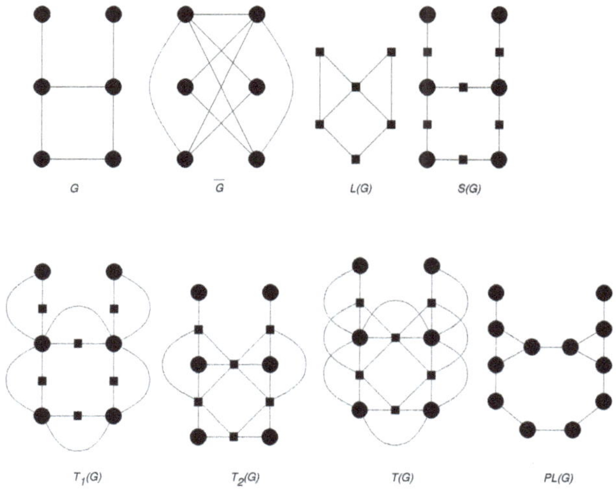

Figure 1. Different graphs derived from G.

The following are some known results on Zagreb indices.

Lemma 1. *[7] If $\overline{G}$ is the complement of G, then:*

$$M_1(\overline{G}) = M_1 + n(n-1)^2 - 4m(n-1).$$

Lemma 2. *[7] If $\overline{G}$ is the complement of G, then:*

$$M_2(\overline{G}) = \frac{2n-3}{2}M_1 - M_2 + \frac{1}{2}n(n-1)^3 - 3m(n-1)^2 + 2m^2.$$

Lemma 3. *[17] If $\overline{G}$ is the complement of G, then:*

$$F(\overline{G}) = 3(n-1)M_1 - F + n(n-1)^3 - 6m(n-1)^2.$$

Lemma 4. *[15] The reformulated first Zagreb index can be written in terms of the Zagreb indices and F-index as:*

$$EM_1 = F - 4M_1 + 2M_2 + 4m.$$

2. Reformulated First Zagreb Index of Some Derived Graphs

In [7] and [8], the authors derived the expressions for Zagreb indices and coindices of those derived graphs, which have been discussed above. In [19], the authors derived the expressions for multiplicative Zagreb indices and coindices of some derived graphs. In this section, we present the expressions for the reformulated first Zagreb index of these derived graphs.

Theorem 1. *If $\overline{G}$ is the complement of G, then:*

$$\begin{aligned} EM_1(\overline{G}) &= 5(n-2)M_1 - 2M_2 - F + 2n(n-1)(n-2)^2 \\ &\quad -4m(n-2)(3n-4) + 4m^2. \end{aligned}$$

Proof. By the definition of the first reformulated Zagreb index,

$$\begin{aligned}
EM_1(\overline{G}) &= \sum_{e\in E(\overline{G})} d_{\overline{G}}(e)^2 \\
&= \sum_{ij\in E(\overline{G})} [d_{\overline{G}}(i) + d_{\overline{G}}(j) - 2]^2 \\
&= \sum_{ij\in E(\overline{G})} [d_{\overline{G}}(i)^2 + d_{\overline{G}}(j)^2] + 4 \sum_{ij\in E(\overline{G})} 1 + 2 \sum_{ij\in E(\overline{G})} d_{\overline{G}}(i)d_{\overline{G}}(j) \\
&\quad -4 \sum_{ij\in E(\overline{G})} [d_{\overline{G}}(i) + d_{\overline{G}}(j)] \\
&= F(\overline{G}) + 4|E(\overline{G})| + 2M_2(\overline{G}) - 4M_1(\overline{G})
\end{aligned}$$

Using Lemma 1, Lemma 2, and Lemma 3 for fourth, third, and first term, respectively, we have:

$$\begin{aligned}
EM_1(\overline{G}) &= 3(n-1)M_1 - F + n(n-1)^3 - 6m(n-1)^2 + 2n(n-1) - 4m \\
&\quad +(2n-3)M_1 - 2M_2 + n(n-1)^3 - 6m(n-1)^2 + 4m^2 \\
&\quad -4M_1 - 4n(n-1)^2 + 16m(n-1) \\
&= 5(n-2)M_1 - 2M_2 - F + 2n(n-1)^3 - 4n(n-1)^2 + 2n(n-1) \\
&\quad -12m(n-1)^2 - 4m + 16m(n-1) + 4m^2 \\
&= 5(n-2)M_1 - 2M_2 - F + 2n(n-1)(n-2)^2 - 4m(n-2)(3n-4) + 4m^2.
\end{aligned}$$

□

Theorem 2. *If $L = L(G)$ is the line graph of G, then:*

$$EM_1(L) = EF + 2EM_2 - 4F + 18M_1 - 8M_2 - 20m.$$

Proof. By the definition of the reformulated first Zagreb index,

$$\begin{aligned}
EM_1(L) &= \sum_{e\in E(L)} d_L(e)^2 \\
&= \sum_{ij\in E(L)} [d_L(i) + d_L(j) - 2]^2 \\
&= \sum_{ij\in E(L)} [d_L(i)^2 + d_L(j)^2] + 4 \sum_{ij\in E(L)} 1 + 2 \sum_{ij\in E(L)} d_L(i)d_L(j) \\
&\quad -4 \sum_{ij\in E(L)} [d_L(i) + d_L(j)] \\
&= F(L) + 4|E(L)| + 2M_2(L) - 4M_1(L).
\end{aligned}$$

Now, by the definition of the line graph, it is clear that:

$$F(L) = EF, M_1(L) = EM_1, M_2(L) = EM_2.$$

Thus:

$$EM_1(L) = EF + 4(\frac{M_1}{2} - m) + 2EM_2 - 4EM_1.$$

Using Lemma 4 for term EM_1,

$$EM_1(L) = EF + 2EM_2 - 4F + 18M_1 - 8M_2 - 20m.$$

□

Now, for subdivision graph S, vertex-semitotal graph T_1, edge-semitotal graph T_2, and total graph T, we can see that there are two types of vertices in these graphs: first, the vertices corresponding to the vertices of G and, second, the vertices corresponding to the edges of G. We call them α-vertices and β-vertices, respectively. Depending on the nature of end vertices, we can divide the edges of these graphs into three types:

1. $\alpha\alpha$-edge: an edge between two α-vertices.
2. $\beta\beta$-edge: an edge between two β-vertices.
3. $\alpha\beta$-edge: an edge between a α-vertex and a β-vertex.

The above idea is very similar to the idea used in [20].

Theorem 3. *If $S = S(G)$ is a subdivision graph of G, then:*

$$EM_1(S) = F.$$

Proof. We can see that for any α-vertex i_α of S:

$$d_S(i_\alpha) = d_G(i_\alpha)$$

and for any β-vertex j_β of S:

$$d_S(j_\beta) = 2.$$

Furthermore, all the edges of S are $\alpha\beta$-edges. Thus:

$$\begin{aligned} EM_1(S) &= \sum_{e_{\alpha\beta}\in E(S)} d_S(e_{\alpha\beta})^2 \\ &= \sum_{i_\alpha j_\beta \in E(S)} [d_S(i_\alpha) + d_S(j_\beta) - 2]^2. \end{aligned}$$

In fact, every α-vertex i_α of S is connected with $d_G(i_\alpha)$ β-vertices, each of degree two. Hence:

$$EM_1(S) = \sum_{i_\alpha \in V(G)} d_G(i_\alpha)[d_G(i_\alpha) + 2 - 2]^2 = F.$$

□

Theorem 4. *If $T_1 = T_1(G)$ is a vertex-semitotal graph of G, then:*

$$EM_1(T_1) = 8(F - M_1 + M_2) + 4m.$$

Proof. First note that for any α-vertex i_α of T_1:

$$d_{T_1}(i_\alpha) = 2d_G(i_\alpha)$$

and for any β-vertex j_β of T_1:

$$d_{T_1}(j_\beta) = 2.$$

Furthermore, any edge of T_1 is either an $\alpha\alpha$-edge or an $\alpha\beta$-edge. Thus:

$$\begin{aligned} EM_1(T_1) &= \sum_{e \in E(T_1)} d_{T_1}(e)^2 \\ &= \sum_{e_{\alpha\alpha} \in E(T_1)} d_{T_1}(e_{\alpha\alpha})^2 + \sum_{e_{\alpha\beta} \in E(T_1)} d_{T_1}(e_{\alpha\beta})^2 \\ &= \sum_{i_\alpha j_\alpha \in E(T_1)} [d_{T_1}(i_\alpha) + d_{T_1}(j_\alpha) - 2]^2 + \sum_{i_\alpha j_\beta \in E(T_1)} [d_{T_1}(i_\alpha) + d_{T_1}(j_\beta) - 2]^2. \end{aligned}$$

For $\alpha\beta$-edges in the second term, it is clear that every α-vertex i_α of T_1 is connected with $d_G(i_\alpha)$ β-vertices, each of degree two. Therefore, corresponding to every vertex i_α in G, there are $d_G(i_\alpha)$ edges in T_1 each of (edge) degree $[2d_G(i_\alpha) + 2 - 2]$. Thus:

$$\begin{aligned} EM_1(T_1) &= \sum_{i_\alpha j_\alpha \in E(G)} [2d_G(i_\alpha) + 2d_G(j_\alpha) - 2]^2 + \sum_{i_\alpha \in V(G)} d_G(i_\alpha)[2d_G(i_\alpha) + 2 - 2]^2 \\ &= 4 \sum_{i_\alpha j_\alpha \in E(G)} [d_G(i_\alpha)^2 + d_G(j_\alpha)^2] + 4 \sum_{i_\alpha j_\alpha \in E(G)} 1 + 8 \sum_{i_\alpha j_\alpha \in E(G)} d_G(i_\alpha) d_G(j_\alpha) \\ &\quad -8 \sum_{i_\alpha j_\alpha \in E(G)} [d_G(i_\alpha) + d_G(j_\alpha)] + 4 \sum_{i_\alpha \in V(G)} d_G(i_\alpha)^3 \\ &= 8(F - M_1 + M_2) + 4m. \end{aligned}$$

□

Theorem 5. *If $T_2 = T_2(G)$ is an edge-semitotal graph of G, then:*

$$EM_1(T_2) = EF + 2EM_2 + 9F - 26M_1 + 16M_2 + 20m.$$

Proof. First note that for any α-vertex i_α of T_2:

$$d_{T_2}(i_\alpha) = d_G(i_\alpha)$$

and for any β-vertex j_β of T_2

$$d_{T_2}(j_\beta) = d_L(j_\beta) + 2.$$

Any edge in T_2 is either an $\alpha\beta$-edge or a $\beta\beta$-edge. In fact,

1. Corresponding to every edge $i_\alpha j_\alpha$ in G, there are two $\alpha\beta$-edges $i_\alpha x_\beta$ and $x_\beta j_\alpha$ in T_2 such that:

$$\begin{aligned} d_{T_2}(i_\alpha) &= d_G(i_\alpha) \\ d_{T_2}(j_\alpha) &= d_G(j_\alpha) \end{aligned}$$

   and:

$$\begin{aligned} d_{T_2}(x_\beta) &= d_L(x_\beta) + 2 = d_G(i_\alpha) + d_G(j_\alpha) - 2 + 2 \\ &= d_G(i_\alpha) + d_G(j_\alpha). \end{aligned}$$

2. $\beta\beta$-edges are the edges corresponding to the edges of $L(G)$.

Thus:

$$\begin{aligned} EM_1(T_2) &= \sum_{e\in E(T_2)} d_{T_2}(e)^2 \\ &= \sum_{e_{\alpha\beta}\in E(T_2)} d_{T_2}(e_{\alpha\beta})^2 + \sum_{e_{\beta\beta}\in E(T_2)} d_{T_2}(e_{\beta\beta})^2 \\ &= \sum_{i_\alpha j_\alpha\in E(G)} [(d_{T_2}(i_\alpha)+d_{T_2}(x_\beta)-2)^2 + (d_{T_2}(x_\beta)+d_{T_2}(j_\alpha)-2)^2] \\ &\quad + \sum_{i_\beta j_\beta\in E(L)} (d_{T_2}(i_\beta)+d_{T_2}(j_\beta)-2)^2 \\ &= \sum_{i_\alpha j_\alpha\in E(G)} [(2d_G(i_\alpha)+d_G(j_\alpha)-2)^2 + (d_G(i_\alpha)+2d_G(j_\alpha)-2)^2] \\ &\quad + \sum_{i_\beta j_\beta\in E(L)} (d_L(i_\beta)+d_L(j_\beta)+2)^2 \\ &= 5F+8m+8M_2-12M_1+F(L)+4|E(L)|+2M_2(L)+4M_1(L). \end{aligned}$$

Now, by the definition of the line graph, it is clear that:

$$F(L) = EF, M_1(L) = EM_1, M_2(L) = EM_2.$$

Thus:

$$EM_1(T_2) = 5F+8m+8M_2-12M_1+EF+4(\frac{M_1}{2}-m)+2EM_2+4EM_1.$$

Using Lemma 4,

$$EM_1(T_2) = EF+2EM_2+9F-26M_1+16M_2+20m.$$

□

Theorem 6. *If $T = T(G)$ is the total graph of G, then:*

$$EM_1(T) = EF+2EM_2+18F-38M_1+28M_2+24m.$$

Proof. First note that for any α-vertex i_α of T:

$$d_T(i_\alpha) = 2d_G(i_\alpha)$$

and for any β-vertex j_β of T:

$$d_T(j_\beta) = d_L(j_\beta)+2.$$

T has all three types of edges. In fact,

1. Corresponding to every edge $i_\alpha j_\alpha$ in G, there is one $\alpha\alpha$-edge, which is $i_\alpha j_\alpha$, and two $\alpha\beta$-edges, which are $i_\alpha x_\beta$ and $x_\beta j_\alpha$ in T such that:

$$\begin{aligned} d_T(i_\alpha) &= 2d_G(i_\alpha) \\ d_T(j_\alpha) &= 2d_G(j_\alpha) \end{aligned}$$

and:

$$\begin{aligned} d_T(x_\beta) &= d_L(x_\beta) + 2 = d_G(i_\alpha) + d_G(j_\alpha) - 2 + 2 \\ &= d_G(i_\alpha) + d_G(j_\alpha). \end{aligned}$$

2. $\beta\beta$-edges are the edges corresponding to the edges of $L(G)$.

Thus:

$$\begin{aligned} EM_1(T) &= \sum_{e \in E(T)} d_T(e)^2 \\ &= \sum_{e_{\alpha\alpha} \in E(T)} d_T(e_{\alpha\alpha})^2 + \sum_{e_{\beta\beta} \in E(T)} d_T(e_{\beta\beta})^2 + \sum_{e_{\alpha\beta} \in E(T)} d_T(e_{\alpha\beta})^2 \\ &= \sum_{i_\alpha j_\alpha \in E(G)} (d_T(i_\alpha) + d_T(j_\alpha) - 2)^2 + \sum_{i_\beta j_\beta \in E(L)} (d_T(i_\beta) + d_T(j_\beta) - 2)^2 \\ &+ \sum_{i_\alpha j_\alpha \in E(G)} [(d_T(i_\alpha) + d_T(x_\beta) - 2)^2 + (d_T(x_\beta) + d_T(j_\alpha) - 2)^2] \\ &= 4 \sum_{i_\alpha j_\alpha \in E(G)} (d_G(i_\alpha) + d_G(j_\alpha) - 1)^2 + \sum_{i_\beta j_\beta \in E(L)} (d_L(i_\beta) + d_L(j_\beta) + 2)^2 \\ &+ \sum_{i_\alpha j_\alpha \in E(G)} [(3d_G(i_\alpha) + d_G(j_\alpha) - 2)^2 + (d_G(i_\alpha) + 3d_G(j_\alpha) - 2)^2] \\ &= 14F - 24M_1 + 20M_2 + 12m + F(L) + 4|E(L)| + 2M_2(L) + 4M_1(L). \end{aligned}$$

Now, using the facts:

$$F(L) = EF, M_1(L) = EM_1, M_2(L) = EM_2,$$

and then using Lemma 4, we get:

$$EM_1(T) = EF + 2EM_2 + 18F - 38M_1 + 28M_2 + 24m.$$

□

Theorem 7. *If $PL = PL(G)$ is a paraline graph of G, then:*

$$EM_1(PL) = 2(^4M_1 + M_1 + M_2) - 5F.$$

Proof. It can be noted that for any vertex $i \in V(G)$, there are $d_G(i)$ vertices in $PL(G)$ having the same degree as i such that all these $d_G(i)$ vertices are connected with each other. In fact, $PL(G)$ can be obtained from G by replacing every vertex i by $K_{d_G(i)}$. Now, the edges of $PL(G)$ can be divided into two categories:

1. The edges in $K_{d_G(i)}$, where $i \in V(G)$.
2. Edges corresponding to edges of G. It can be seen that corresponding to every edge in G, there is an edge in $PL(G)$ of the same degree.

Thus:

$$\begin{aligned} EM_1(PL) &= \sum_{e \in E(PL)} d_{PL}(e)^2 \\ &= \sum_{e \in E(K_{d_G(i)}), i \in V(G)} d_{PL}(e)^2 + \sum_{e \in E(G)} d_G(e)^2. \end{aligned}$$

Now, in each $K_{d_G(i)}$, there are $\frac{d_G(i)(d_G(i)-1)}{2}$ edges, each of degree $d_G(i)+d_G(i)-2=2d_G(i)-2$. Thus:

$$EM_1(PL) = \sum_{i \in V(G)} \frac{d_G(i)(d_G(i)-1)}{2}(2d_G(i)-2)^2 + EM_1.$$

Solving the first term and using Lemma 4 in the second term, we get:

$$\begin{aligned} EM_1(PL) &= 2({}^4M_1 - 3F + 3M_1 - 2m) + F - 4M_1 + 2M_2 + 4m \\ &= 2({}^4M_1 + M_1 + M_2) - 5F. \end{aligned}$$

□

3. Reformulated Second Zagreb Index of Some Derived Graphs

In this section, we derive the expressions for the reformulated second Zagreb index of some derived graphs.

Theorem 8. *If $S = S(G)$ is a subdivision graph of G, then:*

$$EM_2(S) = \frac{1}{2}({}^4M_1 - F) + M_2.$$

Proof. We can divide the pairs of incident edges of $S = S(G)$ into two categories:

1. For any vertex $i \in V(G)$, there are $d_G(i)$ edges, each of degree $d_G(i)$ in S, and all these edges are incident at i. Therefore, for any vertex $i \in V(G)$, the total number of pairs of incident edges lying in this category is $\frac{d_G(i)(d_G(i)-1)}{2}$.
2. Corresponding to every pair of adjacent vertices i and j in G, there is a pair of incident edges of degrees $d_G(i)$ and $d_G(j)$ in S.

Thus:

$$\begin{aligned} EM_2(S) &= \sum_{e \sim f \in E(S)} d_S(e) d_S(f) \\ &= \sum_{i \in V(G)} \frac{d_G(i)(d_G(i)-1)}{2} d_G(i) d_G(i) + \sum_{ij \in E(G)} d_G(i) d_G(j) \\ &= \frac{1}{2}({}^4M_1 - F) + M_2. \end{aligned}$$

□

Before going to the next theorem, we prove another lemma here.

Lemma 5. *For a graph G, the following equality holds.*

$$\sum_{ij \in E(G)} d_G(i) d_G(j)[d_G(i) + d_G(j)] = \frac{1}{3}(EF - {}^4M_1 + 8m) + 2F + 4(M_2 - M_1).$$

Proof.

$$\begin{aligned}\sum_{ij\in E(G)} d_G(i)d_G(j)[d_G(i)+d_G(j)] &= \frac{1}{3}\sum_{ij\in E(G)}[d_G(i)+d_G(j)]^3 - \frac{1}{3}\sum_{ij\in E(G)}[d_G(i)^3+d_G(j)^3] \\ &= \frac{1}{3}\sum_{e\in E(G)}[d_G(e)+2]^3 - \frac{1}{3}{}^4M_1 \\ &= \frac{1}{3}EF + 2EM_1 + 4\sum_{e\in E(G)} d_G(e) + \frac{8}{3}m - \frac{1}{3}{}^4M_1.\end{aligned}$$

Using Lemma 4 and the relation:

$$\sum_{e\in E(G)} d_G(e) = \sum_{i\in V(L)} d_L(i) = 2|E(L)| = M_1 - 2m,$$

we have:

$$\begin{aligned}\sum_{ij\in E(G)} d_G(i)d_G(j)[d_G(i)+d_G(j)] &= \frac{1}{3}EF + 2F - 8M_1 + 4M_2 + 8m + 4M_1 - 8m + \frac{8}{3}m - \frac{1}{3}{}^4M_1 \\ &= \frac{1}{3}(EF - {}^4M_1 + 8m) + 2F + 4(M_2 - M_1).\end{aligned}$$

□

Theorem 9. *If $T_1 = T_1(G)$ is vertex-semitotal graph of G, then:*

$$EM_2(T_1) = \frac{1}{3}(14({}^4M_1) + 4EF + 68m) + 4EM_2 + 6F - 30M_1 + 28M_2.$$

Proof. By definition

$$EM_2(T_1) = \sum_{e\sim f\in E(T_1)} d_{T_1}(e)d_{T_1}(f)$$

We can divide the pairs of incident edges (e, f) of $T_1 = T_1(G)$ into three cases:

Case 1: When $e, f \in E(S)$. Just like Theorem 8, we have two further categories:

1. For any vertex $i \in V(G)$, there are $d_G(i)$ edges, each of degree $2d_G(i)$ in S, and all these edges are incident at i. Therefore, for any vertex $i \in V(G)$, the total number of pairs of incident edges lying in this category is $\frac{d_G(i)(d_G(i)-1)}{2}$.
2. Corresponding to every pair of adjacent vertices i and j in G, there is a pair of incident edges of degrees $2d_G(i)$ and $2d_G(j)$ in S.

Case 2: When $e, f \in E(G)$. For any pair of incident edges e and f in G, e and f are also incident in T_1. Furthermore, for any edge $e = ij$ in G,

$$d_{T_1}(e) = d_{T_1}(i) + d_{T_1}(j) - 2 = 2d_G(i) + 2d_G(j) - 2 = 2(d_G(e) + 1).$$

Case 3: When $e \in E(G), f \in E(S)$. Every edge ij of G has degree $2d_G(i) + 2d_G(j) - 2$ in T. This edge is incident with $d_G(i)$ edges, each of degree $2d_G(i)$ at i, and $d_G(j)$ edges, each of degree $2d_G(j)$ at j.

Thus:

$$
\begin{aligned}
EM_2(T_1) &= \sum_{e\sim f\in E(S)} d_{T_1}(e)d_{T_1}(f) + \sum_{e\sim f\in E(G)} d_{T_1}(e)d_{T_1}(f) + \sum_{e\sim f, e\in E(G), f\in E(S)} d_{T_1}(e)d_{T_1}(f) \\
&= \sum_{i\in V(G)} \frac{d_G(i)(d_G(i)-1)}{2} 2d_G(i)2d_G(i) + \sum_{ij\in E(G)} 2d_G(i)2d_G(j) \\
&+ \sum_{e\sim f\in E(G)} 2(d_G(e)+1)2(d_G(f)+1) \\
&+ \sum_{ij\in E(G)} [d_G(i)(2d_G(i)+2d_G(j)-2)2d_G(i) + d_G(j)(2d_G(i)+2d_G(j)-2)2d_G(j)] \\
&= 2\sum_{i\in V(G)} d_G(i)^4 - 2\sum_{i\in V(G)} d_G(i)^3 + 4M_2 + 4\sum_{e\sim f\in E(G)} d_G(e)d_G(f) \\
&+4\sum_{e\sim f\in E(G)} [d_G(e)+d_G(f)] + 4\sum_{e\sim f\in E(G)} 1 + 4\sum_{ij\in E(G)} [d_G(i)^3 + d_G(j)^3] \\
&-4\sum_{ij\in E(G)} [d_G(i)^2 + d_G(j)^2] + 4\sum_{ij\in E(G)} d_G(i)d_G(j)[d_G(i)+d_G(j)].
\end{aligned}
$$

Using Lemma 5 and the fact:

$$
\sum_{e\sim f\in E(G)} 1 = \sum_{ij\in E(L)} 1 = |E(L)| = \frac{M_1}{2} - m
$$

we have:

$$
\begin{aligned}
EM_2(T_1) &= 2(^4M_1) - 2F + 4M_2 + 4EM_2 + 4EM_1 + 2M_1 - 4m \\
&+4(^4M_1) - 4F + \frac{4}{3}EF - \frac{4}{3}(^4M_1) + \frac{32}{3}m + 8F + 16M_2 - 16M_1.
\end{aligned}
$$

Now, using Lemma 4 and simplifying the above expression, we get:

$$
EM_2(T_1) = \frac{1}{3}(14(^4M_1) + 4EF + 68m) + 4EM_2 + 6F - 30M_1 + 28M_2.
$$

□

Proposition 1. *Let K_n be the complete graph on n vertices, then the total number of pairs of incident edges in K_n is $\frac{n(n-1)(n-2)}{2}$.*

Proof. The total number of pairs of incident edges in K_n:

$$
\begin{aligned}
&= \sum_{e\sim f\in E(K_n)} 1 = \sum_{ij\in E(L(K_n))} 1 = E(L(K_n)) = \frac{M_1(K_n)}{2} - |E(K_n)| \\
&= \frac{n(n-1)^2}{2} - \frac{n(n-1)}{2} = \frac{n(n-1)(n-2)}{2}.
\end{aligned}
$$

□

Theorem 10. *If $PL = PL(G)$ is a paraline graph of G, then:*

$$
EM_2(PL) = 2(^5M_1) + \frac{1}{3}(2EF - 26(^4M_1) + 16m) + 14F - 10M_1.
$$

Proof. As we explained in Theorem 7, $PL(G)$ can be obtained from G by replacing every vertex i by $K_{d_G(i)}$. Therefore, we can divide the pairs of incident edges of $PL = PL(G)$ into two cases:

Case 1: For any vertex $i \in V(G)$, there are $\frac{d_G(i)(d_G(i)-1)}{2}$ edges each of degree $(2d_G(i)-2)$ in the corresponding $K_{d_G(i)}$ of PL, and all these edges are incident with each other. By Proposition 1, for any vertex $i \in V(G)$, the total number of pairs of incident edges in $K_{d_G(i)}$ is:

$$= \frac{d_G(i)(d_G(i)-1)(d_G(i)-2)}{2}.$$

Case 2: Corresponding to every edge ij in G, there is an edge in $PL(G)$ of the same degree that is incident with $(d_G(i)-1)$ edges, each of degree $(2d_G(i)-2)$ at i, and $(d_G(j)-1)$ edges, each of degree $(2d_G(j)-2)$ at j.

Thus:

$$\begin{aligned} EM_2(PL) &= \sum_{e\sim f\in E(PL)} d_{PL}(e)d_{PL}(f) \\ &= \sum_{e\sim f\in E(K_{d_G(i)}),i\in V(G)} d_{PL}(e)d_{PL}(f) \\ &\quad + \sum_{e=ij\in E(G)} [(d_G(i)-1)d_G(e)(2d_G(i)-2)+(d_G(j)-1)d_G(e)(2d_G(j)-2)] \end{aligned}$$

$$\begin{aligned} &= \sum_{i\in V(G)} [\frac{d_G(i)(d_G(i)-1)(d_G(i)-2)}{2}](2d_G(i)-2)(2d_G(i)-2) \\ &\quad +2\sum_{ij\in E(G)} (d_G(i)+d_G(j)-2)[(d_G(i)-1)^2+(d_G(j)-1)^2] \\ &= 2\sum_{i\in V(G)} d_G(i)(d_G(i)-1)^3(d_G(i)-2) \\ &\quad +2\sum_{ij\in E(G)} [(d_G(i)^3+d_G(j)^3)-4(d_G(i)^2+d_G(j)^2)+6(d_G(i)+d_G(j)) \\ &\quad -4d_G(i)d_G(j)+d_G(i)d_G(j)[d_G(i)+d_G(j)]-4]. \end{aligned}$$

Now, using Lemma 5:

$$\begin{aligned} EM_2(PL) &= 2({}^5M_1)-10({}^4M_1)+18F-14M_1+8m+2({}^4M_1)-8F+12M_1-8M_2 \\ &\quad +\frac{2}{3}EF-\frac{2}{3}({}^4M_1)+\frac{16}{3}m+4F+8M_2-8M_1-8m \\ &= 2({}^5M_1)+\frac{1}{3}(2EF-26({}^4M_1)+16m)+14F-10M_1. \end{aligned}$$

□

4. Conclusions

In this note, we obtained some relations for degrees between a derived graph and its parent graph. Using these relations, the expressions for reformulated Zagreb indices of some derived graphs have been found in terms of some topological indices of the parent graph. Finding expressions of derived graphs like these is an open problem for many other topological indices.

Author Contributions: All the authors contributed equally to preparing this article.

Funding: This work is partially supported by the China Postdoctoral Science Foundation under grant No. 2017M621579 and the Postdoctoral Science Foundation of Jiangsu Province under grant No. 1701081B. Project of Anhui Jianzhu University under Grant no. 2016QD116 and 2017dc03. Anhui Province Key Laboratory of Intelligent Building and Building Energy Saving.

Acknowledgments: The authors are grateful to the referees for the useful comments that improved the first version of this article.

Conflicts of Interest: The authors declare no conflict of interest.

References

1. Shang, Y. Perturbation results for the Estrada index in weighted networks. *J. Phys. A Math. Theor.* **2011**, *44*, 075003. [CrossRef]
2. Bondy, J.A.; Murty, U.S.R. *Graph Theory*; Springer: Berlin, Germany, 2008.
3. Gutman, I.; Trinajstić, N. Graph theory and molecular orbitals. III. Total π-electron energy of alternate hydrocarbons. *Chem. Phys. Lett.* **1972**, *17*, 535–538. [CrossRef]
4. Balaban, A.T. *From Chemical Topology to Three-Dimensional Geometry*; Plenum Press: New York, NY, USA, 1997.
5. Devillers, J.; Balaban, A.T. (Eds.) *Topological Indices and Related Descriptors in QSAR and QSPR*; Gordon and Breach: Amsterdam, The Netherlands, 1999.
6. Todeschini, R.; Consonni, V. *Handbook of Molecular Descriptors*; Wiley-VCH: Weinheim, Germany, 2000.
7. Gutman, I.; Furtula, B.; Kovijanić Vukićević, Z.; Popivoda, G. On Zagreb indices and coindices. *Match Commun. Math. Comput. Chem.* **2015**, *74*, 5–16.
8. Basavanagoud, B.; Gutman, I.; Chetana, S.G. On second Zagreb index and coindex of some derived graphs. *Kragujev. J. Sci.* **2015**, *37*, 113–121.
9. Miličević, A.; Nikolić, S.; Trinajstić, N. On reformulated Zagreb indices. *Mol. Divers.* **2004**, *8*, 393–399. [CrossRef] [PubMed]
10. Shang, Y. On the number of spanning trees, the Laplacian eigenvalues, and the Laplacian Estrada index of subdivided-line graphs. *Open Math* **2016**, *14*, 641–648. [CrossRef]
11. De, N. Some bounds of reformulated Zagreb indices. *Appl. Math. Sci.* **2012**, *6*, 5005–5012.
12. Ilić, A.; Zhou, B. On reformulated Zagreb indices. *Discr. Appl. Math.* **2012**, *160*, 204–209. [CrossRef]
13. Ji, S.; Li, X.; Huo, B. On reformulated Zagreb indices with respect to acyclic, unicyclic and bicyclic graphs. *Match Commun. Math. Comput. Chem.* **2014**, *72*, 723–732.
14. Su, G.; Xiong, L.; Xu, L.; Ma, B. On the maximum and minimum first reformulated Zagreb index of graphs with connectivity at most *k*. *Filomat* **2011**, *25*, 75–83. [CrossRef]
15. Zhou, B.; Trinajstić, N. Some properties of the reformulated Zagreb indices. *J. Math. Chem.* **2010**, *48*, 714–719. [CrossRef]
16. Liu, J.-B.; Wang, C.; Wang, S. Zagreb indices and multiplicative Zagreb indices of Eulerian graphs. *Bull. Malays. Math. Sci. Soc.* **2019**, *42*, 67–78. [CrossRef]
17. De, N.; Nayeem, S.M.A.; Pal, A. The F-coindex of some graph operations. *Springer Plus* **2016**, *5*, 221. [CrossRef] [PubMed]
18. Furtula, B.; Gutman, I. A forgotten topological index. *J. Math. Chem.* **2015**, *53*, 1184–1190. [CrossRef]
19. Basavanagoud, B.; Patil, S. Multiplicative Zagreb indices and coindices of some derived graphs. *Opuscula Math.* **2016**, *36*, 287–299. [CrossRef]
20. Shang, Y. Vulnerability of networks: Fractional percolation on random graphs. *Phys. Rev. E* **2014**, *89*, 012813. [CrossRef] [PubMed]

mathematics

MDPI

Article

Wirelength of Enhanced Hypercube into Windmill and Necklace Graphs

Jia-Bao Liu [1], Micheal Arockiaraj [2] and John Nancy Delaila [2,*]

[1] School of Mathematics and Physics, Anhui Jianzhu University, Hefei 230601, China; liujiabaoad@163.com
[2] Department of Mathematics, Loyola College, Chennai 600034, India; marockiaraj@loyolacollege.edu
* Correspondence: nancydelaila20@gmail.com

Received: 20 March 2019; Accepted: 19 April 2019; Published: 26 April 2019

Abstract: An embedding of an interconnection network into another is one of the main issues in parallel processing and computing systems. Congestion, dilation, expansion and wirelength are some of the parameters used to analyze the efficiency of an embedding in which resolving the wirelength problem reduces time and cost in the embedded design. Due to the potential topological properties of enhanced hypercube, it has become constructive in recent years, and a lot of research work has been carried out on it. In this paper, we use the edge isoperimetric problem to produce the exact wirelengths of embedding enhanced hypercube into windmill and necklace graphs.

Keywords: embedding; edge congestion; wirelength; enhanced hypercube

1. Introduction

Graph theory is one of the most interesting branches of mathematics, with wide applications in the domain of computer science, leading to the choice of a network in the development of parallel computers on a commercial basis. An interconnection network can be modeled by a graph, in which processors and links between processors are denoted as vertices and edges of a graph, respectively. Interconnection network has many advantages and inherent applications in the field of system designs such as scheduling of multiprocessor systems and distributed systems. The problem of efficiently implementing a parallel algorithm developed for one network into another can be modeled by a graph-embedding problem [1]. A graph embedding of a guest graph G into a host graph H is defined by a bijective mapping $f : V(G) \to V(H)$ together with a mapping P_f which assigns to each edge (u, v) of G a path between $f(u)$ and $f(v)$ in H [2].

An edge congestion of an embedding of G into H is the maximum number of edges of a graph G that are embedded on any single edge of H. Let $EC_f(e)$ denote the number of edges (u, v) of graph G such that e is in a path $P_f((u, v))$ between $f(u)$ and $f(v)$ of H [2,3]. In other words,

$$EC_f(e) = |\{(u,v) \in E(G) : e \in P_f((u,v))\}|.$$

The wirelength of an embedding f of G into H is computed by

$$WL_f(G,H) = \sum_{(u,v)\in E(G)} | P_f((u,v)) | = \sum_{e\in E(H)} EC_f(e).$$

The wirelength of G into H is defined as

$$WL(G,H) = min\, WL_f(G,H)$$

where the minimum is taken over all embeddings f of G into H. The wirelength problem of a graph embedding emerges from VLSI designs [4], networks for parallel computer systems [5] and structural engineering [6,7].

Wirelength problems have been considered for enhanced hypercube into wounded lobsters [8], r-rooted complete binary tree [1], complete binary tree [9], caterpillar and path [10]. The wirelength of hypercubes into necklace, windmill and snake graphs have been examined in [11]. In this paper, we explore the exact wirelength of enhanced hypercube into necklace and windmill graphs and the main contributions are presented in Theorems 2, 3 and 4.

The rest of the paper is organized as follows. In Section 2, we study edge isoperimetric problem and 2-partition lemma. Enhanced hypercube and its properties will be discussed in Section 3. In Section 4, we compute the minimum wirelength of embedding enhanced hypercube into windmill and necklace graphs. Finally, Section 5 concludes the paper.

2. Edge Isoperimetric Problem

Consider an interesting NP-complete problem [12] namely combinatorial isoperimetric problem which optimizes or selects the best structure among several possibilities and arises naturally in communication engineering, computer science, physical sciences and mathematics [13]. The following two versions of the EIP of a graph G have been considered in the literature [14–16].

EIP (1): Find a subset of the vertices of a given graph such that the edge cut separating this subset from its complement has minimal size among all subsets of the same cardinality.

Mathematically, if for a given m, where $m = 1, 2, \ldots, |V|$, $\theta_G(m) = \min_{A \subseteq V, |A| = m} |\theta_G(A)|$, where $\theta_G(A) = \{(u, v) \in E : u \in A, v \notin A\}$, then the problem is to find $A \subseteq V$ such that $|A| = m$ and $\theta_G(m) = |\theta_G(A)|$. Such subsets are called optimal with respect to EIP (1). If a set of vertices is optimal with respect to EIP (1), then it is trivial that its complement is also optimal to EIP (1).

EIP (2): Find a subset of the vertices of a given graph such that the number of edges in the subgraph induced by this subset is maximal among all induced subgraphs with the same number of vertices.

Mathematically, if for a given m, where $m = 1, 2, \ldots, |V|$, $I_G(m) = \max_{A \subseteq V, |A| = m} |I_G(A)|$, where $I_G(A) = \{(u, v) \in E : u, v \in A\}$, then the problem is to find $A \subseteq V$ such that $|A| = m$ and $I_G(m) = |I_G(A)|$. Such subsets are called optimal with respect to EIP (2).

Clearly, if a subset of vertices is optimal with respect to EIP (2), then its complement is also an optimal set only for regular graphs and moreover, if a subset of vertices is optimal with respect to EIP (2), it is also optimal with respect to EIP (1). In the case of non-regular graphs, if a subset of vertices is optimal with respect to EIP (2), it need not be optimal to EIP (1) and there is no specific condition to optimality [16].

We now state the congestion and partition lemmas which will be used to compute the exact wirelengths in our paper.

Lemma 1. *(Congestion Lemma) [3] Let G be an r-regular graph and f be an embedding of G into H. Let S be an edge cut of H such that the removal of edges of S leaves H into two components H_1 and H_2 and let $G_1 = G[f^{-1}(H_1)]$ and $G_2 = G[f^{-1}(H_2)]$. Also S satisfies the following conditions:*

(i) For every edge $(a, b) \in G_i$, $i = 1, 2$, $P_f((a, b))$ has no edges in S.

(ii) For every edge $(a, b) \in G$ with $a \in G_1$ and $b \in G_2$, $P_f((a, b))$ has exactly one edge in S.

(iii) G_1 is an optimal set.

Then $EC_f(S)$ is minimum and $EC_f(S) = r|V(G_1)| - 2|E(G_1)|$.

Lemma 2. *(2-Partition Lemma) [17] Let $f : G \to H$ be an embedding. Let $[2E(H)]$ denote a collection of edges of H repeated exactly 2 times. Let $\{S_1, S_2, \ldots, S_m\}$ be a partition of $[2E(H)]$ such that each S_i is an edge cut of H. Then*

$$WL_f(G,H) = \frac{1}{2}\sum_{i=1}^{m} EC_f(S_i).$$

3. Properties of Enhanced Hypercubes

The hypercube (Q_n) has received extensive attention in view of its regular structure, small diameter and good connection with a relatively small vertex degree [7,18]. As the effort to improve its efficiency, several variants of Q_n have been proposed.

In many variants of hypercube, the topological structure of enhanced hypercube network $(Q_{n,k})$ is considered to be a significant topology due mainly to its reliability, efficiency and the fault tolerance of $Q_{n,k}$ are better than Q_n, which shows that the enhanced hypercube is an excellent choice of network topology to improve traffic distributions, bandwidth capabilities and performance in parallel processing computer systems [18].

Definition 1 ([19])**.** *The enhanced hypercube $Q_{n,k}$, $1 \le k \le n-1$, is an $(n+1)$-regular graph with vertex set $V(Q_{n,k}) = V(Q_n)$ labeled as $\{0, 1 \ldots, 2^n - 1\}$ and edge set $E(Q_{n,k}) = E(Q_n) \cup \{\ (x_1x_2 \ldots x_kx_{k+1} \ldots x_n,\ x_1x_2 \ldots x_{k-1}\overline{x}_k\overline{x}_{k+1} \ldots \overline{x}_n),\ x_i = 0 \text{ or } 1, 1 \le i \le n\ \}$. The edges of Q_n in $Q_{n,k}$ are called hypercube edges and the remaining edges of $Q_{n,k}$ are called complementary edges.*

Remark 1. $|V(Q_{n,k})| = 2^n$ *and* $|E(Q_{n,k})| = (n+1)2^{n-1}$.

Theorem 1 ([10])**.** *For $1 \le i \le 2^n$, $L_i = \{0, 1, \ldots, i-1\}$ is an optimal set in $Q_{n,k}$.*

Lemma 3 ([10])**.** *For $1 \le i \le 2^n$, $|E(Q_{n,k}[L_i])| = |E(Q_n[L_i])| + \left\lfloor \frac{i}{2^{n-k+1}} \right\rfloor 2^{n-k} + [x - 2^{n-k}]^+$ where $x = i - \left\lfloor \frac{i}{2^{n-k+1}} \right\rfloor 2^{n-k+1}$ and*

$$[x]^+ = \begin{cases} 0 & : \ x < 0 \\ x & : \ x \ge 0. \end{cases}$$

4. Computation of Wirelength

In this section, we compute the exact wirelength of enhanced hypercubes into windmill and necklace graphs. The basic definitions and results to obtain the minimum wirelength are explained as follows.

Lemma 4. *For $i = 1, 2, \ldots, n-1$, $Ncut\ S_i^{2^i} = \{2^i, 2^i+1, \ldots, 2^{i+1}-1\}$ is an optimal set in $Q_{n,k}$.*

Proof. Define $\varphi : Ncut\ S_i^{2^i} \to L_{2^i}$ by $\varphi(2^i + p) = p$. Let the binary representation of $2^i + p$ be $\alpha_1\alpha_2 \ldots \alpha_n$. Then the binary representation of p is $\underbrace{00\ldots00}_{(n-i)\ times}\ \alpha_{n-i+1} \ldots \alpha_n$. To show that $Q_{n,k}[Ncut\ S_i^{2^i}]$ is isomorphic to $Q_{n,k}[L_{2^i}]$, we discuss the following cases for $(x, y) \in E(Q_{n,k}[Ncut\ S_i^{2^i}])$.

Case 1. Let (x, y) be the hypercube edge in $Q_{n,k}[Ncut\ S_i^{2^i}]$. Suppose the binary representations of x and y are

$$x = \alpha_1\alpha_2 \ldots \alpha_{n-i}\beta_1\beta_2 \ldots \beta_k \ldots \beta_i,$$

$$y = \alpha_1\alpha_2 \ldots \alpha_{n-i}\beta_1\beta_2 \ldots \overline{\beta}_k \ldots \beta_i.$$

Then,

$$\varphi(x) = \underbrace{00\ldots00}_{(n-i)\ times}\ \beta_1\beta_2\ldots\beta_k\ldots\beta_i,$$

$$\varphi(y) = \underbrace{00\ldots00}_{(n-i)\ times}\ \beta_1\beta_2\ldots\overline{\beta}_k\ldots\beta_i.$$

hence $(x,y) \in E(Q_{n,k}[Ncut\ S_i^{2^i}]) \Leftrightarrow$ the binary representation of x and y differ in exactly one bit $\Leftrightarrow$ the binary representation of $\varphi(x)$ and $\varphi(y)$ differ in exactly one bit $\Leftrightarrow (\varphi(x),\varphi(y)) \in E(Q_{n,k}[L_{2^i}])$.

Case 2. Let (x,y) be the complementary edge in $Q_{n,k}[Ncut\ S_i^{2^i}]$. Suppose the binary representations of x and y are

$$x = \alpha_1\alpha_2\alpha_3\ldots\alpha_{n-i}\alpha_{n-i+1}\ldots\alpha_{k-1}\alpha_k\alpha_{k+1}\ldots\alpha_{n-1}\alpha_n,$$

$$y = \alpha_1\alpha_2\alpha_3\ldots\alpha_{n-i}\alpha_{n-i+1}\ldots\alpha_{k-1}\overline{\alpha}_k\overline{\alpha}_{k+1}\ldots\overline{\alpha}_{n-1}\overline{\alpha}_n.$$

Then,

$$\varphi(x) = \underbrace{00\ldots00}_{(n-i)\ times}\ \alpha_{n-i+1}\ldots\alpha_{k-1}\alpha_k\alpha_{k+1}\ldots\alpha_n,$$

$$\varphi(y) = \underbrace{00\ldots00}_{(n-i)\ times}\ \alpha_{n-i+1}\ldots\alpha_{k-1}\overline{\alpha}_k\overline{\alpha}_{k+1}\ldots\overline{\alpha}_n.$$

hence $(x,y) \in E(Q_{n,k}[Ncut\ S_i^{2^i}]) \Leftrightarrow$ the binary representations of x and y differ from the kth to n^{th} bits $\Leftrightarrow$ the binary representations of $\varphi(x)$ and $\varphi(y)$ differ from the kth to n^{th} bits $\Leftrightarrow (\varphi(x),\varphi(y)) \in E(Q_{n,k}[L_{2^i}])$. Hence $Q_{n,k}[Ncut\ S_i^{2^i}]$ and $Q_{n,k}[L_{2^i}]$ are isomorphic.
From the above cases and Theorem 1, we infer that $Ncut\ S_i^{2^i}$ is an optimal set in $Q_{n,k}$. □

The following result is an easy consequence of Lemma 4.

Lemma 5. *For* $i = 1,2,\ldots,n-1$, $Ncut\ S_i^{2^i-1} = \{2^i, 2^i+1,\ldots,2^{i+1}-2\}$ *is an optimal set in* $Q_{n,k}$.

Definition 2. *[11] Let* K_{t_i} *be a complete graph on* t_i *vertices,* $1 \le i \le m$. *Let* $t_1 = 2^r$ *and* $t_i = 2^{r+i-2}+1$ *for all* $2 \le i \le m$ *such that* $\biguplus_{i=1}^{m} K_{t_i}$ *has one vertex (s) as common. The resultant graph is called a windmill graph and is denoted by* $WM(K_{t_1}, K_{t_2},\ldots,K_{t_m})$.

Remark 2. *We denote* $w_k = \sum_{i=1}^{k}(t_i-1)+1$, $1 \le k \le m$ *and* $w_0 = t_0 = 0$. *Then the windmill graph has* $w_m = 2^n$ *vertices, see Figure 1.*

Theorem 2. *The wirelength of* $Q_{n,k}$ *into* $WM(K_{t_1}, K_{t_2},\ldots,K_{t_m})$ *is given by* $WL(Q_{n,k}, WM(K_{t_1}, K_{t_2},\ldots,K_{t_m})) = \frac{1}{4}\{(n+1)(2^{n+1}+2^{m+r}-4)\} - \sum_{i=1}^{m}|E(Q_{n,k}[L_{t_i-1}])|$.

Proof. The proof is divided into three parts A, B, and C comprising of the embedding algorithm, proof of correctness, and computation of wirelength, respectively.

Part A:

Label the vertices of $Q_{n,k}$ by lexicographic order from 0 to 2^n-1. Label the vertices of K_{t_1} in $WM(K_{t_1}, K_{t_2},\ldots,K_{t_m})$ as $0,1,2,\ldots,t_1-1$ such that t_1-1 is the label of common vertex s. For $2 \le i \le m$, label the vertices of K_{t_i} (except s) in $WM(K_{t_1}, K_{t_2},\ldots,K_{t_m})$ as $w_{i-1}+j, j = 0,1,2,\ldots,t_i-2$. Define an embedding f of $Q_{n,k}$ into $WM(K_{t_1}, K_{t_2},\ldots,K_{t_m})$ given by $f(x) = x$.

Part B:

We assume that the labels represent the vertices to which they are assigned. Table 1 gives the notations for edge cuts of windmill graph as depicted in Figure 1.

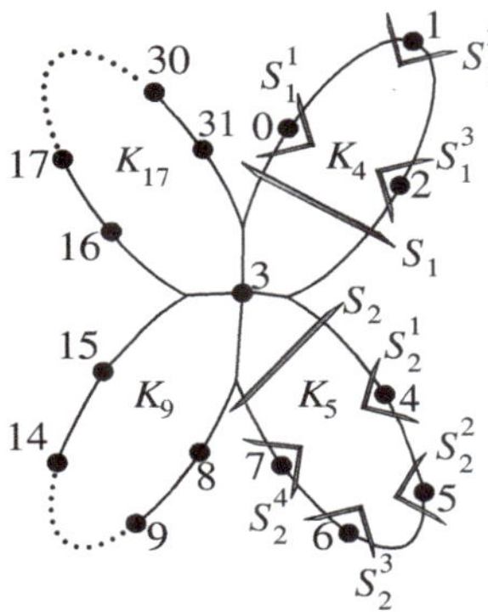

Figure 1. The edge cuts of windmill $WM(K_4, K_5, K_9, K_{17})$.

Table 1. Edge cuts in windmill graph.

Cut Notation	Elements in the Cut	Range
S_i	$\{(w_1 - 1, x) : x \in (V(K_{t_i}) - \{w_1 - 1\})\}$	$1 \le i \le m$
S_i^j	$\{(w_{i-1} + j - 1, x) : x \in (V(K_{t_i}) - \{w_{i-1} + j - 1\})\}$	$1 \le i \le m, 1 \le j \le t_i - 1$

Then $\{S_i : 1 \le i \le m\} \cup \{S_i^j : 1 \le i \le m, 1 \le j \le t_i - 1\}$ is a partition of $[2E(WM(K_{t_1}, K_{t_2}, \ldots, K_{t_m}))]$. The edge cut S_i of $WM(K_{t_1}, K_{t_2}, \ldots, K_{t_m})$ disconnects $WM(K_{t_1}, K_{t_2}, \ldots, K_{t_m})$ into two components X_i and $\overline{X_i}$ where $V(X_i) = \{w_{i-1}, w_{i-1} + 1, \ldots, w_{i-1} + t_i - 2\}$. Let G_i and $\overline{G_i}$ be the preimage of X_i and $\overline{X_i}$ under f respectively. By Lemmas 4 & 5, G_i is an optimal set and each S_i satisfies conditions (i)–(iii) of the Congestion Lemma. Therefore, $EC_f(S_i)$ is minimum.

Similarly, the edge cut S_i^j of $WM(K_{t_1}, K_{t_2}, \ldots, K_{t_m})$ disconnects $WM(K_{t_1}, K_{t_2}, \ldots, K_{t_m})$ into two components X_i^j and $\overline{X_i^j}$ where $V(X_i^j) = \{w_{i-1} + j - 1\}$. Let G_i^j and $\overline{G_i^j}$ be the preimage of X_i^j and $\overline{X_i^j}$ under f respectively. Since G_i^j is an optimal set and each S_i^j satisfies conditions $(i) - (iii)$ of the Congestion Lemma. Therefore, $EC_f(S_i^j)$ is minimum. The 2-Partition Lemma implies that $WL_f(Q_{n,k}, WM(K_{t_1}, K_{t_2}, \ldots, K_{t_m}) = WL(Q_{n,k}, WM(K_{t_1}, K_{t_2}, \ldots, K_{t_m})$.

Part C:

By Part B, we have $EC_f(S_i) = (n+1)(t_i - 1) - 2|E(Q_{n,k}[L_{t_i - 1}])|$, $EC_f(S_i^j) = (n+1)$ for all $1 \le i \le m, 1 \le j \le t_i - 1$. Therefore, the wirelength of enhanced hypercube into windmill graph is given by $WL(Q_{n,k}, WM(K_{t_1}, K_{t_2}, \ldots, K_{t_m}) = \frac{1}{2} \sum\limits_{i=1}^{m} \{(n+1)(t_i - 1) - 2|E(Q_{n,k}[L_{t_i - 1}])|\} + \frac{n+1}{2}(2^n - 1) = \frac{1}{4}\{(n+1)(2^{n+1} + 2^{m+r} - 4)\} - \sum\limits_{i=1}^{m} |E(Q_{n,k}[L_{t_i - 1}])|$. □

Definition 3 ([11]). *Let $K_{1,m}$ be a star graph on $m+1$ vertices (say $v_0, v_1, \ldots, v_m$) and K_{t_i} be complete graphs on t_i vertices, $1 \le i \le m$. Let $t_1 = 2^r$, $t_i = 2^{r+i-2}$ for all $2 \le i \le m-1$ and $t_m = 2^{r+m-2} - 1$ such that $K_{1,m} \uplus (\bigcup\limits_{i=1}^{m} K_{t_i})$ has v_i as common. The resultant graph is called a complete star necklace and is denoted by $SN(K_{1,m}; K_{t_1}, K_{t_2}, \ldots, K_{t_m})$.*

Remark 3. *We denote* $s_k = \sum_{i=0}^{k} t_i$, $0 \leq k \leq m$ *where* $t_0 = 0$. *Then the complete star necklace has* $s_m + 1 = 2^n$ *vertices, see Figure 2.*

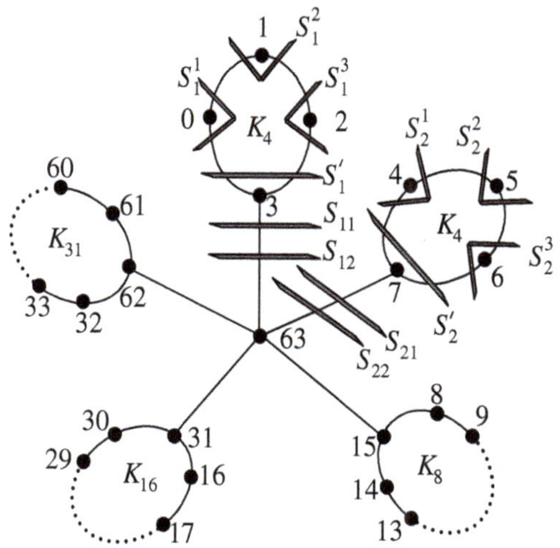

Figure 2. The edge cuts of complete star necklace $SN(K_{1,5}; K_4, K_4, K_8, K_{16}, K_{31})$.

Theorem 3. $WL(Q_{n,k}, SN(K_{1,m}; K_{t_1} K_{t_2}, \ldots, K_{t_m})) = \frac{1}{4}\{(n+1)(2^{n+1} + 3 \cdot 2^{m+r} - 4m - 8)\} - \sum_{i=1}^{m} \{2|E(Q_{n,k}[L_{t_i}])| + |E(Q_{n,k}[L_{t_i - 1}])|\}$.

Proof. The proof technique is similar to Theorem 2 as divided into three parts A, B, and C.

Part A:

Label the vertices of $Q_{n,k}$ by lexicographic order from 0 to $2^n - 1$. For $1 \leq i \leq m$, label the vertices of K_{t_i} in $SN(K_{1,m}; K_{t_1}, K_{t_2}, \ldots, K_{t_m})$ as $s_{i-1} + j, j = 0, 1, 2, \ldots, t_i - 1$ such that $s_i - 1$ is the label of v_i, and v_0 as $2^n - 1$. Define an embedding f of $Q_{n,k}$ into $SN(K_{1,m}; K_{t_1}, K_{t_2}, \ldots, K_{t_m})$ given by $f(x) = x$.

Part B:

We assume that the labels represent the vertices to which they are assigned. Table 2 gives the notations for edge cuts of complete star necklace graph as depicted in Figure 2.

Table 2. Edge cuts in complete star necklace graph.

Cut Notation	Elements in the Cut	Range
S_{i1}	$\{s_i - 1, 2^n - 1\}$	$1 \leq i \leq m$
S_{i2}	$\{s_i - 1, 2^n - 1\}$	$1 \leq i \leq m$
$S_i^{'}$	$\{(s_i - 1, x) : x \in (V(K_{t_i}) - \{s_i - 1\})\}$	$1 \leq i \leq m$
S_i^{j}	$\{(s_{i-1} + j - 1, x) : x \in (V(K_{t_i}) - \{s_{i-1} + j - 1\})\}$	$1 \leq i \leq m, 1 \leq j \leq t_i - 1$

Then $\{S_{i1},\ S_{i2},\ S_i^{'} : 1 \leq i \leq m\} \cup \{S_i^j : 1 \leq i \leq m,\ 1 \leq j \leq t_i - 2\}$ is a partition of $[2E(SN(K_{1,m}; K_{t_1}, K_{t_2}, \ldots, K_{t_m}))]$. The edge cut S_{i1} of $SN(K_{1,m}; K_{t_1}, K_{t_2}, \ldots, K_{t_m})$ disconnects $SN(K_{1,m}; K_{t_1}, K_{t_2}, \ldots, K_{t_m})$ into two components X_i and $\overline{X_i}$ where $V(X_i) = \{s_{i-1}, s_{i-1} + 1, \ldots, s_i - 1\}$. Let G_i and $\overline{G_i}$ be the preimage of X_i and $\overline{X_i}$ under f respectively. By Lemma 4, G_i is an optimal set and each S_{i1} satisfies conditions (i)–(iii) of the Congestion Lemma. Therefore, $EC_f(S_{i1})$ is minimum. Similarly, $EC_f(S_{i2})$ is minimum.

The edge cut $S_i^{'}$ of $SN(K_{1,m}; K_{t_1}, K_{t_2}, \ldots, K_{t_m})$ disconnects $SN(K_{1,m}; K_{t_1}, K_{t_2}, \ldots, K_{t_m})$ into two components $X_i^{'}$ and $\overline{X_i^{'}}$ where $V(X_i^{'}) = \{s_{i-1}, s_{i-1}+1, \ldots, s_i - 2\}$. Let $G_i^{'}$ and $\overline{G_i^{'}}$ be the preimage of $X_i^{'}$ and $\overline{X_i^{'}}$ under f respectively. By Lemma 5, $G_i^{'}$ is an optimal set and each $S_i^{'}$ satisfies conditions (i)–(iii) of the Congestion Lemma. Therefore, $EC_f(S_i^{'})$ is minimum.

The edge cut S_i^j of $SN(K_{1,m}; K_{t_1}, K_{t_2}, \ldots, K_{t_m})$ disconnects $SN(K_{1,m}; K_{t_1}, K_{t_2}, \ldots, K_{t_m})$ into two components X_i^j and $\overline{X_i^j}$ where $V(X_i^j) = \{s_{i-1} + j - 1\}$. Let G_i^j and $\overline{G_i^j}$ be the preimage of X_i^j and $\overline{X_i^j}$ under f respectively. Since G_i^j is an optimal set and each S_i^j satisfies conditions $(i) - (iii)$ of the Congestion Lemma. Therefore, $EC_f(S_j^{'})$ is minimum. The 2-Partition Lemma implies that $WL_f(Q_{n,k}, SN(K_{1,m}; K_{t_1}, K_{t_2}, \ldots, K_{t_m})) = WL(Q_{n,k}, SN(K_{1,m}; K_{t_1}, K_{t_2}, \ldots, K_{t_m}))$.

Part C:

By Part B, we have $EC_f(S_{i1}) = EC_f(S_{i2}) = (n+1)t_i - 2|E(Q_{n,k}[L_{t_i}])|$, $EC_f(S_i^{'}) = (n+1)(t_i - 1) - 2|E(Q_{n,k}[L_{t_i - 1}])|$, $EC_f(S_i^j) = (n+1)$ for all $1 \leq i \leq m$, $1 \leq j \leq t_i - 1$. Therefore, the wirelength of enhanced hypercube into complete star necklace graph is given by $WL(Q_{n,k}, SN(K_{1,m}; K_{t_1}, K_{t_2}, \ldots, K_{t_m})) = \sum_{i=1}^{m} \{(n+1)t_i - 2|E(Q_{n,k}[L_{t_i}])|\} + \frac{1}{2}\sum_{i=1}^{m}\{(n+1)(t_i - 1) - 2|E(Q_{n,k}[L_{t_i - 1}])|\} + \frac{n+1}{2}(2^n - m - 1) = \frac{1}{4}\{(n+1)(2^{n+1} + 3(2^{m+r}) - 4m - 8)\} - \sum_{i=1}^{m}\{2|E(Q_{n,k}[L_{t_i}])| + |E(Q_{n,k}[L_{t_i - 1}])|\}$. □

Definition 4 ([11]). *Let K_m be a complete graph on m vertices (say $v_1, v_2, \ldots, v_m$) and K_{t_i} be complete graphs on t_i vertices, $1 \leq i \leq m$. Let $t_1 = 2^r$ and $t_i = 2^{r+i-2}$ for all $2 \leq i \leq m$ such that $K_m \uplus (\bigcup_{i=1}^{m} K_{t_i})$ has v_i as common. The resultant graph is called a circular necklace graph and is denoted by $CN(K_m; K_{t_1}, K_{t_2}, \ldots, K_{t_m})$.*

Remark 4. *We denote $c_k = \sum_{i=0}^{k} t_i$, $0 \leq k \leq m$ where $t_0 = 0$. Then the circular necklace has $c_m = 2^n$ vertices, see Figure 3.*

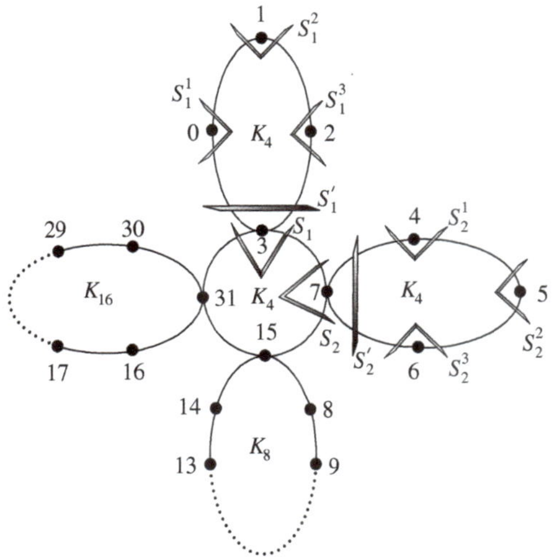

Figure 3. The edge cuts of circular necklace $CN(K_4; K_4, K_4, K_8, K_{16})$.

Theorem 4. *The wirelength of $Q_{n,k}$ into $CN(K_m; K_{t_1}, K_{t_2}, \ldots, K_{t_m})$ is given by $WL(Q_{n,k}, CN(K_m; K_{t_1}, K_{t_2}, \ldots, K_{t_m})) = \frac{1}{2}\{(n+1)(2^{m+r} + 2^n - 2m)\} - \sum_{i=1}^{m}\{|E(Q_{n,k}[L_{t_i}])| + |E(Q_{n,k}[L_{t_i - 1}])|\}$.*

Proof. We label the vertices of $Q_{n,k}$ by lexicographic order from 0 to $2^n - 1$. For $1 \leq i \leq m$, label the vertices of K_{t_i} in $CN(K_m; K_{t_1}, K_{t_2}, \dots, K_{t_m})$ as $c_{i-1} + j, j = 0, 1, 2, \dots, t_i - 1$ such that $c_i - 1$ is the label of v_i. Define an embedding f of $Q_{n,k}$ into $CN(K_m; K_{t_1}, K_{t_2}, \dots, K_{t_m})$ given by $f(x) = x$.

We assume that the labels represent the vertices to which they are assigned. Table 3 gives the notations for edge cuts of circular necklace graph as depicted in Figure 3.

Table 3. Edge cuts in circular necklace graph.

Cut Notation	Elements in the Cut	Range
S_i	$\{(c_i - 1, x) : x \in (V(K_m) - \{c_i - 1\})\}$	$1 \leq i \leq m$
$S_i^{'}$	$\{(c_i - 1, x) : x \in (V(K_{t_i}) - \{c_i - 1\})\}$	$1 \leq i \leq m$
S_i^j	$\{(c_{i-1} + j - 1, x) : x \in (V(K_{t_i}) - \{c_{i-1} + j - 1\})\}$	$1 \leq i \leq m, 1 \leq j \leq t_i - 1$

Then $\{ S_i, S_i^{'} : 1 \leq i \leq m \} \cup \{ S_i^j : 1 \leq i \leq m, 1 \leq j \leq t_i - 2 \}$ is a partition of $[2E(CN(K_m; K_{t_1}, K_{t_2}, \dots, K_{t_m}))]$. The edge cut S_i of $CN(K_m; K_{t_1}, K_{t_2}, \dots, K_{t_m})$ disconnects $CN(K_m; K_{t_1}, K_{t_2}, \dots, K_{t_m})$ into two components X_i and $\overline{X_i}$ where $V(X_i) = \{c_{i-1}, c_{i-1} + 1, \dots, c_i - 1\}$. Let G_i and $\overline{G_i}$ be the preimage of X_i and $\overline{X_i}$ under f respectively. By Lemma 4, G_i is an optimal set and each S_i satisfies conditions $(i) - (iii)$ of the Congestion Lemma. Therefore, $EC_f(S_i)$ is minimum.

The edge cut $S_i^{'}$ of $CN(K_m; K_{t_1}, K_{t_2}, \dots, K_{t_m})$ disconnects $CN(K_m; K_{t_1}, K_{t_2}, \dots, K_{t_m})$ into two components $X_i^{'}$ and $\overline{X_i^{'}}$ where $V(X_i^{'}) = \{c_{i-1}, c_{i-1} + 1, \dots, c_i - 2\}$. Let $G_i^{'}$ and $\overline{G_i^{'}}$ be the preimage of $X_i^{'}$ and $\overline{X_i^{'}}$ under f respectively. By Lemma 5, $G_i^{'}$ is an optimal set and each $S_i^{'}$ satisfies conditions (i)–(iii) of the Congestion Lemma. Therefore, $EC_f(S_i^{'})$ is minimum.

The edge cut S_i^j of $CN(K_m; K_{t_1}, K_{t_2}, \dots, K_{t_m})$ disconnects $CN(K_m; K_{t_1}, K_{t_2}, \dots, K_{t_m})$ into two components X_i^j and $\overline{X_i^j}$ where $V(X_i^j) = \{c_{i-1} + j - 1\}$. Let G_i^j and $\overline{G_i^j}$ be the preimage of X_i^j and $\overline{X_i^j}$ under f respectively. Since G_i^j is an optimal set and each S_i^j satisfies conditions (i)–(iii) of the Congestion Lemma. Therefore $EC_f(S_j^{'})$ is minimum. The 2-Partition Lemma implies that $WL_f(Q_{n,k}, CN(K_m; K_{t_1}, K_{t_2}, \dots, K_{t_m})) = WL(Q_{n,k}, CN(K_m; K_{t_1}, K_{t_2}, \dots, K_{t_m}))$.

Now, we have $EC_f(S_i) = (n + 1)t_i - 2|E(Q_{n,k}[L_{t_i}])|$, $EC_f(S_i^{'}) = (n + 1)(t_i - 1) - 2|E(Q_{n,k}[L_{t_i - 1}])|$, $EC_f(S_i^j) = (n + 1)$ for all $1 \leq i \leq m, 1 \leq j \leq t_i - 1$. Therefore, the wirelength of enhanced hypercube into circular necklace graph is given by $WL(Q_{n,k}, CN(K_m; K_{t_1}, K_{t_2}, \dots, K_{t_m}))$ $= \frac{1}{2} \sum\limits_{i=1}^{m} \{(n + 1)t_i - 2|E(Q_{n,k}[L_{t_i}])|\} + \frac{1}{2} \sum\limits_{i=1}^{m} \{(n + 1)(t_i - 1) - 2|E(Q_{n,k}[L_{t_i - 1}])|\} + \frac{n+1}{2}(2^n - m) =$ $\frac{1}{2}\{(n + 1)(2^{m+r} + 2^n - 2m)\} - \sum\limits_{i=1}^{m} \{|E(Q_{n,k}[L_{t_i}])| + |E(Q_{n,k}[L_{t_i - 1}])|\}$. □

5. Conclusions

In this paper, we have computed the minimum wirelength of embedding enhanced hypercube into host graph such as windmill and necklace graphs by partitioning the edge set of the host graph. On comparing with the wirelength of hypercube into windmill and necklace graphs, we found that the computation varies by degree of enhanced hypercube. The results obtained in this paper would build a great impact on parallel computing systems. Furthermore, it would be an interesting line of research to compute the wirelength of general r-regular graph into windmill and necklace graphs.

Author Contributions: Conceptualization, M.A. and J.N.D.; methodology, M.A. and J.N.D.; investigation, M.A. and J.N.D.; writing—original draft preparation, M.A.and J.N.D.; writing—review and editing, M.A. and J.N.D.; supervision, M.A. and J.-B.L.; funding acquisition, J.-B.L.

Funding: This research was funded by the China Postdoctoral Science Foundation under Grant 2017M621579; the Postdoctoral Science Foundation of Jiangsu Province under Grant 1701081B; Project of Anhui Jianzhu University under Grant no. 2016QD116 and 2017dc03.

Conflicts of Interest: The authors declare no conflict of interest.

References

1. Abraham, J.; Arockiaraj, M. Wirelength of enhanced hypercubes into r-rooted complete binary trees. *Electron. Notes Discrete Math.* **2016**, *53*, 373–382. [CrossRef]
2. Bezrukov, S.L.; Chavez, J.D.; Harper, L.H.; Röttger, M.; Schroeder, U.-P. The congestion of n-cube layout on a rectangular grid. *Discret. Math.* **2000**, *213*, 13–19. [CrossRef]
3. Manuel, P.; Rajasingh, I.; Rajan, B.; Mercy, H. Exact wirelength of hypercubes on a grid. *Discrete Appl. Math.* **2009**, *157*, 1486–1495. [CrossRef]
4. Bhatt, S.N.; Leighton, F.T. A framework for solving VLSI graph layout problems. *J. Comp. Syst. Sci.* **1984**, *28*, 300–343. [CrossRef]
5. Leighton, F.T. *Introduction to Parallel Algorithms and Architectures: Arrays, Trees, Hypercubes*; Morgan Kaufmann Publishers: San Mateo, CA, USA, 1992.
6. Lai, Y.L.; Williams, K. A survey of solved problems and applications on bandwidth, edgesum, and profile of graphs. *J. Graph Theory* **1999**, *31*, 75–94. [CrossRef]
7. Xu, J. *Topological Structure and Analysis of Interconnection Networks*; Springer Publishing Company: Dordrecht, The Netherlands, 2001.
8. Rajasingh, I.; Rajan, B.; Mercy, H.; Manuel, P. Exact wirelength of hypercube and enhanced hypercube layout on wounded lobstars. In Proceedings of the 4th International Multiconference on Computer Science and Information Technology, Amman, Jordan, 5–7 April 2006; pp. 449–454.
9. Manuel, P. Minimum average congestion of enhanced and augmented hypercubes into complete binary trees. *Discret. Appl. Math.* **2011**, *159*, 360–366. [CrossRef]
10. Arockiaraj, M.; Liu, J.B.; Shalini, A.J. Vertex decomposition method for wirelength problem and its applications to enhanced hypercube networks. *IET Comput. Digit. Tech.* **2019**, *13*, 87–92. [CrossRef]
11. Rajasingh, I.; Rajan, B.; Rajan, R.S. Embedding of hypercubes into necklace, windmill and snake graphs. *Inf. Process. Lett.* **2012**, *112*, 509–515. [CrossRef]
12. Garey, M.R.; Johnson, D.S.; Stockmeyer, L. Some simplified NP-complete problems. In Proceedings of the Sixth Annual ACM Symposium on Theory of Computing, Seattle, WA, USA, 30 April–2 May 1974; pp. 47–63.
13. Harper, L.H. *Global Methods for Combinatorial Isoperimetric Problems*; Cambridge University Press: Cambridge, UK, 2004.
14. Boals, A.J.; Gupta, A.K.; Sherwani, N.A. Incomplete hypercubes: Algorithms and embeddings. *J. Supercomput.* **1994**, *8*, 263–294. [CrossRef]
15. Garey, M.R.; Johnson, D.S. *Computers and Intractability, A Guide to the Theory of NP-Completeness*; Freeman: San Francisco, CA, USA, 1979.
16. Bezrukov, S.L.; Das, S.K.; Elsässer, R. An edge-isoperimetric problem for powers of the petersen graph. *Ann. Comb.* **2000**, *4*, 153–169. [CrossRef]
17. Arockiaraj, M.; Manuel, P.; Rajasingh, I.; Rajan, B. Wirelength of 1-fault Hamiltonian graphs into wheels and fans. *Inf. Process. Lett.* **2011**, *111*, 921–925. [CrossRef]
18. Liu, H. The Structural Features of Enhanced Hypercube Networks. In Proceedings of the 5th International Conference on Natural Computation, Tianjin, China, 14–16 April 2009; pp. 345–348.
19. Tzeng, N.F.; Wei, S. Enhanced hypercubes. *IEEE Trans. Comput.* **1991**, *40*, 284–294. [CrossRef]

mathematics

MDPI

Article

On Subtrees of Fan Graphs, Wheel Graphs, and "Partitions" of Wheel Graphs Under Dynamic Evolution

Yu Yang [1,2], An Wang [3,*], Hua Wang [4,5,*], Wei-Ting Zhao [2] and Dao-Qiang Sun [2]

1 School of Mathematical Sciences, Shanghai Jiao Tong University, Shanghai 200240, China; yangyu@sjtu.edu.cn
2 School of Computer Science, Pingdingshan University, Pingdingshan 467000, China; pdsxyzwt@163.com (W.-T.Z.); aqiang_de@163.com (D.-Q.S.)
3 School of Mathematics and Statistics Science, Pingdingshan University, Pingdingshan 467000, China
4 College of Software, Nankai University, Tianjin 300071, China
5 Department of Mathematical Sciences, Georgia Southern University, Statesboro, GA 30460, USA
* Correspondence: kedawangan@163.com (A.W.); hwang@georgiasouthern.edu (H.W.)

Received: 18 April 2019; Accepted: 20 May 2019; Published: 24 May 2019

Abstract: The number of subtrees, or simply the subtree number, is one of the most studied counting-based graph invariants that has applications in many interdisciplinary fields such as phylogenetic reconstruction. Motivated from the study of graph surgeries on evolutionary dynamics, we consider the subtree problems of fan graphs, wheel graphs, and the class of graphs obtained from "partitioning" wheel graphs under dynamic evolution. The enumeration of these subtree numbers is done through the so-called subtree generation functions of graphs. With the enumerative result, we briefly explore the extremal problems in the corresponding class of graphs. Some interesting observations on the behavior of the subtree number are also presented.

Keywords: subtree; generating function; fan graph; wheel graph; "partitions" of wheel graph

1. Introduction

The study of graph invariants or topological indices has been proven to be of crucial importance in various interdisciplinary topics. Generally, the existing known topological indices could be divided into distance-based, degree-based, or structure-based ones; some of the studies on these three categories are referred to in [1–4] and the references therein. The number of subtrees, or simply the subtree number, is one of the counting-based graph invariants.

Finding and/or enumerating special topological structures or graph patterns has become an important problem due to their applications including, to name a few, frequent subgraphs mining [5], network optimization design [6,7], and local network reliability [8,9]. In particular, the subtree number has also been shown to be correlated to phylogenetic reconstruction [10] and various chemical indices such as the Wiener index (closely correlated with the boiling point of paraffin [3]), the Merrifield–Simmons index, and the Hosoya index [11]. Research results also show that there exists an amazing "negative correlation" between the number of subtrees and the Wiener index [12–17]. Therefore, the subtree number index can indirectly characterize the physical–chemical characteristics of molecules.

Various topics related to the subtree number have been explored over the past years, such as extremal problems [12,18,19], the subtree density problem [20–22], generic chemical structures storage problems [23,24], fault tolerant computing and parallel scheduling [25,26], recognition of substructure [27], simplification of the flow networks [28], Markovian queuing systems decomposition [29], and matching problems [30,31].

By using "generating functions", Yan and Yeh [4] presented a linear-time algorithm to evaluate the subtree weight sum of a tree, leading to an algorithmic approach to compute the subtree number of a tree. Following this approach and more recently, Yang et al. [32,33] proposed enumerating algorithms for BC-subtrees of trees, unicyclic, and edge-disjoint bicyclic graphs and computed the subtree number on spiro and polyphenyl hexagonal chains [17]. With structure mapping and weights transferring of cycles, Yang et al. [34] also presented enumerating algorithms for subtrees of hexagonal and phenylene chains. More recently, Chin et al. [35] presented the subtree number of complete graphs, complete bipartite graphs, and theta graphs, as well as the ratio of spanning trees to all subtrees of the above graphs.

Let $K_{1,n} = (V(K_{1,n}), E(K_{1,n}); f, g)$ be a weighted star on $n+1$ vertices with center c_0, vertex weight function $f(v) = y$ for all $v \in V(K_{1,n})$, and edge weight function $g(e) = z$ for all $e \in E(K_{1,n})$. Suppose the n leaves are labeled, in counterclockwise order, $c_1, c_2, \ldots, c_n$, then the wheel graph W_n is obtained from $K_{1,n}$ by adding the edges (c_i, c_{i+1}) for $i = 1, \ldots, n$ (here we let $c_{n+1} = c_1$).

As is well known, star $K_{1,n}$ and wheel graphs W_n are very typical network topologies used in designing and implementing communication networks. The wheel graph W_n is the planar graph with a chromatic number not greater than 4. Specifically, the chromatic number of W_n is 3 for odd n and 4 for even n. Some researchers have tried to apply this property of W_n to prove the famous Four Color Theorem [36]. Moreover, W_n can also be used in wireless ad hoc networks [37]. Therefore, it is worth studying the new structural characteristics of these two graphs, especially from new perspectives.

Inspired by the significant effect of graph surgeries on evolutionary dynamics in [38], we are motivated to study the subtree number in graphs resulting from graph surgeries. In particular, we will consider a class of graphs that "lie in between" the fan and wheel graphs, called the "partitions" (explanation will be given later) of wheel graphs.

Definition 1. *Given the star $K_{1,n}$ on $n+1$ vertices defined above with center vertex c_0 and $0 \le j \le n$, the graph $K_{1,n}^j$ is obtained from $K_{1,n}$ by adding the edges (c_s, c_{s+1}) for all $1 \le s \le n-1$ except for $s = j, 2j, 3j, \ldots$ (see Figure 1a). If $n = j > 0$, we call $K_{1,j}^j$ the fan graph on $j+1$ vertices. It is also easy to see that $K_{1,0}^0$ is the single vertex c_0, and $K_{1,n}^1$ is the star $K_{1,n}$.*

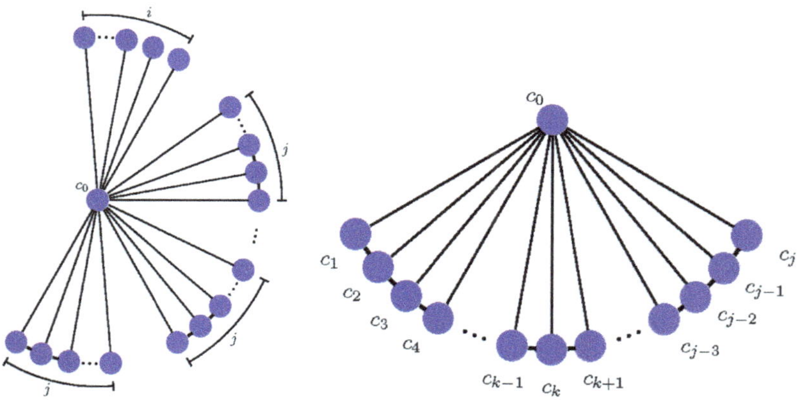

(**a**) The graph $K_{1,n}^j$. (**b**) The fan graph $K_{1,j}^j$.

Figure 1. The graph $K_{1,n}^j$ and the fan graph $K_{1,j}^j$.

From the definitions of the graph $K_{1,n}^j$ and the wheel graph W_n, we see that the graph $K_{1,n}^j$ can be vividly described in terms of W_n through "partitions" (and deletion of edges $\{(c_s, c_{s+1}) | s = j, 2j, 3j, \ldots\}$) perspective.

The reason that we consider our graphs as vertex and edge weighted is to use the so-called subtree generating functions. We assume a graph $G = (V(G), E(G); f, g)$ to be a weighted graph and define the vertex-weight function $f : V(G) \to \Re$ and edge-weight function $g : E(G) \to \Re$, where $\Re$ is the real number. Unless otherwise noted, for a graph G, we initialize its vertex weight function $f(v) = y$ for each $v \in V(G)$ and its edge weight function $g(e) = z$ for each $e \in E(G)$ throughout this paper.

The following notations need to be listed before introducing the main tool:

- $G - X$ denotes the graph obtained from G by removing all elements of X;
- $S(G)$ (resp. $S(G; v)$) denotes the set of subtrees of G (resp. containing v);
- $S(G; (a, b))$ denotes the set of subtrees of G containing the edge (a, b);
- $\omega(T_1)$ denotes the weight of subtree $T_1 \in S(G)$;
- $F(\cdot)$ is the sum of weights of subtrees in $S(\cdot)$;
- $\eta(\cdot)$ is the cardinality (namely the number) of the corresponding $S(\cdot)$ set of subtrees.

We define the weight of a subtree T_s of G, denoted by $\omega(T_s)$, as the product of the weights of the vertices and edges in T_s. The generating function of subtrees of G, denoted by $F(G; f, g)$, is the sum of weights of subtrees of G. Namely, $F(G; f, g) = \sum_{T_1 \in S(G)} \omega(T_1)$. Similarly, we define $F(G; f, g; v_i) = \sum_{T_1 \in S(G; v_i)} \omega(T_1), F(G; f, g; (u, v)) = \sum_{T_1 \in S(G; (u, v))} \omega(T_1)$.

By substituting each vertex weight $y = 1$ and edge weight $z = 1$ in these generating functions, we have the corresponding numbers of subtrees under various constraints, i.e., $\eta(T) = F(T; 1, 1), \eta(T; v_i) = F(T; 1, 1; v_i)$, and $\eta(T; (u, v)) = F(T; 1, 1; (u, v))$.

We introduce the following lemma, which will frequently be used in our work.

Lemma 1. *[4] Let P_n be a path on n vertices, with vertex weight function $f(v) = y$ for all $v \in V(P_n)$ and edge weight function $g(e) = z$ for all $e \in E(P_n)$, then $F(P_n; f, g) = \sum_{t=0}^{n-1} (n - t) y^{t+1} z^t$.*

In Section 2, we will present the subtree generating functions of $K_{1,n}^j (1 \le j \le n)$ and the wheel graph W_n. Through using these generating functions and theoretical analysis, we study the extremal graphs, subtree fitting problems, and subtree density behaviors of these graphs in Section 3. Lastly, in Section 4, we summarize our results and comment on potential topics for future work.

2. Subtree Generating Functions of $K_{1,N}^J (1 \le J \le N)$ and Wheel Graph W_n

In this section, we will establish the subtree generating functions of $K_{1,n}^j (1 \le j \le n)$ and wheel graph W_n and provide the theoretical background for our computational analysis. We start by studying the subtree problem of $K_{1,n}^j (1 \le j \le n)$.

2.1. Subtree Generating Functions and Subtree Numbers of $K_{1,N}^J$

Theorem 1. *Let $K_{1,n}^j$ be the weighted graph defined as above, and n, j be non-negative integers with $0 \le j \le n$ and $n \equiv i \ (mod\ j)$. Then*

$$F(K_{1,n}^j; f, g) = F(K_{1,j}^j; f, g; c_0)^{\frac{n-i}{j}} * y(1 + yz)^i * y^{-\frac{n-i}{j}} + iy + \frac{n-i}{j} \sum_{t=0}^{j-1} (j - t) y^{t+1} z^t, \tag{1}$$

with $F(K_{1,j}^j; f, g; c_0) = F(K_{1,j-1}^{j-1}; f, g; c_0) + \sum_{r=1}^{j} r(yz)^r F(K_{1,j-r}^{j-r}; f, g; c_0)$, *and* $F(K_{1,0}^0; f, g; c_0) = y$, $F(K_{1,1}^1; f, g; c_0) = y + y^2 z$.

Proof. We consider the subtrees of $K_{1,n}^j$ by cases

(i) not containing the center c_0,
(ii) containing the center c_0.

From Lemma 1, we have the subtree generating function of case (i) as

$$\frac{n-i}{j}\sum_{t=0}^{j-1}(j-t)y^{t+1}z^t + iy. \tag{2}$$

With the contraction method of [4] and structure analysis, we have the subtree generating function of case (ii) as

$$F(K_{1,j}^j; f, g; c_0)^{\frac{n-i}{j}} * y(1+yz)^i * y^{-\frac{n-i}{j}}. \tag{3}$$

Denote $e_j = (c_0, c_j)$ and $\widetilde{e_j} = (c_{j-1}, c_j)$, and divide all subtrees for $S(K_{1,j}^j; c_0)$ (see Figure 1b) into four cases $S(K_{1,j}^j; c_0) = \mathcal{S}_1 \cup \mathcal{S}_2 \cup \mathcal{S}_3 \cup \mathcal{S}_4$ where

- $\mathcal{S}_1$ is the collection of subtrees that contain neither e_j nor $\widetilde{e_j}$;
- $\mathcal{S}_2$ is the collection of subtrees that contain e_j, but not $\widetilde{e_j}$;
- $\mathcal{S}_3$ is the collection of subtrees that contain $\widetilde{e_j}$, but not e_j;
- $\mathcal{S}_4$ is the collection of subtrees that contain both e_j and $\widetilde{e_j}$.

From the definitions of subtree weight and subtree generating function, we know that:

(a) $\mathcal{S}_1 = S(K_{1,j-1}^{j-1}; c_0)$;

(b) $\mathcal{S}_2 = \{T_1 + e_j | T_1 \in \mathcal{S}_1\}$, where $T_1 + e_j$ are the trees obtained from T_1 by attaching an edge e_j at vertex c_0;

(c) We can write $\mathcal{S}_3$ as

$$\mathcal{S}_3 = \{T + (c_0, c_{j-k}) + \bigcup_{h=2-k}^{r} (c_{j-k-h+1}, c_{j-k-h+2}) | T \in S(K_{1,j-k-r}^{j-k-r}; c_0)\} \tag{4}$$

for $k = 1, 2, \ldots, j-1$ and $r = 1, 2, \ldots, j-k$;

(d) For each subtree $T_4 \in \mathcal{S}_4$, we know that T_4 must not contain the edge (c_0, c_{j-1}). Consequently, we can further consider the subtrees that contain edges $(c_0, c_j) \bigcup_{r=1}^{k} (c_{j-r}, c_{j-r+1})$ but not (c_{j-k-1}, c_{j-k}) recursively for $k = 1, 2, \ldots, j-1$.

With (a)–(d),we have

$$\sum_{T_1 \in \mathcal{S}_1} \omega(T_1) = F(K_{1,j-1}^{j-1}; f, g; c_0), \tag{5}$$

$$\sum_{T_2 \in \mathcal{S}_2} \omega(T_2) = \sum_{T_1 \in \mathcal{S}_1} f(c_j) g(e_j) \omega(T_1) = yzF(K_{1,j-1}^{j-1}; f, g; c_0), \tag{6}$$

$$\begin{aligned}\sum_{T_3 \in \mathcal{S}_3} \omega(T_3) &= \sum_{k=1}^{j-1}\left(\sum_{r=1}^{j-k}\left[\prod_{h=2-k}^{r}\left(f(c_{j-k-h+1}) g((c_{j-k-h+1}, c_{j-k-h+2}))\right) f(c_j) g(c_0, c_{j-k}) F(K_{1,j-k-r}^{j-k-r}; f, g; c_0)\right]\right) \\ &= \sum_{k=1}^{j-1}\left(\sum_{r=1}^{j-k}(yz)^{r+k} F(K_{1,j-k-r}^{j-k-r}; f, g; c_0)\right),\end{aligned} \tag{7}$$

$$\begin{aligned}\sum_{T_4 \in \mathcal{S}_4} \omega(T_4) &= f(c_j) g(e_j) \sum_{k=1}^{j-1}\left(\prod_{r=1}^{k} f(c_{j-r}) g((c_{j-r}, c_{j-r+1}))\right) F(K_{1,j-k-1}^{j-k-1}; f, g; c_0) \\ &= \sum_{k=1}^{j-1}(yz)^{k+1} F(K_{1,j-k-1}^{j-k-1}; f, g; c_0).\end{aligned} \tag{8}$$

Hence, by Equations (5)–(8), we have

$$\begin{aligned} F(K_{1,j}^{j};f,g;c_0) &= \sum_{T_1\in\mathcal{S}_1}\omega(T_1)+\sum_{T_2\in\mathcal{S}_2}\omega(T_2)+\sum_{T_3\in\mathcal{S}_3}\omega(T_3)+\sum_{T_4\in\mathcal{S}_4}\omega(T_4)\\ &=F(K_{1,j-1}^{j-1};f,g;c_0)+\sum_{r=1}^{j}r(yz)^rF(K_{1,j-r}^{j-r};f,g;c_0) \end{aligned} \tag{9}$$

with $F(K_{1,0}^{0};f,g;c_0)=y$, $F(K_{1,1}^{1};f,g;c_0)=y+y^2z$.

Combining Equations (2), (3), and (9), we have Equation (1) and the theorem follows. □

Following similar arguments to Theorem 1, we can also obtain the subtree generating functions of the graphs obtained from identifying the centers of different fan graphs $K_{1,j_t}^{j_t}$ $(t=1,2,\ldots,l)$. We skip the tedious details for the sake of space.

Theorem 2. *Let $K_{1,j_1}^{j_1},K_{1,j_2}^{j_2},\ldots,K_{1,j_l}^{j_l}$ be different weighted fan graphs, and suppose there are n_{j_t} copies of $K_{1,j_t}^{j_t}$ for $t=1,2,\ldots,l$. Let $G=\bigcup_{t=1}^{l}(K_{1,j_t}^{j_t})^{n_{j_t}}$ be the graph that is constructed by identifying the centers of these $\sum_{t=1}^{l}n_{j_t}$ fan graphs with c_0. Then*

$$F(G;f,g)=y^{1-\sum_{t=1}^{l}n_{j_t}}\prod_{t=1}^{l}F(K_{1,j_t}^{j_t};f,g;c_0)^{n_{j_t}}+\sum_{t=1}^{l}\left(n_{j_t}\sum_{r=0}^{j_t-1}(j_t-r)y^{r+1}z^r\right), \tag{10}$$

with $F(K_{1,j_t}^{j_t};f,g;c_0)=F(K_{1,j_t-1}^{j_t-1};f,g;c_0)+\sum_{r=1}^{j_t}r(yz)^rF(K_{1,j_t-r}^{j_t-r};f,g;c_0)$ and $F(K_{1,0}^{0};f,g;c_0)=y$, $F(K_{1,1}^{1};f,g;c_0)=y+y^2z$.

Actually, a single fan graph is a special case of the above discussion, so we can further obtain the subtree generating function for the subtrees containing a particular vertex. Again, we skip the similar but technical details.

Theorem 3. *Let $K_{1,j}^{j}$ be a weighted fan graph with vertex weigh f and edge weigh g (see Figure 1b), then*

$$F(K_{1,j}^{j};f,g;c_1)=F(K_{1,j}^{j};f,g;c_0)-F(K_{1,j-1}^{j-1};f,g;c_0)+\sum_{t=0}^{j-1}y^{t+1}z^t\,, \tag{11}$$

with $F(K_{1,j}^{j};f,g;c_0)=F(K_{1,j-1}^{j-1};f,g;c_0)+\sum_{r=1}^{j}r(yz)^rF(K_{1,j-r}^{j-r};f,g;c_0)$ and $F(K_{1,0}^{0};f,g;c_0)=y$, $F(K_{1,1}^{1};f,g;c_0)=y+y^2z$.

Adding an edge between any two fan graphs (to construct a bigger fan) of a graph will also increase the number of subtrees. Let G be the weighted graph as defined in Theorem 2, and suppose $K_{1,j_r}^{j_r}$ (with non-center vertices labeled counterclockwise as $c_{j_1}^1,c_{j_2}^1,\ldots,c_{j_r}^1$) and $K_{1,j_s}^{j_s}$ (with non-center vertices labelled clockwise as $c_{j_1}^2,c_{j_2}^2,\ldots,c_{j_s}^2$) are the two sub-fan graphs of G with $j_r\geq 1$. Define $G'=G+(c_{j_r}^1,c_{j_s}^2)$ to be the graph obtained from G by adding one edge $(c_{j_r}^1,c_{j_s}^2)$. Meanwhile, denote $\overline{G}$ the graph of $G-\left(\bigcup_{t=1}^{r}(c_0,c_{j_t}^1)\cup\bigcup_{t=1}^{s}(c_0,c_{j_t}^2)\right)$ that contains c_0, and denote $\overline{S}$ the collection of subtrees in $S(K_{1,j_r+j_s}^{j_r+j_s};c_0)$ that contain $(c_{j_r}^1,c_{j_s}^2)$. By dividing the subtrees of G and G' into two cases of containing center c_0 or not, with Lemma 1, definitions of subtree weight and subtree generating function, and combining structure analysis, we have the following theorem:

Theorem 4. *Let G and G′ be the weighted graphs defined above. Then,*

$$F(G';f,g) = F(G;f,g) + y^{-1}F(\overline{G};f,g;c_0)\sum_{T\in\overline{S}}\omega(T) + \sum_{t=0}^{j_s} ty^{t+1}z^t + j_s\sum_{t=j_s+1}^{j_r} y^{t+1}z^t + \sum_{t=j_r+1}^{j_r+j_s-1}(j_r+j_s-t)y^{t+1}z^t. \quad (12)$$

By letting $y = z = 1$ in the subtree generating functions from the above theorems, we have the corresponding subtree numbers of the various related graphs above.

Corollary 1. *The subtree number of* $S(K_{1,j}^j, c_0)$ *is*

$$\eta(K_{1,j}^j; c_0) = \eta(K_{1,j-1}^{j-1}; c_0) + \sum_{r=1}^{j} r\eta(K_{1,j-r}^{j-r}; c_0), \quad (13)$$

with $\eta(K_{1,0}^0; c_0) = 1, \eta(K_{1,1}^1; c_0) = 2.$

With Corollary 1, we have the number of subtrees of $K_{1,j}^j$ that contain central vertex c_0 as illustrated in Figure 2.

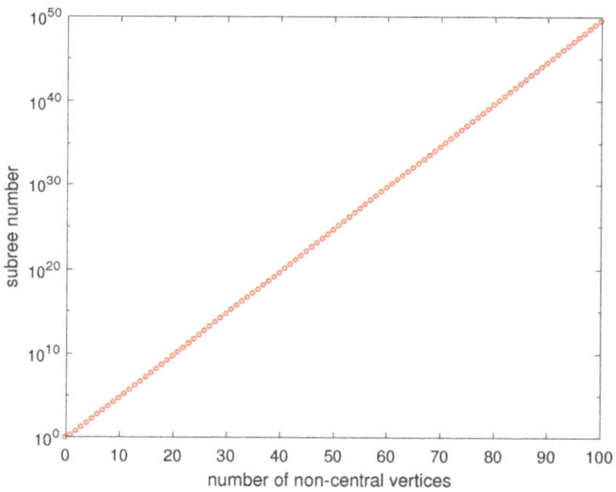

Figure 2. Number of subtrees of $K_{1,j}^j$ that contain central vertex c_0, in semi-log(Log-Y) coordinates.

Corollary 2. *Let n, j be positive integers with* $1 \le j \le n$, *and* $i \equiv n \ (mod\ j)$, *then*

$$\eta(K_{1,n}^j) = \frac{(j+1)(n-i)}{2} + i + 2^i b_j^{\frac{n-i}{j}}, \quad (14)$$

with $b_j = b_{j-1} + \sum_{r=1}^{j} rb_{j-r}$ *and* $b_0 = 1, b_1 = 2.$

With Corollary 2, we have the subtree number of $K_{1,n}^j$, see details in subsection 3.1.

Corollary 3. *Let* $G = \bigcup_{t=1}^{l} (K_{1,j_t}^{j_t})^{n_{j_t}}, j_t, n_{j_t}$ *and l be defined in Theorem 2, then*

$$\eta(G) = \prod_{t=1}^{l} \eta(K_{1,j_t}^{j_t}; c_0)^{n_{j_t}} + \frac{1}{2}\sum_{t=1}^{l} n_{j_t} \times j_t(j_t+1), \quad (15)$$

with $\eta(K_{1,j_t}^{j_t}; c_0) = \eta(K_{1,j_t-1}^{j_t-1}; c_0) + \sum_{r=1}^{j_t} r\eta(K_{1,j_t-r}^{j_t-r}; c_0)$ *and* $\eta(K_{1,0}^0; c_0) = 1$, $\eta(K_{1,1}^1; c_0) = 2$.

Let G_1 be the graph defined in Theorem 2 with $l = 4, j_1 = 2, n_{j_1} = 4; j_2 = 3, n_{j_2} = 3; j_3 = 4, n_{j_3} = 2; j_4 = 5, n_{j_4} = 1$, with Corollary 1 and Corollary 3, we have $\eta(G_1) = 6,048,255,225,665$.

Corollary 4. *The number of subtrees of the fan graph* $K_{1,j}^j$ *containing* c_1 *is*

$$\eta(K_{1,j}^j; c_1) = \eta(K_{1,j}^j; c_0) - \eta(K_{1,j-1}^{j-1}; c_0) + j, \tag{16}$$

with $\eta(K_{1,j}^j; c_0) = \eta(K_{1,j-1}^{j-1}; c_0) + \sum_{r=1}^{j} r\eta(K_{1,j-r}^{j-r}; c_0)$ *and* $\eta(K_{1,0}^0; c_0) = 1$, $\eta(K_{1,1}^1; c_0) = 2$.

Similarly, the number of subtrees of $K_{1,j}^j$ that contain central vertex c_1 (see Figure 3) can be obtained from Corollary 4.

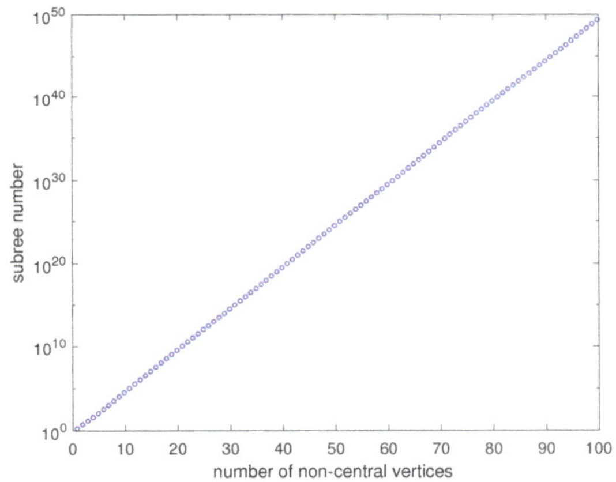

Figure 3. Number of subtrees of $K_{1,j}^j$ that contain the vertex c_1, in semi-log(Log-Y) coordinates.

Corollary 5. *Let G be the graph defined in Theorem 2, and G' be a merged graph from G with $j_r \geq 1$ and $j_s \geq 1$ (Theorem 4), then*

$$\eta(G') = \eta(G) + \eta(\overline{G}; c_0)\eta(K_{1,j_r+j_s}^{j_r+j_s}; c_0, (c_{j_r}^1, c_{j_s}^2)) + j_r j_s, \tag{17}$$

where $\eta(K_{1,j_r+j_s}^{j_r+j_s}; c_0, (c_{j_r}^1, c_{j_s}^2))$ *is the number of subtrees containing both vertex* c_0 *and edge* $(c_{j_r}^1, c_{j_s}^2)$.

Let G, $\overline{G}$, and G' be the graphs defined in Corollary 5 with $l = 2, j_1 = 2, n_{j_1} = 3; j_2 = 3, n_{j_2} = 2$, $j_r = 2$ and $j_s = 3$, with Corollary 1, Corollary 5, and structural analysis, we have $\eta(G)$ = 77,997, $\eta(\overline{G}; c_0) = 684$ $\eta(K_{1,j_r+j_s}^{j_r+j_s}; c_0, (c_{j_r}^1, c_{j_s}^2)) = 75$, and $\eta(G')$ = 129,303.

2.2. Subtree Generating Function and Subtree Number of Wheel Graph W_n

Next we consider the subtree generating function of the weighted wheel graph W_n.

Theorem 5. *Let W_n $(n \geq 3)$ be the weighted wheel graph on $n+1$ vertices with vertex weight function $f \equiv y$ and edge weight function $g \equiv z$. Then*

$$\begin{aligned} F(W_n;f,g) =& y + yzF(W_{n-1};f,g) + (1+2yz)F(K_{1,n-1}^{n-1};f,g;c_0) - 2yzF(K_{1,n-2}^{n-2};f,g;c_0) \\ &+ \sum_{t=1}^{n-1}\Big((1-yz)(n-t) + 2yz\Big)y^t z^{t-1} + 2\sum_{k=1}^{n-1}(yz)^{k+1}F(K_{1,n-1-k}^{n-1-k};f,g;c_0) \\ &+ \sum_{l=0}^{n-3}\sum_{r=0}^{n-3-l}(yz)^{l+r+3}F(K_{1,n-l-r-3}^{n-l-r-3};f,g;c_0), \end{aligned} \tag{18}$$

with $F(W_3;f,g) = 4y + 6y^2z + 12y^3z^2 + 16y^4z^3$, $F(K_{1,j}^{j};f,g;c_0) = F(K_{1,j-1}^{j-1};f,g;c_0) + \sum_{r=1}^{j} r(yz)^r F(K_{1,j-r}^{j-r};f,g;c_0)$, $F(K_{1,0}^{0};f,g;c_0) = y$, *and* $F(K_{1,1}^{1};f,g;c_0) = y + y^2z$.

Proof. For convenience, we let $e_t^* = (c_0, c_t)$ for $t = 1,2,\ldots,n$, $e_n = (c_1, c_n)$ and $e_{n-r} = (c_{n-r}, c_{n-r+1})$ for $r = 1,2,\ldots,n-1$. We also follow the convention that $\bigcup_{r=i}^{j}(c_r, c_{r+1}) = \emptyset$ if $j < i$, and $\sum_{t=i}^{j} b_t = 0$, if $j < i$.

We first consider the subtrees of $W_n (n \geq 3)$ in different cases:
(i) not containing the edge e_n^*,
(ii) containing the edge e_n^*.
The subtrees in case (i) can be further partitioned into four categories. As a result, we have

$$S(W_n - e_n^*) = \overline{S_1}\bigcup\overline{S_2}\bigcup\overline{S_3}\bigcup\overline{S_4},$$

where

- $\overline{S_1}$ is the set of subtrees of $S(W_n - e_n^*)$ that contain neither e_n nor e_{n-1};
- $\overline{S_2}$ is the set of subtrees of $S(W_n - e_n^*)$ that contain e_{n-1}, but not e_n;
- $\overline{S_3}$ is the set of subtrees of $S(W_n - e_n^*)$ that contain e_n but not e_{n-1};
- $\overline{S_4}$ is the set of subtrees of $S(W_n - e_n^*)$ that contain both e_n and e_{n-1}.

From the definition of subtree weight and with structure analysis, we have

$$\sum_{T\in\overline{S_1}}\omega(T) = y + F(K_{1,n-1}^{n-1};f,g), \tag{19}$$

$$\sum_{T\in\overline{S_2}}\omega(T) = \sum_{T\in\overline{S_3}}\omega(T) = yzF(K_{1,n-1}^{n-1};f,g;c_1), \tag{20}$$

and

$$\sum_{T\in\overline{S_4}}\omega(T) = yz\Big(F(W_{n-1};f,g) - F(K_{1,n-1}^{n-1};f,g)\Big). \tag{21}$$

Thus, we have

$$\begin{aligned} \sum_{T\in S(W_n-e_n^*)}\omega(T) =& \sum_{T\in\overline{S_1}}\omega(T) + \sum_{T\in\overline{S_2}}\omega(T) + \sum_{T\in\overline{S_3}}\omega(T) + \sum_{T\in\overline{S_4}}\omega(T) \\ =& y + F(K_{1,n-1}^{n-1};f,g) + yz\Big(2F(K_{1,n-1}^{n-1};f,g;c_1) + F(W_{n-1};f,g) - F(K_{1,n-1}^{n-1};f,g)\Big). \end{aligned} \tag{22}$$

Similarly, for case (ii), we have

$$S(W_n;e_n^*) = \mathcal{S}_1\bigcup\mathcal{S}_2\bigcup\mathcal{S}_3\bigcup\mathcal{S}_4,$$

where

- $\mathcal{S}_1$ is the set of subtrees in $S(W_n; e_n^*)$ that contain neither e_n nor e_{n-1};
- $\mathcal{S}_2$ is the set of subtrees in $S(W_n; e_n^*)$ that contain e_n, but not e_{n-1};
- $\mathcal{S}_3$ is the set of subtrees in $S(W_n; e_n^*)$ that contain e_{n-1} but not e_n;
- $\mathcal{S}_4$ is the set of subtrees in $S(W_n; e_n^*)$ that contain both e_n and e_{n-1}.

Analyzing each case, we have:

(a) $\mathcal{S}_1 = \{T + (c_0, c_n) | T \in S(K_{1,n-1}^{n-1}; c_0)\}$, where $T + (c_0, c_n)$ are the trees obtained from $T (\in S(K_{1,n-1}^{n-1}; c_0))$ by attaching the edge (c_n, c_0) at vertex c_0;

(b) $\mathcal{S}_2 = \{T + e_n^* + (c_1, c_n) + \bigcup_{r=1}^{k-1} (c_r, c_{r+1}) | T \in S(\widetilde{K}_{1,n-1-k}^{n-1-k} c_0)\}$, where $\widetilde{K}_{1,n-1-k}^{n-1-k} (k = 1, 2, \ldots, n-1)$ is the graph of $W_n - \left(e_k \cup e_{n-1} \cup e_n^* \bigcup_{r=1}^{k} e_r^*\right)$ that contains c_0, and obviously, $\widetilde{K}_{1,n-1-k}^{n-1-k} \cong K_{1,n-1-k}^{n-1-k}$;

(c) $\mathcal{S}_3 = \{T + e_n^* + \bigcup_{r=1}^{k} (c_{n-r}, c_{n-r+1}) | T \in S(K_{1,n-1-k}^{n-1-k}; c_0)\}$ for $k = 1, 2, \ldots, n-1$;

(d) In a similar manner as (b) and (c), use variables l and r to count the number of edges trailing off (c_{n-1}, c_n) and (c_1, c_n), respectively, then the subtrees in $\mathcal{S}_4$ are indexed by these two variables, which count extra edges that are on the two sides of the T shape containing the 3 edges required by $\mathcal{S}_4$.

With (a)–(d), we have

$$\sum_{T_1 \in \mathcal{S}_1} \omega(T_1) = \sum_{T \in S(K_{1,n-1}^{n-1}; c_0)} f(c_n) g(e_n^*) \omega(T) = yzF(K_{1,n-1}^{n-1}; f, g; c_0), \tag{23}$$

$$\begin{aligned} \sum_{T_2 \in \mathcal{S}_2} \omega(T_2) = \sum_{T_3 \in \mathcal{S}_3} \omega(T_3) &= f(c_n) g(e_n^*) \sum_{k=1}^{n-1} \prod_{r=1}^{k} (g(e_{n-r}) f(c_{n-r})) F(K_{1,n-1-k}^{n-1-k}; f, g; c_0) \\ &= yz \sum_{k=1}^{n-1} (yz)^k F(K_{1,n-1-k}^{n-1-k}; f, g; c_0), \end{aligned} \tag{24}$$

$$\sum_{T_4 \in \mathcal{S}_4} \omega(T_4) = \sum_{l=0}^{n-3} \sum_{r=0}^{n-3-l} (yz)^{l+r+3} F(K_{1,n-l-r-3}^{n-l-r-3}; f, g; c_0). \tag{25}$$

Now with Equations (23)–(25), we have

$$\begin{aligned} \sum_{T \in S(W_n; (c_0, c_n))} \omega(T) &= \sum_{T_1 \in \mathcal{S}_1} \omega(T_1) + \sum_{T_2 \in \mathcal{S}_2} \omega(T_2) + \sum_{T_3 \in \mathcal{S}_3} \omega(T_3) + \sum_{T_4 \in \mathcal{S}_4} \omega(T_4) \\ &= yzF(K_{1,n-1}^{n-1}; f, g; c_0) + 2 \sum_{k=1}^{n-1} (yz)^{k+1} F(K_{1,n-1-k}^{n-1-k}; f, g; c_0) \\ &\quad + \sum_{l=0}^{n-3} \sum_{r=0}^{n-3-l} (yz)^{l+r+3} F(K_{1,n-l-r-3}^{n-l-r-3}; f, g; c_0). \end{aligned} \tag{26}$$

By Theorem 1, Theorem 3, and Equations (22) and (26), we have

$$\begin{aligned} F(W_n; f, g) &= y + yzF(W_{n-1}; f, g) + (1 + 2yz) F(K_{1,n-1}^{n-1}; f, g; c_0) - 2yzF(K_{1,n-2}^{n-2}; f, g; c_0) \\ &\quad + \sum_{t=1}^{n-1} \left((1 - yz)(n - t) + 2yz \right) y^t z^{t-1} + 2 \sum_{k=1}^{n-1} (yz)^{k+1} F(K_{1,n-1-k}^{n-1-k}; f, g; c_0) \\ &\quad + \sum_{l=0}^{n-3} \sum_{r=0}^{n-3-l} (yz)^{l+r+3} F(K_{1,n-l-r-3}^{n-l-r-3}; f, g; c_0). \end{aligned} \tag{27}$$

Note that W_2 is not a wheel graph. With Theorem 1 and Lemma 1, we have

$$\sum_{T \in S(W_2)} \omega(T) = 3y + 4y^2 z + 5y^3 z^2. \tag{28}$$

Now from Equations (27) and (28), we have

$$F(W_3; f, g) = 4y + 6y^2z + 12y^3z^2 + 16y^4z^3. \tag{29}$$

The subtree generating function of W_n now follows from Equations (27) and (29). □

Letting $y = z = 1$ in Equation (18), we have the following corollary.

Corollary 6. *The subtree number of $W_n (n \geq 3)$ is*

$$\eta(W_n) = \eta(W_{n-1}) + 3\eta(K_{1,n-1}^{n-1}; c_0) + 2n - 1 + 2\sum_{k=2}^{n-1} \eta(K_{1,n-1-k}^{n-1-k}; c_0) + \sum_{l=0}^{n-3}\sum_{r=0}^{n-3-l} \eta(K_{1,n-l-r-3}^{n-l-r-3}; c_0), \tag{30}$$

with $\eta(W_3) = 38$, $\eta(K_{1,j}^{j}; c_0) = \eta(K_{1,j-1}^{j-1}; c_0) + \sum_{r=1}^{j} r\eta(K_{1,j-r}^{j-r}; c_0)$, $\eta(K_{1,0}^{0}; c_0) = 1$ *and* $\eta(K_{1,1}^{1}; c_0) = 2$.

With Corollary 6, the subtree numbers of W_n $(n = 3, 4, \ldots, 50)$ are shown in Table 1.

Table 1. The subtree numbers of wheel graph W_n $(n = 3, 4, \ldots, 50)$.

n	$\eta(W_n)$	n	$\eta(W_n)$	n	$\eta(W_n)$
3	38	19	2,899,980,984	35	269,604,917,347,967,886
4	112	20	9,128,846,611	36	848,689,059,340,934,448
5	332	21	28,736,686,630	37	2,671,587,471,512,527,895
6	1007	22	90,460,187,232	38	8,409,887,625,375,274,755
7	3110	23	284,759,535,167	39	26,473,477,146,304,448,341
8	9704	24	896,394,265,075	40	83,335,833,180,604,495,475
9	30,431	25	2,821,758,641,457	41	262,332,788,908,879,910,034
10	95,643	26	8,882,611,305,147	42	825,797,133,240,010,600,373
11	300,885	27	27,961,563,560,618	43	2,599,525,999,414,007,165,103
12	946,923	28	88,020,178,967,761	44	8,183,045,386,844,876,767,480
13	2,980,538	29	277,078,636,493,555	45	25,759,400,682,377,496,173,050
14	9,382,101	30	872,215,572,630,716	46	81,087,992,568,389,361,552,840
15	29,533,519	31	2,745,646,560,009,062	47	255,256,813,613,269,834,457,576
16	92,968,088	32	8,643,018,158,636,696	48	803,522,677,430,288,749,342,627
17	292,653,642	33	27,207,348,527,149,292	49	2,529,408,261,449,734,855,548,318
18	921,243,536	34	85,645,986,192,695,055	50	7,962,321,827,121,343,008,620,568

As a matter of fact, with the generating function and further structural and theoretical analysis, we can also solve the subtree generation computing problems for the following more generalized types of graphs; here we skip the similar but technical details.

(i) Graph $G = \bigcup_{i=1}^{l} F_{i,n_i}^{t_i}$ is constructed by identifying the center c_0^i of l graphs $F_{i,n_i}^{t_i} (i = 1, 2, \ldots, l)$ to c_0, where $\overline{F_{i,n_i}^{t_i}} (n_i \geq 1, 2 \leq t_i \leq n_i)$ is the graph constructed from a vertex c_0^i and a path $P_{c_1^i c_{n_i}^i} = c_1^i c_2^i \cdots c_{n_i}^i$ by connecting c_0^i with c_1^i, $c_{n_i}^i$, and any other arbitrary $t_i - 2$ vertices on the path $P_{c_1^i c_{n_i}^i}$; for the special case $n_i = 1$, the graph $F_{i,n_i}^{t_i}$ is a path $c_0^i c_1^i$ on vertices c_0^i and c_1^i.

(ii) Graph $G = W_n - \bigcup_{t=1}^{l} (c_0, c_{j_t})(1 \leq l, j_t \leq n)$, namely the graph obtained from wheel W_n by deleting random l different edges $(c_0, c_{j_t})(t = 1, 2, \ldots, l)$ (each j_t is different the others).

3. Behaviors of $K_{1,N}^{J} (1 \leq J \leq N)$ and W_n in Terms of Subtrees

With the subtree generating functions established, we can now analyze the behaviors of the graphs $K_{1,n}^{j} (1 \leq j \leq n)$ and W_n in terms their subtree numbers and other related properties.

3.1. Subtree Numbers and $K^J_{1,N}$

We start with the subtree numbers of the graphs $K^j_{1,n}$ for different n and j. First, we take a quick look at the extremal problems.

Proposition 1. *Among all* $K^j_{1,n}$ $(1 \leq j \leq n)$:

- *the graph* $K^1_{1,n}$ *has* $2^n + n$ *subtrees, fewer than any other* $K^j_{1,n}(j \neq 1)$; *and*
- *the graph* $K^n_{1,n}$ *has* $\frac{n(n+1)}{2} + b_n$ *subtrees (where* $b_n = b_{n-1} + \sum_{r=1}^{n} rb_{n-r}$ *and* $b_0 = 1, b_1 = 2$*), more than any other* $K^j_{1,n}(j \neq n)$.

Proof. Note that adding an edge to a graph will strictly increase the subtree number. The extremal structures $K^1_{1,n}$ and $K^n_{1,n}$ then follow immediately from the fact that the former is a subgraph of any $K^j_{1,n}$ and the latter contains any $K^j_{1,n}$ as a subgraph. □

We are also interested in knowing which $K^j_{1,n}$ $(2 \leq j \leq n-1)$ has the second- or third-largest subtree number. To examine this, we also explore how the subtree numbers of the graphs $K^j_{1,n}$ behave. With Corollary 2 and Matlab computing (see Appendix A for more details), we obtain the subtree number behaviors of $K^j_{1,n}$ as shown in Figure 4.

It seems from Figure 4a,b that among all $K^j_{1,n}$ $(1 \leq j \leq n-1)$:

- $K^{n-1}_{1,n}$ has the second-largest subtree number; and
- $K^{n-2}_{1,n}$ has the third-largest subtree number for odd (or sufficiently large) n.
- the subtree number $K^j_{1,n}$ increment trend meets exponential growth when $j \geq \lceil \frac{n+1}{2} \rceil$.

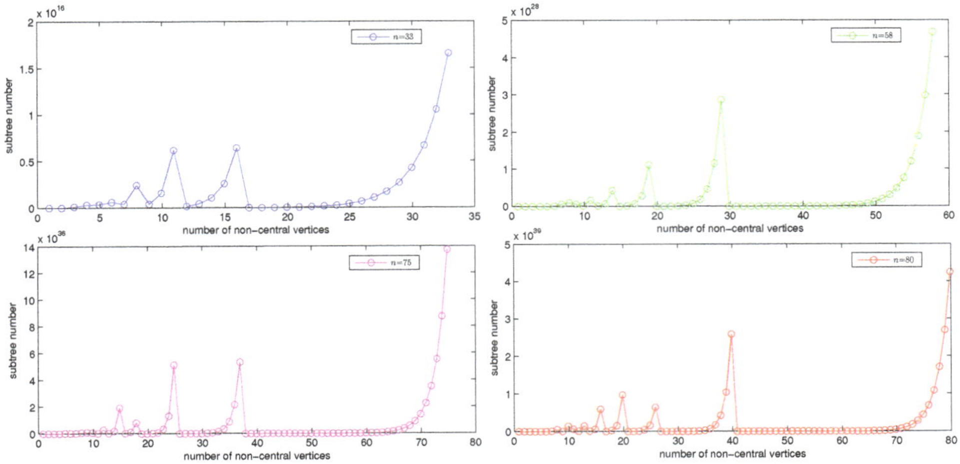

(**a**) Subtree numbers of $K^j_{1,n}$ with $n = 33, 58, 75, 80$, j from 1 to n, in Cartesian coordinates.

Figure 4. *Cont.*

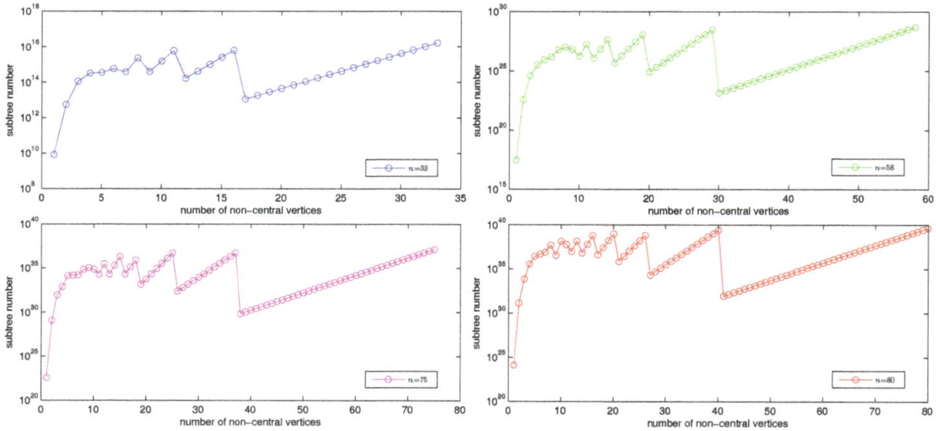

(b) Subtree numbers of $K_{1,n}^{j}$ with $n = 33, 58, 75, 80$, j from 1 to n, in semi-log(Log-Y) coordinates.

Figure 4. Subtree numbers of $K_{1,n}^{j}$ with $n = 33, 58, 75, 80$, j from 1 to n.

In general, it is not difficult to show the following.

Proposition 2. *Suppose $n >> k$ (i.e., n is much larger than k), then among all $K_{1,n}^{j}$ $(1 \leq j \leq n)$, the graph $K_{1,n}^{n-k}$ has the $(k+1)$-th largest subtree number.*

The proof follows from adjusting the "sizes" of the two sub-fan graphs of $K_{1,n}^{j}$. We skip the details here. The same argument can also show the monotonic behavior for large enough j.

Proposition 3. *If $j > \frac{n}{2}$, then $K_{1,n}^{j+1}$ has a larger subtree number than $K_{1,n}^{j}$.*

Understanding the specific behavior of $K_{1,n}^{j}$ for general j in terms of the subtree number seems to be an interesting and nontrivial problem.

Similarly, from Corollary 6, we can obtain the subtree numbers of wheel graph W_n $(n \geq 3)$, as already shown in Table 1. On the other hand, experimental observation shows that the subtree numbers of W_n $(n \geq 3)$ increase very fast and this growth trend seems to fit the linear regression model after doing the logarithmic transformation. Through implementing the linear regression in the MATLAB software for data of the number of subtrees of W_n from $n = 3$ to 602, the appropriate formula for the subtree number of W_n can be stated as

$$\eta(W_n) \approx exp(0.0126 + 1.1466n). \tag{31}$$

As $exp0.0126 \approx 1$, Equation (31) can be rewritten as

$$\eta(W_n) \approx exp1.1466n. \tag{32}$$

With Corollary 6, Equation (32), and logarithmic transformation for each subtree number of W_n and fitted value, we can obtain the subtree number trend of $W_n (n \geq 3)$, as illustrated in Figure 5 where the original and fitted data are respectively marked in red and blue. We believe Equation (32) can be proved or disproved through traditional analytic combinatorial approaches, but we will not pursue the technical analysis here.

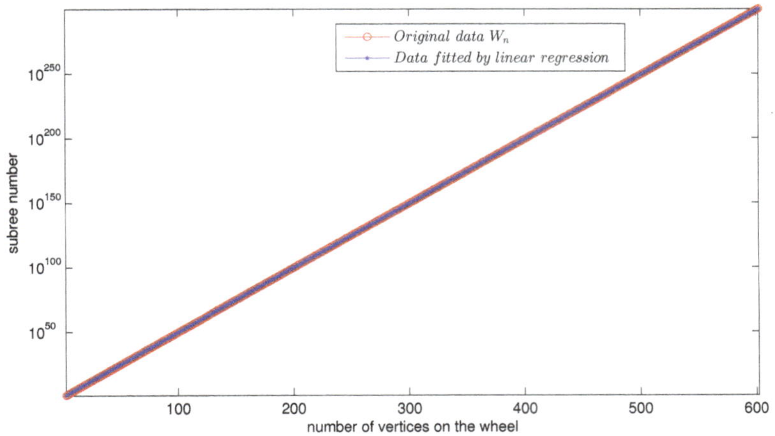

Figure 5. Subtree number trend of wheel graph W_n $(n = 3, 4, \ldots, 602)$, in semi-log(Log-Y) coordinates.

Our work can also be easily applied to more general "partitions" of wheel graphs instead of just the $K_{1,n}^j$s. To illustrate the observations, we introduce some simple notations. Given a positive integer n and partition $\pi = (\pi_1, \pi_2, \ldots, \pi_k)$ of n (with $\pi_1 \geq \pi_2 \geq \ldots \geq \pi_k$ and $n = \sum_{i=1}^{k} \pi_i$), the graph K_π is obtained from $K_{1,n}$ by adding edges (c_j, c_{j+1}) for all $1 \leq j \leq n-1$ except for $j = \sum_{i=1}^{s} \pi_i$, for $s = 1, 2, \ldots, k$. It seems that the subtree numbers of K_πs coincide with some sort of ordering of the partitions π of n. In general, larger and fewer parts in π results in more subtrees in K_π. More precise statements along this line requires further study.

3.2. Subtree Densities of $K_{1,N}^J$ and W_n

The subtree generating function can be used to provide more information than just the number of subtrees. For instance, using the subtree generating function, one can easily obtain the total number of vertices in all subtrees, from which we have the average subtree order in a given graph. The ratio of the average subtree order and the order of the original graph is called the subtree density of the graph. We present the formal subtree density definition as follows:

Definition 2 ([20]). *Suppose G is a graph with n vertices, then $\mu(G) = \frac{1}{k}\sum_{i=1}^{k} n_i$ is the average order of the subtrees of G, where $n_1, n_2, \ldots, n_k$ are the orders of all of G's k non-empty subtrees, and the subtree density of G is defined as $D(G) = \frac{\mu(G)}{n}$.*

It is essentially the probability that a vertex chosen at random from G will belong to a randomly chosen subtree of G. Some of the work related to subtree densities can be found in [20,21,39].

With Theorem 1, Theorem 5, and substituting $z = 1$, we could obtain the vertex generating function of subtrees of $K_{1,n}^j$ and W_n, respectively, i.e., $F(K_{1,n}^j; y, 1)$ and $F(W_n; y, 1)$). The subtree density of $K_{1,n}^j$ and W_n is simply

$$D(G^*) = \frac{\left.\frac{\partial F(G^*; y, 1)}{\partial y}\right|_{y=1}}{F(G^*; 1, 1) * n(G^*)}, \tag{33}$$

where $n(G^*)$ denotes the vertex number of G^*, and G^* can be $K_{1,n}^j$ or W_n.

We will now apply this to find the subtree densities of $K_{1,n}^j$ and W_n.

Clearly, the vertex number of $K_{1,n}^{j}$ and W_n is $n+1$, namely

$$n(K_{1,n}^{j}) = n(W_n) = n+1. \tag{34}$$

With Theorem 1 and letting $z=1$ in Equation (1), we obtain the vertex generating function of subtrees of $K_{1,n}^{j}(1 \le j \le n)$.

$$F(K_{1,n}^{j};f,1) = (1+y)^{i}y^{1-\frac{n-i}{j}} * F(K_{1,j}^{j};f,1;c_0)^{\frac{n-i}{j}} + iy + \frac{n-i}{j}\sum_{t=0}^{j-1}(j-t)y^{t+1}, \tag{35}$$

with $i \equiv n \ (mod\ j)$, $F(K_{1,j}^{j};f,1;c_0) = F(K_{1,j-1}^{j-1};f,1;c_0) + \sum_{r=1}^{j} ry^{r}F(K_{1,j-r}^{j-r};f,1;c_0)$ and $F(K_{1,0}^{0};f,1;c_0) = y$, $F(K_{1,1}^{1};f,1;c_0) = y+y^2$.

Similarly, letting $z=1$ in Equation (18), we obtain the vertex generating function of subtrees of W_n.

$$\begin{aligned} F(W_n;f,1) =& y + yF(W_{n-1};f,1) + (1+2y)F(K_{1,n-1}^{n-1};f,1;c_0) - 2yF(K_{1,n-2}^{n-2};)f,1;c_0) \\ &+ \sum_{t=1}^{n-1}\Big((1-y)(n-t)+2y\Big)y^{t} + 2\sum_{k=1}^{n-1} y^{k+1}F(K_{1,n-1-k}^{n-1-k};f,1;c_0) \\ &+ \sum_{l=0}^{n-3}\sum_{r=0}^{n-3-l} y^{l+r+3}F(K_{1,n-l-r-3}^{n-l-r-3};f,1;c_0), \end{aligned} \tag{36}$$

with $F(W_3;f,1) = 4y+6y^2+12y^3+16y^4$, $F(K_{1,j}^{j};f,1;c_0) = F(K_{1,j-1}^{j-1};f,1;c_0) + \sum_{r=1}^{j} ry^{r}F(K_{1,j-r}^{j-r};f,1;c_0)$ and $F(K_{1,0}^{0};f,1;c_0) = y$, $F(K_{1,1}^{1};f,1;c_0) = y+y^2$.

From Equations (33)–(35), we can obtain the subtree densities of $K_{1,n}^{j}$ $(1 \le j \le n)$ (plotted in Figure 6); related data can be found in Table 2. Similarly, from Equations (33), (34), and (36), we have subtree densities W_n plotted in Figure 7; related data are listed in Table 3.

From Table 2 and Figure 6, it appears that $K_{1,22}^{1}$ and $K_{1,22}^{12}$ have the smallest and second-smallest subtree densities, respectively. $K_{1,22}^{j}(12 \le j \le 22)$ grows linearly with the increase of j. Moreover, from Table 3 and Figure 7, we see that $W_n(3 \le n \le 24)$ increases first in the interval $[3,8]$ and reaches the maximum value when $n=8$, and then decreases gradually in the interval $[9,24]$ and approximates the limit value of 0.8135.

Similarly, as studied in paper [35], we can also discuss the ratio of spanning trees to all subtrees and spanning tree densities of $K_{1,n}^{j}$ and W_n. We skip the details here.

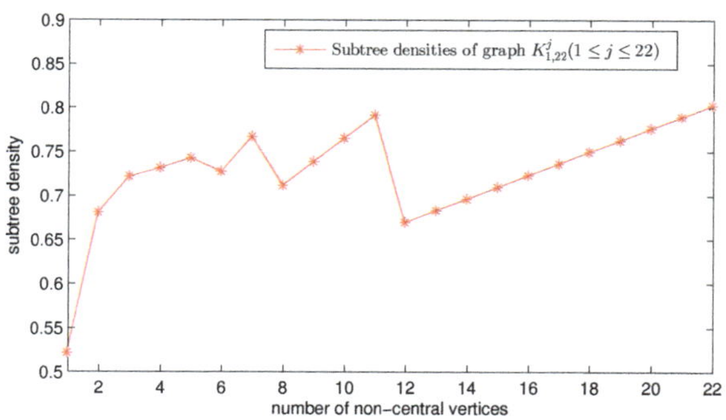

Figure 6. Subtree densities of the graphs $K_{1,22}^{j}$ $(1 \le j \le 22)$.

Table 2. Related data for $K^j_{1,n}$, with $n = 22$ and $j = 1, 2, \ldots, 22$.

j	$P_y(K^j_{1,n})$	$\eta(K^j_{1,n})$	$D(K^j_{1,n})$
1	50,331,670	4,194,326	0.521736621869847
2	5,683,820,588	362,797,089	0.681159363604678
3	29,685,950,982	1,787,743,521	0.721967947748206
4	52,358,400,102	3,110,400,052	0.731884047161090
5	87,226,395,622	5,103,959,426	0.743041170007376
6	56,400,430,972	3,370,318,067	0.727584934777173
7	231,968,014,120	13,141,451,319	0.767462100378713
8	36,440,865,014	2,224,820,302	0.712140856484914
9	93,645,742,414	5,511,577,694	0.738727501280144
10	240,339,417,442	13,653,922,612	0.765314128820302
11	616,080,713,876	33,825,095188	0.791900742502907
12	9,137,267,062	592,843,864	0.670113169892776
13	14,666,890,448	933,106,276	0.683406494463958
14	23,533,982,520	1,468,662,129	0.696699813456106
15	37,748,062,639	2,311,599,999	0.709993128330647
16	60,525,953,078	3,638,341,646	0.723286440182849
17	97,015,668,926	5,726,566,014	0.736579749833999
18	155,453,553,144	9,013,325,743	0.749873057893726
19	249,013,516,065	14,186,519,625	0.763166364810336
20	398,761,664,190	22,328,865,632	0.776459670910826
21	638,376,020,719	35,144,507,194	0.789752976431952
22	1,021,684,176,864	55,315,680,041	0.803046281543951

* $P_y(G)$ stands for $\left.\frac{\partial F(G;y,1)}{\partial y}\right|_{y=1}$.

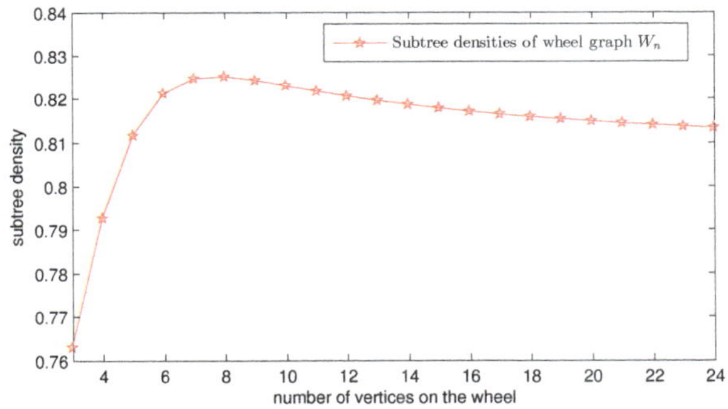

Figure 7. Subtree densities of wheel graph $W_n (3 \le n \le 24)$.

Table 3. Related data for wheel graph W_n $(n = 3, 4 \ldots, 24)$.

k	$P_y(W_k)$	$D(W_k)$	k	$P_y(W_k)$	$D(W_k)$
3	116	0.7631578947368421	14	115,215,423	0.8186895664414613
4	444	0.7928571428571428	15	386,480,089	0.8178844370865525
5	1617	0.8117469879518072	16	1,291,505,336	0.8171718247840813
6	5789	0.8212512413108243	17	4,301,328,493	0.8165375712479791
7	20,519	0.8247186495176849	18	14,282,430,812	0.8159697793983819
8	72,064	0.8251351103783091	19	47,296,291,958	0.815458656779937
9	250,841	0.8242943051493543	20	156,239,476,051	0.8149961728004785
10	865,923	0.8230636087039588	21	514,980,557,554	0.8145757185906111
11	2,967,219	0.8218031806171793	22	1,693,994,724,188	0.8141918205521789
12	10,102,071	0.8206394655271703	23	5,561,968,202,536	0.813839912225434
13	34,200,012	0.8196030381092273	24	18,230,780,418,139	0.8135161559345164

* $P_y(G)$ stands for $\left.\frac{\partial F(G;y,1)}{\partial y}\right|_{y=1}$.

4. Concluding Remarks

In this paper we established the subtree generating functions of the wheel graph W_n and "partitions" $K_{1,n}^j$ $(1 \leq j \leq n)$, along with other related structures such as the fan graphs. With the subtree generating functions, we are able to study the subtree number and subtree densities of these graphs. From our computational analysis, several theoretical conclusions are drawn, especially on the extremal problems with respect to the subtree number. These results provide a fundamental basis for studying subtree problems of multi-cyclic graphs, chemical compounds, and complex graphs.

As one can see, the computational results in Section 3.2 show interesting observations on the characteristics of these structures. Attempting to verify some of them, especially on those related to the subtree densities, and exploring the BC-subtree number index, Wiener index, Harary index, atom-bond connectivity trends of wheel graph W_n, "partitions" $K_{1,n}^j$ $(1 \leq j \leq n)$, and their generalizations would make very interesting projects for future work. Exploring the novel structural properties from the evolutionary perspective of graphs with more complicated cycle structures, particularly some important nanomaterials such as the pentagonal carbon nanocone, is also a very attractive future topic.

Author Contributions: H.W, Y.Y and A.W contributed to supervision, project administration, and the formal analysis. W.-T.Z and D.-Q.S contributed to the methodology and writing of the original draft preparation. The final draft was written by H.W and Y.Y

Funding: This work is supported by the National Natural Science Foundation of China (grant nos. 61702291, 61772102, 61472058,11531001), the China Postdoctoral Science Foundation (grant no. 2018M632095), the Program for Science & Technology Innovation Talents in Universities of Henan Province (grant no. 19HASTIT029), the Joint NSFC-ISF Research Program (jointly funded by the National Natural Science Foundation of China and the Israel Science Foundation (No. 11561141001)), the Key Research Project in Universities of Henan Province (grant nos. 19B110011, 19B630015), the Scientific Research Starting Foundation for High-Level Talents of Pingdingshan University (grant no.PXY-BSQD2017006), and the Simons Foundation (grant no. 245307).

Conflicts of Interest: The authors declare no conflict of interest.

Appendix A. Matlab Code, Python Code, And Data

In what follows, we provide the source codes related to Figure 4. Python code **Pythoncode1** (derived from Corollary 2) can output the data files "n=33.txt", "n=58.txt", "n=75.txt", and "n=80.txt" by inputting the integer 33, 58, 75, 80, respectively. MATLAB codes **MCF1a, MCF1b** can draw the Figure 4a,b, respectively.

Pythoncode1. The Python code for computing the subtree number of $K_{1,n}^j$ $(1 \leq j \leq n)$.

```
n=int(input())
b=[-1 for i in range(n+1)]
b[0]=1
b[1]=2
for j in range(1,n+1):
sum1=0
for r in range(1,j+1):
sum1=sum1+r*b[j-r]
b[j]=b[j-1]+sum1
file_name="n="+str(n)+".txt"
file = open(file_name,'w')
for j in range(1,n+1):
i = n%j
s=(j+1)*(n-i)/2+i+2**(i)*b[j]**((n-i)/j)
file.write(str(s))
file.write('\n')
print(s)
```

MCF1a. MATLAB source code for Figure 1a.

```
clear;
clc;
s1 = load('n=33.txt');
s2 = load('n=58.txt');
s3 = load('n=75.txt');
s4 = load('n=80.txt');
format long;
A =cell(1,10);
A{1,1}='-bo';
A{1,2}='-go';
A{1,3}='-mo';
A{1,4}='-ro';
subplot(2,2,1);
plot(s1,A{1,1});
axis normal;
set(legend('n=33'),'fontsize',8);
xlabel('number of non-central vertices');
ylabel('subtree number');
subplot(2,2,2);
plot(s2,A{1,2});
set(legend('n=58'),'fontsize',8);
xlabel('number of non-central vertices');
ylabel('subtree number');
subplot(2,2,3);
plot(s3,A{1,3});
axis normal;
set(legend('n=75'),'fontsize',8);
xlabel('number of non-central vertices');
ylabel('subtree number');
subplot(2,2,4);
plot(s4,A{1,4});
set(legend('n=80'),'fontsize',8);
xlabel('number of non-central vertices');
ylabel('subtree number');
```

MCF1b: The MATLAB source code for Figure 1b.

```
clear;
clc;
s1 = load('n=33.txt');
s2 = load('n=58.txt');
s3 = load('n=75.txt');
s4 = load('n=80.txt');
format long;
A =cell(1,10);
A{1,1}='-bo';
A{1,2}='-go';
A{1,3}='-mo';
A{1,4}='-ro';
```

```
subplot(2,2,1);
semilogy(s1,A{1,1});
axis normal;
set(legend('n=33','location','southeast'),'fontsize',8);
xlabel('number of non-central vertices');
ylabel('subtree number');
subplot(2,2,2);
semilogy(s2,A{1,2});
set(legend('n=58','location','southeast'),'fontsize',8);
xlabel('number of non-central vertices');
ylabel('subtree number');
subplot(2,2,3);
semilogy(s3,A{1,3});
axis normal;
set(legend('n=75','location','southeast'),'fontsize',8);
xlabel('number of non-central vertices');
ylabel('subtree number');
subplot(2,2,4);
semilogy(s4,A{1,4});
set(legend('n=80','location','southeast'),'fontsize',8);
xlabel('number of non-central vertices');
ylabel('subtree number');
```

References

1. Das, K.C. Atom-bond connectivity index of graphs. *Discret. Appl. Math.* **2010**, *158*, 1181–1188. [CrossRef]
2. Dobrynin, A.A.; Entringer, R.; Gutman, I. Wiener index of trees: Theory and applications. *Acta Appl. Math.* **2001**, *66*, 211–249. [CrossRef]
3. Wiener, H. Structural determination of paraffin boiling points. *J. Am. Chem. Soc.* **1947**, *1*, 17–20. [CrossRef]
4. Yan, W.; Yeh, Y. Enumeration of subtrees of trees. *Theor. Comput. Sci.* **2006**, *369*, 256–268. [CrossRef]
5. Ingalalli, V.; Ienco, D.; Poncelet, P. Mining frequent subgraphs in multigraphs. *Inf. Sci.* **2018**, *451*, 50–66. [CrossRef]
6. Tilk, C.; Irnich, S. Combined column-and-row-generation for the optimal communication spanning tree problem. *Comput. Oper. Res.* **2018**, *93*, 113–122. [CrossRef]
7. Zetina, C.A.; Contreras, I.; Fernández, E.; Luna-Mota, C. Solving the optimum communication spanning tree problem. *Eur. J. Oper. Res.* **2019**, *273*, 108–117. [CrossRef]
8. Chechik, S.; Emek, Y.; Patt-Shamir, B.; Peleg, D. Sparse reliable graph backbones. *Inf. Comput.* **2012**, *210*, 31–39. [CrossRef]
9. Xiao, Y.; Zhao, H.; Liu, Z.; Mao, Y. Trees with large numbers of subtrees. *Int. J. Comput. Math.* **2017**, *94*, 372–385. [CrossRef]
10. Knudsen, B. Optimal multiple parsimony alignment with affine gap cost using a phylogenetic tree. *Algorithms Bioinform. Lect. Notes Comput. Sci.* **2003**, *2812*, 433–446.
11. Wagner, S.G. Correlation of graph-theoretical indices. *SIAM J. Discret. Math.* **2007**, *21*, 33–46. [CrossRef]
12. Székely, L.; Wang, H. On subtrees of trees. *Adv. Appl. Math.* **2005**, *34*, 138–155.
13. Székely, L.; Wang, H. Binary trees with the largest number of subtrees. *Discret. Appl. Math.* **2007**, *155*, 374–385.
14. Zhang, X.; Zhang, X.; Gray, D.; Wang, H. The number of subtrees of trees with given degree sequence. *J. Graph Theory* **2013**, *73*, 280–295. [CrossRef]
15. Wang, H. The extremal values of the Wiener index of a tree with given degree sequence. *Discret. Appl. Math.* **2008**, *156*, 2647–2654. [CrossRef]
16. Deng, H. Wiener indices of spiro and polyphenyl hexagonal chains. *Math. Comput. Model.* **2012**, *55*, 634–644. [CrossRef]

17. Yang, Y.; Liu, H.; Wang, H.; Fu, H. Subtrees of spiro and polyphenyl hexagonal chains. *Appl. Math. Comput.* **2015**, *268*, 547–560. [CrossRef]
18. Czabarka, É.; Székely, L.A.; Wagner, S. On the number of nonisomorphic subtrees of a tree. *J. Graph Theory* **2018**, *87*, 89–95. [CrossRef]
19. Zhang, X.; Zhang, X. The Minimal Number of Subtrees with a Given Degree Sequence. *Graphs Comb.* **2015**, *31*, 309–318. [CrossRef]
20. Jamison, R.E. On the average number of nodes in a subtree of a tree. *J. Comb. Theory Ser. B* **1983**, *35*, 207–223. [CrossRef]
21. Vince, A.; Wang, H. The average order of a subtree of a tree. *J. Comb. Theory Ser. B* **2010**, *100*, 161–170. [CrossRef]
22. Wagner, S.; Wang, H. On the Local and Global Means of Subtree Orders. *J. Graph Theory* **2016**, *81*, 154–166. [CrossRef]
23. Nakayama, T.; Fujiwara, Y. BCT Representation of Chemical Structures. *J. Chem. Inf. Comput. Sci.* **1980**, *20*, 23–28. [CrossRef]
24. Nakayama, T.; Fujiwara, Y. Computer representation of generic chemical structures by an extended block-cutpoint tree. *J. Chem. Inf. Comput. Sci.* **1983**, *23*, 80–87. [CrossRef]
25. Frederickson, G.N.; Hambrusch, S.E. Planar linear arrangements of outerplanar graphs. *IEEE Trans. Circuits Syst.* **1988**, *35*, 323–333. [CrossRef]
26. Wada, K.; Luo, Y.; Kawaguchi, K. Optimal fault-tolerant routings for connected graphs. *Inf. Process. Lett.* **1992**, *41*, 169–174. [CrossRef]
27. Heath, L.; Pemmaraju, S. Stack and queue layouts of directed acyclic graphs: Part II. *SIAM J. Comput.* **1999**, *28*, 1588–1626. [CrossRef]
28. Misiolek, E.; Chen, D.Z. Two flow network simplification algorithms. *Inf. Process. Lett.* **2006**, *97*, 197–202. [CrossRef]
29. Fox, D. Block cutpoint decomposition for markovian queueing systems. *Appl. Stoch. Model. Data Anal.* **1988**, *4*, 101–114. [CrossRef]
30. Barefoot, C. Block-cutvertex trees and block-cutvertex partitions. *Discret. Math.* **2002**, *256*, 35–54. [CrossRef]
31. Mkrtchyan, V. On trees with a maximum proper partial 0-1 coloring containing a maximum matching. *Discret. Math.* **2006**, *306*, 456–459. [CrossRef]
32. Yang, Y.; Liu, H.; Wang, H.; Feng, S. On algorithms for enumerating BC-subtrees of unicyclic and edge-disjoint bicyclic graphs. *Discret. Appl. Math.* **2016**, *203*, 184–203. [CrossRef]
33. Yang, Y.; Liu, H.; Wang, H.; Makeig, S. Enumeration of BC-subtrees of trees. *Theor. Comput. Sci.* **2015**, *580*, 59–74. [CrossRef]
34. Yang, Y.; Liu, H.; Wang, H.; Deng, A.; Magnant, C. On Algorithms for Enumerating Subtrees of Hexagonal and Phenylene Chains. *Comput. J.* **2017**, *60*, 690–710. [CrossRef]
35. Chin, A.J.; Gordon, G.; MacPhee, K.J.; Vincent, C. Subtrees of graphs. *J. Graph Theory* **2018**, *89*, 413–438. [CrossRef]
36. Cahit, I. Spiral chains: A new proof of the four color theorem. *arXiv* **2004**, arXiv: 0408247.
37. Yang, Z.; Liu, Y.; Li, X.Y. Beyond trilateration: On the localizability of wireless ad hoc networks. *IEEE/ACM Trans. Netw. (ToN)* **2010**, *18*, 1806–1814. [CrossRef]
38. Allen, B.; Lippner, G.; Chen, Y.T.; Fotouhi, B.; Momeni, N.; Yau, S.T.; Nowak, M.A. Evolutionary dynamics on any population structure. *Nature* **2017**, *544*, 227. [CrossRef]
39. Haslegrave, J. Extremal results on average subtree density of series-reduced trees. *J. Comb. Theory Ser. B* **2014**, *107*, 26–41. [CrossRef]

mathematics

MDPI

Article

A Study of Regular and Irregular Neutrosophic Graphs with Real Life Applications

Liangsong Huang [1,*], Yu Hu [1], Yuxia Li [1], P. K. Kishore Kumar [2], Dipak Koley [3] and Arindam Dey [4]

1 Key Laboratory for Robot Intelligent Technology of Shandong Province, Shandong University of Science and Technology, Qingdao 266590, China; 13573246239@163.com (Y.H.); yuxiali2004@sdust.edu.cn (Y.L.)
2 Information Technology Department, Al Musanna College of Technology, Sultanate of Oman 111, Oman; kishore2982@gmail.com
3 Department of Computer Application, Techno India Hooghly, Chinsurah 712101, West Bengal, India; dipakkoley973@gmail.com
4 Department of Computer Science and Engineering, Saroj Mohan Institute of Technology, Hooghly 712512, West Bengal, India; arindam84nit@gmail.com
* Correspondence: lshuang@sdust.edu.cn

Received: 23 April 2019; Accepted: 3 June 2019; Published: 17 June 2019

Abstract: Fuzzy graph theory is a useful and well-known tool to model and solve many real-life optimization problems. Since real-life problems are often uncertain due to inconsistent and indeterminate information, it is very hard for an expert to model those problems using a fuzzy graph. A neutrosophic graph can deal with the uncertainty associated with the inconsistent and indeterminate information of any real-world problem, where fuzzy graphs may fail to reveal satisfactory results. The concepts of the regularity and degree of a node play a significant role in both the theory and application of graph theory in the neutrosophic environment. In this work, we describe the utility of the regular neutrosophic graph and bipartite neutrosophic graph to model an assignment problem, a road transport network, and a social network. For this purpose, we introduce the definitions of the regular neutrosophic graph, star neutrosophic graph, regular complete neutrosophic graph, complete bipartite neutrosophic graph, and regular strong neutrosophic graph. We define the d_m- and td_m-degrees of a node in a regular neutrosophic graph. Depending on the degree of the node, this paper classifies the regularity of a neutrosophic graph into three types, namely d_m-regular, td_m-regular, and m-highly irregular neutrosophic graphs. We present some theorems and properties of those regular neutrosophic graphs. The concept of an m-highly irregular neutrosophic graph on cycle and path graphs is also investigated in this paper. The definition of busy and free nodes in a regular neutrosophic graph is presented here. We introduce the idea of the μ-complement and h-morphism of a regular neutrosophic graph. Some properties of complement and isomorphic regular neutrosophic graphs are presented here.

Keywords: neutrosophic graph; complete neutrosophic graph; bipartite neutrosophic graph; complement neutrosophic graph; road transport network; wireless multihop network and social network

PACS: J0101

---

1. Introduction

Graph theory has many real-life applications for problems in computer applications, systems analysis, computer networks, transportation, operations research, and economics. A graph is basically a model of relations, and it is a used to represent the real-life problem consisting of relationships between objects. The vertices and edges of the graph are used to represent the objects and the relations between objects, respectively. In many optimization problems, the available information is inexact or imprecise for various reasons such as the loss of information, a lack of evidence, imperfect statistical data and insufficient information. Generally, the uncertainty in real-life problems may be present in the information that defines the problem.

Classical graph theory uses the basic concept of classical set theory, which was presented by Cantor. In a classical graph, for any vertex or edge, there are two possibilities: it is either in the graph or it is not in the graph. Therefore, classical graphs cannot model uncertain optimization problems. Real-life problems are often uncertain, which are hard to model using classical graphs. The fuzzy set [1] is an extended version of the classical set, where the objects have varying membership degrees. A fuzzy set gives its objects different membership degrees between zero and one. The membership degree is not the same as probability; rather, it describes membership in vaguely-defined sets. The concept of fuzziness in graph theory [2] was described by Kaufmann [3] using the fuzzy relation. Rosenfeld [4] introduced some concepts such as bridges, cycles, paths, trees, and the connectedness of the fuzzy graph and described some of the properties of the fuzzy graph. Gani and Radha [5] presented the notation of the regular fuzzy graph. Many other researchers, such as Samanta and Pal [6], Rashmanlou and Pal [7], Rashmanlou et al. [8,9], Paramik [10], Nandhini [11], Ghorai and Pal [12], Dey et. al [13], and Borzooei [14] presented much work in the domain of the fuzzy graph and its several applications in real life. Samanta and Pal [6] and Rashmanlou and Pal [15] presented the concept of the irregular and regular fuzzy graph. They also described some applications of those graphs.

Atanassov [16–18] presented the idea of the intuitionistic fuzzy set (IFS) as a modified version of the classical fuzzy set. The classical fuzzy set uses only the membership grade (degree), but IFS uses independent membership grade and non-membership grade for any entity, and the only requirement is that the sum of non-membership and membership degree values be no greater than one. Shannon and Atanassov [19] presented the idea of the IFS relation and the intuitionistic fuzzy graph (IFG) and discussed many theorems, proofs, and proprieties [19]. Parvathi et al. [20–22] presented many different operations such as the join, union, and product of two IFGs. Rashmanlou et al. [23] presented some products such as strong, direct, and lexicographic products for two IFGs. For further detailed descriptions of IFGs, please refer to [24–27]. Although the fuzzy graph and IFG have been used to model many real-word problems, uncertainties due to inconsistent and indeterminate information about a problem cannot be represented properly by the fuzzy graph or IFG. Therefore, a new idea is needed to deal with such scenarios.

Smarandache [28] introduced the idea of the neutrosophic set, by modifying the concept of the fuzzy set. The neutrosophic set can work with uncertain, indeterminate, vague, and inconsistent information of any uncertain real-life problem. It is basically a modified version of the crisp set, Type 1 fuzzy set, and IFS. It is described by the truth, indeterminate, and false membership degrees of any object. These three membership degrees are independent and always lie within $]^{-}0, 1^{+}[$, i.e., a nonstandard unit interval. The neutrosophic graph [12] can efficiently model the inconsistent information about any real-life problem. Recently, many researchers have more actively worked on neutrosophic graph theory; for instance, Ye [29], Yang et al. [30], Naz et al. [31], Broumi [15], and Arkam [32–35].

The vertex degree is a significant way to represent the total number of relations of a vertex in a graph, and the vertex degree can be used to analyze the graph. Gani and Lathi [36] proposed the concept of irregularity, total irregularity, and total degree in a fuzzy graph. Maheswari and Sekar [37] proposed the notation of the d_2-vertex in a fuzzy graph and also described several properties on the d_2-vertex degree of a fuzzy graph. Darabian et al. [38] presented the idea of the d_m-regular vague graph, td_m-regular vague graph, m-highly irregular vague graph, and m-highly total irregular vague graph, and they discussed some properties of those graphs. They described some real-life applications (e.g., fullerene molecules, wireless networks, and road networks) of regular vague graphs. Neutrosophic graphs are more effective, precise, flexible, and compatible when modeling uncertain real-life problems compared to fuzzy graphs or vague graphs. Thus, the use of the neutrosophic graph is inevitable for modeling optimization problems [29,34,39–41] in real-life scenarios, and it is essential to present some new properties and theories for neutrosophic graphs. This idea motivates us to introduce different types of neutrosophic graphs (regular, bipartite, isomorphic, and μ complement neutrosophic graphs) and their related theorems.

The concept of the regularity and degree of nodes has a significant role in both theories and applications (e.g., social network analysis, road transpotation network, wireless multihop network, and the assignment problem) in neutrosophic graph theory. The main contributions of this manuscript are as follows.

(i) As far as we know, there exists no research work on the regularity of the neutrosophic graph until now. Therefore, in this manuscript, we present the definition of the regular neutrosophic graph, star neutrosophic graph, regular strong neutrosophic graph, and complete bipartite neutrosophic graph.
(ii) We introduce the two types of degree, d_m and total d_m-degrees, of a node in a neutrosophic graph. The definition of busy and free nodes in a regular neutrosophic graph are presented here.
(iii) Depending on the degree of the node, we introduce the three types of regularity of a neutrosophic graph, d_m-regular, td_m-regular, and m-highly-irregular neutrosophic graphs (HING), and some properties are also described. We also investigate the concept of m-HING on the cycle graph and path graph in this manuscript.
(iv) We have presented the definition of the μ-complement and h-morphism of a neutrosophic graph. Some properties of the complement and isomorphic regular neutrosophic graph are also presented here.
(v) Some real-life applications of the regular neutrosophic graph and complete bipartite neutrosophic graph such as the road transport network, social networks, and assignment problem are described in this paper.

2. Preliminaries

In this portion, we will discuss the single-valued neutrosophic graph (SVNG), adjacent node, path, isolated node, strength of a path, strong SVNG, complement SVNG, and complete SVNG, which are efficient for the present work.

Definition 1. *Let $G = (N, M)$ be an SVNG [42,43] where N and M are represented by two neutrosophic sets on V and E, respectively, which satisfy the following.*

$$T_M(a,b) \le \min\left(T_N(a), T_N(b)\right)$$
$$I_M(a,b) \ge \max\left(I_N(a), I_N(b)\right)$$
$$F_M(a,b) \ge \max\left(F_N(a), F_N(b)\right)$$

Here, a and b are two vertices of G, and $(a,b) \in E$ is an edge of G.

Definition 2. *The two nodes a and b of an SVNG $G = (N, M)$ are defined as the adjacent nodes in G if and only if $(T_M(a,b), I_M(a,b), F_M(a,b)) > 0$. The two nodes a and b are called the neighbor nodes, and (a,b) is incident at a and b.*

Definition 3. *Let $G = (N, M)$ be an SVNG and P be a path of G. P is a collection of different nodes, $a_0, a_1, a_2, ..., a_n$ such that $(T_M(a_{i-1}, a_i), I_M(a_{i-1}, a_i), F_M(a_{i-1}, a_i)) > 0$ for $0 \le i \le n$. Here, n represents the neutrosophic length of the path P. A single neutrosophic vertex, i.e., a_i in G, is also assumed as a path. The path length of a single node a_i is $(0,0,0)$. We define the order pair (a_{i-1}, a_i) as the edge of the path. P is said to be a neutrosophic cycle if $a_0 = a_n$ and $n \ge 3$.*

Definition 4. *A node $a_i \in V$ of an SVNG G is called the isolated node if there exists no incident arc to the vertex a_i.*

Definition 5. *Let $G = (N, M)$ be an SVNG. If G has a path P of path length [42] k from node a to node b in G such as $P = \{a = a_1, (a_1, a_2), a_2, ..., a_{k-1}(a_{k-1}, a_k), a_k = b)\}$, then $T_M^k(a,b)$, $I_M^k(a,b)$, and $F_M^k(a,b)$ are described as follows.*

$$\begin{aligned} T_M^k(a,b) &= \sup\left(T_M(a,a_1) \wedge T_M(a_1,a_2) \wedge ... \wedge T_B(a_{k-1},a_k)\right) \\ I_M^k(a,b) &= \inf\left(I_M(a,a_1) \vee I_M(a_1,a_2) \vee ... \vee I_M(a_{k-1},a_k)\right) \\ F_M^k(a,b) &= \inf\left(F_M(a,a_1) \vee F_M(a_1,a_2) \vee ... \vee F_M(a_{k-1},a_k)\right) \end{aligned}$$

Definition 6. *Let $G = (N, M)$ be an SVNG. The strength of connection of a path P between two nodes a and b is defined by $\left(T_M^\infty(a,b), I_M^\infty(a,b), F_M^\infty(a,b)\right)$ where:*

$$\begin{aligned} T_M^\infty(a,b) &= \sup\{T_M^k(a,b) \mid k = 1,2,...\} \\ I_M^\infty(a,b) &= \inf\{I_M^k(a,b) \mid k = 1,2,...\} \\ F_M^\infty(a,b) &= \inf\{F_B^k(a,b) \mid k = 1,2,...\} \end{aligned}$$

Definition 7. *Let $G = (N, M)$ be an SVNG. G is said to be a connected SVNG [42] if there exists no isolated node in G.*

Definition 8. *Let $G = (N, M)$ be an SVNG and $a \in V$ be node of G. The degree of node a is the sum of the truth membership values, the sum of the indeterminacy membership values, and the sum of the membership values of falsity of all the arcs that are adjacent to node a. The degree of node [35] a is denoted by $d(a) = (d_T(a), d_I(a), d_F(a))$ where:*

$$d_T(a) = \sum_{\substack{a \in V \\ a \neq b}} T_M(a,b), d_I(a) = \sum_{\substack{a \in V \\ a \neq b}} I_M(a,b), d_F(a) = \sum_{\substack{a \in V \\ a \neq b}} F_M(a,b)$$

Here, $d_T(a)$, $d_I(a)$, and $d_F(a)$ are the degree of the truth membership value, the degree of the indeterminacy membership value, and the degree falsity membership value, respectively, of vertex a.

Definition 9. *Let $G = (N, M)$ be an SVNG. Then, the order of G is denoted by $O(G) = (O_T(G), O_I(G), O_F(G))$ where:*

$$O_T(G) = \sum_{a \in V} T_N(a), O_I(G) = \sum_{a \in V} I_N(a), O_F(G) = \sum_{a \in V} F_N(a)$$

Here, $O_T(G)$, $O_I(G)$, and $O_F(G)$ are the order of the membership degree of the truth value, the indeterminacy value, and the falsity value of G, respectively.

Definition 10. *Let $G = (N, M)$ be an SVNG and the size graph of G be described as $S(G) = (S_T(G), S_I(G), S_F(G))$ where:*

$$S_T(G) = \sum_{\substack{a,b \in V \\ a \neq b}} T_M(a,b), S_I(G) = \sum_{\substack{a,b \in V \\ a \neq b}} I_M(a,b), S_F(G) = \sum_{\substack{a,b \in V \\ a \neq b}} F_M(a,b)$$

Here, $S_T(G)$, $S_I(G)$, and $S_F(G)$ are respectively the order of the membership degree of the truth, indeterminacy, and falsity of G.

Definition 11. *Let $G = (N, M)$ be an SVNG. G is said to be a strong SVNG [31,35,44] if:*

$$\begin{aligned} T_M(a,b) &= \min\left(T_N(a), T_N(b)\right) \\ I_M(a,b) &= \max\left(I_N(a), I_N(b)\right) \\ F_M(a,b) &= \max\left(F_N(a), F_N(b)\right), \forall (a,b) \in E \end{aligned}$$

Definition 12. *Let $G = (N, M)$ be an SVNG and G be said to be a complete SVNG [31,35,44] if:*

$$T_M(a,b) = \min\left(T_N(a), T_N(b)\right)$$
$$I_M(a,b) = \max\left(I_N(a), I_N(b)\right)$$
$$F_M(a,b) = \max\left(F_N(a), F_N(b)\right), \forall a,b \in V$$

Definition 13. *Let $G = (N, M)$ be an SVNG. $G^c = (N^c, M^c)$ is the complement of an SVNG [31,35,44] if $N^c = N$ and M^c is computed as below.*

$$T_M^c(a,b) = \min\left(T_N(a), T_N(b)\right) - T_M(a,b)$$
$$I_M^c(a,b) = \max\left(I_N(a), I_N(b)\right) - I_M(a,b)$$
$$F_M^c(a,b) = \max\left(F_N(a), F_N(b)\right) - F_M(a,b), \forall a,b \in V$$

Here, $T_M^c(a,b)$, $I_M^c(a,b)$, and $F_M^c(a,b)$ denote the true, intermediate, and false membership degree for edge (a,b) of G^c.

3. Regular, d_m-Regular, and td_m-Regular Neutrosophic Graphs

In this section, first we define the regular neutrosophic graph, regular strong neutrosophic graph, d_m-degree, and td_m-degree of nodes in a neutrosophic graph. Then, we propose the notions of d_m and td_m-regular neutrosophic graphs and prove the necessary and sufficient conditions, for which under these conditions, d_m-regular with td_m-regular neutrosophic graphs are equivalent.

Definition 14. *Let $G = (N, M)$ be an SVNG. G is a regular neutrosophic graph if it satisfies the following conditions.*

$$\sum_{a \neq b} T_M(a,b) = Constant, \sum_{a \neq b} I_M(a,b) = Constant, \sum_{a \neq b} F_M(a,b) = Constant$$

Definition 15. *Let $G = (N, M)$ be an SVNG. G is a regular strong neutrosophic graph if it satisfies the following conditions.*

$$T_M(a,b) = \min\left(T_N(a), T_N(b)\right) \text{ and } \sum_{a \neq b} T_M(a,b) = Constant$$
$$I_M(a,b) = \max\left(I_N(a), I_N(b)\right) \text{ and } \sum_{a \neq b} I_M(a,b) = Constant$$
$$F_M(a,b) = \max\left(F_N(a), F_N(b)\right) \text{ and } \sum_{a \neq b} F_M(a,b) = Constant$$

Definition 16. *Let $G = (N, M)$ be an SVNG and the d_m-degree of any node a in G be denoted as $d_m(a)$ where:*

$$d_m(a) = \left(\sum_{a \neq b \in V} T_M^m(a,b), \sum_{a \neq b \in V} I_M^m(a,b), \sum_{a \neq b \in V} F_M^m(a,b) \right)$$

Here, the path $a, a_1, a_2, \ldots, a_{m-1}, b$ is the shortest path between the nodes u and v, and the length of this path is m.

Example 1. *We have considered an example of an SVNG G, presented in Figure 1.*

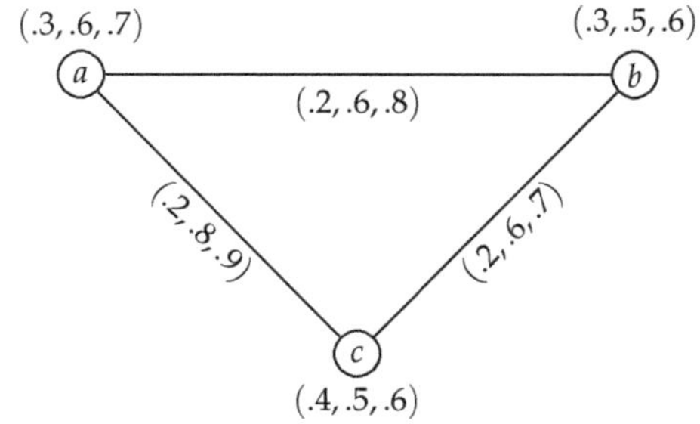

Figure 1. A neutrosophic graph G.

Then, the d_2-degree of the nodes in G is computed as follows.

$$d_2(a) = (0.2+0.2, 0.6+0.8, 0.8+0.9) = (0.4, 1.4, 1.7)$$
$$d_2(b) = (0.2+0.2, 0.8+0.8, 0.9+0.9) = (0.4, 1.6, 1.8)$$
$$d_2(c) = (0.2+0.2, 0.6+0.8, 0.8+0.9) = (0.4, 1.6, 1.8)$$

Definition 17. *Let $G = (N, M)$ be an SVNG and $a \in V$ be a node G. The total d_m-degree (td_m-degree) of node a in G is computed as follows.*

$$td_m(a) = \left(\left(\sum_{a \neq b \in V} T_M^m(a,b) + T_N(a)\right), \left(\sum_{a \neq b \in V} I_M^m(a,b) + I_N(a)\right), \left(\sum_{a \neq b \in V} F_M^m(a,b) + F_N(a)\right)\right)$$

Here, the path $a, a_1, a_2, ..., a_{m-1}, b$ is the shortest path between the nodes u and v, and the length of this path is m.

Example 2. *Let us consider an example of an SVNG, shown in Figure 1. Then, the td_2-degree of the nodes in G is as follows.*

$$td_2(a) = (((0.2+0.2)+0.3), ((0.6+0.8)+0.6), ((0.8+0.9)+0.7)) = (0.7, 2.0, 2.4)$$
$$td_2(b) = (((0.2+0.2)+0.3), ((0.8+0.8)+0.6), ((0.9+0.9)+0.6)) = (0.7, 2.1, 2.4)$$
$$td_2(a) = (((0.2+0.2)+0.4), ((0.8+0.8)+0.5), ((0.9+0.9)+0.6)) = (0.8, 2.1, 2.4)$$

Definition 18. *Let $G = (N, M)$ be a neutrosophic graph. G is said to be a $(m, (d_1, d_2, d_3))$-regular neutrosophic graph or d_m-regular if for all nodes $a \in V$ in G, $d_m(a) = (d_1, d_2, d_3)$.*

Example 3. *An example of an $(m, (d_1, d_2, d_3))$-totally regular neutrosophic graph is pictured in Figure 2. Figure 2 is a (2, (0.1, 0.6,0.9))-regular neutrosophic graph.*

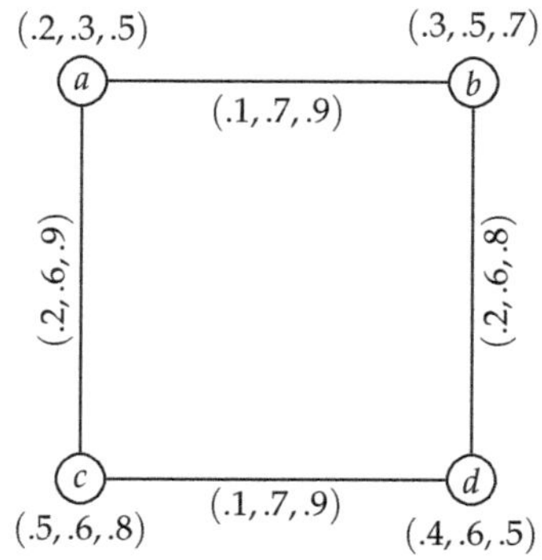

Figure 2. A neutrosophic graph G.

$$d_2(a) = (((0.2+0.2)+0.3), ((0.6+0.8)+0.6), ((0.8+0.9)+0.7)) = (0.7, 2.0, 2.4)$$
$$td_2(b) = (((0.2+0.2)+0.3), ((0.8+0.8)+0.6), ((0.9+0.9)+0.6)) = (0.7, 2.1, 2.4)$$
$$td_2(a) = (((0.2+0.2)+0.4), ((0.8+0.8)+0.5), ((0.9+0.9)+0.6)) = (0.8, 2.1, 2.4)$$

Definition 19. *Let $G = (N, M)$ is a neutrosophic graph. G is an $(m, (k_1, k_2, k_3))$-totally regular neutrosophic graph or td_m-regular neutrosophic graph if for all nodes $a \in V$ in G, $td_m(a) = (k_1, k_2, k_3)$.*

Example 4. *An example of a $(m, (k_1, k_2, k_3))$-totally regular neutrosophic graph is shown in Figure 3. It is a (2, (0.6, 2.0,2.1))-regular neutrosophic graph.*

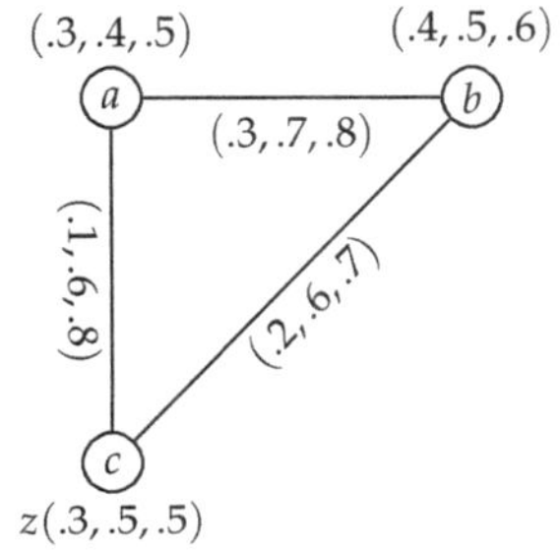

Figure 3. A neutrosophic graph G.

Theorem 1. *Let $G = (N, M)$ be an SVNG. If T_N, $I_N$and F_N are constant functions, then G is a d_m-regular neutrosophic graph if G is a td_m-regular neutrosophic graph (m is a positive integer).*

Proof. Suppose that for every node v in $G, (T_N(v), I_N(v), F_N(v)) = (c_1, c_2, c_3)$ and $d_m(v) = (d_1, d_2, d_3)$. Then:

$$td_m(v) = d_m(v) + (T_N(v), I_N(v), F_N(v)) = (d_1 + c_1, d_2 + c_2, d_3 + c_3);$$

hence, G is a td_m-regular neutrosophic graph. If G is a td_m-regular neutrosophic graph, then the proof is similar to the previous case. □

Theorem 2. *Let $G = (N, M)$ be an $(m, (d_1, d_2, d_3))$-totally regular and an $(m, (k_1, k_2, k_3))$-totally regular neutrosophic graph with n nodes. Then, T_N, I_N and F_N are constant functions, and $O(G) = n(k_1 - d_1, k_2 - d_2, k_3 - d_3)$.*

Proof. If G is an $(m, (d_1, d_2, d_3))$-regular neutrosophic graph and an $(m, (k_1, k_2, k_3))$-totally regular neutrosophic graph respectively, then for all $v \in V$ we get,

$$d_m(u) = (d_1, d_2, d_3) \to (\sum_{u \neq v \in V} T_M^m(uv), \sum_{u \neq v \in V} I_b^m(uv), \sum_{u \neq v \in V} F_b^m(uv)) = (d_1, d_2, d_3)$$

and:

$$td_m(v) = (k_1, k_2, k_3) \to (\sum_{u \neq v \in V} T_B^m(uv), + T_N(v), \sum_{u \neq v \in V} I_M^m(uv), + I_N(v), \sum_{u \neq v \in V} F_M^m(uv), + F_N(v)) = (k_1, k_2, k_3).$$

Hence, by (1), $(d_1, d_2, d_3) = (k_1 - T_N(v), k_2 - I_N(v), k_3 - F_N(v))$ and so $(T_N(v), I_N(v), F_N(v)) = (k_1 - d_1, k_2 - d_2, k_3 - d_3)$. Then, T_N,I_N and F_N are constant functions, and since G has n nodes, we get:

$$O(G) = (\sum_{v\in V} T_N(v), \sum_{v\in V} I_N(v), \sum_{v\in V} F_N(v)) = n(k_1 - d_1, k_2 - d_2, k_3 - d_3).$$

□

Free and Busy Vertex on the Regular Neutrosophic Graph

Definition 20. *Let $G = (N, M)$ be a regular neutrosophic graph and $a \in V$ be a node of G. The neutrosophic busy value of a a is denoted by $D(a) = (D_T(a), D_I(a), D_F(a))$, where:*

$$D_T(a) = \sum \min(T_N(a), T_N(a_i)), D_I(a) = \sum \max(I_N(a), I_N(a_i)), D_F(a) = \sum \max(F_N(a) \vee F_N(a_i))$$

Here, a_i denotes the adjacent vertex of a of G. The busy value of regular neutrosophic graph G is the sum of all busy values of all the nodes of G.

Definition 21. *Let $G = (N, M)$ be a regular neutrosophic graph and $a \in V$ be a node of G. The a of G is a busy node or vertex if:*

$$T_N(a) \leq d_T(a), I_N(a) \leq d_I(a), F_N(a) \geq d_F(a)$$

If the vertex a is not a busy vertex, then it is said to be a free vertex.

Definition 22. *Let $G = (N, M)$ be a regular neutrosophic graph and $(a, b) \in E$ be an arc of G. a, b is said to be an effective edge if:*

$$T_M(a,b) = \min(T_N(a), T_N(b)), I_M(a,b) = \max(I_N(a), I_N(b)), F_M(a,b) = \max(F_N(a), F_N(b))$$

4. Regularity on the Complement and Isomorphic Neutrosophic Graph

Definition 23. *Let $G_1 = (N_1, M_1)$ and $G_2 = (N_2, M_2)$ be two regular neutrosophic graphs.*

1. *A homomorphism h from a regular neutrosophic graph G_1 to another regular neutrosophic graph G_2 is a neutrosophic mapping function $h : V_1 \to V_2$, which always satisfies following conditions:*
   (*i*) $\forall a_1 \in V_1, a_1b_1 \in E_1$
      (*a*) $T_{N_1}(a_1) \leq T_{N_2}(h(a_1)), I_{N_1}(a_1) \leq I_{N_2}(h(a_1)), F_{N_1}(a_1) \geq F_{N_2}(h(a_1))$
      (*b*) $T_{B_1}(a_1b_1) \leq T_{M_2}(h(a_1)h(a_2)), I_{M_1}(a_1b_1) \leq I_{M_2}(h(a_1)h(a_2)),$ $F_{M_1}(a_1b_1) \geq F_{B_2}(h(a_1)h(a_2))$
2. *A isomorphism h from a regular neutrosophic graph G_1 to another regular neutrosophic graph G_2 is a neutrosophic bijective mapping function $h : V_1 \to V_2$, which always satisfies the following:*
   (*i*) $\forall a_1 \in V_1, a_1b_1 \in E_1$
      (*c*) $T_{N_1}(a_1) = T_{N_2}(h(a_1)), I_{N_1}(a_1) = I_{N_2}(h(a_1)), F_{N_1}(a_1) = F_{N_2}(h(a_1))$
      (*d*) $T_{M_1}(a_1b_1) = T_{M_2}(h(a_1)h(a_2)), I_{M_1}(a_1b_1) = I_{M_2}(h(a_1)h(a_2)),$ $F_{M_1}(a_1b_1) = F_{M_2}(h(a_1)h(a_2))$
3. *A weak isomorphism h from a regular neutrosophic graph G_1 to another regular neutrosophic graph G_2 is a neutrosophic bijective mapping function $h : V_1 \to V_2$, which always satisfies the following:*
   (*i*) $\forall a_1 \in V_1$

(e) h is a homomorphism.

(f) $T_{N_1}(a_1) = T_{N_2}(h(a_1)), I_{N_1}(a_1) = I_{N_2}(h(a_1)), F_{N_1}(a_1) = F_{N_2}(h(a_1))$.

4. *A co-weak isomorphism h from a regular neutrosophic graph G_1 to another regular neutrosophic graph G_2 is a neutrosophic bijective mapping function $h : V_1 \to V_2$, which always satisfies the following:*

(i) $\forall a_1b_1 \in E_1$

(g) h is a homomorphism.

(h) $T_{N_1}(a_1b_1) = T_{N_2}(h(a_1b_1)), I_{N_1}(a_1b_1) = I_{N_2}(h(a_1b_1)), F_{N_1}(a_1b_1) = F_{N_2}(h(a_1b_1))$.

Definition 24. *Let $G = (N, M)$ be a connected neutrosophic graph. The neutrosophic graph G is called an HING if and only if all the adjacent nodes of G are always different neutrosophic degrees.*

Theorem 3. *If G_1 and G_2 are two isomorphic highly-irregular neutrosophic graphs (HING), then their size and order are always equal.*

Proof. An isomorphism h is between two HINGs G_1 and G_2 with the underlying node set V_1 and V_2, respectively, then:

(i) $\forall a_1 \in V_1, a_1b_1 \in E_1$

(a) $T_{N_1}(a_1) = T_{N_2}(h(a_1)), I_{N_1}(a_1) = I_{N_2}(h(a_1)), F_{N_1}(a_1) = F_{N_2}(h(a_1))$

(b) $T_{M_1}(a_1b_1) = T_{B_2}(h(a_1)h(a_2)), I_{M_1}(a_1b_1) = I_{M_2}(h(a_1)h(a_2)),$
$F_{M_1}(a_1b_1) = F_{M_2}(h(a_1)h(a_2))$

Therefore, we have:

$$\begin{aligned}
S(G_1) &= \left(\sum_{a_1b_1\in E_1} T_{M_1}(a_1b_1), \sum_{a_1b_1\in E_1} I_{M_1}(a_1b_1), \sum_{a_1b_1\in E_1} F_{M_1}(a_1b_1)\right) \\
&= \left(\sum_{a_1,b_1\in V_1} T_{M_2}(h(a_1),h(b_1)), \sum_{a_1,b_1\in V_1} I_{M_2}(h(a_1),h(b_1)), \sum_{a_1b_1\in V_1} F_{M_2}(h(a_1),h(b_1))\right) \\
&= \left(\sum_{a_2b_2\in E_2} T_{M_2}(a_2b_2), \sum_{a_2b_2\in E_2} I_{M_2}(a_2b_2), \sum_{a_2b_2\in E_2} F_{M_2}(a_2b_2)\right) = S(G_2)
\end{aligned}$$

$$\begin{aligned}
O(G_1) &= \left(\sum_{a_1\in V_1} T_{N_1}(a_1), \sum_{a_1\in V_1} I_{N_1}(a_1), \sum_{a_1\in V_1} F_{N_1}(a_1)\right) \\
&= \left(\sum_{a_1\in V_1} T_{N_2}(h(a_1)), \sum_{a_1\in V_1} I_{N_2}(h(a_1)), \sum_{a_1\in V_1} F_{N_2}(h(a_1))\right) \\
&= \left(\sum_{a_2\in V_2} T_{N_2}(a_2), \sum_{a_2\in V_2} I_{N_2}(a_2), \sum_{a_2\in V_2} F_{N_2}(a_2)\right) = O(G_2)
\end{aligned}$$

□

Theorem 4. *Let G_1 and G_2 be two isomorphic HINGs. Then, the degrees of nodes a and $h(a)$ of G_1 and G_2 are the same.*

Proof. If G_1 and G_2 are two isomorphic HINGs, respectively, then $T_{M_1}(a) = T_{M_1}(h(a_1)h(b_1))$, $I_{M_1}(a) = I_{M_1}(h(a_1)h(b_1))$ and $F_{M_1}(a) = F_{M_1}(h(a_1)h(b_1))$ for all $a_1, b_1 \in V_1$. Therefore,

$$d_T(a_1) = \sum_{a_1,b_1 \in V_1} T_{M_1}(a_1b_1) = \sum_{a_1,b_1 \in V_1} T_{M_2}(h(a_1)h(b_1)) = d_T(h(a_1))$$
$$d_I(a_1) = \sum_{a_1,b_1 \in V_1} I_{M_1}(a_1b_1) = \sum_{a_1,b_1 \in V_1} I_{M_2}(h(a_1)h(b_1)) = d_I(h(a_1))$$
$$d_F(a_1) = \sum_{a_1,b_1 \in V_1} F_{M_1}(a_1b_1) = \sum_{a_1,b_1 \in V_1} F_{M_2}(h(a_1)h(b_1)) = d_F(h(a_1))$$

for all $a_1 \in V_1$. This proves that the degrees of nodes u and $h(u)$ of G_1 and G_2 are preserved. □

Definition 25. *Let $G = (N, M)$ be a connected neutrosophic graph. G is defined as: mm*

1. *A self-weak neutrosophic complementary isomorphic graph if graph G is weak isomorphic with G^c.*
2. *A self-complementary neutrosophic graph if $G^c \cong G$.*

Theorem 5. *Let G be a self-complementary HING. Then:*

$$\sum_{a \neq b} T_M(a,b) \leq \frac{1}{2} \sum_{a \neq b} \min(T_N(a), T_N(b)))$$
$$\sum_{a \neq b} I_M(a,b) \geq \frac{1}{2} \sum_{a \neq b} \max(I_N(a), I_N(b)))$$
$$\sum_{a \neq b} F_M(a,b) \geq \frac{1}{2} \sum_{a \neq b} \max(F_N(a), F_N(b)))$$

Definition 26. *Let $G = (N, M)$ be a neutrosophic graph and $G^\mu = (N^\mu, M^\mu)$ be the μ complement of G where $N = N^\mu$ and $M^\mu = \left(T^\mu_M, I^\mu_M, F^\mu_M\right)$ where:*

$$T^\mu_M = \begin{cases} \min(T_N(a), T_M(b)) - T_M(a,b) & \text{if } T_M(a,b) > 0 \\ 0 & \text{if } T_M(a,b) = 0 \end{cases}$$
$$I^\mu_M = \begin{cases} I_M(a,b) - \max(T_N(a), T_M(b)) & \text{if } I_M(a,b) > 0 \\ 0 & \text{if } I_M(a,b) = 0 \end{cases}$$
$$F^\mu_M = \begin{cases} F_M(a,b) - \max(F_N(a), F_M(b)) & \text{if } F_M(a,b) > 0 \\ 0 & \text{if } F_M(a,b) = 0 \end{cases}$$

Theorem 6. *Let G^μ be the μ-complement of an HING. G need not be highly irregular.*

Proof. All the adjacent nodes of G with different neutrosophic degrees or all the other nodes of G with different neutrosophic degrees may possibly be adjacent nodes with equivalent degrees. This condition opposes the idea of HING. □

Theorem 7. *Let G_1 and G_2 be two HINGs. If G_1 and G_2 are two isomorphic neutrosophic graphs, then G_1^μ and G_2^μ are also isomorphic, and vice versa.*

Proof. Let G_1 and G_2 be two isomorphic neutrosophic graphs; there always exists a map $h : V_1 to V_2$ satisfying $T_{N_1}(a) = T_{N_2}(h(a))$, for all $a \in V_1$ and $T_{M_1}(ab) = T_{M_2}(h(a)h(b))$, $I_{M_1}(ab) = I_{M2}(h(a)h(b))$ and $F_{M1}(ab) = F_{M2}(h(a)h(b))$, for all $ab \in E_1$. By the concept of the μ complement neutrosophic graph, we have $T^\mu_{M_1}(ab) = \min(T_{N1}(a), T_{N1}(b) - T_{M1}(ab) = \min(T_{N2}(h(a)), T_{N2}(h(b)))) - T_{M2}(h(a)h(b))$, $F^\mu_{M_1}(ab) = F_{M_1}(ab) - \max(F_{N1}(a), F_{N1}(b)) = T_{M2}(h(a)h(b)) - \max(F_{N1}(h(a)), F_{N2}(h(b)))$ for all

$ab \in E_1$ ($I^{\mu}_{M_1}$ (ab) values can be calculated using the same method). Hence, $G^{\mu}_1 \cong G^{\mu}_2$. The proof of the converse part is straight forward. □

5. Complete Bipartite Neutrosophic Graphs

Definition 27. *A neutrosophic graph $G = (N, M)$ is called a complete bipartite neutrosophic graph if the vertex set V can be divided into two nonempty sets, such that for every $v_1, v_2 \in V_1$ or V_2 and for every $u \in V_1$ and $v \in V_2$:*

Theorem 8. *Let $G = (N, M)$ be a complete bipartite neutrosophic graph, such that each partition of V has nodes.*

(i) If for every $v \in V_1 \backslash V_2$ or $v \in V_2 \backslash V_1$, $(t_N(v), i_N(v), f_N(v)) = (c_1 , c_2, c_3)$ and for the other nodes as u, $t_N(u) \geq c_1$, $i_N(u) \leq c_2$, $f_N(u) \leq c_3$, , then for all $1 \leq m \leq 2k - 1$, Gis a d_m regular neutrosophic graph.
(ii) Let $W = \{v_1, v_2, ..., v_k\}$ be a set of nodes, such that it contains at least one node of each partition of V.

Proof. Suppose that for every $v_1 \in V_1 \backslash V_2$, $(t_N(v), i_N(v), f_N(v)) = (c_1 , c_2, c_3)$. Therefore, in V_1, t_N and f_N are constant functions, and since for every $v_2 \in V_1 \backslash V_2$, $t_N(v_1) \leq t_N(v_1)$,$i_N(v_1) \geq i_N(v_2)$, $f_N(v_1) \geq f_N(v_{21})$). Then, by the previous definition, we have that for every $uv \in E$, we get $(t_M(uv), i_M(uv), f_M(uv)) = (c_1 , c_2 , c_3)$.

Hence, for every $v \in V$ and $1 \leq m \leq 2k - 1$, if m is odd, then $d_m(v) = k(c_1 , c_2 , c_3)$ and if m is even, then $d_m(v) = (k - 1)(c_1 , c_2 , c_3)$.

Therefore, in both cases, we see that G is a d_m-regular neutrosophic graph. Now, if for every $v_1 \in V_1 \backslash V_2$, $(t_N(v), i_N(v), f_N(v)) = (c_1 , c_2, c_3)$, then the proof is similar to the proof of (i).

(ii) The proof is similar to the previous theorem only by the difference, that every $v \in V$ *and* $k + 1 \leq m \leq 2k - 1$; if m is odd, then and if m is even, then $d_m(v) = (k - 1)(c_1 , c_2 , c_3)$.

Remark 1. *If $V_1 = \{v_1, v_2, ..., v_n\}$ and $V_1 = \{u_1, u_2, ..., u_m\}$ such that $n \neq m$,, then the above theorem does not hold.*

Example 5. *Consider the complete bipartite neutrosophic graph $G = (N, M)$ in Figure 4 shown below.*

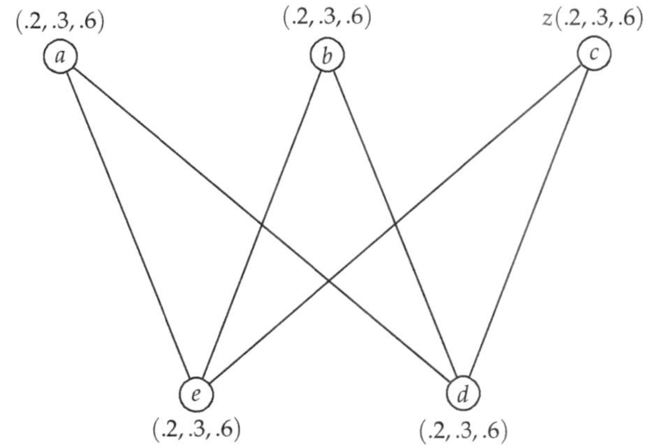

Figure 4. An example of a complete bipartite neutrosophic graph.

Since G is a complete bipartite neutrosophic graph, then for every $uv \in E, (t_M(uv), i_M(uv), f_M(uv)) = (0.2, 0.3, 0.6)$. *Hence:*

$d_1(a) = (0.4, 0.6, 1.2)$, $d_1(b) = (0.4, 0.6, 1.2)$, $d_1(c) = (0.4, 0.6, 1.2)$,$d_1(d) = (0.6, 0.9, 1.8)$, $d_1(e) = (0.6, 0.9, 1.8)$

$$d_2(a) = (0.4, 0.6, 1.2), d_2(b) = (0.4, 0.6, 1.2), d_2(c) = (0.4, 0.6, 1.2),$$
$$d_2(d) = (0.2, 0.3, 0.6), d_2(e) = (0.2, 0.3, 0.6)$$
$$d_3(a) = (0.4, 0.6, 1.2), d_3(b) = (0.4, 0.6, 1.2), d_3(c) = (0.4, 0.6, 1.2),$$
$$d_3(d) = (0.6, 0.9, 1.8), d_3(e) = (0.6, 0.9, 1.8)$$
$$d_4(a) = (0.4, 0.6, 1.2), d_4(b) = (0.4, 0.6, 1.2), d_4(c) = (0.4, 0.6, 1.2),$$
$$d_4(d) = (0, 1, 1), d_4(e) = (0, 1, 1)$$

Since t_N, i_N and f_N are constant functions, we have for all $m = 1, 2, 3, 4$, G does not follow the condition of the d_m-regular neutrosophic graph.

□

Theorem 9. *Let $G = (N, M)$ be a regular neutrosophic graph on cycle graph $G^* = (V, E)$ with $k \geq 3$ nodes. If for $k-(n-1)$ arcs of G as uv, $(T_M(uv), I_M(uv), F_M(uv)) = (c_1, c_2, c_3)$ and for arcs as ab, $T_M(ab) \geq c_1$, $I_M(ab) \geq c_2$, and $F_M(ab) \geq c_3$, then for any $n \leq m \leq k-1$, G is a d_m-regular neutrosophic graph.*

Proof. Let $G = (N, M)$ be a regular neutrosophic graph on cycle graph G^* with k nodes and for $k-(n-1)$ arcs of G as uv, $(T_M(uv), I_M(uv), F_M(uv)) = (c_1, c_2, c_3)$.

Since G^* is a cycle, we have for every node w in G that there exist exactly two distinct nodes w_1 and w_2 such that w and w_1 are connected by means of one path of length m (where $n \leq m \leq k-1$) and each path contains at least one arc uv.

Hence, for two paths,
$\lambda_1 : w_1 = u_0, u_1, u_2, ...u_{m-1}, u_m = w$ and $\lambda_2 : w_1 = v_0, v_1, v_2, ...v_{m-1}, v_m = w$.

Therefore, we obtain
$T_M^m(w_1w) + T_M^m(w_2w) = T_M(w_1u_1) \wedge ... \wedge T_M(u_{m-1}w) + T_M(w_2v_1) \wedge ... \wedge T_M(v_{m-1}w) = c_1 + c_1 = 2c_1$.
$I_M^m(w_1w) + I_M^m(w_2w) = I_M(w_1u_1) \vee ... \vee I_M(u_{m-1}w) + I_M(w_2v_1) \vee ... \vee I_M(v_{m-1}w) = c_1 + c_1 = 2c_1$.
$F_M^m(w_1w) + F_M^m(w_2w) = F_M(w_1u_1) \vee ... \vee F_M(u_{m-1}w) + F_M(w_2v_1) \vee ... \vee F_M(v_{m-1}w) = c_1 + c_1 = 2c_1$.

□

Remark 2. *Let G be a regular neutrosophic graph on cycle graph G^* with k nodes. If T_M, I_M, and F_M are constant functions and T_N, I_N, and F_N are not, then for any $1 \leq m \leq k-1$, G is not a td_m-regular neutrosophic graph in general.*

Theorem 10. *Let $G = (N, M)$ be a neutrosophic graph on cycle graph $G^* = (V, E)$ with $k \geq 3$ nodes. If G is a d_1-regular neutrosophic graph, then T_M, I_M, and F_M are constant functions.*

Proof. Suppose that G is a d_1-regular neutrosophic graph and there exists $uv \in E$ such that $(T_M(uv), I_M(uv), F_M(uv)) = (c_1, c_2, c_3)$, where for every $ab \in E - \{uv\}$, $(T_M(ab), I_M(ab), F_M(ab)) = (c_1, c_2, c_3)$. Then, $T_M(uv) \neq c_1$, $I_M(uv) \neq c_2$, and $F_M(uv) \neq c_3$.

Case 1 : If $T_M(uv) \neq c_1$,
$\sum_{u \neq z \in V} T_M^1(uz) = c_1 + T_M(uv)$, $\sum_{v \neq z \in V} T_M^1(vz) = c_1 + T_M(uv)$.

Since G has $k \geq 3$ nodes, there exists $w \in V$ such that $w \neq v, u$. Hence, we get $\sum_{u \neq z \in V} T_M^1(wz) = 2c_1$, which implies that $T_M^1(wz) \neq T_M^1(vz)$ and $T_M^1(wz) \neq T_M^1(uz)$. Therefore, $d_1(w) \neq d_1(u)$ and $d_1(w) \neq d_1(v)$; hence, G is not a d_1-regular neutrosophic graph, which is a contradiction.

Case 2 : If $I_M(uv) \neq c_2$,
$\sum_{u\neq z\in V} I_M^1(uz) = c_2 + I_M(uv)$, $\sum_{v\neq z\in V} I_M^1(vz) = c_2 + I_M(uv)$.

Since G has $k \geq 3$ nodes, there exists $w \in V$ such that $w \neq v, u$. Hence, we get $\sum_{u\neq z\in V} I_M^1(wz) = 2c_2$, which implies that $I_M^1(wz) \neq I_M^1(vz)$ and $I_M^1(wz) \neq I_M^1(uz)$. Therefore, $d_1(w) \neq d_1(u)$ and $d_1(w) \neq d_1(v)$; hence, G is not a d_1-regular neutrosophic graph, which is a contradiction.

Case 3 : If $F_M(uv) \neq c_3$,
$\sum_{u\neq z\in V} F_M^1(uz) = c_3 + F_M(uv)$, $\sum_{v\neq z\in V} F_M^1(vz) = c_3 + F_M(uv)$.

Since G has $k \geq 3$ nodes, there exists $w \in V$ such that $w \neq v, u$. Hence, we get $\sum_{u\neq z\in V} F_M^1(wz) = 2c_3$, which implies that $F_M^1(wz) \neq F_M^1(vz)$ and $F_M^1(wz) \neq F_M^1(uz)$. Therefore, $d_1(w) \neq d_1(u)$ and $d_1(w) \neq d_1(v)$; hence, G is not a d_1-regular neutrosophic graph, which is a contradiction. Hence the proof. □

Theorem 11. *Let $G = (N, M)$ be a neutrosophic graph on the path graph $G^* = (V, E)$ with $2m$ nodes. If for at least $m+1$ arcs of G as uv, $(T_M(uv), I_M(uv), F_M(uv)) = (c_1, c_2, c_3)$ and for other arcs as ab, $T_M \geq c_1$, $I_M \leq c_2$ and $F_M \leq c_3$, then G is also a d_m-regular neutrosophic graph.*

Proof. Suppose that G^* is a path with $2m$ nodes and for at least $m+1$arcs of G as uv, $(T_M(uv), I_M(uv), F_M(uv)) = (c_1, c_2, c_3)$. Since G^* is a path, we have for every arbitrary node w in G that there exists only one node as z such that w and z are connected by means of one path of length m and this path contains at least one uv. Hence, for the path $\lambda : z = z_0, z_1, z_2, ... z_{m-1}, z_m = w$. We get $T_M^m(zw) = T_M(zz_1) \wedge ... \wedge T_M(z_{m-1}w) = c_1$, $I_M^m(zw) = I_M(zz_1) \vee ... \vee I_M(z_{m-1}w) = c_2$, and $F_M^m(zw) = F_M(zz_1) \vee ... \vee F_M(z_{m-1}w) = c_3$.

Therefore, $d_m(w) = (T_M^m(zw), I_M^m(zw), F_M^m(zw)) = (c_1, c_2, c_3)$, and so, G is a $(m, (c_1, c_2, c_3))$-regular neutrosophic graph. □

Remark 3. *Let $G = (N, M)$ be a neutrosophic graph on the path graph $G^* = (V, E)$ with $2m$ nodes. If T_M, I_M, and F_M are constant functions and T_N, I_N, and F_N are not, then G is not td_m-regular.*

6. m-Highly Irregular Neutrosophic Graph

In this section, we present the notation of m-HING and the m-highly totally irregular neutrosophic graph (HTING). We have proven the necessary and sufficient conditions for which the m-highly irregular and m-HTINGs are equivalent, and also, we provide m-highly irregularity on neutrosophic graph G where G^m is the cycle, path, or star graph.

Definition 28. *Let $G = (N, M)$ be a regular connected neutrosophic graph, then:*

(i) G is said to be m-HING, if each node of G is adjacent to the other nodes with the different values of the d_m-degree.
(ii) G is said to be m-HTING, if each node of G is adjacent to the other nodes with the different values of the td_m-degree.

Theorem 12. *Let $G = (N, M)$ be a neutrosophic graph. If T_M, I_M, and F_M are used to denote the three constant functions and m represents a positive integer number, then G is m-HTING if G is m-HING.*

Proof. Suppose that G is m-HTING. Then, the value of the td_m-degree of all the pairs of adjacent nodes are different. Let two nodes v and u be the pair of adjacent nodes with the distinct td_m-degree. Hence, we have

$$td_m(u) = \left[\sum_{u\neq w\in V} T_M^m(uw) + T_N(u), \sum_{u\neq w\in V} I_M^m(uw) + I_N(u), \sum_{u\neq w\in V} F_M^m(uw) + F_N(u)\right]$$

$$td_m(v) = \left[\sum_{v\neq w\in V} T_M^m(vw) + T_N(v), \sum_{v\neq w\in V} I_M^m(vw) + I_N(v), \sum_{v\neq w\in V} F_M^m(vw) + F_N(v)\right].$$

Since $td_m(u) \neq td_m(v)$, we have, $\sum_{u\neq w\in V} T_M^m(uw) + T_N(u) \neq \sum_{v\neq w\in V} T_M^m(vw) + T_N(v)$, $\sum_{u\neq w\in V} I_M^m(uw) + I_N(u) \neq \sum_{v\neq w\in V} I_M^m(vw) + I_N(v)$ and $\sum_{u\neq w\in V} F_M^m(uw) + F_N(u) \neq \sum_{v\neq w\in V} F_M^m(vw) + F_N(v)$.

Since $T_N(u) = T_N(v)$, $I_N(u) = I_N(v)$, and $F_N(u) = F_N(v)$, hence we get $\sum_{u\neq w\in V} T_M^m(uw) \neq \sum_{v\neq w\in V} T_M^m(vw)$, $\sum_{u\neq w\in V} I_M^m(uw) \neq \sum_{v\neq w\in V} I_M^m(vw)$, and $\sum_{u\neq w\in V} F_M^m(uw) \neq \sum_{v\neq w\in V} F_M^m(vw)$.
Therefore, we have,

$$d_m(u) = \left[\sum_{u\neq w\in V} T_M^m(uw), \sum_{u\neq w\in V} I_M^m(uw), \sum_{u\neq w\in V} F_M^m(uw)\right] \neq \left[\sum_{v\neq w\in V} T_M^m(vw), \sum_{v\neq w\in V} I_M^m(vw), \sum_{v\neq w\in V} F_M^m(vw)\right] = d_m(v).$$

Hence, any pair of adjacent nodes in G has distinct a d_m-degree.
Then, G is m-HING.

Conversely, G is an m-HING. Then, the value of the d_m-degree of each and every pair of adjacent nodes such as u and v is distinct. This implies that $d_m(u) \neq d_m(v)$. Since $T_N(u) = T_N(v)$, $I_N(u) = I_N(v)$, and $F_N(u) = F_N(v)$, hence $td_m(u) = d_m(u) + T_N(u) \neq id_m(u) = d_m(u) + I_N(u) \neq fd_m(u) = d_m(u) + F_N(u)$, and so, any two adjacent nodes in G has the distinct td_m-degree. Therefore, G is an m-HTING.

Hence the proof. □

Theorem 13. *Let $G = (N, M)$ be a neutrosophic graph on cycle graph $G^* = (V, E)$ with $k \geq 3$ nodes and for all $i = 1, 2, ..., k-1$ (where $v_{k+1} = v_1$), $T_M(v_iv_{i+1}) < T_M(v_{i+1}v_{i+2})$or $I_M(v_iv_{i+1}) > I_M(v_{i+1}v_{i+2})$ or $F_M(v_iv_{i+1}) > F_M(v_{i+1}v_{i+2})$. Then, G is a $\left[1, 2, 3, \left[\frac{k}{2}\right]\right]$-HING.*

Proof. Suppose that $v_1, v_2, ..., v_k, v_1$ is the arcs of G, where $T_M(v_1v_2) < T_M(v_2v_3) < ... < T_M(v_iv_{i+1}) < T_M(v_{i+1}v_{i+2}) < ... < T_M(v_kv_1)$ and there exists $v_iv_{i+1} \in E$ such that for $1 \leq m \leq \left[\frac{k}{2}\right]$, $d_m(v_i) = d_m(v_{i+1})$. Therefore, we get $\sum_{v_i\neq v_{i+1}\in V} T_M^m(v_iv_j) = T_M(v_iv_{i+1}) + T_M(v_jv_{j+1})$ and $\sum_{v_{i+1}\neq v_j\in V} T_M^m(v_iv_j) = T_M(v_iv_{i+1}) + T_M(v_{j+1}v_{j+2})$. Since $d_m(v_i) = d_m(v_{i+1})$, $T_M(v_iv_{i+1}) + T_M(v_jv_{j+1}) = T_M(v_iv_{i+1}) + T_M(v_{j+1}v_{j+2})$, thus $T_M(v_ivi+1) = T_M(v_{j+1}vj+2)$, which is a contradiction.

Now, for every $v_1, v_2, ..., v_k, v_1 \in E$, also we have
$I_M(v_1v_2) > I_M(v_2v_3) > ... > I_M(v_iv_{i+1}) > I_M(v_{i+1}v_{i+2}) > ... > I_M(v_kv_1)$,
$F_M(v_1v_2) > F_M(v_2v_3) > ... > F_M(v_iv_{i+1}) > F_M(v_{i+1}v_{i+2}) > ... > F_M(v_kv_1)$.

The proof is similar to the case as true membership, and the contradiction is obtained. Therefore, G is an $\left[1, 2, 3, \left[\frac{k}{2}\right]\right]$-HING. □

Remark 4. *Let $G = (N, M)$ be a neutrosophic graph on cycle graph $G^* = (V, E)$ with $k \geq 3$ nodes and for all $i = 1, 2, ..., k-1$ (where $v_{k+1} = v_1$), $T_M(v_iv_{i+1}) < T_M(v_{i+1}v_{i+2})$or $I_M(v_iv_{i+1}) > I_M(v_{i+1}v_{i+2})$ or $F_M(v_iv_{i+1}) > F_M(v_{i+1}v_{i+2})$. Then, G is not a $\left[1, 2, 3, \left[\frac{k}{2}\right]\right]$-HING in general.*

Example 6. *Consider the neutrosophic graph $G = (N, M)$ on graph $G^* = (V, E)$ shown in the Figure 5.*

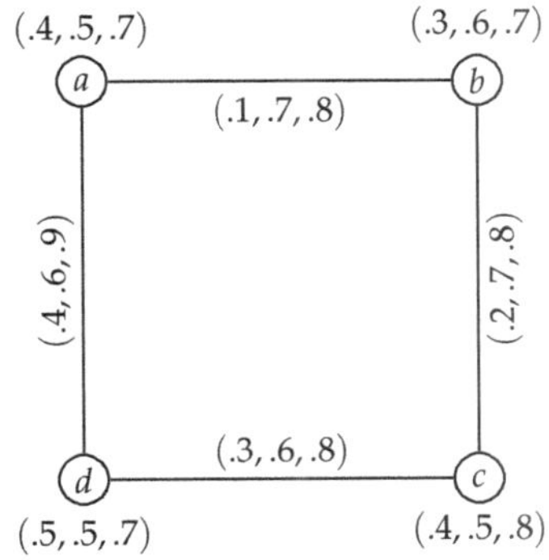

Figure 5. *m*-highly irregular neutrosophic graph.

We see that $T_M(ab) < T_M(bc) < T_B(cd) < T_B(da)$, *and we get*
$td_2(a) = (0.3 + 0.4, 0.6 + 0.4) = (0.7, 1.0)$
$td_2(b) = (0.2 + 0.3, 0.6 + 0.6) = (0.5, 1.2)$
$td_2(c) = (0.3 + 0.4, 0.6 + 0.5) = (0.7, 1.1)$
$td_2(d) = (0.2 + 0.5, 0.6 + 0.4) = (0.7, 1.0)$
Since for two vertices a and b, they are adjacent, $td_2(a) = td_2(b)$. *Then, G is not a two-HTING.*

Theorem 14. *Let* $G = (N, M)$ *be a neutrosophic graph on cycle graph* $G^* = (V, E)$ *with* $k \geq 3$ *nodes and for all* $i = 1, 2, ..., k-1$ *(where* $v_{k+1} = v_1$*),* $T_M(v_iv_{i+1}) < T_M(v_{i+1}v_{i+2})$ *or* $I_M(v_iv_{i+1}) > I_M(v_{i+1}v_{i+2})$ *or* $F_M(v_iv_{i+1}) > F_M(v_{i+1}v_{i+2})$. *Then, G is a* $\left[1, 2, 3, \left[\frac{k}{2}\right]\right]$*-HING.*

Proof. Suppose that $v_1, v_2, ..., v_k, v_1$ is the arcs of G, where $T_M(v_1v_2) < T_M(v_2v_3) < ... < T_M(v_iv_{i+1}) < T_M(v_{i+1}v_{i+2}) < ... < T_M(v_{k-1}v_k)$. Then, for every $1 \leq m \leq \left[\frac{k}{2}\right]$, we get

$$\sum_{v_i \neq v_j \in V} T_M^m(v_iv_j) = \begin{cases} T_M(v_iv_{i+1}), \ if\ i = 1, 2, 3, ..., m \\ T_M(v_iv_{i+1}) + T_M(v_{i-m}v_{i-m+11}, \ if\ i = m+1, ..., k-m \\ T_M(v_{i-m}v_{i-m+1}, \ if\ i = k-m+1, ...k. \end{cases}$$

Hence, for every two adjacent nodes as v_i and v_{i+1} in G, we get $d_m(v_i) \neq d_m(v_{i+1})$. Therefore, G is a $\left[1, 2, 3, \left[\frac{k}{2} - 1\right]\right]$-HING. Now, suppose that $m = \left[\frac{k}{2}\right]$. If k is even, then we have

$$\sum_{v_i \neq v_j \in V} T_M^m(v_iv_j) = \begin{cases} T_M(v_iv_{i+1}), \ if\ i = 1, 2, 3, ..., m \\ T_M(v_{i-m}v), \ if\ i = m+1, ..., k-m \end{cases}$$

If k is odd, then we have

$$\sum_{v_i \neq v_j \in V} T_M^m(v_iv_j) = \begin{cases} T_M(v_iv_{i+1}), \ if\ i = 1, 2, 3, ..., m \\ T_M(v_1v_2) + T_M(v_{i-m}v_{i-m+11}, \ if\ i = m+1 \\ T_M(v_{i-m}v_{i-m+1}, \ if\ i = m+2, ...k. \end{cases}$$

Here, in both cases, we see that for every two adjacent nodes v_i and v_{i+1} in G, we get $d_m(v_i) \neq d_m(v_{i+1})$, and so, G is a $\left[\frac{k}{2}\right]$-HING. Similarly, for the indeterministic and false membership function, for each arc, it is proven.

Hence the proof. □

Example 7. *Consider the the neutrosophic graph* $G = (N, M)$ *on graph* $G^* = (V, E)$ *shown in Figure 6.*

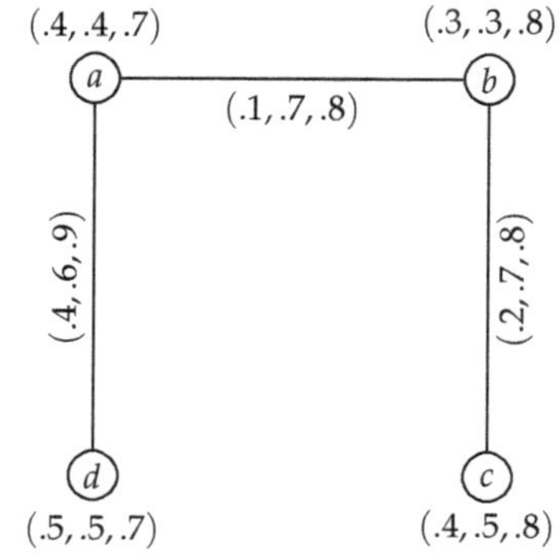

Figure 6. Neutrosophic graph.

We see that $T_M(ab) < T_M(bc) < T_M(cd)$, *and we get*
$td_2(a) = (0.1 + 0.4, 0.4 + 0.4) = (0.5, 0.8)$
$td_2(b) = (0.2 + 0.3, 0.5 + 0.3) = (0.5, 0.8)$
$td_2(c) = (0.1 + 0.4, 0.4 + 0.2) = (0.5, 0.6)$
$td_2(d) = (0.2 + 0.5, 0.5 + 0.4) = (0.7, 0.9)$
Since for two adjacent nodes a and b, $td_2(a) = td_2(b)$. *Then, we see that G is not a two-HTING.*

Definition 29. *Let* $G = (N, M)$ *be a neutrosophic graph. The graph G is said to be a star neutrosophic graph if it is a complete bipartite neutrosophic graph such that one partition of V contains only one node.*

Theorem 15. *Let* $G = (N, M)$ *be a star neutrosophic graph with* $k \geq 4$ *nodes such that the node* v_1 *is incident to the other nodes. If for all* $i = 1, 2, ..., k$, $T_M(v_iv_{i+1}) < T_M(v_{i+1}v_{i+2})$, *or* $I_M(v_iv_{i+1}) > I_M(v_{i+1}v_{i+2})$, *or* $F_M(v_iv_{i+1}) > F_M(v_{i+1}v_{i+2})$. *Then, G is a* $[1, 2]$*-HING.*

Proof. Suppose that $v_1, v_2, ..., v_k, v_1$ is the arcs of G, where $T_M(v_1v_2) < T_M(v_2v_3) < ... < T_M(v_iv_{i+1}) < T_M(v_{i+1}v_{i+2}) < ... < T_M(v_kv_1)$.

Then, $\sum\limits_{v_i \neq v_j \in V} T^1_M(v_1v_j) = \sum\limits_{j=2}^{k} T_M(v_1v_j)$ and $\sum\limits_{v_i \neq v_j \in V} T^1_M(v_1v_j) = T_M(v_1v_j); i = 2, 3, ..., k$.

Hence, $d_1(v_1) \neq d_1(v_2) \neq d_1(v_3) \neq .... \neq d_1(v_k)$. There, G is a one-HING. In addition, we see that $d_1(v_1) = (0, 1)$ and $\sum\limits_{v_i \neq v_j \in V} T^2_M(v_1v_j) = (k - i)t_M(v_1v_i) + \sum\limits_{j=2}^{i-1} T_M(v_1v_j)$. Hence, for every $i = 2, ..., k$, $d_1(v_i) \neq d_1(v_1)$, and so, G is a two-HING. Similarly, the proof is established for the indeterminacy and false membership function of each arc.

Hence the proof. □

7. Some Applications of Regular and Bipartite Neutrosophic Graphs

There are many networks that are used in real-life scenarios. Graph theory plays a vital role in the application parts of these networks. Some networks are designed for distributing commodities or services. The power grid, Internet-related issues, and road transportation networks are some examples of technological networks. There has been a phenomenal change in network research in recent years from a small graphical structure to a wide statistical analysis of large-scale networks.

Most of the real-life networks/structures are ambiguous and vague in nature and not well defined. Indeterministic information exists in almost every real-life scenario. The neutrosophic graph is more flexible and compatible to model the indeterministic information in a network when compared to the fuzzy graph or vague graph. A graph is a classical way of modeling real-life networks, which consist of relationships between entities. In a graph, the entities are expressed by nodes and relationships by arcs.

If there exist uncertainties due to indeterministic or vague information about the entities, or relations, or both, then the neutrosophic graph model is very efficient to design such networks. Neutrosophic graphs are widespread and important in the area of mathematical modeling, pattern recognition, relational mapping, biological, social networks, and information systems. It can be applied to artificial intelligence with neutrosophic rules in the inference system. These graphs can be represented in different data structures in different applications internally. In those real-life applications, the basic mathematical model is that of a neutrosophic graph.

Transforming all those data-carrying networks to a neutrosophic graph using various types of information from all the possible existing levels, it is a proper multilevel method where all possible levels and ties among them are represented and analyzed mathematically at the same time. Neutrosophic graphs have more applications related to information technology and computer science such that these graphs are used to illustrate networks of communication, data optimization, machine learning, chip design, and much more.

7.1. Assignment Problem

The assignment problem is a well-known and fundamental topic of study in the field of operation research. This problem can be modeled as a bipartite neutrosophic graph. Let a neutrosophic graph $G = (V, E)$ where V denotes the nodes and E represents the arcs be represented as a neutrosophic bipartite graph, if it satisfies the condition such that $V = X \cup Y$ and $X \cap Y = \varpi$; where ϖ is a null matrix, and $E \in X \times Y$. A neutrosophic bipartite graph G is described as a weighted bipartite graph if every edge (i.e., E) of it consists of a neutrosophic weight (i.e., $\varpi\,(i,j)$). The neutrosophic weight of one-to-one matching (i.e., M) is described by the following.

$$\omega\,(M) = \sum_{E \in M} \omega\,(E)$$

A matching (i.e., M) is called a perfect matching if each node of X is assigned to one of the Y nodes. Therefore, a linear assignment problem is defined as determining a perfect matching in G that is maximally weighted.

7.2. Road Transportation Network

A road transportation network can be modeled by a neutrosophic graph, due to its imprecise dataset, neutrosophic information for vertices as the geometric location, the grade of significance, and so on; in addition to the arcs as the width, length, quality, traffic, and so on. Therefore, we can represent a road transportation network properly by using neutrosophic graphs where the vertices and the edges represent the cities and routes between two cities, respectively. Dijkstra's algorithm [45,46] is a popular and well-known algorithm to solve the shortest path problem (SPP). It is an algorithmic approach to find the shortest path between two cities within the road transportation network. Floyd also contributed much to the efficient route-finding algorithms, hence the famous Floyd's algorithm to find the shortest paths in any network with weights. The SPP is very useful to model the network routing protocol. In any shortest path algorithm, we consider a node as the source node and find all the shortest paths between the source vertex and destination vertex using all other nodes in the graph. Hence, the proposed d_m-degree of the source node will pave a new way to find the shortest path. When its efficiency is proven, it could be considered as the best one in the road transport network.

7.3. Social Networks

Social networks are platforms on which millions of people are interconnected everywhere in the world. There are many social networks [47] like Facebook, WhatsApp, and Instagram. These networks have reached a rapid progress in recent years and have spread all over the world in several ways. There are also some security issues pointed out recently by ethical hacking teams; hence to maintain the privacy of the user and the data released by them. In the medium, messages could be tampered with easily when a user sends the information to all the other users of his/her group. However, the situation is that a user should be connected to all other users. Social networks can be modeled by a neutrosophic regular graph, since d_m-degree farthest vertices are assumed by paths; therefore, it is a vertex that has at most d_m-degrees; hence, the secrecy of this vertex is maintained, and the user information is secured. If all the vertices of a social network have the same neutrosophic d_m-degree, then the user's secrecy is maintained by adding a lowest arc and producing a d_m-degree in these graphs. Therefore, in these ways, in social networks, the data integrity is maintained, resisting more neighborhood attacks and threats.

The concept of regularity plays an important role in modeling many real-life optimization problems (e.g., social networks, wireless networks, road networks, assignment problems), so the main focus of this study is to present the concept of regularity in neutrosophic graph theory. The degree of a vertex in a graph is a way to find the number of relations of a vertex, so to analyze a network, it is very essential to find the degree of vertices. In this study, we have described the different types of degrees (d_m-degree and td_m-degree) of nodes in a neutrosophic graph. The idea of the neutrosophic degree of a node is very important from different aspects in the real-world. Here, we have presented the definition of the free vertex (node) and busy vertex (node) of a regular neutrosophic graph, which are very useful to find the solutions of many real-life problems like the traffic congestion problem. Finally, we have mentioned some real-life applications of the regular neutrosophic graph, complete bipartite neutrosophic graph, and node d_m-degree of a neutrosophic graph. This study will be useful when the neutrosophic graphs are extremely large. For example, we have described the utility of the regular neutrosophic graph to model a social network. In the Twenty First Century, social networks are platforms on which millions of people are interconnected everywhere in the world. Therefore, as a future study, we will have to consider some large regular neutrosophic fuzzy graphs to model the social networks in real-life scenarios, and the closeness and diameter need to be computed for the analysis of this social network. Furthermore, we will try to propose some algorithmic methods to determine the closeness and diameter of any social network in real life. Despite the demand for future work, this paper is a useful initial contribution to neutrosophic graph theory, and some problems are described under the neutrosophic environment.

8. Conclusions

The main contribution of this manuscript is to introduce the idea of regularity in neutrosophic graph theory. In this paper, we have described the notion of the d_m-degree and td_m-degree of nodes in a neutrosophic graph. Some different types of neutrosophic graphs such as the regular, regular strong, d_m-regular, td_m-regular, and complete bipartite neutrosophic graph were introduced here. We have also provided some sufficient criteria for which the td_m-regular neutrosophic graph and d_m-regular neutrosophic graph are equivalent. We have introduced the definition of the μ-complement and h-morphism of a neutrosophic graph. Some properties of the complement and isomorphic regular neutrosophic graph were also presented here. Finally, some real-life applications of the regular neutrosophic graph and complete bipartite neutrosophic graph were described for the assignment problem, road transportation network, and social networks. In the future, we will focus on the study of neutrosophic intersection graphs, neutrosophic interval graphs, neutrosophic hyper graphs, and so on. The idea of the neutrosophic graph can be used in several areas of expert systems, image processing, computer networks, and social systems.

Author Contributions: L.H., A.D., Y.H., and Y.L. conceived of the presented idea. P.K.K.K., D.K., L.H., and A.D. developed the theory and performed the computations. Y.H. and Y.L. verified the analytical methods. A.D. contributed reagents/materials/analysis tools. L.H., Y.H., and Y.L. wrote the paper.

Funding: The study was supported by the "Major scientific and technological innovation project of Shandong Province, China (Grant No. 2017CXGC0901)" and the "Major basic research project of Shandong natural science foundation, China (Grant No. ZR2018ZC0436)".

Conflicts of Interest: The authors declare no conflict of interest.

References

1. Zadeh, L.A. Fuzzy sets. *Inf. Control* **1965**, *8*, 338–353. [CrossRef]
2. Dey, A.; Pal, A.; Pal, T. Interval type 2 fuzzy set in fuzzy shortest path problem. *Mathematics* **2016**, *4*, 62. [CrossRef]
3. Kaufmann, A. *Introduction to the Theory of Fuzzy Subsets*; Academic Press: Cambridge, MA, USA, 1975; Volume 2.
4. Rosenfeld, A. Fuzzy graphs. In *Fuzzy Sets and Their Applications to Cognitive and Decision Processes*; Elsevier: Amsterdam, The Netherlands, 1975; pp. 77–95.
5. Gani, A.N.; Radha, K. On regular fuzzy graphs. *J. Phys. Sci.* **2008**, *12*, 33–40.
6. Samanta, S.; Pal, M. Irregular bipolar fuzzy graphs. *arXiv* **2012**, arXiv:1209.1682.
7. Rashmanlou, H.; Pal, M. Some properties of highly irregular interval-valued fuzzy graphs. *World Appl. Sci. J.* **2013**, *27*, 1756–1773.
8. Rashmanlou, H.; Samanta, S.; Pal, M.; Borzooei, R. A study on vague graphs. *SpringerPlus* **2016**, *5*, 1234. [CrossRef] [PubMed]
9. Rashmanlou, H.; Samanta, S.; Pal, M.; Borzooei, R.A. A study on bipolar fuzzy graphs. *J. Intell. Fuzzy Syst.* **2015**, *28*, 571–580.
10. Pramanik, T.; Samanta, S.; Pal, M.; Mondal, S.; Sarkar, B. Interval-valued fuzzy ϕ-tolerance competition graphs. *SpringerPlus* **2016**, *5*, 1981. [CrossRef]
11. Nandhini, S.; Nandhini, E. Strongly Irregular Fuzzy Graphs. *Int. J. Math. Arch. EISSN 2229-5046* **2014**, *5*, 2229–5046.
12. Ghorai, G.; Pal, M. Certain types of product bipolar fuzzy graphs. *Int. J. Appl. Comput. Math.* **2017**, *3*, 605–619. [CrossRef]
13. Dey, A.; Son, L.; Kumar, P.; Selvachandran, G.; Quek, S. New concepts on vertex and edge coloring of simple vague graphs. *Symmetry* **2018**, *10*, 373. [CrossRef]
14. Borzooei, R.A.; Rashmanlou, H.; Samanta, S.; Pal, M. New concepts of vague competition graphs. *J. Intell. Fuzzy Syst.* **2016**, *31*, 69–75. [CrossRef]
15. Borzooei, R.A.; Rashmanlou, H.; Samanta, S.; Pal, M. Regularity of vague graphs. *J. Intell. Fuzzy Syst.* **2016**, *30*, 3681–3689. [CrossRef]
16. Atanassov, K.T. Intuitionistic fuzzy sets. *Fuzzy Sets Syst.* **1986**, *20*, 87–96. [CrossRef]
17. Atanassov, K.T. Intuitionistic fuzzy sets. In Proceedings of the VII ITKR's Session, Sofia, Bulgarian, 20–23 June 1983. (In Bulgarian)
18. Atanassov, K.T. Intuitionistic fuzzy sets. In *Intuitionistic Fuzzy Sets*; Springer: Berlin, Germany, 1999; pp. 1–137.
19. Shannon, A.; Atanassov, K. On a generalization of intuitionistic fuzzy graphs. *NIFS* **2006**, *12*, 24–29.
20. Parvathi, R.; Karunambigai, M. Intuitionistic fuzzy graphs. In *Computational Intelligence, Theory and Applications*; Springer: Berlin, Germany, 2006; pp. 139–150.
21. Parvathi, R.; Thamizhendhi, G. Domination in intuitionistic fuzzy graphs. *Notes Intuit. Fuzzy Sets* **2010**, *16*, 39–49.
22. Parvathi, R.; Karunambigai, M.; Atanassov, K.T. Operations on intuitionistic fuzzy graphs. In Proceedings of the 2009 IEEE International Conference on Fuzzy Systems, Jeju Island, Korea, 20–24 August 2009; pp. 1396–1401.
23. Rashmanlou, H.; Samanta, S.; Pal, M.; Borzooei, R.A. Intuitionistic fuzzy graphs with categorical properties. *Fuzzy Inf. Eng.* **2015**, *7*, 317–334. [CrossRef]
24. Rashmanlou, H.; Samanta, S.; Pal, M.; Borzooei, R.A. Bipolar fuzzy graphs with categorical properties. *Int. J. Comput. Intell. Syst.* **2015**, *8*, 808–818. [CrossRef]
25. Akram, M.; Akmal, R. Operations on intuitionistic fuzzy graph structures. *Fuzzy Inf. Eng.* **2016**, *8*, 389–410. [CrossRef]
26. Rashmanlou, H.; Borzooei, R.; Samanta, S.; Pal, M. Properties of interval valued intuitionistic (S, T)—Fuzzy graphs. *Pac. Sci. Rev. A Nat. Sci. Eng.* **2016**, *18*, 30–37. [CrossRef]
27. Sahoo, S.; Pal, M. Different types of products on intuitionistic fuzzy graphs. *Pac. Sci. Rev. A Nat. Sci. Eng.* **2015**, *17*, 87–96. [CrossRef]

28. Smarandache, F. *A Unifying Field in Logic: Neutrosophic Logic. Neutrosophy, Neutrosophic Set, Neutrosophic Probability and Statistics*; Infinite Study; Modern Science Publisher: New York, NY, USA, 2005.
29. Ye, J. Single-valued neutrosophic minimum spanning tree and its clustering method. *J. Intell. Syst.* **2014**, *23*, 311–324. [CrossRef]
30. Yang, H.L.; Guo, Z.L.; She, Y.; Liao, X. On single valued neutrosophic relations. *J. Intell. Fuzzy Syst.* **2016**, *30*, 1045–1056. [CrossRef]
31. Naz, S.; Rashmanlou, H.; Malik, M.A. Operations on single valued neutrosophic graphs with application. *J. Intell. Fuzzy Syst.* **2017**, *32*, 2137–2151. [CrossRef]
32. Akram, M.; Waseem, N.; Dudek, W.A. Certain types of edge m-polar fuzzy graphs. *Iran. J. Fuzzy Syst.* **2017**, *14*, 27–50.
33. Akram, M. Certain bipolar neutrosophic competition graphs. *J. Indones. Math. Soc.* **2017**, *24*, 1–25.
34. Akram, M.; Siddique, S. Neutrosophic competition graphs with applications. *J. Intell. Fuzzy Syst.* **2017**, *33*, 921–935. [CrossRef]
35. Akrama, M. *Single-Valued Neutrosophic Planar Graphs*; Infinite Study; Modern Science Publisher: New York, NY, USA, 2016.
36. Gani, A.N.; Latha, S. On irregular fuzzy graphs. *Appl. Math. Sci.* **2012**, *6*, 517–523.
37. Maheswari, N.S.; Sekar, C. Semi neighbourly irregular graphs. *Int. J. Comb. Graph Theory Appl.* **2015**, *5*, 135–144.
38. Darabian, E.; Borzooei, R.A.; Rashmanlou, H.; Azadi, M. New concepts of regular and (highly) irregular vague graphs with applications. *Fuzzy Inf. Eng.* **2017**, *9*, 161–179. [CrossRef]
39. Dey, A.; Broumi, S.; Bakali, A.; Talea, M.; Smarandache, F. A new algorithm for finding minimum spanning trees with undirected neutrosophic graphs. *Granul. Comput.* **2019**, *4*, 63–69. [CrossRef]
40. Broumi, S.; Bakali, A.; Talea, M.; Smarandache, F.; Dey, A. Spanning tree problem with Neutrosophic edge weights. *Proc. Comput. Sci.* **2018**, *127*, 190–199. [CrossRef]
41. Broumi, S.; Talea, M.; Smarandache, F.; Bakali, A. Decision-making method based on the interval valued neutrosophic graph. In Proceedings of the IEEE 2016 Future Technologies Conference (FTC), San Francisco, CA, USA, 6–7 December 2016; pp. 44–50.
42. Broumi, S.; Talea, M.; Bakali, A.; Smarandache, F. On bipolar single valued neutrosophic graphs. *J. New Theory* **2016**, *11*, 84–102.
43. Karaaslan, F.; Davvaz, B. Properties of single-valued neutrosophic graphs. *J. Intell. Fuzzy Syst.* **2018**, *34*, 57–79. [CrossRef]
44. Broumi, S.; Smarandache, F.; Talea, M.; Bakali, A. *Operations on Interval Valued Neutrosophic Graphs*; Infinite Study; Modern Science Publisher: New York, NY, USA, 2016.
45. Dijkstra, E.W. A note on two problems in connexion with graphs. *Numer. Math.* **1959**, *1*, 269–271. [CrossRef]
46. Cormen, T.H.; Leiserson, C.E.; Rivest, R.L.; Stein, C. *Introduction to Algorithms*, 2nd ed.; MIT Press: Cambridge, MA, USA, 2001.
47. Zuo, C.; Pal, A.; Dey, A. New Concepts of Picture Fuzzy Graphs with Application. *Mathematics* **2019**, *7*, 470. [CrossRef]

mathematics

MDPI

Article

Matching Extendabilities of $G = C_m \vee P_n$

Zhi-hao Hui [1], **Yu Yang** [2,3,*], **Hua Wang** [4,5,*] **and Xiao-jun Sun** [6]

1 School of Mathematics and Statistics Science, Pingdingshan University, Pingdingshan 467000, China; zhihaoli@pdsu.edu.cn
2 School of Computer Science, Pingdingshan University, Pingdingshan 467000, China
3 School of Mathematical Sciences, Shanghai Jiao Tong University, Shanghai 200240, China
4 College of Software, Nankai University, Tianjin 300071, China
5 Department of Mathematical Sciences, Georgia Southern University, Statesboro, GA 30460, USA
6 School of Electrical and Mechanical Engineering, Pingdingshan University, Pingdingshan 467000, China; pdschris@pdsu.edu.cn
* Correspondence: yangyu@pdsu.edu.cn (Y.Y); hwang@georgiasouthern.edu (H.W.)

Received: 2 September 2019; Accepted: 2 October 2019; Published: 11 October 2019

Abstract: A graph is considered to be induced-matching extendable (bipartite matching extendable) if every induced matching (bipartite matching) of G is included in a perfect matching of G. The induced-matching extendability and bipartite-matching extendability of graphs have been of interest. By letting $G = C_m \vee P_n$ ($m \geq 3$ and $n \geq 1$) be the graph join of C_m (the cycle with m vertices) and P_n (the path with n vertices) contains a perfect matching, we find necessary and sufficient conditions for G to be induced-matching extendable and bipartite-matching extendable.

Keywords: perfect matching; k-extendable; induced matching extendable; bipartite matching extendable graph

1. Introduction

Throughout this paper we follow traditional graph theoretical terminologies and only consider simple connected graphs.

Let G be a graph with vertex set $V(G)$ and edge set $E(G)$. For $S \subseteq V(G)$ we define $E(S) = \{uv \in E(G) \mid u, v \in S\}$. Similarly for $N \subseteq E(G)$ we have $V(N) = \{v \in V(G) \mid \exists x \in V(G),\ vx \in N\}$.

A collection of edges $M \subseteq E(G)$ is a matching of G if no two edges in M are adjacent in G. If $V(M) = V(G)$, then M is a perfect matching of G [1]. A matching M is an induced matching of G if no two edges of M are joined by an edge of G [2].

The problem of matching extendability asks if a matching of G is included in a perfect matching of G. First, the concept of k-extendable graphs (Definition 1) was introduced by Plummer [3]. The family of k-extendable graphs has been studied extensively [4–9].

Definition 1. *A connected graph G is called k-extendable if every matching of size k* $(1 \leq k \leq \frac{1}{2}(|V(G)| - 2))$ *extends to a perfect matching in G.*

Along this line the following definitions are also introduced.

Definition 2 ([10])**.** *A graph G is called k-factor-critical if* $G - S$ *has a perfect matching for any* $S \subseteq V(G)$ *with* $|S| = k$.

Definition 3 ([11])**.** *A connected graph G is called induced-matching extendable if any induced matching of G is included in a perfect matching of G.*

Furthermore, A matching M is a bipartite matching if $G[V(M)]$ is a bipartite graph [12]. From the research of k-extendable graphs, induced-matching extendable graphs and k-factor-critical graphs, the important roles of bipartite matching and non-bipartite matching were noticed. Wang et al. proposed the novel concept of bipartite-matching extendable graph in 2008 [12]. More recently, in 2017, Chiarelli et al. presented the sufficient conditions for graphs G and H, under which the lexicographic product $G[H]$ is 2-extendable [8].

Definition 4 ([12])**.** *A connected graph G is bipartite-matching extendable if every bipartite matching of G is included in a perfect matching of G.*

It is easy to see that a graph G is induced-matching extendable if it is bipartite-matching extendable. We also note that bipartite-matching extendability is the same as regular extendability when G itself is bipartite.

In general, matching extendable graphs frequently appear in applications and have been well studied. See, for instance, resonance circle theory in chemical graphs [13,14]. Other work on matching extendable graph can be found in [12,15–17].

In this note we will consider induced-matching extendability and bipartite-matching extendability for another specific class of graphs. Given two graphs G and H, $G \vee H$ is the graph join of G and H, with every vertex of G connected to every vertex of H by an edge. For general notations and facts on graph join one may see [1,4].

Graph joins have been considered in many different topics including edge-colouring [18], the chromatic index [19,20], the total chromatic number [21], the Laplacian spectrum [22], the skewness [23], the thickness [24]. In the remaining of this paper we consider the induced-matching extendability and bipartite-matching extendability of graph join $C_m \vee P_n$ $(m \geq 3,\ n \geq 1)$. In particular, we will show the following main results.

Theorem 1. *Let m and n be positive integers with $m + n \equiv 0 \pmod 2$:*

(i) If $m > n$, then $C_m \vee P_n$ is k-extendable if and only if

$$k \leq \left\lfloor \frac{n+1}{2} \right\rfloor.$$

(ii) If $n \geq m$, then $C_m \vee P_n$ is k-extendable if and only if

$$k \leq \left\lfloor \frac{m}{2} \right\rfloor.$$

Theorem 2. *Let $m \geq 3$ and $n \geq 1$ be two positive integers, then $C_m \vee P_n$ is induced-matching extendable if and only if*

$$m + n \equiv 0 \pmod 2$$

and

$$\frac{1-(-1)^r}{2} + \left\lfloor \frac{n}{3} \right\rfloor \leq m \leq 3n + 5$$

where $n \equiv r \pmod 3$ for some $0 \leq r \leq 2$.

Theorem 3. *Let $m \geq 3$ and $n \geq 1$ be two positive integers, then $C_m \vee P_n$ is bipartite-matching extendable if and only if*

$$m + n \equiv 0 \pmod 2$$

and

$$\left\lfloor \frac{m}{3} \right\rfloor - \frac{1-(-1)^r}{2} \leq n \leq 3m + 2$$

where $m \equiv r \pmod 3$ *for some* $0 \le r \le 2$.

First, in Section 2, we introduce some previously established facts on matching extendability. We then establish Theorem 1 on k-extendability of the graph join $C_m \vee P_n$ in Section 3. Lastly we prove Theorem 2 in Section 4 and Theorem 3 in Section 5.

2. Some Preliminaries

In this section, we list some interesting and useful previous results on matching extendabilities. They build the foundation for our study.

Lemma 1 ([25]). *A graph G has a perfect matching if and only if* $o(G - S) \le |S|$ *for every* $S \subseteq V(G)$.

Here, $o(G)$ is the number of components of G with an odd number of vertices. From Lemma 1 and the definition of bipartite-matching extendability, Wang et al. obtained the following [12]:

Lemma 2 ([12]). *A graph G is bipartite-matching extendable if and only if* $o(G - V(M) - S) \le |S|$ *for every bipartite matching M of G and every* $S \subseteq V(G) \setminus V(M)$.

Lemma 3 ([12]). *A graph G is bipartite-matching extendable if and only if* $o(G - S) \le |S| - 2m_b(S)$ *for any* $S \subseteq V(G)$, *where* $m_b(S)$ *is the number of edges in a maximum bipartite matching of* $G[S]$.

A matching M is called a *forbidden matching* if it is a bipartite matching and $V(M)$ is a vertex cut such that $G - V(M)$ has an odd component [15]. Consequently a graph G is not bipartite-matching extendable if there exists a forbidden matching. The following is also shown in [15].

Lemma 4 ([15]). *If G is bipartite-matching extendable, then*

- *G is 2-connected;*
- *G does not have a forbidden matching;*
- *if* $\{u, v\}$ *is a vertex cut of G and* $uv \notin E(G)$, *then* $G - \{u, v\}$ *has exactly two components and both of them are odd ;*
- *for a bipartite matching M of G and an independent set X of* $G - V(M)$, $|N_{G-V(M)}(X)| \ge |S|$.

It is easy to see, from the definitions and properties of induced-matching extendable and bipartite-matching extendable graphs, that $W_n = C_n \vee K_1 (n \ge 3)$ is bipartite-matching extendable if and only if $n = 3, 5, 7$. We now move on to consider the generalization of K_1 to P_n.

3. On k-Extendable Graph Joins $C_m \vee P_n$

In this section, we examine when is the graph join $C_m \vee P_n$ k-extendable.

First, let $G = C_m \vee P_n$ have a perfect matching. Then we must have $m + n \equiv 0 \pmod 2$. In the rest of this section we only need to consider m and n being both even or both odd. We now consider Case (i) of Theorem 1. Case (ii) is similar.

Case (i) of Theorem 1. For convenience we let

$$C_m = x_1 x_2 \cdots x_{m-1} x_m$$

(with the edge $x_1 x_m$) and

$$P_n = y_1 y_2 \cdots y_{n-1} y_n.$$

We will first show that $C_m \vee P_n$ is not k-extendable for $k > \lfloor \frac{n+1}{2} \rfloor$. By the Definition 1, it suffices to show that $C_m \vee P_n - V(M)$ dose not have a perfect matching, for some matching M of $C_m \vee P_n$ with $|M| > \lfloor \frac{n+1}{2} \rfloor$.

- If m and n are odd:
  - If $n = 1$, note that for the matching $M_0 = \{x_1y_m, x_2x_3\}$ of G, $G - V(M_0)$ has an isolated vertex x_1. Consequently $G - V(M_0)$ does not have a perfect matching. Hence $C_m \vee P_n$ is not 2-extendable.
  - If $n \geq 3$, from the structural characteristics of graph $C_m \vee P_n$, there must exist a matching, say

$$M_1 = \{y_4y_5, y_6y_7, \cdots, y_{n-1}y_n, x_1y_1, x_3y_2, x_4y_3\}$$

of $C_m \vee P_n$ with size $\frac{n+3}{2}$, such that $C_m \vee P_n - V(M_1)$ has an isolated vertex x_2, which imply that $C_m \vee P_n - V(M_1)$ does not have perfect matching, and thus the $C_m \vee P_n$ is not $\frac{n+3}{2}$-extendable. Consequently, $C_m \vee P_n$ is k ($k \leq \frac{n+3}{2} - 1 = \left\lfloor \frac{n+1}{2} \right\rfloor$) extendable follows from the known conclusion that G is k-extendable, it must be a $k-1$ extendable.

- If m and n are even:
  - If $n = 2$, consider the matching $M_0' = \{x_1y_1, x_3y_2\}$. Then $G - V(M_0')$ has an isolated vertex x_2 and consequently without a perfect matching. Hence $C_m \vee P_n$ is not 2-extendable.
  - If $n \geq 4$, consider the matching

$$M_1' = \{y_3y_4, y_5y_6, \cdots, y_{n-1}y_n, x_my_1, x_2y_2\}$$

of size $\frac{n+2}{2}$. Then $C_m \vee P_n - V(M_1')$ has an isolated vertex x_1 and consequently without a perfect matching. Hence $C_m \vee P_n$ is not $\frac{n+2}{2}$-extendable.

We will now show that $C_m \vee P_n$ is indeed k-extendable for smaller values of k. The key idea in the following argument lies in the fact that $C_m \vee P_n$ is highly connected and $C_m \vee P_n - V(M)$ contains a Hamiltonian path of even order for any "small" matching M.

- If m and n are both odd, with $m > n \geq 1$, note that $C_m \vee P_n$ is $(n+2)$-connected. For every matching M of size $\frac{n+1}{2}$ in $C_m \vee P_n$, $C_m \vee P_n - V(M)$ is connected. By the definition of $C_m \vee P_n$, it is not difficult to see that $C_m \vee P_n - V(M)$ is not only connected but also containing a Hamiltonian path of even order. This implies that $C_m \vee P_n - V(M)$ has a perfect matching. With Definition 1, we have $C_m \vee P_n$ is k-extendable for $k \leq \frac{n+1}{2}$.
- Similarly, for even $m > n \geq 2$ and any matching M of size $\frac{n}{2}$, $C_m \vee P_n - V(M)$ contains a Hamiltonian path of even length. This implies that $C_m \vee P_n - V(M)$ has a perfect matching. Hence $C_m \vee P_n$ is k-extendable for $k \leq \frac{n}{2}$.

□

4. Proof of Theorem 2

We note that the $m + n \equiv 0 \pmod 2$ is obvious for the same reason as before. Also as before we let

$$C_m = x_1x_2 \cdots x_{m-1}x_m$$

(with the edge x_1x_m) and

$$P_n = y_1y_2 \cdots y_{n-1}y_n.$$

First we show the bounds for m are necessary for induced-matching extendability. Suppose that G is induced-matching extendable. For the lower bound:

- If $n = 3s$ for some s, consider an induced matching

$$M = \{y_{3t+1}y_{3t+2} : 0 \leq t \leq s-1 = \lfloor \frac{n}{3} \rfloor - 1\}.$$

If $m \leq \lfloor \frac{n}{3} \rfloor - 1$, then

$$o(G - V(M) - V(C_m)) = o(P_n - V(M)) = \lfloor \frac{n}{3} \rfloor > m = |V(C_m)|.$$

This implies, by Lemma 1, that $G - V(M)$ does not have a perfect matching. This is a contradiction with the assumption G is induced-matching extendable. Therefore, $\lfloor \frac{n}{3} \rfloor \leq m$.

- If $n = 3s + 1$, consider an induced matching

$$M = \{y_{3t+2}y_{3t+3} : 0 \leq t \leq s - 1 = \lfloor \frac{n}{3} \rfloor - 1\}.$$

If $m \leq \lfloor \frac{n}{3} \rfloor$, then

$$o(G - V(M) - V(C_m)) = o(P_n - V(M)) = \lfloor \frac{n}{3} \rfloor + 1 > m = |V(C_m)|,$$

again implying, with Lemma 1, that $G - V(M)$ does not have a perfect matching, a contradiction. Therefore, $\lfloor \frac{n}{3} \rfloor + 1 \leq m$.

- If $n = 3s + 2$, consider the induced matching

$$M = \{y_{3t+1}y_{3t+2} : 0 \leq t \leq s - 1 = \lfloor \frac{n}{3} \rfloor - 1\}.$$

If $m \leq \lfloor \frac{n}{3} \rfloor - 1$, then

$$o(G - V(M) - V(C_m)) = o(P_n - V(M)) = \lfloor \frac{n}{3} \rfloor > m = |V(C_m)|,$$

yielding a contradiction as before. Hence $\lfloor \frac{n}{3} \rfloor \leq m$.

For the upper bound, consider the induced matching

$$M = \{x_{3i+1}x_{3i+2} : 0 \leq i \leq \lfloor \frac{m}{3} \rfloor - 1\}.$$

Then we have

$$o(C_m - V(M)) = \begin{cases} \lfloor \frac{m}{3} \rfloor - 1, & m \equiv 1 \pmod 3 \\ \lfloor \frac{m}{3} \rfloor, & m \equiv 0, 2 \pmod 3). \end{cases}$$

Since G is induced-matching extendable, $G - V(M)$ has a perfect matching. By Lemma 1, we have

$$o(G - V(M) - V(P_n)) = o(C_m - V(M)) \leq |V(P_n)| = n.$$

Thus

$$m \leq \begin{cases} 3n, & m \equiv 0 \pmod 3 \\ 3n + 4, & m \equiv 1 \pmod 3 \\ 3n + 5, & m \equiv 2 \pmod 3 \end{cases} \tag{1}$$

We now show that $G = C_m \vee P_n$ is indeed induced-matching extendable under these conditions. Let M be an induced matching:

- If $V(M) \cap V(C_m) = \emptyset$, then $M \subseteq E(P_n)$ and components of $P_n - V(M)$ are either paths or isolated vertices. Assume $M = \{x_2x_3, x_5x_6, \cdots, x_{3s-4}x_{3s-3}, x_{3s-1}x_{3s}\}$ be the induced matching that maximizes the number of odd components of $P_n - V(M)$, therefore the x_1, x_4, x_7, x_{3s-2} and x_{3s+1} are $\lfloor \frac{n}{3} \rfloor + 1$ isolated vertices of $P_n - V(M)$. It is easy to see that

$$o(P_n - V(M)) \leq \lfloor \frac{n}{3} \rfloor + 1$$

if $n \equiv 1 \pmod 3$, and

$$o(P_n - V(M)) \leq \lfloor \frac{n}{3} \rfloor$$

otherwise.

Now let N be a maximum matching of $P_n - V(M)$:

- if $P_n - V(M) \cup V(N)$ has no vertices left, then $G - V(M) \cup V(N)$ is isomorphic to C_m with m being even. Consequently, $G - V(M) \cup V(N)$ has a perfect matching N_1. Now $N \cup N_1$ is a perfect matching of $G - V(M)$.
- if $P_n - V(M) \cup V(N)$ has some vertices left, then $G - V(M) \cup V(N)$ is isomorphic to the join of C_m and some isolated vertices. It is easy to see that $G - V(M) \cup V(N)$ has a perfect matching, say N_2. We now have $N \cup N_2$ as a perfect matching of $G - V(M)$.

- If $V(M) \cap V(C_m) \neq \emptyset$, we consider two cases: $M \subseteq E(G) - E(C_m) \cup E(P_n)$ or $M \subseteq E(C_m)$:
  - If $M \subseteq E(G) - E(C_m) \cup E(P_n)$, then $|M| = 1$ and it is easy to find a perfect matching for $G - V(M)$.
  - If $M \subseteq E(C_m)$, then the components of $C_m - V(M)$ are either paths or isolated vertices. Let N be a maximum matching of $C_m - V(M)$:
    - If $C_m - V(M) \cup V(N)$ has no vertex left, then $G - V(M) \cup V(N)$ is isomorphic to P_n with even number of vertices. Thus $G - V(M) \cup V(N)$ has a perfect matching, say N_1. Consequently, $N \cup N_1$ is a perfect matching of $G - V(M)$.
    - If $C_m - V(M) \cup V(N)$ has some isolated vertices left, then $G - V(M) \cup V(N)$ is isomorphic to the join of P_n and some isolated vertices. With (1) we know $G - V(M) \cup V(N)$ has a perfect matching, say N_2. Now $N \cup N_2$ is a perfect matching of $G - V(M)$.

Therefore $G - V(M)$ has a perfect matching in all cases, implying that G is induced-matching extendable under the given conditions.

5. Proof of Theorem 3

Some of our arguments here are very similar to those of the previous section. Again we note that $m + n \equiv 0 \pmod 2$ is obvious, and label C_m and P_n the same way.

First we show the only if part. Let $G = C_m \vee P_n$ be bipartite-matching extendable. For the lower bound:

- If $m = 3s$ for some s, consider the bipartite matching

$$M = \{x_{3t+1}x_{3t+2} : 0 \leq t \leq s - 1 = \lfloor \frac{m}{3} \rfloor - 1\}.$$

  If $n \leq \lfloor \frac{m}{3} \rfloor - 1$, then

$$o(G - V(M) - V(P_n)) = o(C_m - V(M)) = \lfloor \frac{m}{3} \rfloor > n = |V(P_n)|,$$

  contradiction to Lemma 2 and the bipartite-matching extendability. Therefore $n \geq \lfloor \frac{m}{3} \rfloor$.
- If $m = 3s + 1$, consider the bipartite matching

$$M = \{x_{3t+1}x_{3t+2} : 0 \leq t \leq s - 1 = \lfloor \frac{m}{3} \rfloor - 1\}.$$

If $n \leq \lfloor \frac{m}{3} \rfloor - 2$, then

$$o(G - V(M) - V(P_n)) = o(C_m - V(M)) = \lfloor \frac{m}{3} \rfloor - 1 > n = |V(P_n)|,$$

a contradiction. Hence $n \geq \lfloor \frac{m}{3} \rfloor - 1$.

- If $m = 3s + 2$, consider the bipartite matching

$$M = \{x_{3t+1}x_{3t+2} : 0 \leq t \leq s = \lfloor \frac{m}{3} \rfloor\}.$$

If $n \leq \lfloor \frac{m}{3} \rfloor - 1$, then

$$o(G - V(M) - V(P_n)) = o(C_m - V(M)) = \lfloor \frac{m}{3} \rfloor - 1 > n = |V(P_n)|,$$

a contradiction. Thus $n \geq \lfloor \frac{m}{3} \rfloor$.

For the upper bound, consider the bipartite matching M with $E(M) \subset E(P_n)$. We have

$$o(P_n - V(M)) \leq \begin{cases} \lfloor \frac{n}{3} \rfloor + 1, & n \equiv 1 \pmod 3 \\ \lfloor \frac{n}{3} \rfloor, & n \equiv 0, 2 \pmod 3 \end{cases}$$

Suppose, for comparison, that $n > 3m + 2$. Consider now a bipartite matching

$$M' = \{y_{3i+2}y_{3i+3} : 0 \leq i \leq \lfloor \frac{n}{3} \rfloor - 1\}.$$

Then we have

$$o(G - V(M) - V(C_m)) = o(P_n - V(M)) \geq m = |V(C_m)|,$$

contradicting to Lemma 2 and the bipartite-matching extendability. Therefore we have $n \leq 3m + 2$.

Next we show that $G = C_m \vee P_n$ is indeed bipartite-matching extendable under these conditions. For this purpose we let M be a bipartite matching and we consider two cases:

- If $V(M) \cap V(P_n) = \emptyset$, then $M \subseteq E(C_m)$. Let N be a maximum matching of $C_m - V(M)$:
  - If $C_m - V(M) \cup V(N)$ has no vertices, then $G - V(M) \cup V(N)$ is isomorphic to P_n with even n. Consequently, $G - V(M) \cup V(N)$ has a perfect matching N_1. Now $N \cup N_1$ is a perfect matching of $G - V(M)$.
  - If $C_m - V(M) \cup V(N)$ has some isolated vertices, then $G - V(M) \cup V(N)$ is isomorphic to the join of P_n and some isolated vertices. It is easy to see that $G - V(M) \cup V(N)$ has a perfect matching, say N_2. Then $N \cup N_2$ is a perfect matching of $G - V(M)$.
- If $V(M) \cap V(P_n) \neq \emptyset$, we have $M \subseteq E(G) - E(C_m) \cup E(P_n)$ or $M \subseteq E(P_n)$:
  - If $M \subseteq E(G) - E(C_m) \cup E(P_n)$, we know that the $M-$ saturated vertices in P_n and C_m are not adjacent to each other in P_n, C_m, respectively. Moreover, the components of $P_n - V(M)$ and $C_m - V(M)$ are either path or isolated vertex. Further note that each vertex in components of $P_n - V(M)$ is adjacent to each vertex in components of $C_m - V(M)$, $G - V(M)$, resulting an odd length Hamiltonian path as before. Hence $G - V(M)$ has a perfect matching.
  - If $M \subseteq E(P_n)$, then

$$o(P_n - V(M)) \leq \begin{cases} \lfloor \frac{n}{3} \rfloor + 1, & n \equiv 1 \pmod 3 \\ \lfloor \frac{n}{3} \rfloor, & n \equiv 0, 2 \pmod 3 \end{cases}$$

Let N' be a maximum matching of $P_n - V(M)$. We then argue as before that $G - V(M) \cup V(N')$ has a perfect matching, say N_1'. Then $N' \cup N_1'$ is a perfect matching of $G - V(M)$.

Thus $G - V(M)$ has a perfect matching in all cases. Hence $G = C_m \vee P_n$ is bipartite-matching extendable under the given conditions.

6. Concluding Remarks

Through searching the Hamiltonian path or cycle for the auxiliary substructure of graph join $G = C_m \vee P_n$, we presented the necessary and sufficient conditions for G to be induced-matching extendable and bipartite-matching extendable. Our results provide a fundamental basis that helps study the induced and bipartite matching extendability for general graphs, and will probably be used to analyze the resonance circle properties of the chemical graphs.

As for future work, we plan to explore the correlations between the k-extendable and forbidden subgraphs of graphs. It is also interesting to investigate which of the graphs $C_m \vee C_n (m, n \geq 3)$ are k-extendable, induced-matching extendable or bipartite-matching extendable.

Author Contributions: Z.-h.H. and X.-j.S. contribute for supervision, project administration and formal analysis. Z.-h.H. and Y.Y. contribute for methodology and writing original draft preparation. The final draft we written by H.W. and Y.Y.

Funding: This work is supported by the National Natural Science Foundation of China (grant nos. 661702291, 61472058,11801371); China Postdoctoral Science Foundation (grant no. 2018M632095), the Program for Science & Technology Innovation Talents in Universities of Henan Province (grant no. 19HASTIT029), the Key Research Project in Universities of Henan Province (grant nos. 19B110011, 19B630015), the Scientific Research Starting Foundation for High-Level Talents of Pingdingshan University (grant no.PXY-BSQD2017006), and the Simons Foundation (grant no. 245307).

Conflicts of Interest: The authors declare no conflict of interest.

References

1. Bondy, J.A.; Murty, U.S.R. *Graph Theory with Application*; American Elsevier: New York, NY, USA, 1976.
2. Cameron, K. Induced matching. *Discret. Appl. Math.* **1989**, *24*, 97–102. [CrossRef]
3. Plummer, M. On n-extendable graphs. *Discret. Math.* **1980**, *31*, 201–210. [CrossRef]
4. Lovász, L.; Plummer, M.D. Matching Theory. In *Annals of Discrete Mathematics*; Elsevier: Amsterdam, The Netherlands,1986; Volume 29.
5. Plummer, M.D. Extending matchings in graphs: A survey. *Discret. Math.* **1994**, *127*, 277–292. [CrossRef]
6. Györi, E.; Imrich, W. On the Strong Product of a k-Extendable and an l-Extendable Graph. *Graphs Comb.* **2001**, *17*, 245–253.
7. Lou, D.; Yu, Q. Connectivity of k-extendable graphs with large k. *Discret. Appl. Math.* **2004**, *136*, 55–61. [CrossRef]
8. Chiarelli, N.; Dibek, C.; Ekim, T.; Gözüpek, D.; Miklavič, Š. On matching extendability of lexicographic products. *RAIRO-Oper. Res.* **2017**, *51*, 857–873. [CrossRef]
9. Zhou, J. Characterization of the induced matching extendable graphs with 2n vertices and 3n edges. *Discret. Math.* **2018**, *341*, 1021–1031. [CrossRef]
10. Favaron, O. On k-factor-critical graphs. *Discuss. Math. Graph Theory* **1996**, *1*, 41–51. [CrossRef]
11. Yuan, J.J. Induced matching extendable graphs. *J. Graph Theory* **1998**, *28*, 203–213.
12. Wang, X.; Zhang, Z.; Lin, Y. Bipartite matching extendable graphs. *Discret. Math.* **2008**, *308*, 5334–5341. [CrossRef]
13. Pauling, L. *The Nature of the Chemical Bond*, 3rd ed.; Cornell University Press: Ithaca, NY, USA, 1960.
14. Trinajstić, N. *Chemical Graph Theory*; CRC Press: Boca Raton, FL, USA, 1992.
15. Wang, X.; Feng, A.; Lin, Y. 4-regular bipartite matching extendable graphs. *ARS Comb.* **2013**, *110*, 113–128.
16. Wang, X.; Lin, Y. Three-matching intersection conjecture for perfect matching polytopes of small dimensions. *Theor. Comput. Sci.* **2013**, *482*, 111–114. [CrossRef]
17. Wang, X.; Song, X.; Yuan, J. On matching cover of graphs. *Math. Program.* **2014**, *147*, 499–518. [CrossRef]
18. Simone, C.D.; Mello, C.P.D. Edge-colouring of join graphs. *Theor. Comput. Sci.* **2006**, *335*, 364–370. [CrossRef]

19. Lima, A.R.C.; Garcia, G.; Zatesko, L.M.; de Almeida, S.M. On the chromatic index of cographs and join graphs. *Electron. Notes Discret. Math.* **2015**, *50*, 433–438. [CrossRef]
20. Zorzi, A.; Zatesko, L. On the chromatic index of join graphs and triangle-free graphs with large maximum degree. *Discret. Appl. Math.* **2018**, *245*, 183C189. [CrossRef]
21. Li, G.; Zhang, L. Total chromatic number of one kind of join graphs. *Discret. Math.* **2006**, *306*, 1895–1905. [CrossRef]
22. Sun, L.; Wang, W.; Jiang, Z.; Bu, C. Laplacian spectral characterization of some graph join. *Indian J. Pure Appl. Math.* **2015**, *46*, 279–286. [CrossRef]
23. Chia, G.L.; Sim, K.A. On the skewness of the join of graphs. *Discret. Appl. Math.* **2013**, *161*, 2405–2409. [CrossRef]
24. Chen, Y.; Yang, Y. The thickness of the complete multipartite graphs and the join of graphs. *J. Comb. Optim.* **2017**, *34*, 194–202. [CrossRef]
25. Tutte, W.T. The factorization of linear graphs. *J. Lond. Math. Soc.* **1947**, *22*, 107–111. [CrossRef]

mathematics

MDPI

Article

An Application of Total-Colored Graphs to Describe Mutations in Non-Mendelian Genetics

Raúl M. Falcón [1,*], Óscar J. Falcón [2] and Juan Núñez [2]

1 Department of Applied Mathematics I, University of Seville, 41012 Sevilla, Spain
2 Department of Geometry and Topology, University of Seville, 41012 Sevilla, Spain; oscfalgan@yahoo.es (Ó.J.F.); jnvaldes@us.es (J.N.)
* Correspondence: rafalgan@us.es

Received: 16 September 2019; Accepted: 3 November 2019; Published: 6 November 2019

Abstract: Any gene mutation during the mitotic cell cycle of a eukaryotic cell can be algebraically represented by an isotopism of the evolution algebra describing the genetic pattern of the inheritance process. We identify any such pattern with a total-colored graph so that any isotopism of the former is uniquely related to an isomorphism of the latter. This enables us to develop some results on graph theory in the context of the molecular processes that occur during the S-phase of a mitotic cell cycle. In particular, each monochromatic subset of edges is identified with a mutation or regulatory mechanism that relates any two statuses of the genotypes of a pair of chromatids.

Keywords: evolution theory; evolution algebra; mitotic cell cycle; total-colored graph

MSC: 17D92; 05C15; 92E10

1. Introduction

At present, non-associative algebras are considered an adequate theoretical framework to address important topics in genetics. In fact, there are so many non-associative algebras that have attracted the interest of geneticists that it would be difficult to make an exhaustive list of them. Let us mention, as an example, the following: Mendelian algebras, gametic algebras, zygotic algebras, baric algebras, train algebras, copular algebras, Bernstein algebras, and evolution algebras. These algebras, and generally all the algebras used to model inheritance in genetics, are referred to as "genetic algebras".

In particular, Jin Jung Paul Tian established in his PhD thesis in 2004, later published in 2008, the foundations of the theory of evolution algebras [1]. On page 3 of that thesis, he wrote the following words that in principle could seem like a premonition, but they have been actually confirmed by the numerous publications cited throughout this paper.

> General genetic algebras are the product of interaction between Biology and Mathematics. Mendelian Genetics introduced a new subject to Mathematics: general genetic algebras. The study of these algebras reveals algebraic structures of Mendelian Genetics, which always simplifies and shortens the way to understand genetic and evolutionary phenomena. Indeed, it is the interplay between the purely mathematical structure and the corresponding genetic properties that makes this area so fascinating... Now, non-Mendelian Genetics is a basic language of molecular geneticists. Logically, we can ask what non-Mendelian Genetics offers to Mathematics. The answer is "evolution algebras".

Indeed, evolution algebras were introduced [1–3] to model those probabilistic laws that regulate how the genotypes of a eukaryotic cell undergoing mitosis have influence on the genotypes of the offspring. Specifically, an n-dimensional algebra over a field $\mathbb{K}$ is an evolution algebra if there exists a natural basis $\{e_1, \ldots, e_n\}$ such that $e_ie_j = 0$, for all $i \neq j$. Each basis vector represents a genotype of a given phenotype so that the role of self-replication of genotypes is played by the products $e_ie_i = \sum_{j=1}^{n} c_{ij}e_j$, for all $1 \leq i \leq n$. In the case of dealing with $\mathbb{K} = \mathbb{R}$ as the base field, if $\sum_{j=1}^{n} c_{ij} = 1$ and $c_{i,j} \geq 0$, for all i, j, then each structure constant c_{ij} represents the probability that the genotype e_i becomes e_j in the offspring. In any case, even if there exist some advances in the algebraic study of such a probabilistic meaning [4,5], evolution algebras are usually studied without probabilistic restrictions on their structure constants [6–16].

The n-tuple $(e_1e_1, \ldots, e_ne_n)$ is called the genetic pattern of both the evolution algebra and the mitotic cell cycle with respect to the phenotype under study. The distribution into isotopism classes of two- and three-dimensional evolution algebras over any base field is already known [17,18]. This determines, in turn, the spectrum of genetic patterns of mitotic cell cycles with respect to a phenotype that is respectively associated with two and three genotypes. Being introduced by Albert [19] as a generalization of isomorphisms of algebras, the importance of dealing with isotopisms of algebras in genetics was brought to light by Holgate and Campos [20,21]. They realized how isotopisms can be used to formulate gene mutations in any inheritance process algebraically. See [22] for a recent survey on the theory of isotopisms.

The simplicity of describing any evolution algebra by means of its genetic pattern enables one to find interesting connections with graph theory, which was introduced into genetics by Benzer [23] to show the linear internal structure of genes and how this structure is affected by mutations with no detectable tendency to be reverted. More specifically, given a graph G with a finite set of vertices $\{v_1, \ldots, v_n\}$, Tian [1] defined the evolution algebra of natural basis $\{e_1, \ldots, e_n\}$ such that $e_i^2 = \sum_{j \in \Gamma(v_j)} e_j$, for all $1 \leq i \leq n$. Here, $\Gamma(v_j)$ denotes the set of neighbors of the vertex v_j in G. Such a definition induces a faithful functor among graphs and evolution algebras that maps any isomorphism between two finite graphs to an isomorphism between their related evolution algebras. Based on this fact, Tian contemplated as further work the translation of known results on graph theory to the language of evolution algebras. In this regard, for instance, Cadavid, Rodiño, and Rodríguez [5] recently dealt with the connection between the evolution algebra induced by a random walk on a graph and the evolution algebra derived from the same graph. In a subsequent paper [24], they also characterized the space of derivations of evolution algebras associated with finite graphs.

The reverse problem of describing a graph derived from a given evolution algebra has also been dealt with in the literature. In this regard, Núñez et al. [25] and Elduque and Labra [26,27] associated a weighted digraph to any given finite-dimensional evolution algebra and studied the relationship among them. In particular, in the last two contributions, it was proven that the nonexistence of oriented cycles in such a digraph involves the nilpotency of the corresponding algebra. Further, Cabrera et al. [28] described another directed graph whatever the dimension of the evolution algebra is, from which its annihilator and irreducibility may be determined. Unlike Tian's graphs, however, since all of these proposals depend on the basis of the algebra, isomorphic evolution algebras are not related in general to isomorphic graphs.

This paper delves into this last topic in the case of dealing with evolution algebras over finite fields. Specifically, we introduce a total-colored graph that can be associated with any given evolution algebra over a finite field so that any isomorphism of the former is uniquely related to an isotopism of the latter. The underlying idea behind the proposed graph derives from a previous work of the authors [29] in which a pair of colored graphs was introduced in order to describe faithful functors relating the category of finite-dimensional algebras over finite fields with the category of vertex-colored graphs. They were based in turn on a proposal of McKay et al. [30], who identified the isotopisms of Latin squares with isomorphisms of vertex-colored graphs. A first attempt to approach the graphs introduced in [29] to the theory of evolution algebras was carried out in [31], where a step-by-step construction of an edge-colored graph derived from the genetic pattern of an evolution algebra over a finite field was established. The total-colored graph here introduced not only simplifies this last construction by reducing both the number of vertices and edges under consideration, but also facilitates a relationship among different concepts on evolution algebras, graph theory, and genetics.

The paper is organized as follows. Section 2 shows some preliminary concepts and results on the subjects that are used throughout the paper. Section 3 deals with the description of the total-colored graph that we associate with any given evolution algebra over a finite field. Finally, Section 4 establishes the existence of a faithful functor between both categories of evolution algebras and their total-colored graphs.

2. Preliminaries

This section deals with some basic concepts on genetics, genetic algebras, evolution algebras, graph theory, and isotopisms of algebras that we use throughout the paper. For more details about the last four topics, we refer the reader to [3,19,32,33], respectively. We also refer the reader to [34,35] for different applications of graph theory into genetics and molecular biology.

2.1. Genetics

It is well known that genes are the molecular units of hereditary information within every cell. Disposed in sequential order, they give rise to chromosomes, which are formed by a pair of long twisted strands of deoxyribonucleic acid (DNA) and contain the genetic code to synthesize the proteins that determine each attribute or phenotype of the organism. Genes of a given phenotype always appear in chromosomes at the same position and determine the genotype of the organism with respect to such a phenotype.

In every eukaryotic cell, DNA is mostly contained within its nucleus. The mitotic cell cycle is the series of phases dividing any such cell into two daughter cells with the same DNA. In particular, during the S-phase, both strands of DNA within a nuclear chromosome are separated and serve as templates to synthesize new complementary daughter strands. Up to rare gene mutation, this gives rise to two molecules of identical DNA, called chromatids, which are separated as independent chromosomes into two new nuclei. In turn, the extra-nuclear genetic material of the parent cell is randomly distributed between both daughter cells. Both processes constitute the M-phase of the mitotic cell cycle. During the transition from the S-phase to the M-phase, regulatory mechanisms are produced to check any DNA damage motivated by gene mutations. This stage is known as the G_2-phase.

2.2. Genetic Algebras

General genetic algebras were introduced in the 1920s and 1930s. It was Serebrowski [36] who was the first to interpret algebraically the sign $\times$, which indicated sexual reproduction, and to give a mathematical formulation of Mendel's laws, formulated by G.Mendel in his famous papers of 1865 and 1866 (see [37], for instance). Glivenkov [38] introduced the so-called Mendelian algebras for diploid populations with one locus or two unlinked loci. Independently, Kostitzin [39] also introduced a symbolic multiplication to express Mendel's laws. However, the systematic study of genetic algebras is attributed to Etherington.

Indeed, Mendel's laws were formulated in terms of non-associative algebras by Etherington in several papers in 1939–1941 (and others later), among them [40–42]. In these works, algebraic patterns outlined by Mendel in his famous papers of 1865 and 1866 were developed. Some improvements to the mathematical approach for formulating Mendel's laws were given by Jennings [43] in 1917, by Serebrovski[36] himself in 1934, and by Glivenkov [38] in 1936. Since then, many authors have made relevant contributions to genetics in the context of non-associative algebras: Schafer [44], Gonshör [45], Haldane [46], Holgate [21], Heuch [47], Reiersöl [48], Korol [49], Lyubich [50], Wörz-Busekros [51], and Bertrand [52]. Further information on these algebras can be checked in [51] and the references therein.

2.3. Evolution Algebras

Tian and Vojtechovsky [1,3] showed a complete study of the algebraic properties of an evolution algebra. Let E_1 and E_2 be evolution algebras, and let $B_1 = \{e_1, \ldots, e_n\}$ be a natural basis on E_1. A linear map $\phi : E_1 \mapsto E_2$ is said to be a homomorphism of evolution algebras if $\phi(a\,b) = \phi(a)\,\phi(b)$, for every $a, b \in E_1$. Moreover, if ϕ is bijective, then it is said to be an isomorphism of evolution algebras.

In [15], an evolution algebra was described by identifying the coefficients of inheritance of a bisexual population as the structure constants of the algebra. In this respect, an evolution algebra of a bisexual population is an evolution algebra with a set of females partitioned into finitely many different types and the males having only one type. The idea of treating in a separate way male and female components of the population is due to Etherington [40], and it was later formalized by Holgate [53].

Finally, it is convenient to note that there exist some types of algebras that are not properly evolution algebras in the sense of Tian, but are quite related to them. One of these types is that formed by evolution algebras of a chicken population. Another one is that formed by evolution algebras of a mosquito population (see [54,55], respectively, for further information).

2.4. Graph Theory

A graph is a pair $G = (V, E)$ formed by a set V of vertices and a set E of edges containing two vertices, which are said to be adjacent. Two edges sharing a common vertex are said to be incident. A graph is complete if any pair of its vertices is adjacent. A path between two vertices $v, w \in V$ is any alternating sequence of vertices and edges, $\{v_0, e_0, v_1, \ldots, v_{n-1}, e_{n-1}, v_n\}$, for which:

(a) $v_0 = v$ and $v_n = w$;
(b) each edge e_i contains both vertices v_i and v_{i+1}, for all $i \in \{0, \ldots, n-1\}$; and
(c) all its vertices and all its edges are pairwise distinct.

The number of edges in any path constitutes its length. A graph is said to be connected if there exists a path between any pair of its vertices. The minimum length of a path between two given vertices $v, w \in V$ constitutes their distance $d(v, w)$. The eccentricity $\epsilon(v)$ of a vertex $v \in V$ is the maximum distance from v to any other vertex in V. The radius $\rho(G)$ and the diameter $\delta(G)$ of a graph G are, respectively, the minimum and maximum eccentricity of its vertices.

A graph is total-colored if all its vertices and edges are colored from a given palette of colors. This is proper-edge-colored if no two incident edges are colored with the same color. Two total-colored graphs that are related to the same palette of colors are isomorphic if there exists a bijection between their corresponding sets of vertices that preserves (a) their adjacency, (b) the color of edges, and (c) the fact that two given vertices are colored with the same color. A path between two vertices of a total-colored graph is said to be alternating if it contains at least three vertices and any two adjacent edges have different colors.

2.5. Isotopisms of Algebras

Two n-dimensional algebras A_1 and A_2 are isotopic if there exist three non-singular linear transformations f, g, and h from A_1 to A_2 such that $f(x)g(y) = h(xy)$, for all $x, y \in A_1$. The triple (f, g, h) is called an isotopism from A_1 to A_2. From a strict point of view, the third component h need not be defined for every vector of the algebra A_1, but only for its derived algebra $A_1^2 = \{xy\colon\ x, y \in A_1\}$. The notion of the isotopism of algebras was introduced by Albert [19] in 1942 as a generalization of the classical concept of the isomorphism of algebras (which arises when $f = g = h$). In particular, isotopisms make it possible to gather together non-isomorphic algebras that, at first sight, seemed to be completely different. Since the original manuscript of Albert, isotopisms have been widely used in the literature to study and classify distinct types of non-associative algebras like division algebras, alternative algebras, Jordan algebras, Lie algebras, Malcev algebras, genetic algebras, or evolution algebras, with relevance not only in mathematics, but also in natural sciences and engineering. See [22] for a recent survey in this regard.

In the context of evolution algebras, the transformations f and g represent the gene mutation of each one of the two chromatids in which the nuclear genetic material duplicates during the S-phase of a mitotic cell cycle. On the other hand, the transformation h represents the regulatory mechanisms during the G_2-phase that check any possible DNA damage that is produced during the S-phase. More specifically, for each $\alpha \in \{f, g, h\}$, one has $\alpha(e_i) = \sum_{j=1}^{n} \alpha_{ij} e_j$, for all $i \leq n$, where α_{ij} represents the probability that the genotype e_i mutates (if $\alpha \in \{f, g\}$) or is regulated (if $\alpha = h$) to e_j.

Note that a mitotic cell cycle is accomplished only if the genotypes of both chromatids that have been created after the mutations related to the transformations f and g coincide. Otherwise, there is no offspring. To this end, the regulatory mechanism related to the transformation h is crucial.

3. Description of the Total-Colored Graph

In this section, we describe our proposal to associate a total-colored graph $\mathcal{G}(A)$ with any given n-dimensional evolution algebra A over a finite field $\mathbb{K}$. Let $\{e_1, \ldots, e_n\}$ be the natural basis of such an algebra, and suppose $e_i e_i = \sum_{j=1}^{n} c_{ij} e_j$, with $c_{ij} \in \mathbb{K}$, for all i, j. Since we are interested in mutations among different genotypes, we may suppose $n > 1$. Thus, let $\mathcal{G}(A) := (V(A), E(A))$ be the total-colored graph having:

1. the set of vertices:
$$V(A) := \{v_{x,y} \mid x, y \in A,\ xy \neq 0\}, \tag{1}$$
where $v_{x,y}$ and $v_{x',y'}$ are equally colored in $V(A)$ if and only if $xy = x'y'$;
2. and whose set of edges $E(A)$ is formed by the disjoint union of both monochromatic subsets:
$$E_1(A) := \{v_{x,y} v_{x,z} \mid x, y, z \in A,\ y \neq z\} \tag{2}$$
$$E_2(A) := \{v_{x,y} v_{z,y} \mid x, y, z \in A,\ x \neq z\} \tag{3}$$

In order to illustrate this definition, Figure 1 shows, for instance, the total-colored graphs of the three isotopism classes of non-trivial two-dimensional evolution algebra over the finite field $\mathbb{F}_2$ [17]. Colors of vertices are represented by the symbols •, ▲, and ▼, while colors of edges are represented by solid and dashed lines.

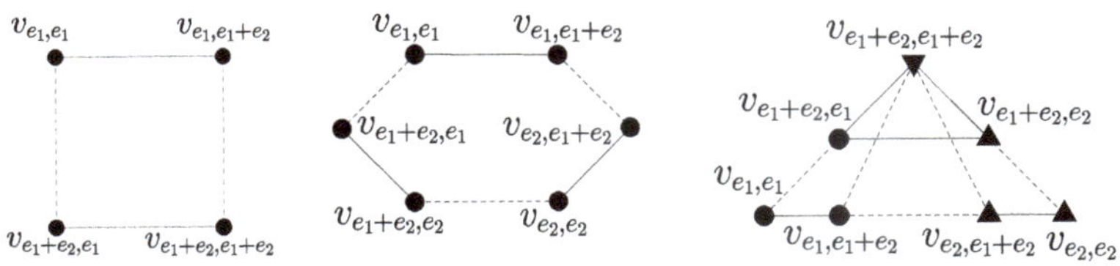

Figure 1. Total-colored graphs of the three two-dimensional evolution algebras over $\mathbb{F}_2$ having respective genetic patterns $(e_1, 0)$, (e_1, e_1), and (e_1, e_2).

The graph $\mathcal{G}(A)$ does not contain multiple edges, because every pair of adjacent vertices is contained in exactly one edge of either $E_1(A)$ or $E_2(A)$. The adjacency within each one of these two sets gives rise to a partition of $V(A)$. More specifically, two vertices $v_{x,y}$ and $v_{x',y'}$ belong to the same class according to the partition derived from $E_1(A)$ (respectively, $E_2(A)$) if and only if $x = x'$ (respectively, $y = y'$). Then, the next result follows readily from the commutativity of the algebra A and the fact that every two vertices within the same equivalence class are adjacent.

Lemma 1. *Both subgraphs $(V(A), E_1(A))$ and $(V(A), E_2(A))$ are isomorphic and constitute a disjoint union of complete graphs within $\mathcal{G}(A)$.*

From a genetic point of view, every vertex in $\mathcal{G}(A)$ represents a probable status of the genotype located in two given chromatids C_1 and C_2 during the S-phase of a mitotic cell cycle, together with the probable status of the genotype of the offspring. More specifically, if $x = \sum_{i=1}^{n} a_i e_i$ and $y = \sum_{i=1}^{n} b_i e_i$ are two vectors in A such that $xy \neq 0$, then the vertex $v_{x,y} \in V(A)$ is related to the following facts.

1. Let $j \in \{1, \ldots, n\}$. If $a_j = 0$ (respectively, $b_j = 0$), then the chromatid C_1 (respectively, C_2) does not contain the genotype e_j. The same occurs for the offspring if $\sum_{i=1}^{n} a_i b_i c_{ij} = 0$, which corresponds to the coefficient of the basis vector e_j in the product xy.
2. Let $j, k \in \{1, \ldots, n\}$ be such that $j \neq k$. If $a_j > a_k$ (respectively, $b_j > b_k$), then it is more probable for the chromatid C_1 (respectively, C_2) to contain the genotype e_j than the genotype e_k. If $a_j = a_k$ (respectively, $b_j = b_k$), then both possibilities are equally probable. The same occurs for the offspring if $\sum_{i=1}^{n} a_i b_i c_{ij} \geq \sum_{i=1}^{n} a_i b_i c_{ik}$.

Furthermore, two different vertices in $\mathcal{G}(A)$ are equally colored if and only if they are associated with the same probable status of the offspring's genotype.

Concerning the edges in $\mathcal{G}(A)$, they represent the regulatory mechanisms produced during the G_2-phase of the mitotic cell cycle. Specifically, every edge $v_{x,y}v_{x,z} \in E_1(A)$ (respectively, $v_{x,y}v_{z,y} \in E_2(A)$) represents a mutation between two probable statuses y and z (respectively, x and z) of the genotype in the chromatid C_2 (respectively, C_1), without changing to this end the probable status of the genotype in C_1 (respectively, C_2). Thus, for instance, the vertex v_{e_1,e_1+e_2} in any of the three graphs in Figure 1 involves the chromatid C_1 containing the genotype e_1, while the chromatid C_2 contains, with the same probability, either the genotype e_1 or e_2. The first case would give rise to an offspring of genotype e_1, because $e_1e_1 = e_1$, whereas the second one would not give rise to any offspring, because $e_1e_2 = 0$. In order to avoid the latter, the regulatory mechanism produced during the G_2-phase of the mitotic cell could mutate the possible genotype e_2 in C_2 into the genotype e_1. This is represented in the graph $\mathcal{G}(A)$ by the edge $v_{e_1,e_1+e_2}v_{e_1,e_1} \in E_1(A)$.

Every path within the graph $\mathcal{G}(A)$ is therefore interpreted as a series of mutations from a pair of probable statuses of the genotype of both chromatids C_1 and C_2 into another. The following result shows that these mutations are possible, whatever the probable statuses under consideration are.

Lemma 2. *The graph $\mathcal{G}(A)$ is connected. Moreover, $\delta(\mathcal{G}(A)) \leq 7$.*

Proof. Let $\{v_{x,y}, v_{z,t}\} \subseteq V(A)$, where, for each $\alpha \in \{x, y, z, t\}$, it is $\alpha = \sum_{i=1}^{n} c_i^{\alpha} e_i$. Since $xy \neq 0 \neq zt$, there exist two positive integers $j, j' \leq n$ such that $\sum_{i=1}^{n} c_i^x c_i^y c_{ij} \neq 0 \neq \sum_{i=1}^{n} c_i^z c_i^t c_{ij'}$. Let $i, i' \in \{1, \ldots, n\}$ be such that $c_i^x c_i^y c_{ij} \neq 0 \neq c_{i'}^z c_{i'}^t c_{i'j'}$. Then, the ordered set of vertices:

$$\{v_{x,y}, v_{x,e_i}, v_{e_i,e_i}, v_{e_i,e_i+e_{i'}}, v_{e_{i'},e_i+e_{i'}}, v_{e_{i'},e_{i'}}, v_{x',e_{i'}}, v_{x',y'}\} \subseteq V(A),$$

determines a path in $\mathcal{G}(A)$ of seven alternating edges in $E_1(A)$ and $E_2(A)$. □

In order to deal with the tightness of the upper bound in Lemma 2, we introduce some new concepts. To this end, let $\{v_{x,y}, v_{z,t}\} \subseteq V(A)$, where, for each $\alpha \in \{x, y, z, t\}$, it is $\alpha = \sum_{i=1}^{n} c_i^{\alpha} e_i$. We say that the vertices $v_{x,y}$ and $v_{z,t}$ are compatible if there exists a positive integer $i \leq n$ such that $c_i^{\alpha} \neq 0$, for all vectors $\alpha \in \{x, y, z, t\}$, and $e_ie_i \neq 0$. Further, a vertex in $V(A)$ is called pure if it has the form $v_{ae_i,be_i} \in V(A)$ for some $a, b \in \mathbb{K} \setminus \{0\}$ and for some i with $e_ie_i \neq 0$. Otherwise, the vertex is called hybrid.

Proposition 1. *The eccentricity of every pure vertex in $\mathcal{G}(A)$ is, at most, three.*

Proof. Let $v_{x,y} \in V(A)$ be a vertex that is not adjacent to a pure vertex $v_{ae_i,be_i} \in V(A)$, where, for each $\alpha \in \{x, y\}$, we have $\alpha = \sum_{i=1}^{n} c_i^{\alpha} e_i$. If $c_i^x \neq 0$ (respectively, $c_i^y \neq 0$), then the ordered set of vertices $\{v_{x,y}, v_{x,be_i}, v_{ae_i,be_i}\} \subseteq V(A)$ (respectively, $\{v_{x,y}, v_{ae_i,y}, v_{ae_i,be_i}\} \subseteq V(A)$) determines a two-length alternating path in $\mathcal{G}(A)$. Otherwise, if $c_i^x = 0 = c_i^y$, then the ordered set of vertices $\{v_{x,y}, v_{x,be_i+y}, v_{ae_i,be_i+y}, v_{ae_i,be_i}\} \subseteq V(A)$ determines a three-length alternating path. □

Proposition 2. *The eccentricity of every hybrid vertex in $\mathcal{G}(A)$ is, at most, three.*

Proof. Let $v_{x,y}$ and $v_{z,t}$ be two non-adjacent vertices in $\mathcal{G}(A)$, where, for each $\alpha \in \{x, y, z, t\}$, we have $\alpha = \sum_{i=1}^{n} c_i^\alpha e_i$. In addition, suppose that the vertex $v_{x,y}$ is hybrid. From Proposition 1, if $v_{z,t}$ were a pure vertex, then $d(v_{x,y}, v_{z,t}) \leq 3$. Thus, from now on, we focus on the case that $v_{z,t}$ is also hybrid.

If the vertices $v_{x,y}$ and $v_{z,t}$ are compatible, then there exists a positive integer $i \leq n$ such that $c_i^x, c_i^y, c_i^z, c_i^t \in \mathbb{K} \setminus \{0\}$ and $e_ie_i \neq 0$. Thus, the ordered set of vertices $\{v_{x,y}, v_{x,e_i}, v_{z,e_i}, v_{z,t}\}$ determines a three-length alternating path.

Otherwise, if the vertices $v_{x,y}$ and $v_{z,t}$ are not compatible, then there exist two different positive integers $i, j \leq n$ such that $c_i^x, c_i^y, c_j^z, c_j^t \in \mathbb{K} \setminus \{0\}$ and $e_ie_i \neq 0 \neq e_je_j$. If $c_j^x \neq 0$, then the ordered set of vertices $\{v_{x,y}, v_{x,e_j}, v_{z,e_j}, v_{z,t}\}$ determines a three-length alternating path. Similar paths of the same length can be found if $c_j^y \neq 0$, $c_i^z \neq 0$, or $c_i^t \neq 0$. On the other hand, if $c_j^x = c_j^y = c_i^z = c_i^t = 0$, then the ordered set of vertices $\{v_{x,y}, v_{x,e_i+e_j}, v_{z,e_i+e_j}, v_{z,t}\}$ determines a three-length alternating path. □

Theorem 1. *If the genetic pattern of the evolution algebra A has more than one non-zero component, then:*

$$2 \leq \rho(\mathcal{G}(A)) \leq 3 = \delta(\mathcal{G}(A)).$$

Otherwise, $\rho(\mathcal{G}(A)) = \delta(\mathcal{G}(A)) = 2$.

Proof. The non-triviality of the algebra involves the existence of a positive integer $i \leq n$ such that $e_ie_i \neq 0$. In addition, let $j \leq n$ be a positive integer such that $i \neq j$. This exists because $n > 1$. Then, since the graph $\mathcal{G}(A)$ is connected and contains, for instance, the edge $v_{e_i,e_i}v_{e_i+e_j,e_i}$, we have that $\rho(\mathcal{G}(A)) > 0$.

Now, suppose that $\rho(\mathcal{G}(A)) = 1$. Then, there exists a vertex $v_{x,y} \in V(A)$ that is adjacent to any other vertex in the graph. Since every evolution algebra is commutative, we have that $yx = xy \neq 0$, and hence, $v_{y,x} \in V(A)$. This is adjacent to $v_{x,y}$, and hence, from (2) and (3), we have that $x = y$. Moreover, since the vertex $v_{e_i,e_i} \in V(A)$ is adjacent to $v_{x,x}$, we have that $x = e_i$. This implies that $e_je_j = 0$. Otherwise, the vertices v_{e_i,e_i} and v_{e_j,e_j} should be adjacent, which is a contradiction with the fact that $e_i \neq e_j$. As a consequence, $(e_i + e_j)(e_i + e_j) = e_ie_i \neq 0$. Thus, the vertices v_{e_i,e_i} and $v_{e_i+e_j,e_i+e_j}$ should be adjacent, and hence, $e_i = e_i + e_j$, which is also a contradiction. Therefore, $\rho(\mathcal{G}(A)) \geq 2$. This fact, together with Propositions 1 and 2, holds the first assertion of the theorem.

Now, in order to prove the second assertion, suppose the existence of a positive integer $i \leq n$ such that $e_ie_i \neq 0$ and $e_je_j = 0$, for all $j \neq i$. Then, every pair of vertices $v_{x,y}, v_{z,t} \in V(A)$ is compatible so that the coefficient of the basis vector e_i in each one of the four vectors $x, y, z, t \in A$ is not zero. In particular, $xt \neq 0$, and hence, the ordered set of vertices $\{v_{x,y}, v_{x,t}, v_{z,t}\}$ determines a two-length alternating path. Therefore, $\delta(\mathcal{G}(A)) \leq 2$, and the result holds readily. □

Note that all the bounds in the first assertion of Theorem 1 are tight. Thus, for instance, the radius of the second graph in Figure 1 is three, whereas the radii of the first and third graphs in the same figure are two. Furthermore, from a genetic point of view, Theorem 1 can be interpreted as the fact that any pair of probable statuses of the genotype of two chromatids can be transformed into another pair by means of at most two or three mutations.

4. A Faithful Functor among Evolution Algebras and Graphs

Apart from the simplicity of describing the role of genes in mutation problems, the total-colored graph that we have just introduced in the previous section becomes also relevant from a categorical point of view. In this regard, we make use in this section of the mentioned graph in order to introduce a faithful functor among the category of evolution algebras on a finite field and the category of total-colored graphs. Unlike the graphs associated with evolution algebras that already exist in the literature (see the introductory section), our description enables one to ensure that isomorphic evolution algebras always give rise to isomorphic total-colored graphs. Moreover, we prove in Proposition 3 that the previous assertion holds indeed for isotopic evolution algebras and not only for isomorphic ones.

From here on, the following notations are used:

- Let $\mathcal{E}$ and $\mathcal{M}$ respectively denote the set of finite-dimensional evolution algebras over a finite field and the set formed by their associated total-colored graphs. For each algebra $A \in \mathcal{E}$, the set of vertices of the graph $\mathcal{G}(A)$ is $V(A) := \{v^A_{x,y} \mid x, y \in A, xy \neq 0\}$.
- Let $\mathrm{Isot}(\mathcal{E})$ and $\mathrm{Isom}(\mathcal{M})$ respectively denote the set of isotopisms of algebras in $\mathcal{E}$ and the set of isomorphisms of graphs in $\mathcal{M}$.
- Let $\mathrm{Isot}(A_1, A_2)$ and $\mathrm{Isom}(G_1, G_2)$ respectively denote the set of isotopisms between two given evolution algebras $A_1, A_2 \in \mathcal{E}$ and the set of isomorphisms between two given graphs $G_1, G_2 \in \mathcal{M}$.
- Let $\mathfrak{E}$ denote the category formed by the sets $\mathcal{E}$ and $\mathrm{Isot}(\mathcal{E})$.
- Let $\mathfrak{M}$ denote the category formed by the sets $\mathcal{M}$ and $\mathrm{Isom}(\mathcal{M})$.

Let us finish this paper with the establishment of a faithful functor between the categories $\mathfrak{E}$ and $\mathfrak{M}$. In this regard, let $F : \mathfrak{E} \to \mathfrak{M}$ be such that:

(a) $F(A) := \mathcal{G}(A)$, for all $A \in \mathcal{E}$.

(b) For each isotopism $\Theta = (f, g, h) \in \mathrm{Isot}(\mathcal{E})$ between two isotopic evolution algebras $A_1, A_2 \in \mathcal{E}$, let $F_\Theta : \mathcal{G}(A_1) \to \mathcal{G}(A_2)$ be such that:

$$F_\Theta(v^{A_1}_{x,y}) := v^{A_2}_{f(x),g(y)}, \text{ for all } v^{A_1}_{x,y} \in V(A_1).$$

The non-singularity of the three linear transformations f, g, and h involves F_Θ being a well-defined bijection from $V(A_1)$ to $V(A_2)$. Observe to this end that, for each $x, y \in A$, one has that $f(x)g(y) = h(xy) \neq 0$ if and only if $xy \neq 0$.

Proposition 3. *The map $F : \mathfrak{E} \to \mathfrak{M}$ is a faithful functor.*

Proof. Let $A_1, A_2 \in \mathcal{E}$ and $\Theta \in \mathrm{Isot}(A_1, A_2)$. Then, we define the map:

$$\begin{array}{rcl} \mathrm{Isot}(A_1, A_2) & \to & \mathrm{Isom}(\mathcal{G}(A_1), \mathcal{G}(A_2)) \\ \Theta & \to & F_\Theta \end{array} \quad (4)$$

Let us prove that this is well defined and injective. Suppose to this end that $\Theta = (f, g, h) \in \mathrm{Isot}(A_1, A_2)$. The non-singularity of both linear transformations f and g implies that $F_\Theta : V(A_1) \to V(A_2)$ maps $E_1(A_1)$ and $E_2(A_1)$ to $E_1(A_2)$ and $E_2(A_2)$, respectively. Furthermore, the non-singularity of the linear transformation h implies that $F_\Theta(V(A_1)) = V(A_2)$. Hence, $F_\Theta \in \mathrm{Isom}(\mathcal{G}(A_1), \mathcal{G}(A_2))$.

Now, suppose the existence of a pair of isotopisms $\Theta_1 = (f_1, g_1, h_1)$ and $\Theta_2 = (f_2, g_2, h_2)$ in the set $\mathrm{Isot}(A_1, A_2)$ such that $F_{\Theta_1} = F_{\Theta_2}$. Then, $v^{A_2}_{f_1(x),g_1(y)} = F_{\Theta_1}(v^{A_1}_{x,y}) = F_{\Theta_2}(v^{A_1}_{x,y}) = v^{A_2}_{f_2(x),g_2(y)}$, for all $x, y \in A_1$ such that $xy \neq 0$. As a consequence, $f_1 = f_2$ and $g_1 = g_2$, and hence, $h_1(xy) = f_1(x)g_1(y) = f_2(x)g_2(y) = h_2(xy)$. Therefore, both isotopisms, Θ_1 and Θ_2, coincide. □

5. Conclusions and Further Work

This paper dealt with the description of a total-colored graph associated with the genetic pattern of any finite-dimensional evolution algebra over a finite field. We proved that both categories formed by such algebraic and combinatorial structures are related by a faithful functor. The faithfulness of the functor implies that any isotopism of evolution algebras is uniquely related to an isomorphism of total-colored graphs. As a consequence, isotopisms of evolution algebras can be thought from now on as isomorphisms of total-colored graphs satisfying some extra properties. An explicit determination and characterization of the latter were established as further work in order to identify under which circumstances the functor under consideration is indeed fully faithful. In any case, these extra properties derive straightforwardly from the linearity of any isotopism of algebras.

Further, throughout the paper, we interpreted some basic properties and results of the described total-colored graph in the context of the genetic processes that occur during a mitotic cell cycle. Keeping in mind that the spectrum of possible patterns is currently known only for those evolution algebras that are involved with two or three genotypes, we established as further work the study of graph invariants of the proposed total-colored graph in order to deal with a higher number of genotypes.

It is also required to delve into the translation of algebraic concepts and results on evolution algebras to the language of total-colored graphs. A particular focus of interest in this regard is the associator $\mathcal{A}(x,y,z) := (xy)z - x(yz)$, which constitutes, according to Goze and Remms [56], an interesting way of classifying nonassociative algebras over the group algebra $\mathbb{K}[\Sigma_3]$ of the symmetric group Σ_3 over a field $\mathbb{K}$ and also for studying their algebraic properties. In order to illustrate such a classification in the case of dealing with evolution algebras, let A be an n-dimensional evolution algebra over the field $\mathbb{K}$, with natural basis $\{e_1,\ldots,e_n\}$ and such that $e_ie_i = \sum_{j=1}^n c_{ij}e_j$, with $c_{ij} \in \mathbb{K}$, for all i,j. Then,

$$\mathcal{A}(e_i,e_i,e_j) + \mathcal{A}(e_j,e_i,e_i) = c_{ij}e_je_j - c_{ij}e_je_j = 0, \text{ for all } i \neq j,$$

and hence,

$$\mathcal{A}(e_i,e_k,e_j) + \mathcal{A}(e_k,e_i,e_j) + \mathcal{A}(e_j,e_i,e_k) + \mathcal{A}(e_j,e_k,e_i) = 0, \text{ for all } i,j,k.$$

According to the notation that is used in [56], the previous identity becomes:

$$\mathcal{A} \circ \Phi_v = 0.$$

Here, Φ_v denotes the right action on the group algebra $\mathbb{K}[\Sigma_3]$ by means of the vector:

$$v := \mathrm{Id} + \tau_{12} + c_2 + \tau_{13},$$

where, from here on, τ_{ij} denotes the transposition $(i,j) \in \Sigma_3$ and c_2 is the cycle of order three $(1,3,2) \in \Sigma_3$. The orbit of this vector v for the natural action of Σ_3 on the group algebra $\mathbb{K}[\Sigma_3]$ is:

$$\mathcal{O}(v) := \{v, \mathrm{Id} + \tau_{13} + \tau_{23} + c_1, \tau_{23} + c_1 + c_2 + \tau_{12}\},$$

where c_1 denotes the cycle of order three $(1,2,3) \in \Sigma_3$. Then, $\mathbb{K}(\mathcal{O}(v))$ is an invariant irreducible space of the group algebra $\mathbb{K}[\Sigma_3]$. In particular,

$$2\mathrm{Id} + 2\tau_{12} + 2\tau_{13} + 2\tau_{23} + 2c_1 + 2c_2 \in \mathbb{K}(\mathcal{O}(v)).$$

According to the terminology in [56], this means that evolution algebras are $\mathbb{K}[\Sigma_3]$-associative. This implies that a development of algebraic properties (like operad, homology, cohomology, amongst others) could be done as further work.

Author Contributions: Investigation, R.M.F., Ó.J.F., and J.N.; methodology, R.M.F.; supervision, R.M.F. and J.N.; validation, R.M.F.; writing, original draft, R.M.F. and Ó.J.F.; writing, review and editing, R.M.F. and J.N.

Funding: This research received no external funding.

Acknowledgments: The authors want to express their gratitude to the anonymous referees for the comprehensive reading of the paper and their pertinent comments and suggestions, which helped improve the manuscript. In particular, we are very grateful to the anonymous reviewer who suggested that we delve into the possible description of the non-associativity of evolution algebras according to Goze's classification [56]. This work is partially supported by the Research Projects FQM-016 and FQM-326 from Junta de Andalucía.

Conflicts of Interest: The authors declare no conflict of interest.

References

1. Tian, J.P. *Evolution Algebras and their Applications*; Lecture Notes in Mathematics; Springer-Verlag: Berlin, Germany, 2008; Volume 1921.
2. Tian, J.P. Evolution Algebra Theory. Ph.D. Thesis, University of California, Riverside, CA, USA, 2004.
3. Tian, J.P.; Vojtechovsky, P. Mathematical concepts of evolution algebras in non-mendelian Genetics. *Quasigroups Relat. Syst.* **2006**, **14**, 111–122.
4. Khudoyberdiyev, A.; Omirov, B.A.; Qaralleh,I. Few remarks on evolution algebras. *J. Algebra Appl.* **2015**, *14*, 1550053. [CrossRef]
5. Cadavid, P.; Rodiño, M.L.; Rodríguez, P.M. The connection between evolution algebras, random walks and graphs. *J. Algebra Appl.* **2019**. [CrossRef]
6. Cabrera Casado, Y.; Siles Molina, M.; Velasco, M.V. Classification of three-dimensional evolution algebras. *Linear Algebra Appl.* **2017**, *524*, 68–108. [CrossRef]
7. Camacho, L.M.; Gómez, J.R.; Omirov, B.A.; Turdibaev, R.M. Some properties of evolution algebras. *Bull. Korean Math. Soc.* **2013**, *50*, 1481–1494. [CrossRef]
8. Camacho, L.M.; Gómez, J.R.; Omirov, B.A.; Turdibaev, R.M. The derivations of some evolution algebras. *Linear Multilinear A* **2013**, 61, 309–322. [CrossRef]
9. Casas, J.M.; Ladra, M.; Rozikov, U.A. A chain of evolution algebras. *Linear Algebra Appl.* **2011**, *435*, 852–870. [CrossRef]
10. Casas, J.M.; Ladra, M.; Omirov, B.A.; Rozikov, U.A. On nilpotent index and dibaricity of evolution algebras. *Linear Algebra Appl.* **2013**, *439*, 90–05. [CrossRef]
11. Casas, J.M.; Ladra, M.; Omirov, B.A.; Rozikov, U.A. On Evolution Algebras. *Algebra Colloq.* **2014**, *21*, 331–342. [CrossRef]
12. Dzhumadildaev, A.; Omirov, B.A.; Rozikov, U.A. Constrained evolution algebras and dynamical systems of a bisexual population. *Linear Algebra Appl.* **2016**, *496*, 351–380. [CrossRef]
13. Labra, A.; Ladra, M.; Rozikov, U.A. An evolution algebra in population Genetics. *Linear Algebra Appl.* **2014**, *457*, 348–362. [CrossRef]
14. Ladra, M.; Omirov, B.A.; Rozikov, U.A. Dibaric and evolution algebras in biology. *Lobachevskii J. Math.* **2014**, *35*, 198–210. [CrossRef]
15. Ladra, M.; Rozikov, U.A. Evolution algebra of a bisexual population. *J. Algebra* **2013**, *378*, 153–172. [CrossRef]
16. Omirov, B. A.; Rozikov, U.A.; Tulenbayev, K.M. On real chains of evolution algebras. *Linear Multilinear Algebra* **2015**, *63*, 586–600. [CrossRef]
17. Falcón, O.J.; Falcón, R.M.; Núñez, J. Classification of asexual diploid organisms by means of strongly isotopic evolution algebras defined over any field. *J. Algebra* **2017**, *472*, 573–593. [CrossRef]
18. Falcón, O.J.; Falcón, R.M.; Núñez, J. Algebraic computation of genetic patterns related to three-dimensional evolution algebras. *Appl. Math. Comput.* **2018**, *319*, 510–517. [CrossRef]
19. Albert, A.A. Non-Associative Algebras: I. Fundamental Concepts and Isotopy. *Ann. Math. Second. Ser.* **1942**, *43*, 685–707. [CrossRef]
20. Campos, T.M.M.; Holgate, P. Algebraic Isotopy in Genetics. *IMA J. Math. Appl. Med. Biol.* **1987**, *4*, 215–222. [CrossRef]
21. Holgate, P. Genetic algebras associated with polyploidy. *Proc. Edinb. Math. Soc.* **1966**, *15*, 1–9. [CrossRef]
22. Falcón, R.M.; Falcón, O.J.; Núñez, J. A Historical Perspective of the Theory of Isotopisms. *Symmetry* **2018**, *10*, 1–21. [CrossRef]
23. Benzer, S. On the topology of genetic fine structure. *Proc. Nat. Acad. Sci. USA* **1959**, *45*, 1607–1620. [CrossRef] [PubMed]
24. Cadavid, P.; Rodiño, M.L.; Rodríguez, P.M. Characterization theorems for the spaces of derivations of evolution algebras associated to graphs. *Linear Multilinear A* **2019**. [CrossRef]
25. Núñez, J.; Silvero, M.; Villar, M.T. Mathematical tools for the future: Graph Theory and graphicable algebras. *Appl. Math. Comput.* **2013**, *219*, 6113–6125. [CrossRef]
26. Elduque, A.; Labra, A. Evolution algebras and graphs. *J. Algebra Appl.* **2015**, *14*, 1550103. [CrossRef]
27. Elduque, A.; Labra, A. On nilpotent evolution algebras. *Linear Algebra Appl.* **2016**, *505*, 11–31. [CrossRef]

28. Cabrera Casado, Y.; Siles Molina, M.; Velasco, M.V. Evolution algebras of arbitrary dimension and their decompositions. *Linear Algebra Appl.* **2016**, *495*, 122–162. [CrossRef]
29. Falcón, O.J.; Falcón, R.M.; Núñez, J.: Pacheco, A.M.; Villar, M.T. Computation of isotopisms of algebras over finite fields by means of graph invariants. *J. Comput. Appl. Math.* **2017**, *318*, 307–315. [CrossRef]
30. McKay, B.D.; Meynert, A.; Myrvold, W. Small Latin Squares, Quasigroups and Loops. *J. Combin. Des.* **2007**, *15*, 98–119. [CrossRef]
31. Falcón, O.J.; Falcón, R.M.; Núñez, J. Mutation graphs of asexual diploid organisms. In *X Encuentro Andaluz de Matemática Discreta*; Universidad de Cádiz: Cádiz, Spain, 2017; pp. 103–106.
32. Brooker, R. *Genetics: Analysis and Principles*; McGraw-Hill Education: New York, NY, USA, 2018.
33. Harary, F. *Graph Theory*; Addison Wesley: Reading, MA, USA, 1969.
34. Jones, N.C.; Pevzner, P.A. *An introduction to Bioinformatics Algoritms*; MIT Press: Cambridge, MA, USA, 2004.
35. Pevzner, P.A. *Computational Molecular Biology*; MIT Press: Cambridge, MA, USA, 2000.
36. Serebrowsky, A. On the properties of the Mendelian equations. *Dokl. Akad. Nauk. SSSR* **1934**, *2*, 33–36.
37. Mendel, G. *Experiments in Plant-Hybridization*; Classic Papers in Genetics; Prentice-Hall Inc.: Upper Saddle River, NJ, USA, 1959.
38. Glivenkov, V. Algebra Mendelienne. *C. R. Acad. Sci. URSS* **1936**, *4*, 385–386.
39. Kostitzin, V.A. Sur les coefficients mendeliens d'heredite. *C. R. Acad. Sci. URSS* **1938**, *206*, 883–885.
40. Etherington, I.M.H. Genetic algebras. *Proc. R. Soc. Edinb.* **1939**, *59*, 242–258. [CrossRef]
41. Etherington, I.M.H. Duplication of linear algebras. *Proc. Edinb. Math. Soc.* **1941**, *6*, 222–230. [CrossRef]
42. Etherington, I.M.H. Non-associative algebra and the symbolism of Genetics. *Proc. R. Soc. Edinb. Sect. B* **1941**, *61*, 24–42. [CrossRef]
43. Jennings, H.S. Observed changes in hereditary characters in relation to evolution. *J. Wash. Acad. Sci.* **1917**, *7*, 281–301.
44. Schafer, R.D. Structure of genetic algebras. *Am. J. Math.* **1949**, *71*, 121–135. [CrossRef]
45. Gonshör, H. Contributions to genetic algebras. *Proc. Edinb. Math. Soc.* **1971**, *17*, 289–298. [CrossRef]
46. Haldane, J.B.S. A mathematical theory of natural and artificial selection. *Bull. Math. Biol.* **1990**, *52*, 209–240.
47. Heuch, I. Sequences in genetic algebras for overlapping generations. *Proc. Edinb.* **1972**, *16*, 291–294. [CrossRef]
48. Reiersöl, O. Genetic algebras studied recursively and by means of differential operators. *Math. Scand.* **1962**, *10*, 25–44. [CrossRef]
49. Korol, A.B.; Preygel, I.A.; Preygel, S.I. *Recombination Variability and Evolution*; Chapman & Hall: London, UK, 1994.
50. Lyubich, Y.I. *Mathematical Structures in Population Genetics*; Springer: Berlin, Germany, 1992.
51. Wörz-Busekros, A. *Algebras in Genetics*; Springer: Berlin/Heidelberg, Germany, 1980.
52. Bertrand, M. *Algèbres non Associatives et Algèbres Gènètiques*; Gauthier-Villars Editeur: Paris, France, 1966.
53. Holgate, P. Characterizations of genetic algebras. *J. Lond. Math. Soc.* **1972**, *2*, 169–174. [CrossRef]
54. Dzhumadil'daev, A.S.; Omirov, B.A.; Rozikov, U.A. On a class of evolution algebras of "chicken" population. *Int. J. Math.* **2014**, *25*, 1450073. [CrossRef]
55. Rozikov, U.A.; Velasco, M.V. A discrete-time dynamical system and an evolution algebra of mosquito population. *J. Math. Biol.* **2019**, *78*, 1225–1244. [CrossRef] [PubMed]
56. Goze, M.; Remm, E. A class of nonassociative algebras. *Algebra Colloq.* **2007**, *14*, 313–326. [CrossRef]

mathematics

MDPI

Article

f-Polynomial on Some Graph Operations

Walter Carballosa [1], José Manuel Rodríguez [2], José María Sigarreta [3,*] and Nodari Vakhania [4]

1 Department of Mathematics and Statistics, Florida International University, 11200 SW 8th Street, Miami, FL 33199, USA; waltercarb@gmail.com
2 Departamento de Matemáticas, Universidad Carlos III de Madrid, Avenida de la Universidad 30, 28911 Leganés, Spain; jomaro@math.uc3m.es
3 Facultad de Matemáticas, Universidad Autónoma de Guerrero, Carlos E. Adame No. 54 Col. Garita, 39650 Acapulco, Mexico
4 Centro de Investigación en Ciencias, Universidad Autónoma del Estado de Morelos, Av. Universidad 1001 Col. Chamilpa, 62209 Cuernavaca, Mexico; nodari@uaem.mx
* Correspondence: jsmathguerrero@gmail.com

Received: 20 September 2019; Accepted: 3 November 2019; Published: 8 November 2019

Abstract: Given any function $f : \mathbb{Z}^+ \to \mathbb{R}^+$, let us define the f-index $I_f(G) = \sum_{u \in V(G)} f(d_u)$ and the f-polynomial $P_f(G,x) = \sum_{u \in V(G)} x^{1/f(d_u)-1}$, for $x > 0$. In addition, we define $P_f(G,0) = \lim_{x \to 0^+} P_f(G,x)$. We use the f-polynomial of a large family of topological indices in order to study mathematical relations of the inverse degree, the generalized first Zagreb, and the sum lordeg indices, among others. In this paper, using this f-polynomial, we obtain several properties of these indices of some classical graph operations that include corona product and join, line, and Mycielskian, among others.

Keywords: inverse degree index; generalized first Zagreb index; sum lordeg index; corona product; join of graphs; line graph; Mycielskian graph; polynomials in graphs

1. Introduction

A topological index is a single number that represents a chemical structure via the molecular graph, in graph theoretical terms, whenever it correlates with a molecular property. Hundreds of topological indices have been recognized to be useful tools in research, especially in chemistry. Topological indices have been used to understand physicochemical properties of compounds. They usually enclose topological properties of a molecular graph in a single real number. Several topological indices were introduced by the seminal work by Wiener [1]. They have been studied and generalized by several researchers since then. In particular, topological indices based on end-vertex degrees of edges have been studied over almost 50 years (see, e.g., [2–11]).

A graph, usually denoted $G(V(G);E(G))$, consists of a set of vertices $V(G)$ together with a set $E(G)$ of unordered pairs of vertices called edges. The number of vertices in a graph is usually denoted $n = |V(G)|$, while the number of edges is usually denoted $m = |E(G)|$; these two basic parameters are called the order and size of G, respectively. Miličević and Nikolić defined in reference [12] the first variable Zagreb index as

$$M_1^\alpha(G) = \sum_{u \in V(G)} d_u^\alpha,$$

where d_u is the degree of the vertex u and $\alpha \in \mathbb{R}$.

Note that M_1^2 is the first Zagreb index M_1, M_1^{-1} is the inverse index ID, M_1^3 is the forgotten index F, etc.

The harmonic index, defined in reference [13] as

$$H(G) = \sum_{uv \in E(G)} \frac{2}{d_u + d_v},$$

has been studied in the last years (see, e.g., [14–19]).

In reference [20], the harmonic polynomial of a graph G is defined as

$$H(G,x) = \sum_{uv \in E(G)} x^{d_u + d_v - 1},$$

and the harmonic polynomials of some graphs are computed. The harmonic polynomials of the line of some graphs are computed in reference [21].

This polynomial owes its name to the fact that $2 \int_0^1 H(G,x)\, dx = H(G)$.

The inverse degree index $ID(G)$ of a graph G is defined by

$$ID(G) = \sum_{u \in V(G)} \frac{1}{d_u} = \sum_{uv \in E(G)} \left(\frac{1}{d_u^2} + \frac{1}{d_v^2} \right).$$

The inverse degree index first attracted attention through numerous conjectures (see [13]). This index has been studied in reference [22–26].

The inverse degree polynomial of a graph G was defined in reference [27] as

$$ID(G,x) = \sum_{u \in V(G)} x^{d_u - 1}.$$

We have $\int_0^1 ID(G,x)\, dx = ID(G)$.

Given any function $f : \mathbb{Z}^+ \to \mathbb{R}^+$, let us define the f-index

$$I_f(G) = \sum_{u \in V(G)} f(d_u)$$

and the f-polynomial

$$P_f(G,x) = \sum_{u \in V(G)} x^{1/f(d_u) - 1},$$

for $x > 0$. In addition, we define $P_f(G,0) = \lim_{x \to 0^+} P_f(G,x)$. Note that $P_f(G,x) = ID(G,x)$ when $f(t) = 1/t$.

The degrees of the vertices are graph invariants, i.e., if two graphs are isomorphic, then the corresponding vertices by any isomorphism have the same degrees. Hence, the f-polynomial is also a graph invariant, i.e., two isomorphic graphs have the same ID polynomial.

Polynomials have proved to be useful in the study of several topological indices (see, e.g., [27–29]).

There are many papers studying several topological indices of graph operations (see, e.g.,[27–30]).

Throughout this paper, $G = (V(G), E(G))$ denotes a (non-oriented) finite simple (without multiple edges and loops) graph without isolated vertices. The main aim of this paper is to study mathematical relations of the inverse degree, the generalized first Zagreb, and the sum lordeg indices, among others. In order to do that, we use the f-polynomial of a large family of topological indices, introduced in reference [31]. We obtain inequalities (and even closed formulas in the case of the ID polynomial) involving the f-polynomial of many classical graph operations, which include corona product, join, line and Mycielskian, among others. These results allow us to obtain new inequalities for the inverse degree, the generalized first Zagreb, and the sum lordeg indices of these graph operations.

2. Definitions and Background

In the following sections, we obtain inequalities for the f-polynomial of many classical graph operations, which include corona product, join, line and Mycielskian, among others. The f-polynomial of other graph operations (Cartesian product, lexicographic product, and Cartesian sum) is studied in reference [31]. The different kinds of graph operations are an important research topic (see [32] and the references therein). Some large graphs are composed from some existing smaller ones by using several graph operations, and many properties of such large graphs are strongly associated with those of the corresponding smaller ones.

Let us recall the definitions of some classical products in graph theory.

The join $G_1 + G_2$ is defined as the graph obtained by taking one copy of G_1 and one copy of G_2 and joining by an edge each vertex of G_1 with each vertex of G_2.

The corona product $G_1 \circ G_2$ is the graph obtained by taking $|V(G_1)|$ copies of G_2 and joining each vertex of the i-th copy with the vertex $v_i \in V(G_1)$.

The ID polynomial is related to the polynomials associated to some topological indices.

In reference [33], Shuxian defined the following polynomial related to the first Zagreb index:

$$M_1^*(G,x) := \sum_{u \in V(G)} d_u x^{d_u}.$$

Note that $x(xID(G,x))' = M_1^*(G,x)$.

The ID polynomial is also related to other polynomials, like the harmonic polynomial (see [27] and Propositions 7–10 in this paper).

The following result states some of the main properties of P_f.

Proposition 1. *If G is a graph with order n and $f : \mathbb{Z}^+ \to \mathbb{R}^+$, then:*

- *$P_f(G,x)$ is a polynomial if and only if $1/f(d_u) \in \mathbb{Z}^+$ for every $u \in V(G)$,*
- *$P_f(G,x)$ is a positive C^∞ function on $(0,\infty)$,*
- *$P_f(G,x)$ is a continuous function on $[0,\infty)$ if and only if $P_f(G,0) < \infty$,*
- *$P_f(G,x)$ is a continuous function on $[0,\infty)$ if and only if $f(d_u) \leqslant 1$ for every $u \in V(G)$,*
- *$P_f(G,x)$ is an integrable function on $[0,A]$ for every $A > 0$, and $\int_0^1 P_f(G,x)\,dx = I_f(G)$,*
- *$P_f(G,x)$ is increasing on $(0,\infty)$ if and only if $f(d_u) \leqslant 1$ for every $u \in V(G)$,*
- *$P_f(G,x)$ is strictly increasing on $(0,\infty)$ if and only if $f(d_u) \leqslant 1$ for every $u \in V(G)$, and $f(d_v) \neq 1$ for some $v \in V(G)$,*
- *$P_f(G,x)$ is convex on $(0,\infty)$ if $f(d_u) \in (0,1/2] \cup [1,\infty)$ for every $u \in V(G)$,*
- *$P_f(G,x)$ is strictly convex on $(0,\infty)$ if $f(d_u) \in (0,1/2] \cup [1,\infty)$ for every $u \in V(G)$, and $f(d_v) \notin \{1/2,1\}$ for some $v \in V(G)$,*
- *$P_f(G,x)$ is concave on $(0,\infty)$ if $f(d_u) \in [1/2,1]$ for every $u \in V(G)$,*
- *$P_f(G,x)$ is strictly concave on $(0,\infty)$ if $f(d_u) \in [1/2,1]$ for every $u \in V(G)$, and $f(d_v) \notin \{1/2,1\}$ for some $v \in V(G)$,*
- *$P_f(G,1) = n$.*

Proof. The first statement is direct, since $P_f(G,x)$ is a polynomial if and only if $1/f(d_u) - 1 \in \mathbb{Z}$ for every $u \in V(G)$.

The second and third statements are direct.

The fourth statement holds since $P_f(G,x)$ is a continuous function on $[0,\infty)$ if and only if $1/f(d_u) - 1 \geqslant 0$ for every $u \in V(G)$.

Since $f > 0$, $1/f(d_u) - 1 > -1$ for every $u \in V(G)$, and $P_f(G,x)$ is an integrable function on $[0,A]$ for every $A > 0$. Thus, a simple computation gives $\int_0^1 P_f(G,x)\,dx = I_f(G)$.

If there exists $u \in V(G)$ with $f(d_u) > 1$, then $1/f(d_u) - 1 < 0$ and $\lim_{x\to 0^+} x^{1/f(d_u)-1} = \infty$; thus, $\lim_{x\to 0^+} P_f(G,x) = \infty$, and $P_f(G,x)$ is not increasing on $(0,\infty)$. If $f(d_u) \leqslant 1$ for every $u \in V(G)$, then $1/f(d_u) - 1 \geqslant 0$ for every $u \in V(G)$, and so $P_f(G,x)$ is increasing on $(0,\infty)$. If this is the case, $P_f(G,x)$

is strictly increasing on $(0,\infty)$ if $1/f(d_v)-1>0$ for some $v\in V(G)$, i.e., $f(d_v)\neq 1$ for some $v\in V(G)$; if $f(d_u)=1$ for every $u\in V(G)$, then $P_f(G,x)$ is constant, and so it is not strictly increasing on $(0,\infty)$.

If $f(d_u)\in(0,1/2]\cup[1,\infty)$ for every $u\in V(G)$, then $1/f(d_u)-1\in(-1,0]\cup[1,\infty)$ for every $u\in V(G)$, and $P_f(G,x)$ is convex on $(0,\infty)$. If this is the case, $P_f(G,x)$ is strictly convex on $(0,\infty)$ if $1/f(d_v)-1\in(-1,0)\cup(1,\infty)$ for some $v\in V(G)$, i.e., $f(d_v)\notin\{1/2,1\}$ for some $v\in V(G)$.

If $f(d_u)\in[1/2,1]$ for every $u\in V(G)$, then $1/f(d_u)-1\in[0,1]$ for every $u\in V(G)$, and $P_f(G,x)$ is concave on $(0,\infty)$. If this is the case, $P_f(G,x)$ is strictly concave on $(0,\infty)$ if $1/f(d_v)-1\in(0,1)$ for some $v\in V(G)$, i.e., $f(d_v)\notin\{1/2,1\}$ for some $v\in V(G)$.

Finally, $P_f(G,1)=\sum_{u\in V(G)}1=n$. □

In particular, if $\alpha\in\mathbb{R}$ and $f(t)=t^\alpha$, then Proposition 1 gives $\int_0^1 P_f(G,x)\,dx=M_1^\alpha(G)$.

In particular, we have the following properties.

Proposition 2. *If G is a k-regular graph with order n and $f:\mathbb{Z}^+\to\mathbb{R}^+$, then $P_f(G,x)=nx^{1/f(k)-1}$.*

The following result shows the polynomial P_f for well-known graphs, such as: K_n (complete graph), C_n (cycle graph), Q_n (hypercube graph), $K_{n_1;n_2}$ (complete bipartite graph), S_n (star), P_n (path graph), and W_n (wheel graph).

Proposition 3. *If $f:\mathbb{Z}^+\to\mathbb{R}^+$, then*

$$P_f(K_n,x)=nx^{1/f(n-1)-1},\qquad P_f(C_n,x)=nx^{1/f(2)-1},$$

$$P_f(Q_n,x)=2^n x^{1/f(n)-1},\qquad P_f(K_{n_1,n_2},x)=n_1x^{1/f(n_2)-1}+n_2x^{1/f(n_1)-1},$$

$$P_f(S_n,x)=x^{1/f(n-1)-1}+(n-1)x^{1/f(1)-1},\qquad P_f(P_n,x)=(n-2)x^{1/f(2)-1}+2x^{1/f(1)-1},$$

$$P_f(W_n,x)=x^{1/f(n-1)-1}+(n-1)x^{1/f(3)-1}.$$

Fix $\delta\in\mathbb{Z}^+$ and $f:\mathbb{Z}^+\to\mathbb{R}^+$. We say that f satisfies the δ-additive property 1 (and we write $f\in AP_1(\delta)$) if

$$\frac{1}{f(x+y)}\geqslant\frac{1}{f(x)}+\frac{1}{f(y)}$$

for every $x,y\in\mathbb{Z}^+$ with $x,y\geqslant\delta$.

f satisfies the δ-additive property 2 (and we write $f\in AP_2(\delta)$) if

$$\frac{1}{f(x+y)}\leqslant\frac{1}{f(x)}+\frac{1}{f(y)}$$

for every $x,y\in\mathbb{Z}^+$ with $x,y\geqslant\delta$.

Finally, f satisfies the δ-additive property 3 (and we write $f\in AP_3(\delta)$) if

$$\frac{1}{f(x+y)}\leqslant\min\left\{\frac{1}{f(x)},\frac{1}{f(y)}\right\}$$

for every $x,y\in\mathbb{Z}^+$ with $x,y\geqslant\delta$.

Remark 1. *Note that if $f\in AP_j(\delta)$ for some $1\leqslant j\leqslant 3$, then $f\in AP_j(\delta')$ for every $\delta'\geqslant\delta$.*

If f is an increasing function on $[\delta,\infty)$, then $f\in AP_3(\delta)$.

Note that $f(t)=1/t$ satisfies the 1-additive properties 1 and 2, i.e., $f\in AP_1(1)\cap AP_2(1)$.

The following result appears in reference [31].

Theorem 1. *Let $\alpha\in\mathbb{R}$ and $f(t)=t^\alpha$.*

(1) *If $\alpha\leqslant -1$, then $f\in AP_1(1)$.*

(2) *If* $\alpha \in [-1, 0]$, *then* $f \in AP_2(1)$.
(3) *If* $\alpha \geqslant 0$, *then* $f \in AP_3(1)$.

Next, we present two useful improvements (for convex functions) of Chebyshev's inequality.

Lemma 1 ([34]). *If* $f_1, \ldots, f_k$ *are non-negative convex functions on* $[a, b]$, *then*

$$\frac{1}{b-a}\int_a^b \prod_{i=1}^k f_i(x)\,dx \geqslant \frac{2^k}{k+1}\prod_{i=1}^k \frac{1}{b-a}\int_a^b f_i(x)\,dx.$$

Lemma 2 ([35], Corollary 5.2). *If* $f_1, \ldots, f_k$ *are non-negative convex functions on* $[a, b]$, *then*

$$\int_a^b \prod_{i=1}^k f_i(x)\,dx \leqslant \frac{2}{k+1}\left(\prod_{i=1}^k \int_a^b f_i(x)\,dx\right)^{1/k}\left(\prod_{i=1}^k \big(f_i(a)+f_i(b)\big)\right)^{1-1/k}.$$

3. Join of Graphs

Theorem 2. *Let* $\delta \in \mathbb{Z}^+$ *and* G_1 *and* G_2 *be two graphs with order* n_1 *and* n_2, *respectively, and minimum degree of at least* δ, *and* $f : \mathbb{Z}^+ \to \mathbb{R}^+$. *The* f*-polynomial of the join* $G_1 + G_2$ *satisfies the following inequalities for* $x \in (0, 1]$.

(1) *If* $f \in AP_1(\delta)$, *then*

$$P_f(G_1+G_2,x) \leqslant x^{1/f(n_2)}P_f(G_1,x) + x^{1/f(n_1)}P_f(G_2,x).$$

(2) *If* $f \in AP_2(\delta)$, *then*

$$P_f(G_1+G_2,x) \geqslant x^{1/f(n_2)}P_f(G_1,x) + x^{1/f(n_1)}P_f(G_2,x).$$

(3) *If* $f \in AP_3(\delta)$, *then*

$$P_f(G_1+G_2,x) \geqslant P_f(G_1,x) + P_f(G_2,x).$$

Proof. If $u \in V(G_1)$ (respectively, $u \in V(G_2)$), then its degree in $G_1 + G_2$ is $d_u + n_2$ (respectively, $d_u + n_1$).

Assume first that $f \in AP_1(\delta)$. Since $d_u \geqslant \delta$ for every $u \in V(G_1) \cup V(G_2)$, $f \in AP_1(\delta)$, and $x \in (0, 1]$,

$$\begin{aligned} P_f(G_1+G_2,x) &= \sum_{u\in V(G_1)} x^{1/f(d_u+n_2)-1} + \sum_{v\in V(G_2)} x^{1/f(d_v+n_1)-1} \\ &\leqslant \sum_{u\in V(G_1)} x^{1/f(d_u)-1}x^{1/f(n_2)} + \sum_{v\in V(G_2)} x^{1/f(d_v)-1}x^{1/f(n_1)} \\ &= x^{1/f(n_2)}P_f(G_1,x) + x^{1/f(n_1)}P_f(G_2,x). \end{aligned}$$

If $f \in AP_2(\delta)$, then a similar argument allows us to obtain the corresponding inequality.

Assume now that $f \in AP_3(\delta)$. We have

$$\begin{aligned} P_f(G_1+G_2,x) &= \sum_{u\in V(G_1)} x^{1/f(d_u+n_2)-1} + \sum_{v\in V(G_2)} x^{1/f(d_v+n_1)-1} \\ &\geqslant \sum_{u\in V(G_1)} x^{1/f(d_u)-1} + \sum_{v\in V(G_2)} x^{1/f(d_v)-1} = P_f(G_1,x) + P_f(G_2,x). \end{aligned}$$

□

Theorems 1 and 2 have the following consequence.

Theorem 3. *Let G_1 and G_2 be two graphs with order n_1 and n_2, respectively, $\alpha \in \mathbb{R}$, and $f(t) = t^\alpha$. The f-polynomial of the join $G_1 + G_2$ satisfies the following inequalities for $x \in (0,1]$.*

(1) *If $\alpha \leqslant -1$, then*

$$P_f(G_1 + G_2, x) \leqslant x^{n_2^{-\alpha}} P_f(G_1, x) + x^{n_1^{-\alpha}} P_f(G_2, x).$$

(2) *If $\alpha \in [-1, 0]$, then*

$$P_f(G_1 + G_2, x) \geqslant x^{n_2^{-\alpha}} P_f(G_1, x) + x^{n_1^{-\alpha}} P_f(G_2, x).$$

(3) *If $\alpha \geqslant 0$, then*

$$P_f(G_1 + G_2, x) \geqslant P_f(G_1, x) + P_f(G_2, x).$$

Theorem 3 has the following consequence.

Corollary 1. *Given two graphs G_1 and G_2, with order n_1 and n_2, respectively, the ID polynomial of the join $G_1 + G_2$ is*

$$ID(G_1 + G_2, x) = x^{n_2} ID(G_1, x) + x^{n_1} ID(G_2, x).$$

Since $f(t) = t\sqrt{\log t} \in AP_3(2)$, Theorem 2 has the following consequence.

Corollary 2. *Let G_1 and G_2 be two graphs without pendant vertices and with order n_1 and n_2, respectively. If $f(t) = t\sqrt{\log t}$, then the f-polynomial of the join $G_1 + G_2$ satisfies for $x \in (0,1]$*

$$P_f(G_1 + G_2, x) \geqslant P_f(G_1, x) + P_f(G_2, x).$$

Next, we obtain bounds for $I_f(G_1 + G_2)$ by using the previous inequalities for $P_f(G_1 + G_2, x)$.

Proposition 4. *Let $\delta \in \mathbb{Z}^+$ and let G_1 and G_2 be two graphs with order n_1 and n_2, respectively, and a minimum degree of at least δ. If $f \in AP_3(\delta)$, then*

$$I_f(G_1 + G_2) \geqslant I_f(G_1) + I_f(G_2).$$

Proof. Theorem 2 gives

$$P_f(G_1 + G_2, x) \geqslant P_f(G_1, x) + P_f(G_2, x)$$

for every $x \in (0,1]$. Thus, Proposition 1 gives

$$I_f(G_1 + G_2) = \int_0^1 P_f(G_1 + G_2, x)\, dx \geqslant \int_0^1 P_f(G_1, x)\, dx + \int_0^1 P_f(G_2, x)\, dx = I_f(G_1) + I_f(G_2).$$

□

Theorem 4. *Let $\delta \in \mathbb{Z}^+$, and let G_1 and G_2 be two graphs with order n_1 and n_2, respectively, and a minimum degree of at least δ, and $a > 0$. If $f : \mathbb{Z}^+ \cap [\delta, \infty) \to (0, a/2]$, then the following inequalities hold.*

(1) *If $f \in AP_1(\delta)$, then*

$$I_f(G_1 + G_2) \leqslant \frac{2}{3}\Big(\frac{a\, n_1 f(n_2)}{a + f(n_2)}\, I_f(G_1)\Big)^{1/2} + \frac{2}{3}\Big(\frac{a\, n_2 f(n_1)}{a + f(n_1)}\, I_f(G_2)\Big)^{1/2}.$$

(2) *If $f \in AP_2(\delta)$, then*

$$I_f(G_1 + G_2) \geqslant \frac{4}{3}\Big(\frac{f(n_2)}{a + f(n_2)}\, I_f(G_1) + \frac{f(n_1)}{a + f(n_1)}\, I_f(G_2)\Big).$$

Proof. Let us define the function $g = f/a$. Thus, $g : \mathbb{Z}^+ \cap [\delta, \infty) \to (0, 1/2]$ and Proposition 1 gives that $P_g(G_1, x)$ and $P_g(G_2, x)$ are convex functions on $(0, \infty)$ and continuous on $[0, \infty)$; hence, they are convex on $[0, 1]$.

If $f \in AP_1(\delta)$, then $f/a \in AP_1(\delta)$ and Theorem 2 gives

$$P_{f/a}(G_1 + G_2, x) \leqslant x^{a/f(n_2)} P_{f/a}(G_1, x) + x^{a/f(n_1)} P_{f/a}(G_2, x).$$

Note that $f/a \leqslant 1/2$ gives $a/f - 1 \geqslant 1$, and so $P_{f/a}(G_i, 0) = 0$ and $P_{f/a}(G_i, 1) = n_i$ for $i = 1, 2$. Since $a/f(n_2) \geqslant 2$, we have that $x^{a/f(n_2)}$ is also a convex function on $[0, 1]$, and Lemma 2 gives

$$\begin{aligned}\int_0^1 x^{a/f(n_2)} P_{f/a}(G_1, x)\, dx &\leqslant \frac{2}{3}\Big(\int_0^1 x^{a/f(n_2)} dx \int_0^1 P_{f/a}(G_1, x)\, dx\Big)^{1/2} \\ &\quad \cdot \Big((0+1)\big(P_{f/a}(G_1, 0) + P_{f/a}(G_1, 1)\big)\Big)^{1/2} \\ &= \frac{2}{3}\Big(n_1 \frac{f(n_2)}{a + f(n_2)}\, I_{f/a}(G_1)\Big)^{1/2} \\ &= \frac{2}{3}\Big(\frac{n_1 f(n_2)}{a + f(n_2)}\, \frac{1}{a}\, I_f(G_1)\Big)^{1/2}.\end{aligned}$$

We obtain in a similar way

$$\int_0^1 x^{a/f(n_1)} P_{f/a}(G_2, x)\, dx \leqslant \frac{2}{3}\Big(\frac{n_2 f(n_1)}{a + f(n_1)}\, \frac{1}{a}\, I_f(G_2)\Big)^{1/2}.$$

Hence,

$$\begin{aligned}\frac{1}{a}\, I_f(G_1 + G_2) = I_{f/a}(G_1 + G_2) &= \int_0^1 P_{f/a}(G_1 + G_2, x)\, dx \\ &\leqslant \frac{2}{3}\Big(\frac{n_1 f(n_2)}{a + f(n_2)}\, \frac{1}{a}\, I_f(G_1)\Big)^{1/2} + \frac{2}{3}\Big(\frac{n_2 f(n_1)}{a + f(n_1)}\, \frac{1}{a}\, I_f(G_2)\Big)^{1/2}.\end{aligned}$$

If $f \in AP_2(\delta)$, then $f/a \in AP_2(\delta)$ and Theorem 2 gives

$$P_{f/a}(G_1 + G_2, x) \geqslant x^{a/f(n_2)} P_{f/a}(G_1, x) + x^{a/f(n_1)} P_{f/a}(G_2, x).$$

Thus, Lemma 1 gives

$$\begin{aligned}\frac{1}{a}\, I_f(G_1 + G_2) = I_{f/a}(G_1 + G_2) &= \int_0^1 P_{f/a}(G_1 + G_2, x)\, dx \\ &\geqslant \frac{4}{3}\int_0^1 x^{a/f(n_2)} dx \int_0^1 P_{f/a}(G_1, x)\, dx + \frac{4}{3}\int_0^1 x^{a/f(n_1)} dx \int_0^1 P_{f/a}(G_2, x)\, dx \\ &= \frac{4}{3}\, \frac{f(n_2)}{a + f(n_2)}\, \frac{1}{a}\, I_f(G_1) + \frac{4}{3}\, \frac{f(n_1)}{a + f(n_1)}\, \frac{1}{a}\, I_f(G_2).\end{aligned}$$

□

Corollary 3. *Let $\delta \in \mathbb{Z}^+$, and let G_1 and G_2 be two graphs with order n_1 and n_2, respectively, and a minimum degree of at least δ, and $a > 0$. If $f : \mathbb{Z}^+ \cap [\delta, \infty) \to (0, a/2]$ and $f \in AP_1(\delta)$, then*

$$I_f(G_1 + G_2) \leqslant \frac{2}{3}\Big(\frac{n_1 a}{3}\, I_f(G_1)\Big)^{1/2} + \frac{2}{3}\Big(\frac{n_2 a}{3}\, I_f(G_2)\Big)^{1/2}.$$

Proof. Since $F(t) = at/(a+t)$ is an increasing function on $t \in [0,\infty)$ and $f \leqslant a/2$,

$$n_i \frac{af(n_j)}{a+f(n_j)} \leqslant n_i \frac{a\,\frac{a}{2}}{a+\frac{a}{2}} = \frac{n_i a}{3},$$

and Theorem 4 gives the desired inequality. □

Next, we obtain inequalities for several topological indices of joins of graphs. Theorems 1 and 4 (with $a = 2$) and Proposition 4 have the following consequence.

Corollary 4. *Let G_1 and G_2 be two graphs with order n_1 and n_2, respectively, and $\alpha \in \mathbb{R}$.*

(1) *If $\alpha \leqslant -1$, then*

$$M_1^{\alpha}(G_1+G_2) \leqslant \frac{2}{3}\Big(\frac{2n_1 n_2^{\alpha}}{2+n_2^{\alpha}}\, M_1^{\alpha}(G_1)\Big)^{1/2} + \frac{2}{3}\Big(\frac{2n_2 n_1^{\alpha}}{2+n_1^{\alpha}}\, M_1^{\alpha}(G_2)\Big)^{1/2}.$$

(2) *If $\alpha \in [-1,0]$, then*

$$M_1^{\alpha}(G_1+G_2) \geqslant \frac{4}{3}\Big(\frac{n_2^{\alpha}}{2+n_2^{\alpha}}\, M_1^{\alpha}(G_1) + \frac{n_1^{\alpha}}{2+n_1^{\alpha}}\, M_1^{\alpha}(G_2)\Big).$$

(3) *If $\alpha \geqslant 0$, then*

$$M_1^{\alpha}(G_1+G_2) \geqslant M_1^{\alpha}(G_1) + M_1^{\alpha}(G_2).$$

Corollary 4 gives the following result.

Corollary 5. *If G_1 and G_2 are two graphs with order n_1 and n_2, respectively, then*

$$\begin{aligned}\frac{4}{3}\Big(\frac{1}{2n_2+1}\, ID(G_1) + \frac{1}{2n_1+1}\, ID(G_2)\Big) &\leqslant ID(G_1+G_2)\\ &\leqslant \frac{2}{3}\Big(\frac{2n_1}{2n_2+1}\, ID(G_1)\Big)^{1/2} + \frac{2}{3}\Big(\frac{2n_2}{2n_1+1}\, ID(G_2)\Big)^{1/2}.\end{aligned}$$

The sum lordeg index

$$SL(G) = \sum_{u \in V(G)} d_u \sqrt{\log d_u}$$

is one of the Adriatic indices introduced in reference [36].

Since $f(t) = t\sqrt{\log t}$ is an increasing function on $[1,\infty)$, $f \in AP_3(2)$. Thus, we have the following:

Lemma 3. *If $f(t) = t\sqrt{\log t}$, then $f \in AP_3(2)$.*

Lemma 3 and Proposition 4 have the following consequence.

Corollary 6. *If G_1 and G_2 are graphs without pendant vertices and with order n_1 and n_2, respectively, then*

$$SL(G_1+G_2) \geqslant SL(G_1) + SL(G_2).$$

4. Corona Products

Theorem 5. *Let G_1 and G_2 be two graphs with order n_1 and n_2, respectively, and $f : \mathbb{Z}^+ \to \mathbb{R}^+$. The f-polynomial of the corona product $G_1 \circ G_2$ satisfies the following inequalities for $x \in (0,1]$.*

(1) *If $f \in AP_1(1)$, then*

$$P_f(G_1 \circ G_2, x) \leqslant x^{1/f(n_2)} P_f(G_1, x) + n_1 x^{1/f(1)} P_f(G_2, x).$$

(2) *If $f \in AP_2(1)$, then*

$$P_f(G_1 \circ G_2, x) \geqslant x^{1/f(n_2)} P_f(G_1, x) + n_1 x^{1/f(1)} P_f(G_2, x).$$

(3) *If $\delta \in \mathbb{Z}^+$, G_1 and G_2 have a minimum degree of at least δ, and f is increasing on $\mathbb{Z}^+ \cap [\delta, \infty)$, then*

$$P_f(G_1 \circ G_2, x) \geqslant P_f(G_1, x) + n_1 P_f(G_2, x).$$

Proof. If $u \in V(G_1)$, then its degree in $G_1 \circ G_2$ is $d_u + n_2$. If $G_2^{(i)}$ is a copy of G_2 in $G_1 \circ G_2$ and $v \in V(G_2^{(i)})$, then its degree in $G_1 \circ G_2$ is $d_v + 1$.

Assume first that $f \in AP_1(1)$. Since $f \in AP_1(1)$ and $x \in (0,1]$,

$$\begin{aligned} P_f(G_1 \circ G_2, x) &= \sum_{u \in V(G_1)} x^{1/f(d_u+n_2)-1} + n_1 \sum_{v \in V(G_2)} x^{1/f(d_v+1)-1} \\ &\leqslant \sum_{u \in V(G_1)} x^{1/f(d_u)-1} x^{1/f(n_2)} + n_1 \sum_{v \in V(G_2)} x^{1/f(d_v)-1} x^{1/f(1)} \\ &= x^{1/f(n_2)} P_f(G_1, x) + n_1 x^{1/f(1)} P_f(G_2, x). \end{aligned}$$

If $f \in AP_2(1)$, then a similar argument allows us to obtain the corresponding inequality.

Finally, assume that G_1 and G_2 have a minimum degree of at least δ, and f is increasing on $\mathbb{Z}^+ \cap [\delta, \infty)$. We have

$$\begin{aligned} P_f(G_1 \circ G_2, x) &= \sum_{u \in V(G_1)} x^{1/f(d_u+n_2)-1} + n_1 \sum_{v \in V(G_2)} x^{1/f(d_v+1)-1} \\ &\geqslant \sum_{u \in V(G_1)} x^{1/f(d_u)-1} + n_1 \sum_{v \in V(G_2)} x^{1/f(d_v)-1} = P_f(G_1, x) + n_1 P_f(G_2, x). \end{aligned}$$

□

Theorems 1 and 5 have the following consequence.

Corollary 7. *Let G_1 and G_2 be two graphs with order n_1 and n_2, respectively, and let $\alpha \in \mathbb{R}$ and $f(t) = t^\alpha$. The f-polynomial of the corona product $G_1 \circ G_2$ satisfies the following inequalities for $x \in (0,1]$.*

(1) *If $\alpha \leqslant -1$, then*

$$P_f(G_1 \circ G_2, x) \leqslant x^{n_2^{-\alpha}} P_f(G_1, x) + n_1 x P_f(G_2, x).$$

(2) *If $\alpha \in [-1, 0]$, then*

$$P_f(G_1 \circ G_2, x) \geqslant x^{n_2^{-\alpha}} P_f(G_1, x) + n_1 x P_f(G_2, x).$$

(3) *If $\alpha \geqslant 0$, then*

$$P_f(G_1 \circ G_2, x) \geqslant P_f(G_1, x) + n_1 P_f(G_2, x).$$

Corollary 7 has the following consequence.

Corollary 8. *Given two graphs G_1 and G_2 with order n_1 and n_2, respectively, the ID polynomial of the corona product $G_1 \circ G_2$ is*

$$ID(G_1 \circ G_2, x) = x^{n_2} ID(G_1, x) + n_1 x ID(G_2, x).$$

Theorem 5 has the following consequence.

Corollary 9. *Let G_1 and G_2 be two graphs without pendant vertices and with order n_1 and n_2, respectively. If $f(t) = t\sqrt{\log t}$, then the f-polynomial of the corona product $G_1 \circ G_2$ satisfies for $x \in (0,1]$*

$$P_f(G_1 \circ G_2, x) \geqslant P_f(G_1, x) + n_1 P_f(G_2, x).$$

Next, we obtain bounds for $I_f(G_1 \circ G_2)$ by using the previous inequalities for $P_f(G_1 \circ G_2, x)$.

Proposition 5. *Let $\delta \in \mathbb{Z}^+$, and let G_1 and G_2 be two graphs with order n_1 and n_2, respectively, and a minimum degree of at least δ. If f is increasing on $\mathbb{Z}^+ \cap [\delta, \infty)$, then*

$$I_f(G_1 \circ G_2) \geqslant I_f(G_1) + n_1 I_f(G_2).$$

Proof. Theorem 5 gives

$$P_f(G_1 \circ G_2, x) \geqslant P_f(G_1, x) + n_1 P_f(G_2, x)$$

for every $x \in (0,1]$. Thus, Proposition 1 gives the desired inequality. □

Theorem 6. *Let G_1 and G_2 be two graphs with order n_1 and n_2, respectively, and $a > 0$. If $f : \mathbb{Z}^+ \to (0, a/2]$, then the following inequalities hold.*

(1) *If $f \in AP_1(1)$, then*

$$I_f(G_1 \circ G_2) \leqslant \frac{2}{3}\Big(\frac{a n_1 f(n_2)}{a + f(n_2)}\, I_f(G_1)\Big)^{1/2} + \frac{2n_1}{3}\Big(\frac{a n_2 f(1)}{a + f(1)}\, I_f(G_2)\Big)^{1/2}.$$

(2) *If $f \in AP_2(1)$, then*

$$I_f(G_1 \circ G_2) \geqslant \frac{4}{3}\Big(\frac{f(n_2)}{a + f(n_2)}\, I_f(G_1) + \frac{n_1 f(1)}{a + f(1)}\, I_f(G_2)\Big).$$

Proof. Let us define the function $g = f/a$. Thus, $g : \mathbb{Z}^+ \to (0, 1/2]$ and Proposition 1 gives that $P_g(G_1, x)$ and $P_g(G_2, x)$ are convex functions on $(0,\infty)$ and continuous on $[0,\infty)$; hence, they are convex on $[0,1]$.

If $f \in AP_1(1)$, then $f/a \in AP_1(1)$ and Theorem 5 gives

$$P_{f/a}(G_1 \circ G_2, x) \leqslant x^{a/f(n_2)} P_{f/a}(G_1, x) + n_1 x^{a/f(1)} P_{f/a}(G_2, x).$$

Note that $f/a \leqslant 1/2$ gives $a/f - 1 \geqslant 1$, and so $P_{f/a}(G_i, 0) = 0$ and $P_{f/a}(G_i, 1) = n_i$ for $i = 1, 2$. Since $a/f(n_2) \geqslant 2$, we have that $x^{a/f(n_2)}$ is also a convex function on $[0,1]$, and Lemma 2 gives

$$\begin{aligned}\int_0^1 x^{a/f(n_2)} P_{f/a}(G_1, x)\, dx &\leqslant \frac{2}{3}\Big(\int_0^1 x^{a/f(n_2)} dx \int_0^1 P_{f/a}(G_1, x)\, dx\Big)^{1/2} \\ &\quad \cdot \Big((0+1)\big(P_{f/a}(G_1, 0) + P_{f/a}(G_1, 1)\big)\Big)^{1/2} \\ &= \frac{2}{3}\Big(n_1 \frac{f(n_2)}{a + f(n_2)} \frac{1}{a} I_f(G_1)\Big)^{1/2}.\end{aligned}$$

We obtain, in a similar way,

$$n_1 \int_0^1 x^{a/f(1)} P_{f/a}(G_2, x)\, dx \leqslant \frac{2n_1}{3}\Big(\frac{n_2 f(1)}{a + f(1)} \frac{1}{a} I_f(G_2)\Big)^{1/2}.$$

Hence,

$$\frac{1}{a}\, I_f(G_1 \circ G_2) = I_{f/a}(G_1 \circ G_2) = \int_0^1 P_{f/a}(G_1 \circ G_2, x)\, dx$$
$$\leqslant \frac{2}{3}\Big(\frac{n_1 f(n_2)}{a + f(n_2)}\, \frac{1}{a}\, I_f(G_1) \Big)^{1/2} + \frac{2n_1}{3}\Big(\frac{n_2 f(1)}{a + f(1)}\, \frac{1}{a}\, I_f(G_2) \Big)^{1/2}.$$

If $f \in AP_2(\delta)$, then $f/a \in AP_2(\delta)$ and Theorem 5 gives

$$P_{f/a}(G_1 \circ G_2, x) \geqslant x^{a/f(n_2)} P_{f/a}(G_1, x) + n_1 x^{a/f(1)} P_{f/a}(G_2, x).$$

Thus, Lemma 1 gives

$$\frac{1}{a}\, I_f(G_1 \circ G_2) = I_{f/a}(G_1 \circ G_2) = \int_0^1 P_{f/a}(G_1 \circ G_2, x)\, dx$$
$$\geqslant \frac{4}{3}\int_0^1 x^{a/f(n_2)} dx \int_0^1 P_{f/a}(G_1, x)\, dx + \frac{4n_1}{3}\int_0^1 x^{a/f(1)} dx \int_0^1 P_{f/a}(G_2, x)\, dx$$
$$= \frac{4}{3}\, \frac{f(n_2)}{a + f(n_2)}\, \frac{1}{a}\, I_f(G_1) + \frac{4}{3}\, \frac{n_1 f(1)}{a + f(1)}\, \frac{1}{a}\, I_f(G_2).$$

□

Next, we obtain inequalities for several topological indices of the corona product of graphs. Theorems 1 and 6 (with $a = 2$) and Proposition 5 have the following consequence.

Corollary 10. *Let G_1 and G_2 be two graphs with order n_1 and n_2, respectively, and $\alpha \in \mathbb{R}$.*

(1) *If $\alpha \leqslant -1$, then*

$$M_1^\alpha(G_1 \circ G_2) \leqslant \frac{2}{3}\Big(\frac{2n_1 n_2^\alpha}{2 + n_2^\alpha}\, M_1^\alpha(G_1) \Big)^{1/2} + \frac{2n_1}{3}\Big(\frac{2n_2}{3}\, M_1^\alpha(G_2) \Big)^{1/2}.$$

(2) *If $\alpha \in [-1, 0]$, then*

$$M_1^\alpha(G_1 \circ G_2) \geqslant \frac{4}{3}\Big(\frac{n_2^\alpha}{2 + n_2^\alpha}\, M_1^\alpha(G_1) + \frac{n_1}{3}\, M_1^\alpha(G_2) \Big).$$

(3) *If $\alpha \geqslant 0$, then*

$$M_1^\alpha(G_1 \circ G_2) \geqslant M_1^\alpha(G_1) + n_1 M_1^\alpha(G_2).$$

Corollary 10 gives the following result.

Corollary 11. *If G_1 and G_2 are two graphs with order n_1 and n_2, respectively, then*

$$\frac{4}{3}\Big(\frac{1}{2n_2 + 1}\, ID(G_1) + \frac{n_1}{3}\, ID(G_2) \Big) \leqslant ID(G_1 \circ G_2)$$
$$\leqslant \frac{2}{3}\Big(\frac{2n_1}{2n_2 + 1}\, ID(G_1) \Big)^{1/2} + \frac{2n_1}{3}\Big(\frac{2n_2}{3}\, ID(G_2) \Big)^{1/2}.$$

Proposition 5 has the following consequence.

Corollary 12. *If G_1 and G_2 are graphs without pendant vertices and with order n_1 and n_2, respectively, then*

$$SL(G_1 \circ G_2) \geqslant SL(G_1) + n_1 SL(G_2).$$

5. Mycielskian Graphs

Given a graph G with $V(G) = \{v_1, \ldots, v_n\}$, its Mycielskian graph $\mu(G)$ contains G itself as a subgraph, together with $n+1$ additional vertices $\{u_1, \ldots, u_n, w\}$. Each vertex u_i is connected by an edge to w. In addition, for each edge v_iv_j of G, the Mycielskian graph includes two edges, u_iv_j and v_iu_j. Thus, if G has order n and size m, then $\mu(G)$ has $2n+1$ vertices and $3m+n$ edges. In addition,

$$d_{v_i,\mu(G)} = 2\,d_{v_i}, \qquad d_{u_i,\mu(G)} = d_{v_i} + 1, \qquad d_{w,\mu(G)} = n,$$

for each $i = 1, \ldots, n$, where d_v and $d_{v,\mu(G)}$ denote the degree of the vertex v in G and $\mu(G)$, respectively.

Mycielskian graphs are a construction for embedding any graph into a larger graph with a higher chromatic number while avoiding the creation of additional triangles (see [37]). Mycielskian graphs have been used also in mathematical chemistry; see, e.g., [38–41].

Theorem 7. *Let G be a graph with order n, $a > 0$, $\alpha \in \mathbb{R}$, $f(t) = t^\alpha/a$, and $x \in (0,1]$.*

(1) *If $\alpha \leqslant -1$, then*

$$P_{f/a}(\mu(G), x) \leqslant x^{a2^{-\alpha}-1}\, P_f(G, x^{a2^{-\alpha}}) + x^a P_{f/a}(G, x) + x^{an^{-\alpha}-1}.$$

(2) *If $\alpha \in [-1, 0]$, then*

$$P_{f/a}(\mu(G), x) \geqslant x^{a2^{-\alpha}-1}\, P_f(G, x^{a2^{-\alpha}}) + x^a P_{f/a}(G, x) + x^{an^{-\alpha}-1}.$$

(3) *If $\alpha \geqslant 0$, then*

$$P_{f/a}(\mu(G), x) \geqslant x^{a2^{-\alpha}-1}\, P_f(G, x^{a2^{-\alpha}}) + P_{f/a}(G, x) + x^{an^{-\alpha}-1}.$$

Proof. Assume first that $\alpha \leqslant -1$. Thus, $-\alpha \geqslant 1$ and $(d_{v_i}+1)^{-\alpha} \geqslant d_{v_i}^{-\alpha} + 1$. Since $x \in (0,1]$, we have

$$\begin{aligned}
P_{f/a}(\mu(G), x) &= \sum_{i=1}^{n} x^{a(d_{v_i,\mu(G)})^{-\alpha}-1} + \sum_{i=1}^{n} x^{a(d_{u_i,\mu(G)})^{-\alpha}-1} + x^{a(d_{w,\mu(G)})^{-\alpha}-1} \\
&= \sum_{i=1}^{n} x^{a2^{-\alpha}(d_{v_i}^{-\alpha}-1)+a2^{-\alpha}-1} + \sum_{i=1}^{n} x^{a(d_{v_i}+1)^{-\alpha}-1} + x^{an^{-\alpha}-1} \\
&\leqslant x^{a2^{-\alpha}-1} \sum_{i=1}^{n} \left(x^{a2^{-\alpha}}\right)^{d_{v_i}^{-\alpha}-1} + \sum_{i=1}^{n} x^{ad_{v_i}^{-\alpha}+a-1} + x^{an^{-\alpha}-1} \\
&= x^{a2^{-\alpha}-1}\, P_f(G, x^{a2^{-\alpha}}) + x^a P_{f/a}(G, x) + x^{an^{-\alpha}-1}.
\end{aligned}$$

If $\alpha \in [-1, 0]$, then $(d_{v_i}+1)^{-\alpha} \leqslant d_{v_i}^{-\alpha} + 1$, and so

$$\begin{aligned}
P_{f/a}(\mu(G), x) &= \sum_{i=1}^{n} x^{a2^{-\alpha}(d_{v_i}^{-\alpha}-1)+a2^{-\alpha}-1} + \sum_{i=1}^{n} x^{a(d_{v_i}+1)^{-\alpha}-1} + x^{an^{-\alpha}-1} \\
&\geqslant x^{a2^{-\alpha}-1} \sum_{i=1}^{n} \left(x^{a2^{-\alpha}}\right)^{d_{v_i}^{-\alpha}-1} + \sum_{i=1}^{n} x^{ad_{v_i}^{-\alpha}+a-1} + x^{an^{-\alpha}-1} \\
&= x^{a2^{-\alpha}-1}\, P_f(G, x^{a2^{-\alpha}}) + x^a P_{f/a}(G, x) + x^{an^{-\alpha}-1}.
\end{aligned}$$

If $\alpha \geqslant 0$, then $(d_{v_i}+1)^{-\alpha} \leqslant d_{v_i}^{-\alpha}$, and so

$$\begin{aligned} P_{f/a}(\mu(G),x) &= \sum_{i=1}^{n} x^{a2^{-\alpha}(d_{v_i}^{-\alpha}-1)+a2^{-\alpha}-1} + \sum_{i=1}^{n} x^{a(d_{v_i}+1)^{-\alpha}-1} + x^{an^{-\alpha}-1} \\ &\geqslant x^{a2^{-\alpha}-1} \sum_{i=1}^{n} \left(x^{a2^{-\alpha}}\right)^{d_{v_i}^{-\alpha}-1} + \sum_{i=1}^{n} x^{ad_{v_i}^{-\alpha}-1} + x^{an^{-\alpha}-1} \\ &= x^{a2^{-\alpha}-1} P_f(G, x^{a2^{-\alpha}}) + P_{f/a}(G,x) + x^{an^{-\alpha}-1}. \end{aligned}$$

□

Theorem 7 has the following consequences.

Corollary 13. *If G is a graph with order n, then*

$$ID(\mu(G),x) = x\,ID(G,x^2) + x\,ID(G,x) + x^{n-1}.$$

Next, we obtain bounds for $M_1^\alpha(\mu(G))$ by using the previous inequalities for $P_f(\mu(G),x)$.

Theorem 8. *Let G be a graph with order n and $\alpha \in \mathbb{R}$.*

(1) *If $\alpha \leqslant -1$, then*

$$M_1^\alpha(\mu(G)) \leqslant 2^\alpha M_1^\alpha(G) + \frac{2}{3}\Big(\frac{2}{3}\, n\, M_1^\alpha(G)\Big)^{1/2} + n^\alpha.$$

(2) *If $\alpha \in [-1,0]$, then*

$$M_1^\alpha(\mu(G)) \geqslant \Big(2^\alpha + \frac{4}{9}\Big) M_1^\alpha(G) + n^\alpha.$$

(3) *If $\alpha \geqslant 0$, then*

$$M_1^\alpha(\mu(G)) \geqslant (2^\alpha + 1) M_1^\alpha(G) + n^\alpha.$$

Proof. Let us consider the function $f(t) = t^\alpha$. Assume first that $\alpha \geqslant 0$. Theorem 7 with $a = 1$ gives

$$P_f(\mu(G),x) \geqslant x^{2^{-\alpha}-1} P_f(G, x^{2^{-\alpha}}) + P_f(G,x) + x^{n^{-\alpha}-1}$$

for $x \in (0,1]$. Thus,

$$\begin{aligned} M_1^\alpha(\mu(G)) &= \int_0^1 P_f(\mu(G),x)\,dx \\ &\geqslant \int_0^1 x^{2^{-\alpha}-1} P_f(G, x^{2^{-\alpha}})\,dx + \int_0^1 P_f(G,x)\,dx + \int_0^1 x^{n^{-\alpha}-1} dx \\ &= \int_0^1 \frac{1}{2^{-\alpha}} P_f(G,t)\,dt + M_1^\alpha(G) + \Big[\frac{x^{n^{-\alpha}}}{n^{-\alpha}}\Big]_0^1 \\ &= (2^\alpha+1) M_1^\alpha(G) + n^\alpha. \end{aligned}$$

Note that $f(t) \leqslant 1$ for every $t \in \mathbb{Z}^+$ if $\alpha \leqslant 0$. Thus, $2/f - 1 \geqslant 1$, and $P_{f/2}(\mu(G),x)$ is a convex function on $[0,1]$ with $P_{f/2}(\mu(G),0) = 0$.

Assume now that $\alpha \leqslant -1$. Theorem 7 with $a = 2$ gives

$$P_{f/2}(\mu(G),x) \leqslant x^{2^{1-\alpha}-1} P_f(G, x^{2^{1-\alpha}}) + x^2 P_{f/2}(G,x) + x^{2n^{-\alpha}-1}$$

for $x \in (0,1]$. Thus,

$$
\begin{aligned}
M_1^\alpha(\mu(G)) &= 2\, I_{f/2}(\mu(G)) = 2\int_0^1 P_{f/2}(\mu(G),x)\,dx \\
&\leqslant 2\int_0^1 x^{2^{1-\alpha}-1}\, P_f(G,x^{2^{1-\alpha}})\,dx + 2\int_0^1 x^2 P_{f/2}(G,x)\,dx + 2\int_0^1 x^{2n^{-\alpha}-1}dx \\
&= \int_0^1 \frac{1}{2^{-\alpha}}\, P_f(G,t)\,dt + 2\int_0^1 x^2 P_{f/2}(G,x)\,dx + \left[\frac{x^{2n^{-\alpha}}}{n^{-\alpha}}\right]_0^1 \\
&= 2^\alpha M_1^\alpha(G) + 2\int_0^1 x^2 P_{f/2}(G,x)\,dx + n^\alpha.
\end{aligned}
$$

Since x^2 and $P_{f/2}(\mu(G),x)$ are non-negative convex functions on $[0,1]$, Lemma 2 gives

$$
\begin{aligned}
\int_0^1 x^2 P_{f/2}(G,x)\,dx &\leqslant \frac{2}{3}\left(\int_0^1 x^2 dx \int_0^1 P_{f/2}(G,x)\,dx\right)^{1/2} \left(P_{f/2}(G,0)+P_{f/2}(G,1)\right)^{1/2} \\
&= \frac{2}{3}\left(\frac{1}{3}\cdot\frac{1}{2}\,M_1^\alpha(G)\,n\right)^{1/2} = \frac{1}{3}\left(\frac{2}{3}\,n\,M_1^\alpha(G)\right)^{1/2}.
\end{aligned}
$$

Finally, assume that $\alpha \in [-1,0]$. Theorem 7 with $a=2$ gives

$$
P_{f/2}(\mu(G),x) \geqslant x^{2^{1-\alpha}-1}\, P_f(G,x^{2^{1-\alpha}}) + x^2 P_{f/2}(G,x) + x^{2n^{-\alpha}-1}
$$

for $x \in (0,1]$. Hence, the previous argument gives

$$
M_1^\alpha(\mu(G)) \geqslant 2^\alpha M_1^\alpha(G) + 2\int_0^1 x^2 P_{f/2}(G,x)\,dx + n^\alpha.
$$

Since x^2 and $P_{f/2}(\mu(G),x)$ are non-negative and convex on $[0,1]$, Lemma 1 gives

$$
2\int_0^1 x^2 P_{f/2}(G,x)\,dx \geqslant \frac{8}{3}\int_0^1 x^2 dx \int_0^1 P_{f/2}(G,x)\,dx = \frac{4}{9}\,M_1^\alpha(G).
$$

□

Theorem 8 has the following consequence for the inverse degree index.

Corollary 14. *If G is a graph with order n, then*

$$
\frac{17}{18}\,ID(G) + \frac{1}{n} \leqslant ID(\mu(G)) \leqslant \frac{1}{2}\,ID(G) + \frac{2}{3}\left(\frac{2}{3}\,n\,ID(G)\right)^{1/2} + \frac{1}{n}.
$$

6. ID Polynomials of Other Graph Operations

Let us recall the definition of other graph operations.

Let G be a graph. Given an edge $e = uv$ of G, let $V(e) = \{u, v\}$. Now we can define the following five graph operations.

The line graph, denoted by $\mathcal{L}(G)$, is the graph whose vertices correspond to the edges of G, and two vertices are adjacent if and only if the corresponding edges in G share a vertex.

The subdivision graph, denoted by $S(G)$, is the graph obtained from G by replacing each of its edges by a path of length two, or equivalently, by inserting an additional vertex into each edge of G.

The total graph, denoted by $T(G)$, has as its vertices the edges and vertices of G. Adjacency in $T(G)$ is defined as adjacency or incidence for the corresponding elements of G.

The graph $R(G)$ is obtained from G by adding a new vertex corresponding to each edge of G and then joining each new vertex to the end vertices of the corresponding edge. Another way to describe $R(G)$ is to replace each edge of G by a triangle.

The graph $Q(G)$ is the graph obtained from G by inserting a new vertex into each edge of G and by joining edges the pairs of these new vertices that lie on adjacent edges of G.

Given a graph G, we may define the following sets:

$$EE(G) := \{\{e,e'\} : e,e' \in E(G), e \neq e', |V(e) \cap V(e')| = 1\},$$
$$EV(G) := \{\{e,v\} : e \in E(G), v \in V(e)\}.$$

We may then write these five graph operations as follows:

$$\mathcal{L}(G) := (E(G), EE(G)),$$
$$S(G) := (V(G) \cup E(G), EV(G)),$$
$$T(G) := (V(G) \cup E(G), E(G) \cup EV(G) \cup EE(G)),$$
$$R(G) := (V(G) \cup E(G), E(G) \cup EV(G)),$$
$$Q(G) := (V(G) \cup E(G), EV(G) \cup EE(G)).$$

The following result is elementary.

Proposition 6. *If G is a graph with m edges, then*

$$ID(S(G),x) = ID(G,x) + mx, \qquad ID(S(G)) = ID(G) + \frac{1}{2}m.$$

The computation of the ID polynomials of other graph operations involves harmonic polynomials.

As we have seen, the line graph $\mathcal{L}(G)$ of G is a graph that has a vertex $w_e \in V(\mathcal{L}(G))$ for each edge $e \in E(G)$, and an edge joining w_{e_i} and w_{e_j} when e_i and e_j share a vertex (i.e., $\mathcal{L}(G)$ is the intersection graph of $E(G)$). It is easy to check that if $uv \in E(G)$, then the degree of $w_{uv} \in V(\mathcal{L}(G))$ is $d_u + d_v - 2$.

Line graphs were initially introduced in the papers [1,42], although the terminology of line graph was used in reference [43] for the first time. They are an active topic of research at this moment. In particular, several papers study some topological indices on line graphs (see, e.g., [44,45]).

Proposition 7. *If G is a graph, then*

$$H(G,x) = x^2 ID(\mathcal{L}(G),x), \qquad ID(\mathcal{L}(G)) = \int_0^1 \frac{H(G,x)}{x^2}\,dx.$$

Proof. We have

$$H(G,x) = \sum_{uv\in E(G)} x^{d_u+d_v-1} = \sum_{w\in V(\mathcal{L}(G))} x^{d_w+1} = x^2 \sum_{w\in V(\mathcal{L}(G))} x^{d_w-1} = x^2 ID(\mathcal{L}(G),x),$$
$$ID(\mathcal{L}(G)) = \int_0^1 ID(\mathcal{L}(G),x)\,dx = \int_0^1 \frac{H(G,x)}{x^2}\,dx.$$

□

Proposition 8. *If G is a graph, then*

$$ID(T(G),x) = xID(G,x^2) + H(G,x), \qquad ID(T(G)) = \frac{1}{2}ID(G) + \frac{1}{2}H(G).$$

Proof. We have

$$\begin{aligned} ID(T(G),x) &= \sum_{u\in V(G)} x^{2d_u-1} + \sum_{uv\in E(G)} x^{d_u+d_v-1} = x \sum_{u\in V(G)} (x^2)^{d_u-1} + H(G,x) \\ &= xID(G,x^2) + H(G,x), \\ ID(T(G)) &= \frac{1}{2}\int_0^1 2xID(G,x^2)\,dx + \int_0^1 H(G,x)\,dx = \frac{1}{2}\int_0^1 ID(G,t)\,dt + \frac{1}{2}H(G) \\ &= \frac{1}{2}ID(G) + \frac{1}{2}H(G). \end{aligned}$$

□

Proposition 9. *If G is a graph with m edges, then*

$$ID(R(G),x) = xID(G,x^2) + mx, \qquad ID(R(G)) = \frac{1}{2}ID(G) + \frac{1}{2}m.$$

Proof. We have

$$\begin{aligned} ID(R(G),x) &= \sum_{u\in V(G)} x^{2d_u-1} + \sum_{uv\in E(G)} x^{2-1} = xID(G,x^2) + mx, \\ ID(R(G)) &= \frac{1}{2}\int_0^1 2xID(G,x^2)\,dx + \int_0^1 mx\,dx = \frac{1}{2}ID(G) + \frac{1}{2}m. \end{aligned}$$

□

Proposition 10. *If G is a graph, then*

$$ID(Q(G),x) = ID(G,x) + H(G,x), \qquad ID(Q(G)) = ID(G) + \frac{1}{2}H(G).$$

Proof. We have

$$\begin{aligned} ID(Q(G),x) &= \sum_{u\in V(G)} x^{d_u-1} + \sum_{uv\in E(G)} x^{d_u+d_v-1} = ID(G,x) + H(G,x), \\ ID(Q(G)) &= \int_0^1 ID(G,x)\,dx + \int_0^1 H(G,x)\,dx = ID(G) + \frac{1}{2}H(G). \end{aligned}$$

□

Corollary 15. *If G is a graph with m edges, then*

$$\begin{aligned} ID(S(G)) &= ID(R(G)) + \frac{1}{2}ID(G) = 2\,ID(R(G)) - \frac{1}{2}m, \\ ID(Q(G)) &= ID(T(G)) + \frac{1}{2}ID(G). \end{aligned}$$

Author Contributions: Conceptualization, J.M.R.; Data curation, N.V.; Formal analysis, W.C. and J.M.S. The authors contributed equally to this work.

Funding: J.M.R. and J.M.S.: Supported in part by two grants from the Ministerio de Economía y Competititvidad, Agencia Estatal de Investigación (AEI), and Fondo Europeo de Desarrollo Regional (FEDER) (MTM2016-78227-C2-1-P and MTM2017-90584-REDT), Spain.

Acknowledgments: We would like to thank the referees for their careful reading of the manuscript and for some helpful suggestions that have improved the paper.

Conflicts of Interest: The authors declare no conflict of interest.

References

1. Wiener, H. Structural determination of paraffin boiling points. *J. Am. Chem. Soc.* **1947**, *69*, 17–20. [CrossRef] [PubMed]
2. Gutman, I.; Furtula, B. (Eds.) *Recent Results in the Theory of Randić Index*; Univ. Kragujevac: Kragujevac, Serbia, 2008.
3. Li, X.; Gutman, I. *Mathematical Aspects of Randić Type Molecular Structure Descriptors*; Univ. Kragujevac: Kragujevac, Serbia, 2006.
4. Li, X.; Shi, Y. A survey on the Randić index. *MATCH Commun. Math. Comput. Chem.* **2008**, *59*, 127–156.
5. Randić, M. On characterization of molecular branching. *J. Am. Chem. Soc.* **1975**, *97*, 6609–6615. [CrossRef]
6. Rodríguez-Velázquez, J.A.; Sigarreta, J.M. On the Randić index and condicional parameters of a graph. *MATCH Commun. Math. Comput. Chem.* **2005**, *54*, 403–416.
7. Rodríguez-Velázquez, J.A.; Tomás-Andreu, J. On the Randić index of polymeric networks modelled by generalized Sierpinski graphs. *MATCH Commun. Math. Comput. Chem.* **2015**, *74*, 145–160.
8. Borovicanin, B.; Furtula, B. On extremal Zagreb indices of trees with given domination number. *Appl. Math. Comput.* **2016**, *279*, 208–218. [CrossRef]
9. Das, K.C. On comparing Zagreb indices of graphs. *MATCH Commun. Math. Comput. Chem.* **2010**, *63*, 433–440.
10. Furtula, B.; Gutman, I.; Ediz, S. On difference of Zagreb indices. *Discret. Appl. Math.* **2014**, *178*, 83–88. [CrossRef]
11. Liu, M. A simple approach to order the first Zagreb indices of connected graphs. *MATCH Commun. Math. Comput. Chem.* **2010**, *63*, 425–432.
12. Miličević, A.; Nikolić, S. On variable Zagreb indices. *Croat. Chem. Acta* **2004**, *77*, 97–101.
13. Fajtlowicz, S. On conjectures of Graffiti-II. *Congr. Numer.* **1987**, *60*, 187–197.
14. Deng, H.; Balachandran, S.; Ayyaswamy, S.K.; Venkatakrishnan, Y.B. On the harmonic index and the chromatic number of a graph. *Discret. Appl. Math.* **2013**, *161*, 2740–2744. [CrossRef]
15. Favaron, O.; Mahéo, M.; Saclé, J.F. Some eigenvalue properties in graphs (conjectures of Graffiti-II). *Discret. Math.* **1993**, *111*, 197–220. [CrossRef]
16. Rodríguez, J.M.; Sigarreta, J.M. New Results on the Harmonic Index and Its Generalizations. *MATCH Commun. Math. Comput. Chem.* **2017**, *78*, 387–404.
17. Shwetha Shetty, B.; Lokesha, V.; Ranjini, P.S. On the harmonic index of graph operations. *Trans. Comb.* **2015**, *4*, 5–14. [CrossRef]
18. Wu, R.; Tang, Z.; Deng, H. A lower bound for the harmonic index of a graph with minimum degree at least two. *Filomat* **2013**, **27**, 51–55. [CrossRef]
19. Zhong, L.; Xu, K. Inequalities between vertex-degree-based topological Indices. *MATCH Commun. Math. Comput. Chem.* **2014**, *71*, 627–642.
20. Iranmanesh, M.A.; Saheli, M. On the harmonic index and harmonic polynomial of Caterpillars with diameter four. *Iran. J. Math. Chem.* **2014**, *5*, 35–43. [CrossRef]
21. Nazir, R.; Sardar, S.; Zafar, S.; Zahid, Z. Edge version of harmonic index and harmonic polynomial of some classes of graphs. *J. Appl. Math. Inform.* **2016**, *34*, 479–486. [CrossRef]
22. Dankelmann, P.; Hellwig, A.; Volkmann, L. Inverse degree and edge-connectivity. *Discret. Math.* **2008**, *309*, 2943–2947. [CrossRef]
23. Zhang, Z.; Zhang, J.; Lu, X. The relation of matching with inverse degree of a graph. *Discret. Math.* **2005**, *301*, 243–246. [CrossRef]
24. Erdös, P.; Pach, J.; Spencer, J. On the mean distance between points of a graph. *Congr. Numer.* **1988**, *64*, 121–124.
25. Entringer, R. Bounds for the average distance-inverse degree product in trees. In *Combinatorics, Graph Theory, and Algorithms*; New Issues Press: Kalamazoo, MI, USA, 1996; Volume I–II, pp. 335–352.
26. Rodríguez, J.M.; Sánchez, J.L.; Sigarreta, J.M. Inequalities on the inverse degree index. *J. Math. Chem.* **2019**, *57*, 1524–1542. [CrossRef]
27. Hernández, J.C.; Méndez-Bermúdez, J.A.; Rodríguez, J.M.; Sigarreta, J.M. Harmonic Index and Harmonic Polynomial on Graph Operations. *Symmetry* **2018**, *10*, 456. [CrossRef]
28. Bindusree, A.R.; Naci Cangul, I.; Lokesha, V.; Sinan Cevik, A. Zagreb Polynomials of Three Graph Operators. *Filomat* **2016**, *30*, 1979–1986. [CrossRef]

29. Loghman, A. PI polynomials of product graphs. *Appl. Math. Lett.* **2009**, *22*, 975–979. [CrossRef]
30. Khalifeh, M.H.; Yousefi-Azari, H.; Ashrafi, A.R. The first and second Zagreb indices of some graph operations. *Discr. Appl. Math.* **2009**, *157*, 804–811. [CrossRef]
31. Bermudo, S.; Carballosa, W.; Rodríguez, J.M.; Sigarreta, J.M. f-polynomial on some products of graphs. **2019**, submitted.
32. Imrich, W.; Klavžar, S. *Product Graphs: Structure and Recognition*; John Wiley & Sons: New York, NY, USA, 2000.
33. Shuxian, L. Zagreb polynomials of thorn graphs. *Kragujev. J. Sci.* **2011**, *33*, 33–38.
34. Anderson, B.J. An inequality for convex functions. *Nord. Mat. Tidskr.* **1958**, *6*, 25–26.
35. Csiszár, V.; Móri, T.F. Sharp integral inequalities for products of convex functions. *J. Ineq. Pure Appl. Math.* **2007**, *8*, 94.
36. Vukičević, D.; Gašperov, M. Bond additive modeling 1. Adriatic indices. *Croat. Chem. Acta* **2010**, *83*, 243–260.
37. Mycielski, J. Sur le coloriage des graphes. *Colloq. Math.* **1955**, *3*, 161–162. [CrossRef]
38. Ali, A.; Zhong, L.; Gutman, I. Harmonic Index and its Generalizations: Extremal Results and Bounds. *MATCH Commun. Math. Comput. Chem.* **2019**, *81*, 249–311.
39. Eliasi, M.; Raeisi, G.; Taeri, B. Wiener index of some graph operations. *Discret. Appl. Math.* **2012**, *160*, 1333–1344. [CrossRef]
40. Hua, H.; Das, K.C.; Wang, H. On atom-bond connectivity index of graphs. *J. Math. Anal. Appl.* **2019**, *479*, 1099–1114. [CrossRef]
41. Hua, H.; Reza Ashrafi, A.; Zhang, L. More on Zagreb coindices of graphs. *Filomat* **2012**, *26*, 1210–1220. [CrossRef]
42. Krausz, J. Démonstration nouvelle d'un théorème de Whitney sur les réseaux. *Mat. Fiz. Lapok* **1943**, *50*, 75–85.
43. Harary, F.; Norman, R.Z. Some properties of line digraphs. *Rend. Circ. Math. Palermo* **1960**, *9*, 161–169. [CrossRef]
44. Pestana, D.; Sigarreta, J.M.; Tourís, E. Geometric-arithmetic index and line graph. *J. Math. Chem.* **2019**, *57*, 1427–1447. [CrossRef]
45. Su, G.; Xu, L. Topological indices of the line graph of subdivision graphs and their Schur bounds. *Appl. Math. Comput.* **2015**, *253*, 395–401. [CrossRef]

mathematics

MDPI

Article

Some Results on $(s-q)$-Graphic Contraction Mappings in b-Metric-Like Spaces

Manuel De la Sen [1,*], Nebojša Nikolić [2], Tatjana Došenović [3], Mirjana Pavlović [4] and Stojan Radenović [5]

1 Institute of Research and Development of Porocesses, University of the Basque, 48080 Bilbao, Spain
2 Faculty of Organizational Sciences, University of Belgrade, 11000 Belgrade, Serbia; nebojsa.nikolic@fon.bg.ac.rs
3 Faculty of Technology, University of Novi Sad, Bulevar cara Lazara 1, 21000 Novi Sad, Serbia; tatjanad@tf.uns.ac.rs
4 Department of Mathematics and Informatics, Faculty of Science, University of Kragujevac, Radoja Domanovića 12, 34000 Kragujeva, Serbia; mpavlovic@kg.ac.rs
5 Faculty of Mechanical Enginering, University of Belgrade, Kraljice Marije 16, 11120 Beograd, Serbia; radens@beotel.rs or sradenovic@mas.bg.ac.rs
* Correspondence: manuel.delasen@ehu.eus

Received: 15 November 2019; Accepted: 2 December 2019; Published: 4 December 2019

Abstract: In this paper we consider $(s-q)$-graphic contraction mapping in b-metric like spaces. By using our new approach for the proof that a Picard sequence is Cauchy in the context of b-metric-like space, our results generalize, improve and complement several approaches in the existing literature. Moreover, some examples are presented here to illustrate the usability of the obtained theoretical results.

Keywords: b-metric space; b-metric-like space; general contractive mappings; graphic contraction mappings

JEL Classification: 47H10; 54H25

---

1. Introduction and Preliminaries

First, we present some definitions and basic notions of partial-metric, metric-like, b-metric, partial b-metric and b-metric-like spaces as the generalizations of standard metric spaces. After that, we give a process diagram, where arrows stand for generalization relationships.

Definition 1. *[1] Let X be a nonempty set. A mapping $p_{pm}: X\times X\to[0,+\infty)$ is said to be a p-metric if the following conditions hold for all $x,y,z\in X$:*

$(p_{pm}1)$ $x=y$ *if and only if* $p_{pm}(x,x)=p_{pm}(x,y)=p_{pm}(y,y)$;
$(p_{pm}2)$ $p_{pm}(x,x)\le p_{pm}(x,y)$;
$(p_{pm}3)$ $p_{pm}(x,y)=p_{pm}(y,x)$;
$(p_{pm}4)$ $p_{pm}(x,y)\le p_{pm}(x,z)+p_{pm}(z,y)-p_{pm}(z,z)$.

Then, the pair (X,p_{pm}) is called a partial metric space.

Definition 2. *[2] Let X be a nonempty set. A mapping $b_{ml}: X\times X\to[0,+\infty)$ is said to be metric-like if the following conditions hold for all $x,y,z\in X$:*

(b_l1) $b_{ml}(x,y)=0$ *implies* $x=y$;
(b_l2) $b_{ml}(x,y)=b_{ml}(y,x)$;
(b_l3) $b_{ml}(x,z)\le b_{ml}(x,y)+b_{ml}(y,z)$.

In this case, the pair (X,b_{ml}) is called a metric-like space.

Definition 3. *[3,4] Let X be a nonempty set and $s \geq 1$ a given real number. A mapping $b : X \times X \to [0, +\infty)$ is called a b-metric on the set X if the following conditions hold for all $x, y, z \in X$:*

$(b1)$ $b(x,y) = 0$ *if and only if* $x = y$;

$(b2)$ $b(x,y) = b(y,x)$;

$(b3)$ $b(x,z) \leq s[b(x,y) + b(y,z)]$.

In this case, the pair (X, b) is called a b-metric space (with coefficient $s \geq 1$).

Definition 4. *[5,6] Let X be a nonempty set and $s \geq 1$. A mapping $b_{pb} : X \times X \to [0, +\infty)$ is called a partial b-metric on the set X if the following conditions hold for all $x, y, z \in X$:*

$\left(b_{pb}1\right)$ $x = y$ *if and only if* $p_{pb}(x,x) = p_{pb}(x,y) = p_{pb}(y,y)$;

$\left(b_{pb}2\right)$ $b_{pb}(x,x) \leq b_{pb}(x,y)$;

$\left(b_{pb}3\right)$ $b_{pb}(x,y) = b_{pb}(y,x)$;

$\left(b_{pb}4\right)$ $b_{pb}(x,y) \leq s\left[b_{pb}(x,z) + b_{pb}(z,y)\right] - b_{pb}(z,z)$.

Then, the pair $\left(X, b_{pb}\right)$ is called a partial b-metric space.

Definition 5. *[7] Let X be a nonempty set and $s \geq 1$. A mapping $b_{bl} : X \times X \to [0, +\infty)$ is called b-metric-like on the set X if the following conditions hold for all $x, y, z \in X$:*

$(b_{bl}1)$ $b_{bl}(x,y) = 0$ *implies* $x = y$;

$(b_{bl}2)$ $b_{bl}(x,y) = b_{bl}(y,x)$;

$(b_{bl}3)$ $b_{bl}(x,z) \leq s[b_{bl}(x,y) + b_{bl}(y,z)]$.

In this case, the pair (X, b_{bl}) is called a b-metric-like space with coefficient $s \geq 1$.

Now, we give the process diagram of the classes of generalized metric spaces that were introduced earlier:

Metric space	→	Partial metric space	→	Metric-like space
↓		↓		↓
b-Metric space	→	Partial b-metric space	→	b-Metric-like space

For more details on other generalized metric spaces see [8–14].

The next proposition helps us to construct some more examples of b-metric (respectively partial b-metric, b-metric-like) spaces.

Proposition 1. *Let (X, d) (resp. $\left(X, p_{pm}\right), (X, b_{ml})$) be a metric (resp. partial metric, metric-like) space and $D(x,y) = (d(x,y))^k$ (resp. $P_{pm}(x,y) = \left(p_{pm}(x,y)\right)^k$, $B_{ml}(x,y) = (b_{ml}(x,y))^k$), where $k > 1$ is a real number. Then D (resp. P_{pm}, B_{pm}) is b-metric (resp. partial b-metric, b-metric-like) with coefficient $s = 2^{k-1}$.*

Proof. The proof follows from the fact that

$$u^k + v^k \leq (u+v)^k \leq (a+b)^k \leq 2^{k-1}\left(a^k + b^k\right),$$

for all nonnegative real numbers a, b, u, v with $u + v \leq a + b$. □

It is clear that each metric-like space, i.e., each partial b-metric space, is a b-metric-like space, while the converse is not true. For more such examples and details see [1,2,5–7,15–27]. Moreover, for various metrics in the context of the complex domain see [28,29].

The definitions of convergent and Cauchy sequences are formally the same in partial metric, metric-like, partial b-metric and b-metric like spaces. Therefore, we give only the definition of convergence and Cauchyness of the sequences in b-metric-like space. Moreover, these two notions are formally the same in metric and b-metric spaces.

Definition 6. *[7] Let $\{x_n\}$ be a sequence in a b-metric-like space (X, b_{bl}) with coefficient s.*

(i) The sequence $\{x_n\}$ is said to be convergent to x if $\lim_{n\to\infty} b_{bl}(x_n, x) = b_{bl}(x, x)$;

(ii) The sequence $\{x_n\}$ is said to be b_{bl}-Cauchy in (X, b_{bl}) if $\lim_{n,m\to\infty} b_{bl}(x_n, x_m)$ exists and is finite;

(iii) One says that a b-metric-like space (X, b_{bl}) is b_{bl}-complete if for every b_{bl}-Cauchy sequence $\{x_n\}$ in X there exists an $x \in X$, such that $\lim_{n,m\to\infty} b_{bl}(x_n, x_m) = b_{bl}(x, x) = \lim_{n\to\infty} b_{bl}(x_n, x)$.

Remark 1. *In a b-metric-like space the limit of a sequence need not be unique and a convergent sequence need not be a b_{bl}-Cauchy sequence (see Example 7 in [18]). However, if the sequence $\{x_n\}$ is b_{bl}-Cauchy with $\lim_{n,m\to\infty} b_{bl}(x_n, x_m) = 0$ in the b_{bl}-complete b-metric-like space (X, b_{bl}) with coefficient $s \geq 1$, then the limit of such a sequence is unique. Indeed, in such a case if $x_n \to x$ $(b_{bl}(x_n, x) \to b_{bl}(x, x))$ as $n \to \infty$ we get that $b_{bl}(x, x) = 0$. Now, if $x_n \to x$ and $x_n \to y$ where $x \neq y$, we obtain that:*

$$\frac{1}{s} b_{bl}(x, y) \leq b_{bl}(x, x_n) + b_{bl}(x_n, y) \to b_{bl}(x, x) + b_{bl}(y, y) = 0 + 0 = 0. \tag{1}$$

From $(b_{bl}1)$ it follows that $x = y$, which is a contradiction. The same is true as well for partial metric, metric like and partial b-metric spaces.

The next definition and the corresponding proposition are important in the context of fixed point theory.

Definition 7. *[30] The self-mappings $f, g : X \to X$ are weakly compatible if $f(g(x)) = g(f(x))$, whenever $f(x) = g(x)$.*

Proposition 2. *[30] Let T and S be weakly compatible self-maps of a nonempty set X. If they have a unique point of coincidence $w = f(u) = g(u)$, then w is the unique common fixed point of f and g.*

In this paper we shall use the following result to prove that certain Picard sequences are Cauchy. The proof is completely identical with the corresponding in [31] (see also [25]).

Lemma 1. *Let $\{x_n\}$ be a sequence in a b-metric-like space (X, b_{bl}) with coefficient $s \geq 1$ such that*

$$b_{bl}(x_n, x_{n+1}) \leq \lambda b_{bl}(x_{n-1}, x_n) \tag{2}$$

for some $\lambda, 0 \leq \lambda < \frac{1}{s}$, and each $n = 1, 2, \ldots$.Then $\{x_n\}$ is a b_{bl}-Cauchy sequence in (X, b_{bl}) such that $\lim_{n,m\to\infty} b_{bl}(x_n, x_m) = 0$.

Remark 2. *It is worth noting that the previous lemma holds in the context of b-metric-like spaces for each $\lambda \in [0, 1)$. For more details see [6,32].*

2. Main Results

In line with Jachymski [33], let (X, b_{bl}) be a b-metric-like space and $\mathcal{D}$ denote the diagonal of the Cartesian product $X \times X$. Consider a directed graph G such that the set $V(G)$ of its vertices coincides with X, and the set $E(G)$ of its edges contains all loops, i.e., $E(G) \supseteq \mathcal{D}$. We also assume that G has no parallel edges, so we can identify G with the pair $(V(G), E(G))$. Moreover, we may treat G as a weighted graph by assigning the distance between its vertices to each edge (see [33]).

By G^{-1} we denote the conversion of a graph G, i.e., the graph obtained from G by reversing the direction of edges. Thus, we have

$$E\left(G^{-1}\right) = \{(x, y) \in X \times X : (y, x) \in E(G)\}. \tag{3}$$

The letter $\widetilde{G}$ denotes the undirected graph obtained from G by ignoring the direction of edges. Actually, it will be more convenient for us to treat $\widetilde{G}$ as a directed graph for which the set of its edges is symmetric under the convention

$$E\left(\widetilde{G}\right) = E\left(G\right) \cup E\left(G^{-1}\right). \tag{4}$$

If x and y are vertices in a graph G, then a path in G from x to y of length N ($N \in \mathbb{N}$) is a sequence $\{x_i\}_{i=0}^{N}$ of $N+1$ vertices such that $x_0 = x, x_N = y$ and $(x_{i-1}, x_i) \in E(G)$ for $i = 1, ..., N$. A graph G is connected if there is a path between any two vertices. G is weakly connected if $\widetilde{G}$ is connected.

Recently, some results have appeared providing sufficient conditions for a self mapping of X to be a Picard operator when (X, d) is endowed with a graph. The first result in this direction was given by Jachymski [33]. Moreover, see [34–36].

Definition 8. [33] *We say that a mapping $f : X \to X$ is a Banach G-contraction or simply a G-contraction if f preserves edges of G, i.e.,*

$$\text{for all } x, y \in X : (x, y) \in E(G) \text{ implies } (f(x), f(y)) \in E(G) \tag{5}$$

and f decreases the weights of edges of G as for all $x, y \in X$, there exists $\lambda \in (0,1)$, such that

$$(x, y) \in E(G) \text{ implies } d(f(x), f(y)) \le \lambda d(x, y). \tag{6}$$

Definition 9. [37] *A mapping $g : X \to X$ is called orbitally continuous, if given $x \in X$ and any sequence $\{k_n\}$ of positive integers,*

$$g^{k_n}(x) \to y \text{ as } n \to \infty \text{ implies } g\left(g^{k_n}(x)\right) \to g(y) \text{ as } n \to \infty. \tag{7}$$

Definition 10. [33] *A mapping $g : X \to X$ is called G-continuous, if for any given $x \in X$ and any sequence $\{x_n\}_{n\in\mathbb{N}} \subset X$ with the properties that for all $n \in \mathbb{N}$ the pair $(x_n, x_{n+1}) \in E(G)$ and that $x_n \to x$ as $n \to \infty$ it follows that $g(x_n) \to g(x)$.*

Definition 11. [33] *A mapping $g : X \to X$ is called orbitally G-continuous, if given $x, y \in X$ and any sequence $\{k_n\}$ of positive integers for all $n \in \mathbb{N}$,*

$$g^{k_n}x \to y \text{ and } \left(g^{k_n}(x), g^{k_n+1}(x)\right) \in E(G) \text{ implies } g\left(g^{k_n}(x)\right) \to g(y) \text{ as } n \to \infty. \tag{8}$$

In this section, we consider self-mappings $f, g : X \to X$ with $f(X) \subset g(X)$. Let $x_0 \in X$ be an arbitrary point, then there exists $x_1 \in X$ such that $z_0 = f(x_0) = g(x_1)$. By repeating this step we can build a sequence $\{z_n\}$ such that $z_n = f(x_n) = g(x_{n+1})$ and the following property:

The property $G_{f,g(x_n)}$. If $\{g(x_n)\}_{n\in\mathbb{N}}$ is a sequence in X such that $(g(x_n), g(x_{n+1})) \in E(G)$ for all $n \ge 1$ and $g(x_n) \to x$, then there is a subsequence $\{g(x_{n_i})\}_{i\in\mathbb{N}}$ of $\{g(x_n)\}_{n\in\mathbb{N}}$ such that $(g(x_{n_i}), x) \in E(G)$ for all $i \ge 1$. Note that the property $G_{f,g(x_n)}$ depends only on the pair of mappings f and g, and does not depend on the sequence $\{x_n\}$. Here, we use notation G_{gf} in the following sense: $x \in X$ belongs to G_{gf} if and only if there exists a sequence $\{x_n\}_{n\in\mathbb{N}}$ in X such that $x_0 = x$, $f(x_{n-1}) = g(x_n)$ for $n \in \mathbb{N}$, and $(g(x_n), g(x_m)) \in E(G)$ for all $m, n \in \mathbb{N}$.

Now, we present the first result of this section.

Theorem 1. *(Hardy-Rogers) Let $f, g : X \to X$ be self-mappings defined on a b-metric-like space (X, b_{bl}) (with coefficient $s \geq 1$) endowed with a graph G, and which satisfy*

$$\begin{aligned} s^q b_{bl}(f(x), f(y)) \leq\ & c_1 b_{bl}(g(x), g(y)) + c_2 b_{bl}(g(x), f(x)) + c_3 b_{bl}(g(y), f(y)) \\ & + c_4 b_{bl}(g(x), f(y)) + c_5 b_{bl}(g(y), f(x)), \end{aligned} \tag{9}$$

for all $x, y \in X$ with $(g(x), g(y)) \in E(G)$ where $q \geq 2, c_i \geq 0, i = 1, \ldots, 5$ and either

$$c_1 + c_2 + c_3 + 2c_4 + 2c_5 < \frac{1}{s} \tag{10}$$

or

$$c_1 + 2c_2 + 2c_3 + c_4 + c_5 < \frac{1}{s}. \tag{11}$$

Suppose that $f(X) \subset g(X)$ and at least one of $f(X), g(X)$ is b_{bl}-complete subspace of (X, b_{bl}). Then:

(i) If the pair (f, g) has property $G_{f,g(x_n)}$ and $G_{gf} \neq \emptyset$, then f and g have a point of coincidence in X.

(ii) If x and y in X are points of coincidence of f and g such that $(x, y) \in E(G)$, then $x = y$. Hence, points of coincidence of f and g are unique in X. Moreover, if the pair (f, g) is weakly compatible, then f and g have a unique common fixed point in X.

Proof. (i) Assume that $G_{gf} \neq \emptyset$, there exists $x_0 \in G_{gf}$. Since $f(X) \subset g(X)$, there exists $x_1 \in X$ such that $f(x_0) = g(x_1)$, again we can find $x_2 \in X$ such that $f(x_1) = g(x_2)$. Repeating this step, we can build a sequence $z_n = f(x_n) = g(x_{n+1})$ such that $(z_n, z_m) \in E(G)$. If $z_k = z_{k+1}$ for some $k \in \mathbb{N}$, then $f(x_{k+1}) = g(x_{k+1})$ is a point of coincidence of f and g. Therefore, let $z_n \neq z_{n+1}$ for all $n \in \mathbb{N}$. By Condition (9), we can get that

$$\begin{aligned} b_{bl}(z_n, z_{n+1}) \leq\ & s^q b_{bl}(z_n, z_{n+1}) = s^q b_{bl}(f(x_n), f(x_{n+1})) \\ \leq\ & c_1 b_{bl}(g(x_n), g(x_{n+1})) + c_2 b_{bl}(g(x_n), f(x_n)) + c_3 b_{bl}(g(x_{n+1}), f(x_{n+1})) \\ & + c_4 b_{bl}(g(x_n), f(x_{n+1})) + c_5 b_{bl}(g(x_{n+1}), f(x_n)). \end{aligned} \tag{12}$$

Since $z_n = f(x_n) = g(x_{n+1})$ then Condition (12) becomes

$$\begin{aligned} b_{bl}(z_n, z_{n+1}) \leq\ & c_1 b_{bl}(z_{n-1}, z_n) + c_2 b_{bl}(z_{n-1}, z_n) + c_3 b_{bl}(z_n, z_{n+1}) \\ & + c_4 b_{bl}(z_{n-1}, z_{n+1}) + c_5 b_{bl}(z_n, z_n) \\ \leq\ & c_1 b_{bl}(z_{n-1}, z_n) + c_2 b_{bl}(z_{n-1}, z_n) + c_3 b_{bl}(z_n, z_{n+1}) + s c_4 b_{bl}(z_{n-1}, z_n) \\ & + s c_4 b_{bl}(z_n, z_{n+1}) + 2 s c_5 b_{bl}(z_{n-1}, z_n), \end{aligned} \tag{13}$$

or equivalently:

$$b_{bl}(z_n, z_{n+1}) \leq \lambda b_{bl}(z_{n-1}, z_n), \tag{14}$$

where $\lambda = \frac{c_1 + c_2 + s c_4 + 2 s c_5}{1 - c_3 - s c_4}$. Since, $c_1 + c_2 + c_3 + s c_4 + 2 s c_5 \leq s c_1 + s c_2 + s c_3 + 2 s c_4 + 2 s c_5 < 1$, it follows that $\lambda < 1$. Therefore, by Remark 2 of Lemma 1, the sequence $z_n = f(x_n) = g(x_{n+1})$ is a b_{bl}-Cauchy sequence. The b_{bl}-completeness of $f(X)$ leads to $u \in f(X) \subset g(X)$ such that $z_n \to u = g(v)$ for some $v \in X$. As $z_0 \in G_{gf}$, this implies that $(z_n, z_m) \in E(G)$ for $n, m = 1, 2, \ldots$ and so $(z_n, z_{n+1}) \in E(G)$.

By property $G_{f,g(x_n)}$, there is a subsequence $\{z_{n_i}\}_{i\in\mathbb{N}}$ of $\{z_n\}_{n\in\mathbb{N}}$ such that $(z_{n_i}, u) \in E(G)$. Applying $(b_{bl}3)$, we get

$$\begin{aligned} b_{bl}(f(v), g(v)) &\leq sb_{bl}(f(v), f(x_{n_i})) + sb_{bl}(f(x_{n_i}), g(v)) \\ &\leq s^q b_{bl}(f(v), f(x_{n_i})) + sb_{bl}(f(x_{n_i}), g(v)) \\ &\leq c_1 b_{bl}(g(v), g(x_{n_i})) + c_2 b_{bl}(g(v), f(v)) + c_3 b_{bl}(g(x_{n_i}), f(x_{n_i})) \\ &\leq +c_4 b_{bl}(g(v), f(x_{n_i})) + c_5 b_{bl}(g(x_{n_i}), f(v)) + sb_{bl}(f(x_{n_i}, g(v))) \\ &= c_1 b_{bl}(g(v), z_{n_i-1}) + c_2 b_{bl}(g(v), f(v)) + c_3 b_{bl}(z_{n_i-1}, z_{n_i}) \\ &\quad +c_4 b_{bl}(g(v), z_{n_i}) + c_5 b_{bl}(z_{n_i-1}, f(v)) + sb_{bl}(z_{n_i}, g(v)). \end{aligned} \tag{15}$$

Since $b_{bl}(z_{n_i-1}, f(v)) \leq sb_{bl}(z_{n_i-1}, g(v)) + sb_{bl}(g(v), f(v))$, Condition (15) becomes

$$\begin{aligned} &(1 - c_2 - c_5 s) b_{bl}(f(v), g(v)) \\ \leq\ & c_1 b_{bl}(g(v), z_{n_i-1}) + c_3 b_{bl}(z_{n_i-1}, z_{n_i}) + c_4 b_{bl}(g(v), z_{n_i}) \\ &+c_5 s b_{bl}(z_{n_i-1}, g(v)) + sb_{bl}(z_{n_i}, g(v)). \end{aligned} \tag{16}$$

Taking the limit in Condition (16) as $i \to \infty$ we obtain that $b_{bl}(f(v), g(v)) = 0$, because $c_2 + c_5 s \leq c_1 s + c_2 s + c_3 s + 2c_4 s + 2c_5 s < 1$. That is, $f(v) = g(v) = u$ is a point of coincidence for the mappings f and g, i.e., (i) is proved in the case if $f(X)$ is b_{bl}-complete. The proof for the case if $g(X)$ is b_{bl}-complete is similar.

(ii) Assume that x and y are two different points of coincidence of f and g with $(x, y) \in E(G)$. This means that there are different points x_1 and y_1 from X such that: $f(x_1) = g(x_1) = x$ and $f(y_1) = g(y_1) = y$. Now, according to Condition (9) we get

$$\begin{aligned} sb_{bl}(x, y) &\leq s^q b_{bl}(x, y) = s^q b_{bl}(f(x_1), f(y_1)) \\ &\leq c_1 b_{bl}(g(x_1), g(y_1)) + c_2 b_{bl}(g(x_1), f(y_1)) + c_3 b_{bl}(g(y_1), f(y_1)) \\ &\quad +c_4 b_{bl}(g(x_1), f(y_1)) + c_5 b_{bl}(g(y_1), f(x_1)) \\ &= c_1 b_{bl}(x, y) + c_2 b_{bl}(x, y) + c_3 b_{bl}(y, y) \\ &\quad +c_4 b_{bl}(x, y) + c_5 b_{bl}(y, x) \\ &\leq (c_1 + c_2 + 2c_3 s + c_4 + c_5) b_{bl}(y, x) \\ &\leq (c_1 s + 2c_2 s + 2c_3 s + c_4 s + c_5 s) b_{bl}(y, x) < b_{bl}(y, x). \end{aligned} \tag{17}$$

Hence, if $x \neq y$ we get a contradiction.

If f and g are weakly compatible, then by Proposition 2 f and g have a unique common fixed point. □

Example 1. *Let $X = [0, +\infty)$ and $f, g : X \to X$ be the mappings such that*

$$f(x) = e^x - 1 \quad \text{and} \quad g(x) = e^{4x} - 1.$$

Consider b-metric-like space (X, b_{bl}) under the distance $b_{bl}(x, y) = (x + y)^2$ with coefficient $s = 2$, and the graph $G = (V, E)$ with $V = X$ and $E = \{(x, x) : x \in X\} \cup \{(0, x) : x \in X\}$. Assume that $c_1 = \frac{1}{4}$ and $c_2 = c_3 = c_4 = c_5 = \frac{1}{25}$ for which Inequalities (10) and (11) hold. Note that $(g(x), g(y)) \in E$ if and only if $x = y, x \geq 0$ or $x = 0, y > 0$ or $y = 0, x > 0$. For $q = 2$ let us check whether Condition (9) holds in these cases.

Case 1: $x = y, x \geq 0$;

$$\begin{aligned}
& c_1 b_{bl}(g(x), g(x)) + c_2 b_{bl}(g(x), f(x)) + c_3 b_{bl}(g(x), f(x)) + c_4 b_{bl}(g(x), f(x)) + c_5 b_{bl}(g(x), f(x)) \\
= \; & c_1 \left(e^{4x} - 1 + e^{4x} - 1\right)^2 + (c_2 + c_3 + c_4 + c_5) \left(e^{4x} - 1 + e^x - 1\right)^2 \\
= \; & 4c_1 \left(e^x - 1\right)^2 \left(e^{3x} + e^{2x} + e^x + 1\right)^2 + (c_2 + c_3 + c_4 + c_5) \left(e^x - 1\right)^2 \left(e^{3x} + e^{2x} + e^x + 2\right)^2 \\
\geq \; & 4c_1 \left(e^x - 1\right)^2 4^2 + (c_2 + c_3 + c_4 + c_5) \left(e^x - 1\right)^2 5^2 = \left(\tfrac{1}{4} \cdot 64 + \tfrac{4}{25} \cdot 25\right) \left(e^x - 1\right)^2 \\
> \; & 4 \left(e^x - 1 + e^x - 1\right)^2 = s^q b_{bl}(f(x), f(x)).
\end{aligned}$$

Case 2: $x = 0, y > 0$ *(similarly for* $y = 0, x > 0$*);*

$$\begin{aligned}
& c_1 b_{bl}(g(0), g(y)) + c_2 b_{bl}(g(0), f(0)) + c_3 b_{bl}(g(y), f(y)) + c_4 b_{bl}(g(0), f(y)) + c_5 b_{bl}(g(y), f(0)) \\
= \; & c_1 \left(e^{4y} - 1\right)^2 + c_2 (0 + 0)^2 + c_3 \left(e^{4y} - 1 + e^y - 1\right)^2 + c_4 \left(e^y - 1\right)^2 + c_5 \left(e^{4y} - 1\right)^2 \\
= \; & (c_1 + c_5) \left(e^y - 1\right)^2 \left(e^{3y} + e^{2y} + e^y + 1\right)^2 + c_3 \left(e^y - 1\right)^2 \left(e^{3y} + e^{2y} + e^y + 2\right)^2 + c_4 \left(e^y - 1\right)^2 \\
> \; & (c_1 + c_5) \left(e^y - 1\right)^2 4^2 + c_3 \left(e^y - 1\right)^2 5^2 + c_4 \left(e^y - 1\right)^2 = \left(\tfrac{29}{100} \cdot 16 + \tfrac{1}{25} \cdot 25 + \tfrac{1}{25}\right) \left(e^y - 1\right)^2 \\
> \; & 4 \left(e^y - 1\right)^2 = s^q b_{bl}(f(0), f(y)).
\end{aligned}$$

Hence, f and g satisfy Condition (9) for all $x, y \in X$ such that $(g(x), g(y)) \in E$.

Moreover, there is $x_1 = \frac{x_0}{4}$ such that $g(x_1) = f(x_0)$, $x_2 = \frac{x_0}{4^2}$ such that $g(x_2) = f(x_1)$, and so on. In this way, we can built the sequence $x_n = \frac{x_0}{4^n}, n \in \mathbb{N}$ such that $g(x_n) = f(x_{n-1})$. For $x_0 \neq 0$ it is clear that $(g(x_n), g(x_m)) \notin E$. For $x_0 = 0$, $x_n = 0, n \in \mathbb{N}$ is obtained. Thus, the constant sequence $x_n = 0$ is only convergent sequence such that $(g(x_n), g(x_m)) = (0,0) \in E$, and for each subsequence $(g(x_{n_i}))_{i \in \mathbb{N}}$ of $(g(x_n))_{n \in \mathbb{N}}$ holds $(g(x_{n_i}), 0) = (0,0) \in E$. This means that $x_0 \in G_{gf} \neq \emptyset$ and the pair (f, g) possesses the property $G_{f,g(x_n)}$.

It is obvious that $f(X) \subset g(X)$ and $g(X) = X$ is b_{bl}-complete. Since the mappings f and g are weakly compatible at $x = 0$ ($f(0) = g(0)$ implies $g(f(0)) = f(g(0))$), all conditions of Theorem 1 are satisfied. So, 0 is the unique common fixed point of mappings f and g in X.

Example 2. *Now consider the same b-metric-like space (X, b_{bl}) endowed with the graph G as in Example 1, and the mappings $f, g : X \to X$ such that*

$$f(x) = \begin{cases} e^x - 1, & x \neq 0 \\ 1, & x = 0 \end{cases} \quad \text{and} \quad g(x) = \begin{cases} e^{4x} - 1, & x \neq 0 \\ 2, & x = 0 \end{cases}.$$

In this case we have $G_{gf} = \emptyset$. Namely, for $x_0 = 0$, $x_n = \frac{1}{4^n} \ln 2, n \in \mathbb{N}$ is now obtained, and $(g(x_n), g(x_m)) \notin E$. Hence, the conditions of Theorem 1 are not satisfied. Moreover, we can easily see that the mappings f and g have no coincidence point nor common fixed points.

As corollaries of our Theorem 1, we obtain the next results in the context of b-metric-like spaces endowed with a graph:

Corollary 1. *(Jungck) Let $f, g : X \to X$ be self-mappings defined on a b-metric-like space (X, b_{bl}) (with coefficient $s \geq 1$) endowed with a graph G, and satisfy*

$$s^q b_{bl}(f(x), f(y)) \leq c_1 b_{bl}(g(x), g(y)) \tag{18}$$

for all $x, y \in X$ with $(g(x), g(y)) \in E(G)$ when $c_1 < \frac{1}{s}$. Suppose that $f(X) \subset g(X)$ and at least one of $f(X), g(X)$ is a b_{bl}-complete subspace of (X, b_{bl}). Then

(i) If the property $G_{f,g(x_n)}$ is satisfied and $G_{gf} \neq \emptyset$, then f and g have a point of coincidence in X.

(ii) If x and y in X are points of coincidence of f and g such that $(x, y) \in E(G)$, then $x = y$. Hence, points of coincidence of f and g are unique in X. Moreover, if the pair (f, g) is weakly compatible, then f and g have a unique common fixed point in X.

Corollary 2. *(Kannan) Let $f, g : X \to X$ be self-mappings defined on a b-metric-like space (X, b_{bl}) (with coefficient $s \geq 1$) endowed with a graph G, and satisfy*

$$s^q b_{bl}(f(x), f(y)) \leq c_2 b_{bl}(g(x), f(x)) + c_3 b_{bl}(g(y), f(y)) \tag{19}$$

for all $x, y \in X$ with $(g(x), g(y)) \in E(G)$ when

$$c_2 + c_3 < \frac{1}{2s}. \tag{20}$$

Suppose that $f(X) \subset g(X)$ and at least one of $f(X), g(X)$ is a b_{bl}-complete subspace of (X, b_{bl}). Then

(i) If the property $G_{f,g(x_n)}$ is satisfied and $G_{gf} \neq \emptyset$, then f and g have a point of coincidence in X.

(ii) If x and y in X are points of coincidence of f and g such that $(x, y) \in E(G)$, then $x = y$. Hence, points of coincidence of f and g are unique in X. Moreover, if the pair (f, g) is weakly compatible, then f and g have a unique common fixed point in X.

Corollary 3. *(Chatterjea) Let $f, g : X \to X$ be self-mappings defined on a b-metric-like space (X, b_{bl}) (with coefficient $s \geq 1$) endowed with a graph G, and satisfy*

$$s^q b_{bl}(f(x), f(y)) \leq c_4 b_{bl}(g(x), f(y)) + c_5 b_{bl}(g(y), f(x)), \tag{21}$$

for all $x, y \in X$ with $(g(x), g(y)) \in E(G)$ when

$$c_4 + c_5 < \frac{1}{2s}. \tag{22}$$

Suppose that $f(X) \subset g(X)$ and at least one of $f(X), g(X)$ is a b_{bl}-complete subspace of (X, b_{bl}). Then

(i) If the property $G_{f,g(x_n)}$ is satisfied and $G_{gf} \neq \emptyset$, then f and g have a point of coincidence in X.

(ii) If x and y in X are points of coincidence of f and g such that $(x, y) \in E(G)$, then $x = y$. Hence, points of coincidence of f and g are unique in X. Moreover, if the pair (f, g) is weakly compatible, then f and g have a unique common fixed point in X.

Corollary 4. *(Reich) Let $f, g : X \to X$ be self-mappings defined on a b-metric-like space (X, b_{bl}) (with coefficient $s \geq 1$) endowed with a graph G, and satisfy*

$$s^q b_{bl}(f(x), f(y)) \leq c_1 b_{bl}(g(x), g(y)) + c_2 b_{bl}(g(x), f(x)) + c_3 b_{bl}(g(y), f(y)) \tag{23}$$

for all $x, y \in X$ with $(g(x), g(y)) \in E(G)$ when

$$c_1 + 2c_2 + 2c_3 < \frac{1}{s} \tag{24}$$

Suppose that $f(X) \subset g(X)$ and at least one of $f(X), g(X)$ is a b_{bl}-complete subspace of (X, b_{bl}). Then

(i) If the property $G_{f,g(x_n)}$ is satisfied and $G_{gf} \neq \emptyset$, then f and g have a point of coincidence in X.

(ii) If x and y in X are points of coincidence of f and g such that $(x, y) \in E(G)$, then $x = y$. Hence, points of coincidence of f and g are unique in X. Moreover, if the pair (f, g) is weakly compatible, then f and g have a unique common fixed point in X.

Now, we announce our last result in this section in the context of *b*-metric-like spaces endowed with the graph. The proof is similar enough with the corresponding proof of Theorem 1 and therefore we omit it.

Theorem 2. *(Das-Naik-Ćirić) Let $f, g : X \to X$ be self-mappings defined on a b-metric-like space (X, b_{bl}) (with coefficient $s \geq 1$) endowed with a graph G, and satisfy*

$$s^q b_{bl}(f(x), f(y)) \leq \lambda \max\{b_{bl}(g(x), g(y)), b_{bl}(g(x), f(x)), b_{bl}(g(y), f(y)), b_{bl}(g(x), f(y)), b_{bl}(g(y), f(x))\} \tag{25}$$

for all $x, y \in X$ with $(g(x), g(y)) \in E(G)$ when $\lambda \in [0, \frac{1}{s})$. Suppose that $f(X) \subset g(X)$ and at least one of $f(X), g(X)$ is a b_{bl}-complete subspace of (X, b_{bl}). Then

(i) If the property $G_{f,g(x_n)}$ is satisfied and $G_{gf} \neq \emptyset$, then f and g have a point of coincidence in X.

(ii) If x and y in X are points of coincidence of f and g such that $(x, y) \in E(G)$, then $x = y$. Hence, points of coincidence of f and g are unique in X. Moreover, if the pair (f, g) is weakly compatible, then f and g have a unique common fixed point in X.

Author Contributions: Conceptualization, T.D. and S.R.; Data curation, M.P.; Formal analysis, N.N. and T.D.; Investigation, N.N.; Methodology, M.D.l.S., M.P., N.N. and T.D.; Project administration, M.D.l.S.; Supervision, M.D.l.S. and S.R.

Funding: This research was funded by the Basque Government, grant number IT1207-19; the Serbian Ministry of Science and Technology, grant number OI 174010 and the Ministry of Education and Science and Technological Development of the Republic of Serbia, Grant number -174002 . The APC was funded by the Basque Government through grant IT1207-19.

Acknowledgments: The first author thanks the Basque Government for its support of this work through Grant IT1207-19, the second author is supported by the Serbian Ministry of Science and Technology OI 174010, the third author wish to thank the projects MNTRRS-174009, the fourth author is grateful for the financial support from the Ministry of Education and Science and Technological Development of the Republic of Serbia-174002.

Conflicts of Interest: The authors declare no conflict of interest.

References

1. Matthews, S.G. Partial metric topology. In Proceedings of the 8th Summer Conference on General Topology and Applications, Queens College, New York, NY, USA, 18–20 June 1992; Annals of the New York Academy of Sciences: New York, NY, USA, 1994; pp. 183–197.
2. Harandi, A.A. Metric-like spaces, partial metric spaces and fixed points. *Fixed Point Theory Appl.* **2012**, *2012*, 1–10.
3. Bakhtin, I.A. The contraction mapping principle in quasi-metric spaces. *Funct. Anal.* **1989**, *30*, 26–37.
4. Czerwik, S. Contraction mappings in b-metric spaces. *Acta Math. Inform. Univ. Ostrav.* **1993**, *1*, 5–11.
5. Shukla, S. Partial b-metric spaces and fixed point theorems. *Mediterranean J. Math.* **2014**, *11*, 703–711. [CrossRef]
6. Vujaković, J.; Aydi, H.; Radenović, S.; Mukheimer, A. Some remarks and new results in ordered partial b-metric spaces. *Mathematics* **2019**, *7*, 334. [CrossRef]
7. Alghmandi, M.A.; Hussain, N.; Salimi, P. Fixed point and coupled fixed point theorems on b-metric-like spaces. *J. Inequal. Appl.* **2013**, *402*, 1–25.
8. Agarwal, R.P.; Karapinar, E.; O'Regan, D.; Roldan-Lopez-de-Hierro, A.F. *Fixed Point Theory in Metric Type Spaces*; Springer International Publishing: Cham, Switzerland, 2015. [CrossRef]
9. Banach, S. Sur les opérations dans les ensembles abstraits et leur application aux équations intégrales. *Fundam. Math.* **1922**, *3*, 133–181. [CrossRef]
10. Gaba, Y.U.; Agyingi, C.A.; Leko, D.J. Chatterjea type fixed point in partial b-metric spaces. *arXiv* 2019, arXiv:1902.03108v1.
11. George, R.; Nabwey, H.A.; Ramaswamy, R.; Radenović, S. Some generalized contraction classes and common fixed points in b-metric spaces endowed with a graph. *Mathematics* **2019**, *7*, 754. [CrossRef]
12. George, R.; Tamrakar, E.; Vujaković, J.; Pathak, H.K.; Velusamy, S. (C, ψ^*, G) class of contractions and fixed points in a metric space endowed with a graph. *Mathematics* **2019**, *7*, 482. [CrossRef]
13. Kirk, W.A.; Shahzad, N. *Fixed Point Theory in Distance Spaces*; Springer International Publishing: Cham, Switzerland, 2014. [CrossRef]

14. Rhoades, B.E. Comparison of Various Definitions of Contractive Mappings. *Trans. Am. Math. Soc.* **1977**, *226*, 257–290. [CrossRef]
15. Alsulami, H.; Gulyaz, S.; Karapinar, E.; Erha, I.M. An Ulam stability result on quasi-*b*-metric like spaces. *Open Math.* **2016**, *14*, 1–15. [CrossRef]
16. Aydi, H.; Felhi, A.; Sahmim, S. On common fixed points for (α, ψ)-contractions and generalized cyclic contractions in *b*-metric-like spaces and consequences. *J. Nonlinear Sci. Appl.* **2016**, *9*, 2492–2510. [CrossRef]
17. Aydi, H.; Felhi, A.; Sahmim, S. Common fixed points via implicit contractions on *b*-metric-like spaces. *J. Nonlinear Sci. Appl.* **2017**, *10*, 1524–1537. [CrossRef]
18. Hammad, H.A.; De la Sen, M. Generalized contractive mappings and related results in *b*-metric-like spaces with an application. *Symmetry* **2019**, *11*, 667. [CrossRef]
19. Hussain, N.; Roshan, J.R.; Parvaneh, V.; Kadelburg, Z. Fixed points of contractive mappings in *b*-metric-like spaces. *Sci. World J.* **2014**, 471827. [CrossRef] [PubMed]
20. Mukheimer, A. $\alpha - \psi - \varphi$ contractive mappings in ordered partial *b*-metric spaces. *J. Nonlinear Sci. Appl.* **2014**, *7*, 168–179. [CrossRef]
21. Mustafa, Z.; Jaradat, M.M.M.; Jaradat, H.M. Some common fixed point results of graph on *b*-metric space. *J. Nonlinear Sci. Appl.* **2016**, *9*, 4838–4851. [CrossRef]
22. Nashine, H.K.; Kadelburg, Z. Existence of solutions of Cantilever Beam Problem via α-β-FG-contractions in *b*-metric-like spaces. *Filomat* **2017**, *31*, 3057–3074. [CrossRef]
23. Padhan, S.K.; Jagannadha Rao, G.V.V.; Nashine, H.K.; Agarwal, R.P. Some fixed point results for $(\beta$-ψ_1-$\psi_2)$-contractive conditions in ordered *b*-metric-like spaces. *Filomat* **2017**, *31*, 4587–4612. [CrossRef]
24. Shukla, S.; Radenović, S.; Rajić, V. Ć. Some common fixed point theorems in 0-σ-complete metric-like spaces. *Vietnam J. Math.* **2013**, *41*, 341–352. [CrossRef]
25. Singh, S.L.; Czerwik, S.; Krol, K.; Singh, A. Coincidences and fixed points of hybrid contractions. *Tamsui Oxf. J. Math. Sci.* **2008**, *24*, 401–416.
26. Zoto, K.; Radenović, S.; Ansari, A.H. On some fixed point results for (s, p, α)-contractive mappings in *b*-metric-like spaces and applications to integral equations. *Open Math.* **2018**, *16*, 235–249. [CrossRef]
27. Zoto, K.; Rhoades, B.E.; Radenović, S. Some generalizations for $(\alpha$-$\psi, \phi)$-contractions in *b*-metric-like spaces and an application. *Fixed Point Theory Appl.* **2017**, *26*. [CrossRef]
28. Todorčević, V. *Harmonic Quasiconformal Mappings and Hyperbolic Type Metrics*; Springer Nature: Cham, Switzerland, 2019. [CrossRef]
29. Todorčević, V. Subharmonic behavior and quasiconformal mappings. *Anal. Math. Phys.* **2019**, *9*, 1211–1225. [CrossRef]
30. Abbas, M.; Jungck, G. Common fixed point results for noncommuting mappings without continuity in cone metric spaces. *J. Math. Anal. Appl.* **2008**, *341*, 416–420. [CrossRef]
31. Jovanović, M.; Kadelburg, Z.; Radenović, S. Common fixed point results in metric-type spaces. *Fixed Point Theory Appl.* **2010**, 978121. [CrossRef]
32. Aleksić, S.; Mitrović, Z.D.; Radenović, S. Picard sequences in *b*-metric spaces. *Fixed Point Theory* 2019, accepted.
33. Jachymski, J. The contraction principle for mappings on a metric space with a graph. *Proc. Am. Math. Soc.* **2008**, *1*, 1359–1373. [CrossRef]
34. Beg, I.; But, A.R.; Radojević, S. The contraction principle for set valued mappings on a metric space with a graph. *Comput. Math. Appl.* **2010**, *60*, 1214–1219. [CrossRef]
35. Bojor, F. Fixed point of ϕ-contraction in metric spaces endowed with a graph. *Ann. Univ. Craiova Ser. Math. Inform.* **2010**, *37*, 85–92.
36. Bojor, F. Fixed point theorems for Reich type contractions on metric spaces with graph. *Nonlinear Anal.* **2012**, *75*, 3895–3901. [CrossRef]
37. Ćirić, Lj. *Some Recent Results in Metrical Fixed Point Theory*; University of Belgrade: Belgrade, Serbia, 2003.

mathematics

MDPI

Article

Complexity Analysis and Stochastic Convergence of Some Well-known Evolutionary Operators for Solving Graph Coloring Problem

Raja Marappan * and Gopalakrishnan Sethumadhavan

Department of Computer Applications, School of Computing, SASTRA Deemed University, Tirumalaisamudram 613401, India; sgk@mca.sastra.edu

* Correspondence: raja_csmath@cse.sastra.edu

Received: 20 January 2020; Accepted: 23 February 2020; Published: 25 February 2020

Abstract: The graph coloring problem is an *NP*-hard combinatorial optimization problem and can be applied to various engineering applications. The chromatic number of a graph G is defined as the minimum number of colors required to color the vertex set $V(G)$ so that no two adjacent vertices are of the same color, and different approximations and evolutionary methods can find it. The present paper focused on the asymptotic analysis of some well-known and recent evolutionary operators for finding the chromatic number. The asymptotic analysis of different crossover and mutation operators helps in choosing the better evolutionary operator to minimize the problem search space and computational complexity. The choice of the right genetic operators facilitates an evolutionary algorithm to achieve faster convergence with lesser population size N through an adequate distribution of promising genes. The selection of an evolutionary operator plays an essential role in reducing the bounds for *minimum color* obtained so far for some of the benchmark graphs. This research also focuses on the necessary and sufficient conditions for the global convergence of evolutionary algorithms. The stochastic convergence of recent evolutionary operators for solving graph coloring is newly analyzed.

Keywords: approximation methods; chromatic number; combinatorial optimization; complexity analysis; evolutionary approach; genetic algorithm; graph coloring; *NP*-hard; stochastic convergence

1. Introduction

A simple graph $G = (V, E)$ consists of a set of vertices $V = \{v_1, v_2, v_3 \ldots v_n\}$ and a set of edges $E = \{e_1, e_2, e_3 \ldots e_m\}$ such that each e_i $(1 \le i \le m)$ is uniquely associated with an unordered pair of vertices (v_j, v_k) $(1 \le j, k \le n)$ and $j \ne k$ [1,2]. Its adjacency matrix is denoted by $A(G)$, an $n \times n$ symmetric matrix where $A(j, k) = 1$ $(1 \le j, k \le n)$ if $(v_j, v_k) \in E(G)$; and $A(j, k) = 0$ $(1 \le j, k \le n)$ otherwise [2].

$\chi(G)$, the chromatic number of G, is defined as the minimum number of colors required for $V(G)$ so that no two adjacent vertices are of the same color [2]. The graph coloring problem (GCP) finds the value for $\chi(G)$ and applies it to register allocation, channel assignment, image segmentation, resource utilization, and scheduling [3–10]. With the increasing values of n, the complexity of determining $\chi(G)$ also increases. GCP is an *NP*-hard combinatorial optimization problem. Hence, a fast evolutionary and approximation method is expected to reduce the problem search space by maximizing the number of successful runs [1,11–16]. Tabu search [17], backtracking [18], branch and bound [11], evolutionary algorithm [19–21], branch and cut [22], particle swarm optimization (PSO) [23–26], ant colony optimization (ACO) [27], local and cuckoo search [28,29] are some existing methods for finding $\chi(G)$. Some recent applications of GCP are selective graph coloring [30,31], signed graphs coloring [32,33], scheduling, and resource allocation [7–10,34–38]. In comparison to other methods, GA is useful for solving multi-objective optimization problems with vast search space. Hence, it is expected to design new evolutionary operators to achieve faster stochastic convergence with minimal generations. Some recent evolutionary operators have achieved near-optimal solutions for some benchmark graphs [4,39–41]. Moreover, some recent genetic operators have increased the number of successful runs and reduced expected generations $\overline{g}$, expected crossovers $\overline{c}$, and the expected mutations $\overline{m}$. The notations $\overline{g}$, $\overline{c}$, and $\overline{m}$ indicate respectively the average number of generations, the average number of crossovers, and the average number of mutations performed to obtain the near-optimal solution for the specified number of total runs and generations during the execution of the program. This paper focused on the asymptotic analysis and stochastic convergence of some well-known and current evolutionary operators.

Some of the recent recombination operators are similar to mutation and are executed in place of recombination operation [2]. Some of the well-known and recent recombination operators operate on multiple parent gene sequences [2]. The order-based uniform crossover operator [42] uniformly selects genes from various parents. The penalty based crossover operator [43] generates offspring, which results in a minimum penalty. The one-point and two-point crossover operators generate random numbers and inherit genes from their parents. The graph adapted crossover generates genes based on conflicting and neighboring vertices. The union independent set (UIS) crossover [44] operator unifies the pairs of independent color sets from the selected parents. The greedy partition crossover (GPX) [45] operator divides $V(G)$ into k-sets and generates offspring. The multi-parent crossover [46] operator is the extension of GPX, which creates offspring from multiple parents. The well-informed partition crossover (WIPX) [47] operator selects the color classes from the randomly chosen parents based on a scoring function. The penalty-based color partitioning crossover (PCPX) [48] operator generates offspring based on the partition of $V(G)$ and its penalty. The degree based crossover (DBX) [48] operator applies the heuristics of the largest degree ordering (LDO) to order-based crossover or permutation one point crossover (POP) [40]. POP crossover performs an order-based permutation operation. The multi-point guided crossover (MPGX) [48] operator incorporates problem-specific knowledge. The merging crossover (MOX) [40] operator merges two-parent gene sequences and generates order-based offspring. The merging independent sets (MIS) [40] crossover groups the color sets of parents and generates offspring.

A recent crossover operator, the single parent conflict gene crossover (SPCGX), similar to mutation, is applied to the single parent to identify the conflicting genes and generate better offspring [49]. SPCGX is combined with the conflict gene removal (CGR) procedure to identify and remove some conflicting genes [50]. SPCGX is also connected with the advanced local guided search (ALGS) operator [2] to fine-tune the offspring further and also to reduce the computational complexity. The single parent conflict gene extended crossover (SPCGEX) operator is applied to the selected single parent for fixed iterations. Extended SPCGEX (ESPCGEX) extends SPCGEX with CGR for fixed iterations to produce better offspring. The multipoint SPCGX (MSPCGX) performs several crossovers for conflicting genes. Some of the recent genetic operators have achieved the better solution for some large DIMACS benchmark graphs in the reasonable expected generations [2].

The mutation operators operate on offspring generated by crossover operations. The random mutation operator [43] randomly swaps the color of each vertex with low mutation probability p_m. The polynomial mutation operator rounds off the floating-point color values. The swap mutation is applied to offspring generated by the DBX operator [51]. Problem-specific mutation [51] operators operate on the offspring generated by the MPGX operator. The order mutation (OM) [52] operator generates a random number r and performs r interchanges between vertices. The block mutation (BM) [52] performs the translation of blocks of successive vertices. The color spread mutation (CSM) [52] randomly selects the conflicting edges and moves the genes to new positions. The bad edge stretch mutation (BESM) [52] reduces the number of conflicting edges. The conflict gene mutation (CGM) [2] assigns the conflict-free integers to the offspring and performs fixed iterations. The CGM-CGR operator further reduces the number of conflicts in the offspring. The extended CGM (ECGM) operator fine-tunes the offspring by extending CGM for finite iterations. The multipoint CGM (MCGM) performs multipoint mutations at the conflicting vertices to reduce the number of conflicts.

Section 2 explains the need for complexity analysis and stochastic convergence for evolutionary operators. Section 3 focuses on the general structure of the genetic algorithm for solving GCP. Sections 4 and 5 explain some of the well-known and recent recombination and mutation operators. The necessary and sufficient condition for the global convergence of evolutionary algorithms and the stochastic convergence of recent genetic operators are newly analyzed in Section 6. Conclusions and future research directions are delineated in Section 7.

2. The Need for Complexity Analysis and Stochastic Convergence of Evolutionary Operators

GCP is very challenging for genetic algorithms because of its vast solution space. Hence, designing and choosing the right genetic operators in population-based methods are important for the following reasons [53–56]:

1. Solution space of GCP consists of n^n candidate solutions, and n^n rapidly grows with n. The computational complexity of finding $\chi(G)$ is proportional to m and n. If the solutions are represented using $\chi(G) < n$, then the solution space contains $\chi(G)^n$ assignments. Therefore, the total number of different assignments in the solution space is $\chi(G)!$
2. The computational complexity of GCP is reduced only when the values of $\overline{c}$, $\overline{m}$, and $\overline{g}$ are minimized.
3. The evolutionary operators are expected to improve the performance of the genetic algorithm by reducing the search space as well as by increasing the successful runs.
4. The evolutionary operators should improve the near-optimal solution obtained from the existing methods [11,22,25–27,57–59].
5. The evolutionary operators should avoid additional memory resources and complex computational operations.
6. The performance of operators depends on how many parent gene sequences are considered for crossover and mutation. It also depends on the update of gene sequences [4,39–41].
7. A smaller population size ($N \leq 15$) is expected to reduce the complexity and to achieve a sufficient convergence rate. The crossover and mutation operations can be embedded with the right search operators to reduce the value of $\overline{g}$ further.
8. The evolutionary operators should effectively distribute the promising genes from one generation to the next generation in order to reduce the fitness function values.
9. The graph characteristics, graph instances such as n, m, and graph density, can be considered for choosing the right genetic operators [60].
10. The evolutionary operators should quickly converge with the reduced computational complexity.

3. The General Structure of Genetic Algorithm for Solving GCP

Notations

1. Degree, $\delta(G)$ and $\Delta(G)$ The number of edges incident on v_i ($1 \leq i \leq n$) is the degree, $d(v_i)$ in G. Let $\Delta(G) = \max_{1\leq i\leq n}\{d(v_i)|v_i \in V(G)\}$ and $\delta(G) = \min_{1\leq i\leq n}\{d(v_i)|v_i \in V(G)\}$ be the maximum and minimum degrees of G respectively.
2. Encoding of Gene Sequence The color values of $V(G)$ are encoded as $(g_1, g_2, g_3 \cdots g_n)$, where g_j ($1 \leq j \leq n$) is a non-zero integer.
3. Density The density of G is $\frac{2m}{n(n-1)}$.
4. Population Size, Crossover, Mutation Probability p_c is the crossover probability, and the mutation probability is denoted as p_m. P_g is the population, which represents the collection of gene sequences for generation g. $N = |P_g|$ defines the population size for g.
5. Fitness Function $f(G)$, the general fitness function of G, is the number of distinct genes in the gene sequence [2]. The main objective of GCP is to minimize $f(G)$, hence, $\chi(G) - f(G) = 0$.

6. Conflicting Genes When $g_a = g_b$ and $(v_a, v_b) \in E(G)$, the genes g_a $(1 \le a \le n)$ and g_b $(1 \le b \le n)$ corresponding to vertices v_a and v_b conflict with the gene sequence $(g_1, g_2, g_3 \cdots g_n)$.

The general structure of the genetic algorithm for solving GCP is presented below.

Step 1: Initialization

First, initialize $g = 0$. The initialization of P_g randomly or heuristically generates values in [1, *minimum color*].

Step 2: Evaluation of Fitness Function and Selection of Individuals

The fitness of individuals in the population is evaluated, and one or more parent gene sequence(s) are selected for the generation of the subsequent gene sequences. Different selection methods can be applied to select gene sequences for crossover operation.

Step 3: Recombination or Crossover Operation

The recombination operation is performed on the selected gene sequence(s) with probability p_c (high value of p_c, for example, $p_c = 0.8$) to generate offspring.

Step 4: Performing Mutation

The crossover gene sequences or offspring are mutated with the chosen mutation probability p_m (low value of p_m, for example, $p_m = 0.2$) to generate new offspring.

Step 5: Termination

Find F_g (offspring) for the updated gene sequence. If F_g (offspring) = 0, the generation of offspring stops; otherwise, P_g is updated ($g = g + 1$). If F_g (offspring) < F_{g-1} (worst), P_g is updated by replacing the worst gene with offspring. Then transfer the control to Step 2 for the next generation.

The pseudo-code of the general genetic algorithm is presented below.

```
// Pseudo code of the Genetic Algorithm
Initialization of individuals;
While termination conditions are not satisfied, do
    Evaluate the fitness function and select better individuals;
    Perform the recombination operation;
    Perform the mutation operation;
    Apply elitism operation;
Print the near-optimal results;
```

4. Asymptotic Analysis of Some Existing Well-known and Recent Crossover Operators

Some of the existing well-known and recent crossover or recombination operators for solving GCP are presented in this section. The asymptotic complexity of these operators is also analyzed in Table 1. The simple control constructs if and if-else statements that can be implemented in a constant complexity, $\Theta(c)$. The control constructs *for*, *while*, *do-while*, and *repeat-until* can be implemented in a linear complexity, $\Theta(n)$. The quadratic complexity is represented in $\Theta(n^2)$. For example, color assignments of $V(G)$ can be verified in $\Theta(n^2)$ complexity.

1. Order-based Uniform Crossover

The order-based uniform crossover operator uniformly selects the genes from multiple parents. The procedure can be described in the following steps.

a. Select two parent gene sequences, say p_1 and p_2.
b. Generate the offspring by selecting the first *n*/2 length of genes from p_1 and the next *n*/2 length of genes from p_2.
c. Randomly select any two vertices, v_i and v_j in the offspring, and scramble all vertices between v_i and v_j.

Analysis: Steps (a), (b) and (c) take $\Theta(n)$ complexity. Thus the order-based uniform crossover takes $\Theta(n)$ complexity.

2. Penalty Based Crossover

The penalty based crossover operator generates the offspring, which results in a minimum penalty. The procedure is described in the following steps.

a. Select two parent gene sequences, say p_1 and p_2.
b. Generate the offspring by determining the color for each vertex in every parent by assigning colors, which results in a minimum penalty.

Analysis: Step (a) takes $\Theta(n)$ complexity. Step (b) takes $\Theta(n^2\ \chi(G))$ complexity. The computational complexity of penalty based crossover is $\Theta(n^2\ \chi(G))$.

3. One-point Crossover

The one-point crossover operator generates random numbers and inherits genes from their parents. The procedure is described in the following steps.

a. Select two parent gene sequences, say p_1 and p_2.
b. Choose a random point r in between $[1, n]$.
c. Generate the offspring based on r by inheriting genes from both parents.

Analysis: Step (a) takes $\Theta(n)$ complexity. Step (b) can be implemented in constant running time. Step (c) requires $\Theta(n)$ complexity. Thus the one-point crossover takes $\Theta(n)$ complexity.

4. Two-point Crossover

The two-point crossover operator generates random numbers and inherits genes from their parents. The procedure is described in the following steps.

a. Select two parent gene sequences, say p_1 and p_2.
b. Choose two random points, r and s in between $[1, n]$.
c. Generate the offspring based on r and s by inheriting genes from both parents.

Analysis: Step (a) takes $\Theta(n)$ complexity. Step (b) can be implemented in constant running time. Step (c) requires $\Theta(n)$ complexity. Thus the two-point crossover takes $\Theta(n)$ complexity.

5. Graph Adapted Crossover

The graph adapted crossover generates genes based on the conflicting and neighboring vertices. The procedure is described in the following steps.

a. Select two parent gene sequences, say p_1 and p_2.
b. Choose a random vertex v in $V(G)$.
c. Color the vertices with respect to the parents that have no conflicts for vertex v.
d. If p_1 and p_2 have conflicts with vertex v then assign the least color to its neighboring vertices.

Analysis: Step (a) takes $\Theta(n)$ complexity. Step (b) requires constant running time. Step (c) requires $\Theta(n^2)$ complexity. Step (d) requires $\Theta(n^2)$ complexity. The complexity of graph adapted crossover is $\Theta(n^2)$.

6. UIS Crossover

The UIS crossover operator unifies a pair of independent color sets from the selected parents. The procedure is described in the following steps.

a. Select two parent gene sequences, say p_1 and p_2.
b. Find the largest conflict-free subset of vertices having a color c in p_1.
c. Find the largest conflict-free subset in p_2 such that a common number of vertices in these subsets are maximal.
d. Repeat the steps (b) and (c) for every color.
e. Generate the offspring o_1 and o_2 by performing set union operation.

Analysis: Steps (a), (b) and (c) require $\Theta(n)$ complexity. Step (d) requires $\Theta(n\ \chi(G))$ complexity. Step (e) requires $\Theta(n)$ complexity. Thus the complexity of UIS crossover is $\Theta(n\ \chi(G))$.

7. Multi-Parent Crossover

Multi-parent crossover is an extension of GPX, which creates offspring from multiple parents. The procedure is described in the following steps.

a. Select the parent color classes from multiple parents.
b. Select the largest color class c in the population.
c. Add the color class c to the offspring.
d. Remove all vertices in c from all parent color classes.
e. Repeat the operation until the required number of color classes has reached.
f. Add the remaining vertices randomly to the color class.

Analysis: Steps (a), (b), (c) require $\Theta(n)$ complexity. Steps (d) and (e) require $\Theta(n^2)$ complexity. Step (f) requires $\Theta(n)$ complexity. Thus the complexity of multi-parent crossover is $\Theta(n^2)$.

8. WIPX Crossover

The WIPX operator selects the color classes from the randomly selected parents based on a scoring function. The procedure is described in the following steps.

a. Select the color classes from the randomly chosen parents based on a scoring function.
b. Determine the score of the color set as:

$$\text{Number of conflicts} - \frac{1}{|\mathrm{V}|}(\text{classSize} + \text{classDegree}/(\mathrm{n} * \mathrm{m}))$$

where classSize is the cardinality of the color set, and classDegree is the sum of the degrees of the color class vertices.
c. Apply the multi-parent crossover operation.

Analysis: Step (a) requires $\Theta(n)$ complexity. Step (b) requires $\Theta(n^2)$ complexity. Step (c) requires $\Theta(n^2)$ complexity. Thus the complexity of WIPX crossover is $\Theta(n^2)$.

9. GPX Crossover

GPX operator partitions $V(G)$ into k sets and generates the offspring. The procedure is described in the following steps.

a. Choose two parent configurations $p_1 = ({v_1}^1, {v_2}^1, {v_3}^1 \ldots {v_k}^1)$ and $p_2 = ({v_1}^2, {v_2}^2, {v_3}^2 \ldots {v_k}^2)$ where k is the number of color classes.

b. For each j ($1 \leq j \leq k$) do If j is odd then s =1, else s = 2; Select i such that $v_i{}^s$ has maximum vertices; $v_j = v_i{}^s$; Remove v_j from both p_1 and p_2
c. Choose a random color class and assign $V-(v_1 \cup v_2 \cup v_3 \ldots \cup v_k)$.
d. Generate an offspring $o_1 = (v_1, v_2, v_3 \ldots v_k)$.

Analysis: Step (a) requires $\Theta(n)$ complexity. Step(b) requires $\Theta(n^2)$ complexity. Steps (c) and (d) require $\Theta(n)$ complexity. Thus GPX crossover takes $\Theta(n^2)$ complexity.

10. PCPX Crossover

PCPX crossover operator generates offspring based on partitions of $V(G)$ and its penalty. The procedure is described in the following steps.

a. Select two parent gene sequences, say p_1 and p_2.
b. Consider one parent at a time.
c. Extract a subpartition $V'(G)$, which results in a minimum penalty among $V(G)$ of the main partition.
d. Color $V'(G)$ for the selected parent.
e. Remove $V'(G)$ in both parents.
f. Find the largest color partition for the next parent.
g. Repeat the procedure until all vertices in the offspring are colored.

Analysis: Step (a) requires $\Theta(n)$ complexity. Step(b) takes constant running time. Steps (c), (d), (e). (f) and (g) take $\Theta(n^2)$ complexity. Thus PCPX crossover takes $\Theta(n^2)$ complexity.

11. DBX Crossover

DBX crossover operator applies the heuristics of LDO in order based crossover or POP. The procedure is described in the following steps.

a. Apply the heuristics of LDO in order based crossover or POP operator.
b. Select a pair of random individuals, say P_1 and P_2.
c. Select a random crossover point r in P_1.
d. Copy the vertex colors starting at an initial point to r in P_1 into the offspring O_1;
e. Choose a vertex u in P_2 and v in O_1.
f. If u is not assigned an order in O_1 and if degree(v) < degree(u) then Find the vertex v having minimum order in O_1 with degree(v) < degree(u); Increment the vertex orders by 1 for all vertices having order $\geq v$; Assign vertex u to the order of vertex v; Else assign the least available order to vertex u.

Analysis: Step (a) requires $\Theta(n^2)$ complexity. The selection and copy operations in steps (b), (d), and (e) require $\Theta(n)$ complexity. Step(c) takes a constant running time. Step (f) requires $\Theta(n^2)$ complexity. Thus DBX takes $\Theta(n^2)$ complexity.

12. MPGX Crossover

MPGX operator incorporates problem-specific knowledge. The procedure is described in the following steps.

a. Select a pair of random individuals, say P_1 and P_2;
b. Choose a random crossover point;
c. Assign count = $Pc * n$;
d. While count $\neq 0$ do Select a random vertex i; c_1= color[i] in P_1; c_2 = color[i] in P_2; If $c_1 \neq c_2$ and if penalty of P_1 reduces with c_2 then replace c_1 by c_2; If $c_1 \neq c_2$ and if penalty of P_2 reduces with c_1 then replace c_2 by c_1; count = count − 1

e. Generate offspring.

Analysis: Step (a) requires $\Theta(n)$ complexity. Steps (b), (c), and (e) take a constant running time. Step (d) takes $\Theta(n^2)$ complexity. Thus MPGX crossover takes $\Theta(n^2)$ complexity.

13. POP Crossover

POP crossover performs an order based permutation operation. The procedure is described in the following steps.

a. Select two parent gene sequences, say p_1 and p_2.
b. Choose a random point r in between $[1, n]$.
c. Apply order based crossover.
d. Swap the first portion of strings and apply permutation on the remaining unused vertices copied in the sequence.

Analysis: Steps (a) and (c) require $\Theta(n)$ complexity. Step (b) requires a constant running time. Step (d) requires $\Theta(n^2)$ complexity. Thus the complexity of POP crossover is $\Theta(n^2)$.

14. MOX Crossover

The MOX operator performs merging of two-parent gene sequences and generates the order based offspring. The procedure is described in the following steps.

a. Select two parent gene sequences, say p_1 and p_2.
b. Merge p_1 and p_2 randomly into a single list, which consists of $2n$ elements.
c. Assign the first occurrence of each gene to offspring$_1$.
d. Assign the second occurrence of each gene to offspring$_2$.

Analysis: Steps (a) and (b) require $\Theta(n)$ complexity. Steps (c) and (d) take $\Theta(n^2)$ complexity. The complexity of MOX crossover is $\Theta(n^2)$.

15. MIS Crossover

The MIS crossover operator groups the color sets of parents and generates the offspring. The procedure is described in the following steps.

a. Select two parent gene sequences, say p_1 and p_2.
b. Group the color sets in both p_1 and p_2.
c. Copy the whole color sets from p_1 and p_2 into a single merged list.
d. Assign the first occurrence of each gene into offspring$_1$.
e. Assign the second occurrence of each gene into offspring$_2$.

Analysis: Steps (a), (b) and (c) require $\Theta(n)$ complexity. Steps (d) and (e) take $\Theta(n^2)$ complexity. The complexity of MIX crossover is $\Theta(n^2)$.

16. SPCGX Crossover

SPCGX operation is applied to the selected gene sequences, i and j, to generate two new gene sequences as offspring. The offspring, i' and j' are generated by a chosen crossover probability p_c. SPCGX operator is defined as follows [50]:

a. Let $i = (i_1, i_2, i_3 \ldots i_q \ldots i_r \ldots i_n)$ and $j = (j_1, j_2, j_3 \ldots j_s \ldots j_t \ldots j_n)$ be the two selected sequences.
b. Generate random probability p_{cr}.
c. If $p_{cr} > p_c$ go to step (d).

d. Find all pairs of conflicting edges in i and j.
e. Let (q, r) and (s, t) be such two pairs of conflicting edges (i.e., $i_q = i_r$ and $j_s = j_t$) in i and j, then generate two offspring: $(i_1, i_2, i_3 \ldots i_q \ldots i_r{+}1 \ldots i_n)$ and $(j_1, j_2, j_3 \ldots j_s \ldots j_t{+}1 \ldots j_n)$.
f. Repeat step (e) of crossover operation to all the identified conflicting pairs in step (d) of crossover operation and generate updated gene sequences $i' = (i_{1'}, i_{2'}, i_{3'} \ldots i_{f'} \ldots i_{g'} \ldots i_{n'})$ and $j' = (j_{1'}, j_{2'}, j_{3'} \ldots j_{k'} \ldots j_{l'} \ldots, j_{n'})$.

Analysis: Step (a) requires $\Theta(n)$ complexity. Steps (b), (c), and (e) take a constant running time. Steps (d) and (f) require $\Theta(n^2)$ complexity. SPCGX takes $\Theta(n^2)$ complexity.

17. SPCGX-CGR Crossover

SPCGX is applied with a CGR procedure with a chosen p_c. SPCGX-CGR is given in the following steps [50]:

a. Apply the steps (a) to (e) of SPCGX crossover.
b. Check if the new genes at r and t (i.e., i_r and j_t) corresponding to an edge (r, t) are in conflict with its corresponding adjacent genes. If so, generate $i_r = i_r + 1$ and $j_t = j_t + 1$ such that $i_r + 1 \leq \chi(G)$ and $j_t + 1 \leq \chi(G)$.
c. Repeat step (f) of SPCGX with CGR procedure to all the conflicting pairs and generate updated gene sequences: $i' = (i_{1'}, i_{2'}, i_{3'} \ldots i_{n'})$ and $j' = (j_{1'}, j_{2'}, j_{3'} \ldots j_{n'})$.

Analysis: Steps (a) and (c) require $\Theta(n^2)$ complexity. Step (b) takes a constant running time. SPCGX-CGR takes $\Theta(n^2)$ complexity.

18. SPCGEX Crossover

The SPCGEX operator is applied on the selected gene sequences $i = (i_1, i_2, i_3 \ldots i_n)$ and $j = (j_1, j_2, j_3 \ldots j_n)$ as follows [2]:

a. Randomly select any two most conflicting vertices, r, and t in gene sequences i and j.
b. Generate offspring $(i_1, i_2, i_3 \ldots i_r{+}1 \ldots i_n)$ and $(j_1, j_2, j_3 \ldots j_t{+}1 \ldots j_n)$ such that $i_r + 1 \leq$ minimum color and $1 + j_t \leq$ *minimum color*.
c. Repeat the steps (a) and (b) for fixed iterations to obtain the updated offspring i' and j' as $(i_{1'}, i_{2'}, i_{3'} \ldots i_{n'})$ and $(j_{1'}, j_{2'}, j_{3'} \ldots j_{n'})$.

Analysis: Steps (a) and (c) require $\Theta(n^2)$ complexity. Step (b) takes a constant running time. SPCGEX takes $\Theta(n^2)$ complexity.

19. ESPCGEX Crossover

The steps in ESPCGEX crossover are defined as follows [2].

a. Perform SPCGEX crossover steps (a) and (b).
b. If the new genes i_r and j_t are conflicting, then apply SPCGEX at the conflicting edges (q, r) and (s, t).
c. Repeat the steps (a) and (b) for fixed iterations to obtain the updated offspring i' and j' as $(i_{1'}, i_{2'}, i_{3'} \ldots i_{n'})$ and $(j_{1'}, j_{2'}, j_{3'} \ldots j_{n'})$.

Analysis: Steps (a), (b) and (c) require $\Theta(n^2)$ complexity. Hence ESPCGEX takes $\Theta(n^2)$ complexity.

20. MSPCGX Crossover

The steps in MSPCGX crossover are defined as follows [2].

a. Find the conflicting edges that count to every vertex in gene sequences j and i.

b. Determine the conflicting vertices in j and i.
c. Select arbitrarily conflicting vertices, r in i and t in j.
d. Update i and j as ($i_1, i_2, i_3 \ldots i_r+1 \ldots i_n$) and ($j_1, j_2, j_3 \ldots j_t+1 \ldots j_n$) such that *minimum color* $\geq i_r + 1$ and *minimum color* $\geq j_t + 1$.
e. If the new genes i_r and j_t are conflicting, then generate $i_r = i_r + 1$ and $j_t = j_t + 1$ such that minimum color $\geq i_r + 1$ and minimum color $\geq j_t + 1$.
f. Repeat the steps (d)–(e) for remaining conflicting vertices and update i' and j' as ($i_{1'}, i_{2'}, i_{3'} \ldots i_{n'}$) and ($j_{1'}, j_{2'}, j_{3'} \ldots j_{n'}$).

Analysis: Steps (a), (b), (e) and (f) require $\Theta(n^2)$ complexity. Step (c) takes $\Theta(n)$ complexity. Step (d) takes a constant running time. MSPCGX takes $\Theta(n^2)$ complexity.

Table 1. Asymptotic Complexity of Well-known and Recent Crossover Operators.

SNo	Crossover Operator	Asymptotic Complexity
1	Order-based Uniform Crossover	$\Theta(n)$
2	Penalty Based Crossover	$\Theta(n^2\ \chi(G))$
3	One-point Crossover	$\Theta(n)$
4	Two-point Crossover	$\Theta(n)$
5	Graph Adapted Crossover	$\Theta(n^2)$
6	UIS Crossover	$\Theta(n\ \chi(G))$
7	Multi-Parent Crossover	$\Theta(n^2)$
8	WIPX Crossover	$\Theta(n^2)$
9	GPX Crossover	$\Theta(n^2)$
10	PCPX Crossover	$\Theta(n^2)$
11	DBX Crossover	$\Theta(n^2)$
12	MPGX Crossover	$\Theta(n^2)$
13	POP Crossover	$\Theta(n^2)$
14	MOX Crossover	$\Theta(n^2)$
15	MIS Crossover	$\Theta(n^2)$
16	SPCGX Crossover	$\Theta(n^2)$
17	SPCGX-CGR Crossover	$\Theta(n^2)$
18	SPCGEX Crossover	$\Theta(n^2)$
19	ESPCGEX Crossover	$\Theta(n^2)$
20	MSPCGX Crossover	$\Theta(n^2)$

5. Asymptotic Analysis of Some Existing Well-known and Recent Mutation Operators

Some of the existing well-known and recent mutation operators for solving GCP are presented in this section. The asymptotic complexity of these operators is also analyzed. The asymptotic complexity of these operators is also analyzed in Table 2.

1. Random Mutation

The random mutation operator randomly swaps the color of each vertex with a low mutation probability p_m. The procedure is described in the following steps.

a. Select crossover offspring o_1.
b. Randomly swap the color class of each vertex with a low p_m.

Analysis: Step (a) requires a constant running time. Step (b) takes $\Theta(n)$ complexity. Thus the random mutation takes $\Theta(n)$ complexity.

2. Polynomial Mutation

The polynomial mutation operator rounds off real values of colors to the nearest integer. The procedure is described in the following steps.

a. Select crossover offspring o_1.
b. Choose the random point and round off real values of colors to the nearest integer.

Analysis: Steps (a) and (b) require a constant running time. Thus the polynomial mutation takes $\Theta(c)$ complexity.

3. Problem-Specific Mutation 1

The problem-specific mutation operator is defined with problem-specific knowledge. It is operating on the offspring generated by MPGX operator. The procedure is described in the following steps.

a. Select crossover offspring o_1.
b. Find the missing color c_1 in the range of maximum color [1, maximum color] used.
c. Identify all the vertices $V'(G)$ assigned with maximum color.
d. Replace colors of $V'(G)$ by c_1.

Analysis: Step (a) requires a constant running time. Steps (b) and (c) take $\Theta(n^2)$ complexity. Step (d) takes $\Theta(n)$ complexity. This mutation takes $\Theta(n^2)$ complexity.

4. Problem-Specific Mutation 2

The problem-specific mutation operator is defined with problem-specific knowledge. It operates on the offspring generated by MPGX operator. The procedure is described in the following steps.

a. Choose color c_1 randomly within one less than [1, maximum color].
b. Identify all the vertices $V'(G)$ assigned to a maximum color.
c. Replace colors of $V'(G)$ by c_1.

Analysis: Step (a) requires a constant running time. Step (b) takes $\Theta(n^2)$ complexity. Step (c) takes $\Theta(n)$ complexity. This mutation takes $\Theta(n^2)$ complexity.

5. Problem-Specific Mutation 3

The problem-specific mutation operator is defined with problem-specific knowledge. It operates on the offspring generated by MPGX operator. The procedure is described in the following steps.

a. Choose a vertex v randomly.
b. Choose a color c randomly for v.
c. If the penalty is reduced to v with color c, then set color[v] = c.

Analysis: Steps (a) and (b) require a constant running time. Step (c) takes $\Theta(n^2)$ complexity. This mutation takes $\Theta(n^2)$ complexity.

6. OM

The OM operator generates a random number r and performs r interchanges between vertices. The procedure is described in the following steps.

a. Let $(v_1, v_2, v_3 \ldots v_n)$ be the crossover offspring.
b. Generate a random number r;
c. Generate the offspring by performing r interchanges between vertices.

Analysis: Steps (a) and (b) require a constant running time. Performing r interchanges in Step(c) can also be done in linear time. The OM mutation takes $\Theta(n)$ complexity.

7. BM

The BM operation performs a translation of blocks of $k > 0$ successive vertices. The procedure is described in the following steps.

a. Let $(v_1, v_2, v_3 \ldots v_n)$ be the crossover offspring.
b. If $k = 2$ and $i \in [1, n-1]$, $j \in [1, n]$ are randomly generated then generate the offspring $(v_1, \ldots .v_{i-1}, v_{i+2} \ldots v_j, v_i, v_{i+1}, v_{j+1} \ldots)$.

Analysis: Step (a) requires a constant running time. Step (b) can take $\Theta(n)$ complexity. Thus BM operation takes $\Theta(n)$ complexity.

8. CSM

The CSM randomly selects the conflicting edge and moves the genes to new positions. The procedure is described in the following steps.

a. Let $(v_1, v_2, v_3 \ldots v_n)$ be the crossover offspring.
b. Choose a randomly selected bad edge $e = (v_i, v_j)$ and $e' = (v_k, v_l)$ be the next bad edge such that $(i < j < k < l)$.
c. Move the vertices from positions $i + 1$ to $l - 1$ to randomly chosen new positions.

Analysis: Step (a) requires a constant running time. Steps (b) and (c) need $\Theta(n)$ complexity.

9. BESM

The BESM operation reduces the number of conflicting edges. The procedure is described in the following steps.

a. Let $(v_1, v_2, v_3 \ldots v_n)$ be the crossover offspring.
b. Choose a random bad edge $e = (v_i, v_j)$.
c. Choose a direction either left-to-right or right-to-left randomly.
d. Let v_k be the farthest vertex adjacent to v_i so that $k > j$.
e. Let $e' = (v_l, v_m)$ be the first bad edge so that $l > k$.
f. If there is no vertex v_k or edge e' with the specified properties, the mutation is void. Otherwise, the vertex v_i is moved to the position $m + 1$.

Analysis: Steps (a), (c) require a constant running time. Step (b) takes $\Theta(n^2)$ complexity. Steps (d), (e) and (f) take $\Theta(n)$ complexity. Hence BESM takes $\Theta(n^2)$ complexity.

10. CGM

The CGM operates on offspring i' and j'. The steps are given below [49]:

a. Select any two conflicting genes $i'_{f'} = i'_{g'}$ and $j'_{k'} = j'_{l'}$ in i' and j'.
b. Update i' and j' as $(i_{1'}, i_{2'}, i_{3'} \ldots i_{f'}{-}1 \ldots i_{n'})$ and $(j_{1'}, j_{2'}, j_{3'} \ldots j_{k'}{-}1 \ldots j_{n'})$ such that $1 \leq i_{f'}{-}1$ and $1 \leq j_{k'}{-}1$.
c. Update i'' and j'' as $(i_{1''}, i_{2''}, i_{3''} \ldots i_{n''})$ and $(j_{1''}, j_{2''}, j_{3''} \ldots j_{n''})$ by repeating steps (a)-(b) for fixed iterations.

Analysis: Steps (a) and (c) require $\Theta(n^2)$ complexity. Step (b) requires a constant running time. Hence CGM mutation takes $\Theta(n^2)$ complexity.

11. CGM-CGR Mutation

The CGM-CGR mutation is applied to the offspring i' and j' and its steps are given below [50]:

a. Perform CGM steps (a)–(b).
b. If the new genes $i_{f'}$ and $j_{k'}$ are conflicting then update $i_{f'} = i_{f'}-1$ and $j_{k'} = j_{k'}-1$ such that $1 \leq i_{f'}-1$ and $1 \leq j_{k'}-1$.
c. Update i'' and j'' as $(i_{1''}, i_{2''}, i_{3''} \ldots i_{n''})$ and $(j_{1''}, j_{2''}, j_{3''} \ldots j_{n''})$ by repeating steps (a)–(b) for fixed iterations.

Analysis: Steps (a), (b) and (c) require $\Theta(n^2)$ complexity. Hence CGM-CGR mutation takes $\Theta(n^2)$ complexity.

12. ECGM

The ECGM operation is defined as follows [2]:

a. Perform CGM steps (a) and (b).
b. If new genes $i'_{f'}$ and $j'_{k'}$ are conflicting, then perform CGM at f' and k'.
c. Update i'' and j'' as $(i_{1''}, i_{2''}, i_{3''} \ldots i_{n''})$ and $(j_{1''}, j_{2''}, j_{3''} \ldots j_{n''})$ by repeating steps (a) and (b) for fixed iterations.

Analysis: Steps (a), (b) and (c) require $\Theta(n^2)$ complexity. Hence ECGM takes $\Theta(n^2)$ complexity.

13. MCGM

The MCGM operation is defined as follows [2]:

a. Find the conflict edges count to every vertex in i' and j'.
b. Determine the conflict vertices in i' and j'.
c. Select arbitrarily two conflicting vertices f' in i' and k' in j'.
d. Update i' and j' as $(i_1', i_2', i_3' \ldots i_f'-1 \ldots i_n')$ and $(j_1', j_2', j_3' \ldots j_k'-1 \ldots j_n')$ such that $1 \leq i_f'-1$ and $1 \leq j_k'-1$.
e. If new genes $i'_{f'}$ and $j'_{k'}$ are conflicting then perform CGM at f' and k'.
f. Update i'' and j'' as $(i_{1''}, i_{2''}, i_{3''} \ldots i_{n''})$ and $(j_{1''}, j_{2''}, j_{3''} \ldots j_{n''})$ by repeating steps (d) and (e) for fixed iterations.

Analysis: Steps (a), (b), (e) and (f) require $\Theta(n^2)$ complexity. Steps (c) and (d) take $\Theta(c)$ complexity. Hence MCGM takes $\Theta(n^2)$ complexity.

Table 2. Asymptotic Complexity of Well-known and Recent Mutation Operators.

SNo	Mutation Operator	Asymptotic Complexity
1	Random Mutation	$\Theta(n)$
2	Polynomial Mutation	$\Theta(c)$
3	Problem-Specific Mutation 1	$\Theta(n^2)$
4	Problem-Specific Mutation 2	$\Theta(n^2)$
5	Problem-Specific Mutation 3	$\Theta(n^2)$
6	OM	$\Theta(n)$
7	BM	$\Theta(n)$
8	CSM	$\Theta(n)$
9	BESM	$\Theta(n^2)$
10	CGM	$\Theta(n^2)$
11	CGM-CGR	$\Theta(n^2)$
12	ECGM	$\Theta(n^2)$
13	MCGM	$\Theta(n^2)$

6. Global Convergence of Evolutionary Algorithms and Stochastic Convergence of Recent Genetic Operators

The necessary and sufficient condition for the global convergence of evolutionary algorithms is proved in this section [61,62]. The stochastic convergence of some recent genetic operators for solving GCP is also proved in the present section [2].

6.1. Markov Chain Model of Evolutionary Algorithms

Consider the following optimization problem:

Optimize $\{f(x); x \in S\}$ where S is a measurable space, $f(x)$ is the objective or fitness function where the absolute value of $(f(x))$ is $< +\infty$.

Represent the optimal solution set by $S^* = \{x^*$ such that $f(x^*) =$ optimize $f(x), x \in S\}$.

Let $\mu(S)$ be a measure to space S with $\mu(S^*) > 0$.

Consider the set $S^*_\delta = \{x$ such that $f(x^*) - f(x) < \delta\}$ where $\delta > 0$ is a small number.

Choose an appropriate δ to satisfy $\mu(S^*_\delta) > 0$.

The space S is called an individual space where each state s in S is called an individual.

N represents population size.

$X = S^N$ is the population space.

Denote the elements in x as $x = \{x_1, x_2, x_3 \ldots x_N\}$.

Denote the optimal solution set $X^* = \{x^*$ such that $\exists\, x_i \in x$: $x_i \in S^*\}$.

An evolutionary algorithm to solve this optimization problem is formulated as follows:

a. Initialization: Generate initial population P_0 at generation $g = 0$.
b. Crossover operation: Update the new population P_t by a crossover operator $P_C(g)$.
c. Mutation operation: Update the new population P_t by a mutation operator $P_M(g)$.
d. Selection operation: Select a new population from $P_M(g)$ by a selection operator $P_S(g)$.
e. Termination: If termination conditions hold, then stop; otherwise $g = g + 1$ and update P_g.

Since the state of P_{g+1} is only dependent on the state of P_g, then $\{P_g$: $g = 0, 1, 2 \ldots\}$ can be modeled by a Markov chain with the following transition function [61,62]:

$$P(g;\ x,\ dy) \;=\; P(P_g{+}_1 = dy \text{ such that } P_g = x) = \int_{u \in X}\int_{v \in X} P_C(g;x,\ du)P_M(g;u,\ dv)P_S(g;v,\ dy) \quad (1)$$

Then $\{P_g$: $g = 0, 1, 2 \ldots\}$ converges to X^*, if for any initial population P_0,

$$\lim_{g \to +\infty} \mu_g(X^*) = 1 \text{ where } \mu_g(X^*) = \mu(P_g \in X^*).$$

6.2. Convergence Conditions of Evolutionary Algorithms

The necessary and sufficient condition for the convergence of evolutionary algorithms is as follows [61,62]:

Let $\{P_g$: $g = 0, 1, 2 \ldots\}$ be the Markov chain given by (1). Define $\alpha(g)$ as the difference between flow from the non-optimal solution set to the optimal solution set and vice versa.

$$\alpha(g) = \int_{x \notin X^*} P(P_{g+1} \in X^* | \, P_g = x)\ \mu_g(dx) - \int_{x \in X^*} P(P_{g+1} \notin X^* | \, P_g = x)\ \mu_g(dx)$$

Then $\{P_g$: $g = 0, 1, 2 \ldots\}$ converges to the optimal solution set X^* iff

$$\sum_{g=0}^{+\infty} \alpha(g) = 1 - \mu_0(X^*)$$

Proof: For any given generation g, we have

$$\mu_{g+1}(X^*) = \int\limits_{x \notin X^*} P(P_{g+1} \in X^* | P_g = x)\, \mu_g(dx) + \int\limits_{x \in X^*} P(P_{g+1} \notin X^* | P_g = x)\, \mu_g(dx)$$

$$\mu_{g+1}(X^*) = \mu_g(X^*) - \int\limits_{x \in X^*} P(P_{g+1} \notin X^* | P_g = x)\, \mu_g(dx) + \int\limits_{x \notin X^*} P(P_{g+1} \in X^* | P_g = x)\, \mu_g(dx)$$

$$\mu_{g+1}(X^*) - \mu_g(X^*) = \alpha(g)$$

$$\mu_{g+1}(X^*) - \mu_0(X^*) = \sum_{k=0}^{g} \alpha(k)$$

$$\lim_{g \to +\infty} \mu_{g+1}(X^*) = 1 \iff \sum_{k=0}^{+\infty} \alpha(k) = 1 - \mu_0(X^*)$$

The initial population P_0 is not optimal, that is, $\mu_0(X^*) = 0$. Then

$$\sum_{g=0}^{+\infty} \alpha(g) = 1$$

6.3. Stochastic Convergence of Recent Evolutionary Operators

Some of the recent genetic operators like SPCGX, SPCGEX, ESPCGEX, ECGM, MSPCGX, and MCGM modify P_g through successive stochastic transformations, which can be analyzed by the Markovian model. The subsequent value of P_{g+1} is generated stochastically, and the evolutionary algorithm converges when the reachability and monotone conditions are fulfilled [2,41,63]. Some of the recent genetic operators converge stochastically.

6.3.1. Analyzing the Reachability Condition from j′

If j and j'' be any two individuals in the finite search space S, then j'' is reachable from individual j. That is, $j' = (j_{1'}, j_{2'}, j_{3'} \ldots j_{n'})$ and $j''= (j_{1''}, j_{2''}, j_{3''} \ldots j_{n''})$ are the two gene sequences generated from the gene sequences $j = (j_1, j_2, j_3 \ldots j_n)$ and $j_{'} = (j_{1'}, j_{2'}, j_{3'} \ldots j_{n'})$ respectively. If (0 < Probability {j'' = (*crossover* (j) & *mutation* (j'))} < 1) then the individual j'' is reachable from individual j' [2].

First, consider the effect of crossover operation. Let S be the finite search space of the problem. Consider a gene sequence $j \in S$ such that $j = (j_1, j_2, j_3 \ldots j_n)$. The proposed crossover operators (incremental operators) SPCGX and SPCGEX produce a new gene as $j_r = j_r + 1$ if *minimum color* $\geq j_r + 1$. Hence the new gene takes one of the values in (j_r, j_r+1). The CGR operator further checks the conflict at vertex r. If there is a conflict, then the new gene at vertex r is generated as $j_r = j_r + 1$ such that *minimum color* $\geq j_r + 1$. Clearly, j_r assumes the value in (j_r, $j_r + 1$, $j_r + 2$). Thus the required probability of assigning a new gene at the conflicting vertex is 1/3 for a single crossover operation. Assume that the crossover operation is performed for k times.

⇨ Probability {j' = *crossover* (j)} = $1/3^k > 0$ ($1 \leq j, j' \leq N$), k > 0 (2)

Now consider the effect of mutation operation. The recent mutation operators set a value of the conflicting gene in [1, f(G)]. Thus the required probability of assigning a gene at the conflicting vertex is 1/f(G) for a single mutation operation. Assume that the mutation operation is performed for l times.

⇨ Probability {j'' = *mutation* (j')} = $1/f(\mathrm{G})^l > 0, l > 0$. (3)

For the chosen crossover probability p_c and mutation probability p_m, the probability of generating new gene sequence j'' is computed using the multiplication theorem of probability. The crossover and mutation operations are mutually independent. Hence the required probability of generating j'' is computed as follows:

Probability of {j'' = (*crossover* (j) & *mutation* (j'))} ≥

(Crossover probability) × (Probability of generating j' from j) × (Mutation probability) × (Probability of generating j'' from j')

That is

$$\text{Probability of } \{j'' = (\text{crossover } (j) \ \& \ \text{mutation } (j'))\} \geq p_c.\ \text{Probability } \{j' = \text{Crossover } (j)\}.p_m.\ \text{Probability } \{j'' = \text{Mutation } (j')\} \tag{4}$$

$$\text{Probability of } (j'' = (crossover(j) \ \& \ mutation(j'))\} \geq p_c\, p_m \frac{1}{3^k} \frac{1}{(f(G))^1} > 0$$

Hence the gene sequence j'' is reachable from j'.

6.3.2. Analyzing whether P_g is Monotone

During the iterations of an evolutionary algorithm, P_g is modified by successive probabilistic transformations. The proportionate fitness selection selects one worst and two better gene sequences in every generation as follows:

a. Evaluate $F_g(a)$ for each gene sequence a ($1 \leq a \leq N$) in P_g.
b. Determine $p(a)$ using

$$p(a) = F_g(a) / \sum_{a=1}^{N} F_g(a)$$

c. Compute $E(a) = N\,p(a)$ for all a ($1 \leq a \leq N$).
d. Select the better gene sequences i and j and the worst gene sequence w.

The probability of selection of i and j is always greater than zero. P_g is then updated by performing the elitism operation.

The recombination and mutation operations are performed based on the values of p_{cr} and p_{mr}. For small graphs, some of the recent operators monotonically reduce F_g (i) and F_g (j) during the modification of population by a finite number of probabilistic transformations [2]:

1. The values of F_g (i) and F_g (j) monotonically decreases in subsequent generations and converge to a better near-optimal solution.

   That is, either $F_0(i) \geq F_1(i) \geq F_2(i) \geq \ldots \geq (F_q(i) = 0)$ or $F_0(j) \geq F_1(j) \geq F_2(j) \geq \ldots \geq (F_q(j) = 0)$.

2. F_g (i) and F_g (j) monotonically decrease for a finite number of generations (say t).

i.e., $F_0(i) \geq F_1(i) \geq F_2(i) \geq \ldots \geq F_t(i)$ and $F_0(j) \geq F_1(j) \geq F_2(j) \geq \ldots \geq F_t(j)$.

Then, based on the values of p_{cr} and p_{mr}, the fitness-function values again increase for a finite number of generations (say s).

i.e., $F_t(i) \leq F_{t+1}(i) \leq \ldots \leq F_s(i)$ and $F_t(j) \leq F_{t+1}(j) \leq \ldots \leq F_s(j)$.

For complex graphs, the successive probabilistic transformations find a better near-optimal solution after a finite number of generations. Hence, the successive probabilistic changes utilize P_g as a monotone in some of the subsequent generations to achieve a better near-optimal solution. These successive probabilistic changes, applying for over a large number of generations, for example, 1000 runs each with 5,000,000 generations of the genetic algorithm, achieve a better near-optimal solution [2].

7. Conclusions and Future Work

GCP is an *NP*-hard combinatorial optimization problem and can be applied to innumerable engineering problems. $\chi(G)$ can be obtained using different approximations and evolutionary methods. In comparison to other methods, genetic algorithms are useful in solving multi-objective optimization problems with vast search space [2]. Hence, it is expected to design new evolutionary operators to achieve faster stochastic convergence with a minimal number of generations.

This paper explored the asymptotic analysis of some well-known and recent evolutionary operators for finding chromatic numbers. The asymptotic analysis of different crossover and mutation operators helps in minimizing problem search space and computational complexity. The choice of the right genetic operators facilitates an evolutionary algorithm to achieve faster convergence with lesser population size (N) through an effective distribution of promising genes. The selection of evolutionary operators also plays an essential role in reducing the bounds for *minimum color* obtained for the benchmark graphs [2]. The necessary and sufficient condition for global convergence of the evolutionary algorithm and the stochastic convergence of some recent evolutionary operators for solving GCP are analyzed in this paper.

Our current work enlightens some new research directions [64–66].

i. The recent genetic operators can be combined with better local search strategies in order to further reduce the computational complexity of GCP.
ii. The self-adaptive evolutionary operators with heuristics can be designed to find chromatic numbers as well as can be applied to channel allocation problem (CAP) and scheduling problems. CAP is an extension of GCP, which assigns channels to mobile stations.
iii. The performances of some recent operators can be evaluated under different parameter values, such as graph density, N, p_c, and p_m.
iv. Some of the recent operators can be combined to reduce $\overline{g}$.
v. The stochastic convergence of genetic algorithms can be extended to multi-objective optimization problems, and the necessary conditions for the global convergence of multi-objective optimization problems can be analyzed mathematically.
vi. An algebraic framework that may lead to a unification of different evolutionary operators for combinatorial problems can be designed and analyzed [64–66].

Author Contributions: Methodology, Software, Writing-original draft and formal analysis, R.M.; formal analysis, writing-review and editing, G.S. All authors have read and agreed to the published version of the manuscript.

Funding: This research received no external funding.

Acknowledgments: The authors would like to acknowledge the support rendered by the management of SASTRA Deemed University for providing the financial support and BRNS, India for providing the workstation, which was procured for the project No. 2013/36/40-BRNS/2305 for execution [2]. The authors would like to thank Gary Lewandowski and Michael Trick for uploading the graph repository in W3C [2,60].

Conflicts of Interest: The authors declare no conflict of interest.

References

1. Balakrishnan, R.; Ranganathan, K. *A Textbook of Graph Theory*, 1st ed.; Springer-Verlag Publisher: New York, NY, USA, 2000.
2. Marappan, R.; Sethumadhavan, G. Solution to Graph Coloring using Genetic and Tabu Search Procedures. *Arab. J. Sci. Eng.* **2018**, *43*, 525–542. [CrossRef]
3. Garey, M.R.; Johnson, D.S. *Computers and Intractability: A Guide to the Theory of NP-Completeness*; Freeman and Company: San Francisco, CA, USA, 1979.
4. Maitra, T.; Pal, A.J.; Bhattacharyya, D.; Kim, T.H. Noise Reduction in VLSI Circuits using Modified GA Based Graph Coloring. *Int. J. Control Autom.* **2010**, *3*, 37–44.
5. Yoshino, J.; Ohtomo, I. Study on efficient channel assignment method using the genetic algorithm for mobile communication systems. *Soft Comput.* **2005**, *9*, 143–148. [CrossRef]

6. Chen, X.; Zu, Y.; Xu, J.; Wang, Z.; Yao, B. Vertex-Distinguishing E-Total Colorings of Graphs. *Arab. J. Sci. Eng.* **2011**, *36*, 1485–1500. [CrossRef]
7. Abdelfattah, M.; Shawish, A. Automated Academic Schedule Builder for University's Faculties. In Proceedings of the World Congress on Engineering 2016, London, UK, 29 June–1 July 2016.
8. Saharan, S.; Kumar, R. Graph Coloring based Optimized Algorithm for Resource Utilization in Examination Scheduling. *Appl. Math. Inf. Sci.* **2016**, *10*, 1193–1201. [CrossRef]
9. Tawfiq, F.M.O.; Al-qahtani, K.K.S. Graph Coloring Applied to Medical Doctors Schedule. In Proceedings of the 10th International Conference on Advanced Engineering Computing and Applications in Sciences, Venice, Italy, 9–13 October 2016.
10. Thevenin, S.; Zufferey, N.; Potvin, J.Y. Graph multi-coloring for a job scheduling application. *CIRRELT* 2016. [CrossRef]
11. Mehrotra, A.; Trick, M.A. A Column Generation Approach for Graph Coloring. *Inf. J. Comput.* **1995**, *8*, 344–354. [CrossRef]
12. Szép, T.; Mann, Z.Á. Graph coloring: The more colors, the better? In Proceedings of the 11th International Symposium on Computational Intelligence and Informatics (CINTI), Budapest, Hungary, 18–20 November 2010.
13. Galinier, P.; Hertz, A. A survey of local search methods for graph coloring. *Comput. Oper. Res.* **2006**, *33*, 2547–2562. [CrossRef]
14. Johnson, D.S.; Aragon, C.R.; McGeoch, L.A.; Schevon, C. An Experimental Evaluation; Part II, Graph Coloring and Number Partitioning. *Oper. Res.* **1991**, *39*, 378–406. [CrossRef]
15. Johnson, D.S.; Trick, M.A. *Cliques, Coloring, and Satisfiability*; American Mathematical Society: Providence, RI, USA, 1993; Volume 26.
16. Mizuno, K.; Nishihara, S. Constructive generation of very hard 3-colorability instances. *Discret. Appl. Math.* **2008**, *156*, 218–229. [CrossRef]
17. Hertz, A.; de Werra, D. Using tabu search techniques for graph coloring. *Computing* **1987**, *39*, 345–351. [CrossRef]
18. Monasson, R. On the Analysis of Backtrack Procedures for the Coloring of Random Graphs. *Lect. Notes Phys.* **2004**, *650*, 235–254.
19. Eiben, A.E.; Van Der Hauw, J.K.; Van Hemert, J.I. Graph Coloring with Adaptive Evolutionary Algorithms. *J. Heuristics* **1998**, *4*, 25–46. [CrossRef]
20. Kumar, R.; Tolay, P.; Tiwary, S. Enhancing solution quality of the biobjective graph coloring problem using hybridization of EA: Biobjective graph coloring problem. In Proceedings of the Genetic and Evolutionary Computation Conference (GECCO), Atlanta, GA, USA, 12–16 July 2008; ACM Press: New York, NY, USA, 2008; pp. 547–554.
21. Deb, K. *Multi-Objective Optimization Using Evolutionary Algorithms*; John Wiley & Sons: Hoboken, NJ, USA, 2001.
22. Méndez-Díaz, I.; Zabala, P. A Branch and Cut algorithm for graph coloring. *Discret. Appl. Math.* **2006**, *154*, 826–847. [CrossRef]
23. Cases, B.; Hernandez, C.; Grana, M.; D'Anjou, A. On the ability of Swarms to compute the 3-coloring of graphs. In *Artificial Life*; MIT Press: Cambridge, MA, USA, 2008.
24. Graña, M.; Cases, B.; Hernandez, C.; D'Anjou, A. Further Results on Swarms Solving Graph Coloring. In Proceedings of the ICCSA 2010: Computational Science and Its Applications, Fukuoka, Japan, 23–26 March 2010; pp. 541–551.
25. Hsu, L.Y.; Horng, S.J.; Fan, P.; Khan, M.K.; Wang, Y.R.; Run, R.S.; Lai, J.L.; Chen, R.J. MTPSO algorithm for solving planar graph coloring problem. *Expert Syst. Appl.* **2011**, *38*, 5525–5531. [CrossRef]
26. Cui, G.; Qin, L.; Liu, S.; Wang, Y.; Zhang, X.; Cao, X. Modified PSO algorithm for solving planar graph coloring problem. *Prog. Nat. Sci.* **2008**, *18*, 353–357. [CrossRef]
27. Bui, T.N.; Nguyen, T.H.; Patel, C.M.; Phan, K.A.T. An ant-based algorithm for coloring graphs. *Discret. Appl. Math.* **2008**, *156*, 190–200. [CrossRef]
28. Zhou, Y.; Zheng, H.; Luo, Q.; Wu, J. An improved Cuckoo Search Algorithm for Solving Planar Graph Coloring Problem. *Appl. Math. Inf. Sci.* **2013**, *7*, 785–792. [CrossRef]
29. Prestwich, S. Generalised graph colouring by a hybrid of local search and constraint programming. *Discret. Appl. Math.* **2008**, *156*, 148–158. [CrossRef]

30. Demange, M.; Ekim, T.; Ries, B.; Tanasescu, C. On some applications of the selective graph coloring problem. *Eur. J. Oper. Res.* **2015**, *240*, 307–314. [CrossRef]
31. Demange, M.; Ekim, T.; Ries, B. On the minimum and maximum selective graph coloring problems in some graph classes. *Discret. Appl. Math.* **2016**, *204*, 77–89. [CrossRef]
32. Takeshita, L. Coloring Signed Graphs. 2016, pp. 1–7. Available online: https://math.mit.edu/~{}apost/courses/18.204-2016/18.204_Lynn_Takeshita_final_paper.pdf (accessed on 20 January 2020).
33. Macajov, E.; Raspaud, A.; Skoviera, M. The Chromatic Number of a Signed Graph. Cornell University Library. *arXiv* **2016**, arXiv:1412.6349.
34. Zhou, B. On the Maximum Number of Dominating Classes in Graph Coloring. *Open J. Discret. Math.* **2016**, *6*, 70–73. [CrossRef]
35. Gaspers, S.; Lee, E.J. Faster Graph Coloring in Polynomial Space. Cornell University Library. *arXiv* **2016**, arXiv:1607.06201.
36. Angelini, P.; Bekos, M.A.; De Luca, F.; Didimo, W.; Kaufmann, M.; Kobourov, S.; Montecchiani, F.; Raftopoulou, C.N.; Roselli, V.; Symvonis, A. Vertex-Coloring with Defects. *J. Graph Algorithms Appl.* **2017**, *21*, 313–340. [CrossRef]
37. Galán, S.F. Simple decentralized graph coloring. *Comput. Optim. Appl.* **2017**, *66*, 163–185. [CrossRef]
38. Aslan, M.; Baykan, N.A. A Performance Comparison of Graph Coloring Algorithms. *Int. J. Intell. Syst. Appl. Eng.* **2016**, 1–7. [CrossRef]
39. Fleurent, C.; Jacques, A. Ferland: Genetic and hybrid algorithms for graph coloring. *Ann. Oper. Res.* **1996**, *63*, 437–461. [CrossRef]
40. Mumford, C.L. New Order Based Crossovers for the Graph Coloring Problem. In *Parallel Problem Solving from Nature*; Springer: Berlin/Heidelberg, Germany, 2006; Volume 4193, pp. 880–889.
41. Han, L.; Han, Z. A Novel Bi-objective Genetic Algorithm for the Graph Coloring Problem. In Proceedings of the 2nd International Conference on Computer Modeling and Simulation, Sanya, China, 22–24 January 2010.
42. Hajduk, J.O. An Analysis of Tabu Search for the Graph Coloring Problem. Master's Thesis, Utrecht University, Utrecht, The Netherlands, 2010. Available online: www.cs.uu.nl/education/scripties/pdf.php?SID=INF/SCR-2009-095 (accessed on 20 January 2020).
43. Costa, D.; Hertz, A.; Dubuis, C. Embedding a sequential procedure within an evolutionary algorithm for coloring problems in graphs. *J. Heuristics* **1995**, *1*, 105–128. [CrossRef]
44. Dorne, R.; Hao, J.K. A new genetic local search algorithm for graph coloring. In Proceedings of the International Conference on Parallel Problem Solving from Nature, Amsterdam, The Netherlands, 27–30 September 1998; Springer: Berlin/Heidelberg, Germany, 1998; pp. 745–754.
45. Galinier, P.; Hao, J.K. Hybrid Evolutionary Algorithms for Graph Coloring. *J. Comb. Optim.* **1999**, 3, 379–397. [CrossRef]
46. Lu, Z.; Hao, J.K. A memetic algorithm for graph coloring. *Eur. J. Oper. Res.* **2010**, *203*, 241–250. [CrossRef]
47. Porumbel, D.C.; Hao, J.K.; Kuntz, P. An evolutionary approach with diversity guarantee and well-informed grouping recombination for graph coloring. *Comput. Oper. Res.* **2010**, *37*, 1822–1832. [CrossRef]
48. Saha, S.; Kumar, R.; Baboo, G. Characterization of graph properties for improved Pareto fronts using heuristics and EA for bi-objective graph coloring problem. *Appl. Soft Comput.* **2013**, *13*, 2812–2822. [CrossRef]
49. Marappan, Raja.; Sethumadhavan, Gopalakrishnan. A new genetic algorithm for graph coloring. In Proceedings of the 5th International Conference on Computational Intelligence, Modelling and Simulation, Seoul, Korea, 24–26 September 2013; pp. 49–54.
50. Sethumadhavan, G.; Marappan, R. A Genetic Algorithm for Graph Coloring using Single Parent Conflict Gene Crossover and Mutation with Conflict Gene Removal Procedure. In Proceedings of the IEEE International Conference on Computational Intelligence and Computing Research 2013, Madurai, India, 26–28 December 2013; pp. 350–355.
51. Ryan, E.; Azad, R.M.A.; Ryan, C. On the performance of genetic operators and the random key representation. In Proceedings of the European Conference on Genetic Programming, Coimbra, Portugal, 5–7 April 2004; Springer: Berlin/Heidelberg, Germany, 2004; pp. 162–173.
52. Croitoru, C.; Luchian, H.; Gheorghie, O.; Apetrei, A. A New Genetic Graph Coloring Heuristic. 2002. Available online: http://citeseerx.ist.psu.edu/viewdoc/summary?doi=10.1.1.14.19612002 (accessed on 20 January 2020).

53. Lewis, R.M.R. *A Guide to Graph Coloring, Algorithms and Applications*; Springer: Berlin/Heidelberg, Germany, 2016.
54. Mann, Z.Á. Complexity of Coloring Random Graphs: An Experimental Study of the Hardest Region. *J. Exp. Algorithmics* **2018**, *23*, 1–19. [CrossRef]
55. Mostafaie, T.; Khiyabani, F.M.; Navimipour, N.J. A systematic study on meta-heuristic approaches for solving the graph coloring problem. *Comput. Oper. Res.* **2019**. [CrossRef]
56. Srivastava, S.; Sundararaghavan, V. Graph Coloring Approach to Mesh Generation in Multiphase Media with Smooth Boundaries. *AIAA J.* **2020**, *58*, 198–205. [CrossRef]
57. Dukanovic, I.; Rendl, F. A semidefinite programming-based heuristic for graph coloring. *Discret. Appl. Math.* **2008**, *156*, 180–189. [CrossRef]
58. Hernández, H.; FrogSim, B.C. Distributed graph coloring in wireless ad hoc networks. *Telecommun. Syst.* **2014**, *55*, 211–223. [CrossRef]
59. Segundo, P.S. A new DSATUR-based algorithm for exact vertex coloring. *Comput. Oper. Res.* **2012**, *39*, 1724–1733. [CrossRef]
60. The Graph Coloring Instances. Available online: http://mat.gsia.cmu.edu/COLOR/instances.html (accessed on 20 January 2020).
61. Back, T. *Evolutionary Algorithms in Theory and Practice*; Oxford University Press: New York, NY, USA, 1996; pp. 21–28.
62. He, J.; Yu, X. Conditions for the Convergence of Evolutionary Algorithms. *J. Syst. Archit.* **2001**, *47*, 601–612. [CrossRef]
63. Rudolph, G. Finite Markov Chain Results in Evolutionary Computation: A Tour d'Horizon. *Fundam. Inform.* **1998**, *35*, 67–89. [CrossRef]
64. Marco, B.; Alfredo, M. Automatic algebraic evolutionary algorithms. In *Italian Workshop on Artificial Life and Evolutionary Computation*; Springer: Berlin/Heidelberg, Germany, 2017; pp. 271–283.
65. Marco, B.; Alfredo, M. Learning bayesian networks with algebraic differential evolution. In *International Conference on Parallel Problem Solving from Nature*; Springer: Berlin/Heidelberg, Germany, 2018; pp. 436–448.
66. Marco, B.; Alfredo, M. MOEA/DEP: An algebraic decomposition-based evolutionary algorithm for the multiobjective permutation flow shop scheduling problem. In *European Conference on Evolutionary Computation in Combinatorial Optimization*; Springer: Berlin/Heidelberg, Germany, 2018; pp. 132–145.

mathematics

MDPI

Article

Competition-Independence Game and Domination Game

Chalermpong Worawannotai [1] and Watcharintorn Ruksasakchai [2,*]

[1] Department of Mathematics, Faculty of Science, Silpakorn University, Nakhon Pathom 73000, Thailand; worawannotai_c@silpakorn.edu

[2] Department of Mathematics, Statistics and Computer Science, Faculty of Liberal Arts and Science, Kasetsart University, Kamphaeng Saen Campus, Nakhon Pathom 73140, Thailand

* Correspondence: faaswtr@ku.ac.th

Received: 31 January 2020; Accepted: 17 February 2020; Published: 5 March 2020

Abstract: The domination game is played on a graph by two players, Dominator and Staller, who alternately choose a vertex of G. Dominator aims to finish the game in as few turns as possible while Staller aims to finish the game in as many turns as possible. The game ends when all vertices are dominated. The game domination number, denoted by $\gamma_g(G)$ (respectively $\gamma'_g(G)$), is the total number of turns when both players play optimally and when Dominator (respectively Staller) starts the game. In this paper, we study a version of this game where the set of chosen vertices is always independent. This version turns out to be another game known as the competition-independence game. The competition-independence game is played on a graph by two players, Diminisher and Sweller. They take turns in constructing maximal independent set M, where Diminisher tries to minimize $|M|$ and Sweller tries to maximize $|M|$. Note that, actually, it is the domination game in which the set of played vertices is independent. The competition-independence number, denoted by $I_d(G)$ (respectively $I_s(G)$) is the optimal size of the final independent set in the competition-independence game if Diminisher (respectively Sweller) starts the game. In this paper, we check whether some well-known results in the domination game hold for the competition-independence game. We compare the competition-independence numbers to the game domination numbers. Moreover, we provide a family of graphs such that many parameters are equal. Finally, we present a realization result on the competition-independence numbers.

Keywords: domination game; competition-independence game

1. Introduction

A *dominating set* of a graph G is a set S of vertices of G such that every vertex in G is an element in S or is adjacent to an element in S. The *domination number* of G, denoted by $\gamma(G)$, is the cardinality of a minimum dominating set of G. A set S is *independent* if no two vertices in S are adjacent. The *independence number* of a graph G, denoted by $\alpha(G)$, is the cardinality of a maximum independent set of G. An *independent dominating set* of a graph G is a dominating set of G which is independent. The *independent domination number* of G, denoted by $i(G)$, is the cardinality of a minimum independent dominating set of G.

In 2010, Brešar, Klavžar, and Rall [1] first introduced the domination game. The *domination game* is played on a graph G by two players, Dominator and Staller, who alternately choose a vertex of G in such a way that at least one new vertex is dominated. The game ends when all vertices are dominated. Dominator aims to finish the game in as few turns as possible while Staller aims to finish the game in as many turns as possible. The *game domination number*, denoted by $\gamma_g(G)$ (respectively $\gamma'_g(G)$), is the total number of turns when both players play optimally and when Dominator (respectively Staller) starts the game.

Also, they investigated the relationship between domination number and game domination numbers of a graph. They proved that $\gamma(G) \leq \gamma_g(G) \leq 2\gamma(G) - 1$ for any graph G. In References [1,2], the authors showed that the two game domination numbers of a graph can differ by at most one. The domination game is being studied extensively since it was introduced. Kinnersley, West, and Zamani [2] posted the 3/5 conjecture: for any isolate-free graph G, $\gamma_g(G) \leq 3|V(G)|/5$, and it has attracted several researches (See References [3–7]). In addition, there are researchers studying the domination game numbers on several classes of graphs such as paths, cycles, forests, disjoint union of graphs, split graphs, etc. (See References [4,8–11] for examples).

In 2015, a variation of the domination game called *total domination game* was introduced by Henning, Klavžar, and Rall [12]. In this version of the game, domination is replaced by total domination where a chosen vertex dominates its neighbors but not itself (See References [13–15]). The concept of total domination naturally gives rise to a variety of domination games such as Z-domination game, L-domination game, and LL-domination game (See Reference [16]). Many results in the domination game also hold for any of the above variations. For example, the difference between the number of moves in Dominator-start game and in Staller-start game is at most 1 for any of the above games.

In this paper, we are interested in studying the domination game such that the set of played vertices must be independent. In other words, a player can only play on an undominated vertex. Since an independent dominating set is a maximal independent set and vice versa, this independent version is the same as the competition-independence game which was introduced by Philips and Slater [17] in 2001.

The *competition-independence game* is played on a graph by two players, Diminisher and Sweller. They take turns in constructing maximal independent set M, where Diminisher tries to minimize $|M|$ and Sweller tries to maximize $|M|$. The *competition-independence number*, denoted by $I_d(G)$ (respectively $I_s(G)$), is the optimal size of the final independent set in the competition-independence game if Diminisher (respectively Sweller) starts the game. For the rest of the paper, we will use Dominator instead of Diminisher and Staller instead of Sweller.

Philips and Slater [18] provided the competition-independence numbers of a path. Consequently, the competition-independence numbers of a cycle of n vertices established since the first move in C_n produces P_{n-3}. In 2018, the competition-independence game in trees was studied by Goddard and Henning [19]. They provided the maximum and minimum values of the competition-independence game for trees of maximum degree 3.

In this paper, we check whether some well-known results in the domination game hold for the competition-independence game. We compare the competition-independence numbers to the game domination numbers. Also, we give some classes of graphs in which many parameters are equal. Finally, we establish a realization result of the competition-independence numbers.

2. Relationship between the Competition-Independence Numbers and Other Parameters

A fundamental tool for analyzing the original domination game is the Continuation Principle, which is proved in Reference [2]. We start this section by showing that the Continuation Principle for the domination game does not hold for the competition-independence game. Next, we compare the competition-independence numbers to the game domination numbers. Also, we provide a family of graphs such that many parameters are equal.

Theorem 1 ([2] (Continuation Principle)). *Let G be a (partially dominated) graph and let A and B be subsets of $V(G)$. Let G_A and G_B be partially dominated graphs in which the sets A and B have already been dominated, respectively. If $B \subseteq A$, then $\gamma_g(G_A) \leq \gamma_g(G_B)$ and $\gamma_g'(G_A) \leq \gamma_g'(G_B)$.*

This result is very intuitive and natural. The more coverage the chosen vertices have, the fewer vertices are needed to be chosen to dominate the remaining undominated vertices. This result also holds for total domination game, Z-domination game, L-domination game, and LL-domination game [16]. However, it does not hold for the competition-independence game as the restriction of independence sometimes prevents players from making good moves. We present two simple counterexamples here.

Example 2. *Consider the complete bipartite graph $K_{1,3}$ with vertex set $\{u, v_1, v_2, v_3\}$ and edge set $\{uv_1, uv_2, uv_3\}$. If $A = \{u, v_1\}$ and $B = \emptyset$, then $I_d(G_A) = 2$ and $I_d(G_B) = 1$.*

Example 3. *Let G be the graph obtained from the complete bipartite graph $K_{3,3}$ with partite sets V_1 and V_2 by deleting a perfect matching. If $A = V_1$ and $B = \emptyset$, then $I_s(G_A) = 3$ and $I_s(G_B) = 2$.*

Now, we compute the competition-independence numbers of complete multipartite graphs and complete bipartite graphs minus perfect matchings. These families of graphs will appear as examples throughout this paper.

Lemma 4. *Let $n_1, n_2, ..., n_m$ be positive integers. Then, $I_d(K_{n_1,n_2,...,n_m}) = \min\{n_1, n_2, ..., n_m\}$ and $I_s(K_{n_1,n_2,...,n_m}) = \max\{n_1, n_2, ..., n_m\}$.*

Proof. Let $V_{n_1}, V_{n_2}, ..., V_{n_m}$ be the partite sets of $K_{n_1,n_2,...,n_m}$ with $|V_{n_i}| = n_i$ for $1 \leq i \leq m$. If a player starts the game by playing a vertex in V_{n_i} for some $1 \leq i \leq m$, then both players must alternately choose vertices in the set V_{n_i} so the game will end in n_i moves. Hence, $I_d(K_{n_1,n_2,...,n_m}) = \min\{n_1, n_2, ..., n_m\}$ and $I_s(K_{n_1,n_2,...,n_m}) = \max\{n_1, n_2, ..., n_m\}$. □

Lemma 5. *Let G be the graph obtained from $K_{n,n}$ by deleting a perfect matching where $n \geq 2$. Then, $I_d(G) = n$ and $I_s(G) = 2$.*

Proof. Let $A = \{a_1, a_2, ..., a_n\}$ and $B = \{b_1, b_2, ..., b_n\}$ be the partite sets of the $K_{n,n}$, where a_i and b_i are not adjacent in G for $i \in \{1, 2, ..., n\}$.

We first show that $I_d(G) = n$. Without loss of generality, assume that Dominator starts the game by playing vertex $a_1 \in A$. Then, each vertex in $B - \{b_1\}$ is dominated. Staller plays another vertex in A. Now all vertices in B are dominated. Then, Dominator and Staller must alternately play all vertices in A. Consequently, $I_d(G) = n$.

We next show that $I_s(G) = 2$. Without loss of generality, Staller starts the game by playing vertex $a_1 \in A$. Then, Dominator finishes the game by playing b_1. Hence, $I_s(G) = 2$. □

The following four theorems show that the difference between the competition-independence numbers and the game domination numbers of a graph can be arbitrarily large.

Theorem 6. *For a nonnegative integer n, there is a graph G such that $I_d(G) - \gamma_g(G) = n$.*

Proof. Consider $K_{n+3,t}$, where $t \geq n + 3$. By Lemma 4, $I_d(K_{n+3,t}) = n + 3$. Note that $\gamma_g(K_{n+3,t}) = 3$. Thus, $I_d(K_{n+3,t}) - \gamma_g(K_{n+3,t}) = n$. □

Theorem 7. *For a positive integer n, there is a graph G such that $\gamma_g(G) - I_d(G) = n$.*

Proof. Let u be a vertex and $P_{9n+1} = x_1y_1z_1x_2y_2z_2...x_{3n}y_{3n}z_{3n}x_{3n+1}$ be a path. Define a graph G by $V(G) = \{u\} \cup V(P_{9n+1})$ and $E(G) = \{ux_j | 1 \leq j \leq 3n + 1\} \cup E(P_{9n+1})$. See Figure 1 for example. We first show that $I_d(G) = 3n + 1$. To show that $I_d(G) \leq 3n + 1$, we present a strategy for Dominator. Dominator starts the game by playing the vertex u first. Since $ux_j \in E(G)$ for all $1 \leq j \leq 3n + 1$, Staller and Dominator cannot play any vertex x_j for all $1 \leq j \leq 3n + 1$. Thus, after the game ends, for each j, exactly one of y_j or z_j is played. Therefore, $3n + 1$ vertices are played and so $I_d(G) \leq 3n + 1$. To show that $I_d(G) \geq 3n + 1$, we present a strategy for Staller.

Case 1: Dominator starts the game by playing the vertex u.

Then, Staller plays y_1, and thus by similar arguments as above, after the game ends, $3n + 1$ vertices are played. Therefore, the number of moves in this case is at least $3n + 1$.

Case 2: Dominator starts the game by playing the vertex y_i or z_i for some $1 \leq i \leq 3n$.

Then, Staller plays the vertex u, and thus by similar arguments as above, after the game ends, $3n + 1$ vertices are played. Therefore, the number of moves in this case is at least $3n + 1$.

Case 3: Dominator starts the game by playing the vertex x_i for some $1 \leq i \leq 3n+1$.

Without loss of generality, we may assume that $1 \leq i \leq \frac{3n+1}{2}$. Then, Staller plays his first move on vertex y_{i+1}. Consider the set $X := \{x_1, x_2, x_3, ..., x_{3n+1}\} - \{x_i, x_{i+1}\}$. Notice that each $x \in X$ is not dominated. In order to dominate X, an additional $3n-1$ vertices are played. Thus, after the game ends, at least $3n+1$ vertices are played.

From the above cases, $I_d(G) \geq 3n+1$. Hence, $I_d(G) = 3n+1$.

We next show that $\gamma_g(G) = 4n+1$. To show that $\gamma_g(G) \geq 4n+1$, we present a strategy for Staller.

Case 1: Dominator starts the game by playing the vertex u.

In each turn of Staller, he plays a vertex which dominates only one more undominated vertex. Since the vertex x_j is dominated for all $1 \leq j \leq 3n+1$, it follows that in each turn of Dominator, he can dominate at most two more undominted vertices. Consequently, in each round of the game played by Staller and then Dominator, there are at most three new dominated vertices. Since after the first move of Dominator at u there are $9n+2-(3n+1)-1 = 6n$ undominated vertices, it follows that the number of moves in this case is at least $\frac{2}{3} \cdot 6n + 1 = 4n+1$.

Case 2: Dominator starts the game by playing a vertex v where $v \neq u$.

Then, v dominates at most four vertices including v itself. Staller plays his first move on the vertex u, and after that, he plays to dominate one new undominated vertex at a time. Thus, for each $1 \leq j \leq 3n+1$, x_j is dominated. After that, for each round of the game, Dominator can dominate at most two more undominated vertices. Consequently, in each round of the game played by Dominator and then Staller, there are at most three new dominated vertices. After u and v are played, there are at least $(9n+2) - 4 - (3n+1) + 1 = 6n-2$ undominated vertices. Note that $(6n-2) = (6n-3)+1$. Therefore, the number of moves in this case is at least $\frac{2}{3} \cdot (6n-3) + 1 + 2 = 4n+1$.

From the above cases, $\gamma_g(G) \geq 4n+1$.

To show that $\gamma_g(G) \leq 4n+1$, we present a strategy for Dominator. Dominator starts the game by playing the vertex u. For each turn of Dominator, he plays a vertex to dominate two new leftmost vertices if possible. Note that if Dominator cannot dominate two new vertices in his turn, that is he can dominate only one new vertex, then it means that Staller played a vertex to dominate two new vertices earlier. Therefore, on average, a move by Staller and a move by Dominator dominates at least three new vertices. Thus, $\gamma_g(G) \leq \frac{2}{3}(6n) + 1 = 4n+1$. Consequently, $\gamma_g(G) = 4n+1$.

Therefore, $\gamma_g(G) - I_d(G) = (4n+1) - (3n+1) = n$, as required. □

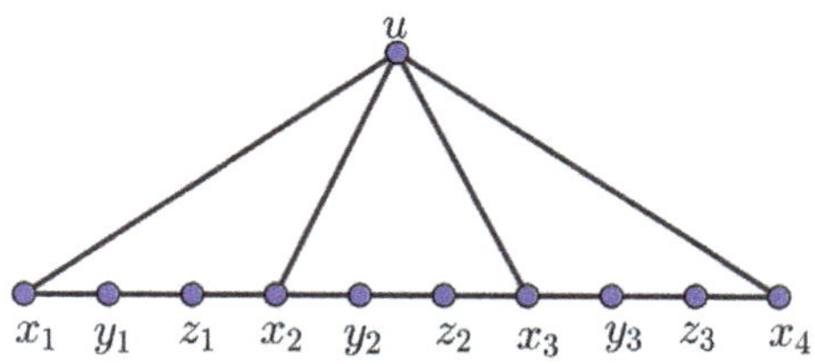

Figure 1. The graph G in the proof of Theorem 7, where $n = 1$.

Theorem 8. *For a nonnegative integer n, there is a graph G such that* $I_s(G) - \gamma'_g(G) = n$.

Proof. Consider $K_{1,n+2}$. By Lemma 4, $I_s(K_{1,n+2}) = n+2$. Note that $\gamma'_g(K_{1,n+2}) = 2$. Thus, $I_s(K_{1,n+2}) - \gamma'_g(K_{1,n+2}) = n$. □

Theorem 9. *For a positive integer n, there is a graph G such that* $\gamma'_g(G) - I_s(G) = n$.

Proof. Let $P_{9n+1} = x_1y_1z_1x_2y_2z_2...x_{3n}y_{3n}z_{3n}x_{3n+1}$ be a path and K_{3n+1} be a complete graph with $V(K_{3n+1}) = \{u_1, u_2, ..., u_{3n+1}\}$.

Define a graph G by $V(G) = V(P_{9n+1}) \cup V(K_{3n+1})$ and $E(G) = E(P_{9n+1}) \cup E(K_{3n+1}) \cup \{u_jx_k | i \neq k$ and $1 \leq j, k \leq 3n+1\}$. See Figure 2 for example.

We first show that $\gamma'_g(G) = 4n+2$. To show that $\gamma'_g(G) \leq 4n+2$, we present a strategy for Dominator.

Case 1: Staller starts the game by playing the vertex u_i for some $1 \leq i \leq 3n+1$.

Without loss of generality, we may assume that $1 \leq i \leq \frac{3n+1}{2}$. Dominator responds by playing the vertex y_i. At this point, y_i, z_i and all u_k's $(1 \leq k \leq 3n+1)$ are dominated. After that, in each move of Dominator, he plays the leftmost vertex that can dominate two new vertices if possible. Note that, if Dominator cannot dominate two new vertices in his turn, that is he can dominate only one new vertex, then it means that Staller played a vertex to dominate two new vertices earlier. Therefore, excluding the first two moves, on average, a move by Staller and a move by Dominator dominate at least three new vertices. After Dominator plays y_i, there are $(12n+2) - 2(3n+1) - 2 = 6n-2$ undominated vertices. Notice that $6n-2 = (6n-3)+1 = 3(2n-1)+1$. Hence, the number of moves in this case is at most $\frac{2}{3}(6n-3)+1+2 = 4n+1 < 4n+2$.

Case 2: Staller starts the game by playing the vertex y_i or z_i for some $1 \leq i \leq 3n$.

Without loss of generality, we may assume that Staller starts the game by playing the vertex y_i for some $1 \leq i \leq 3n$. Dominator responds by playing the vertex u_i and so each vertex x_j is dominated for $1 \leq j \leq 3n+1$ and that each vertex u_k is dominated for $1 \leq k \leq 3n+1$. By similar arguments as in case 1, after Dominator plays u_i, there are $(12n+2) - 2(3n+1) - 2 = 6n-2$ undominated vertices and so the number of moves in this case is at most $4n+2$.

Case 3: Staller starts the game by playing the vertex x_i for some $1 \leq i \leq 3n$.

Without loss of generality, we may assume that $1 \leq i \leq \frac{3n+1}{2}$. Dominator responds by playing the vertex u_i and so each vertex x_j is dominated for $1 \leq j \leq 3n+1$. By similar arguments as in case 1, after Dominator plays u_i, there are at most $(12n+2) - 2(3n+1) - 1 = 6n-1$ undominated vertices. Note that $6n-1 = (6n-3)+2 = 3(2n-1)+2$. Hence, the number of moves in this case is at most $\frac{2}{3}(6n-3)+2+2 = 4n+2$.

From the above cases, $\gamma'_g(G) \leq 4n+2$.

To show that $\gamma'_g(G) \geq 4n+2$, we present a strategy for Staller. Staller starts the game by playing the vertex x_1. Then, each vertex u_j is dominated for $2 \leq j \leq 3n+1$ and y_1 is also dominated.

Case 1: Dominator responds by playing the vertex u_1.

In each round of Staller's turn, he plays the leftmost vertex that can dominate one new vertex. Note that, in each Dominator's turn, he can dominate at most two new vertices. Therefore, excluding the first two moves, on average, a move by Staller and a move by Dominator dominate at most three new vertices. After Dominator plays u_1, there are $(12n+2) - 2(3n+1) - 1 = 6n-1$ undominated vertices. Notice that $6n-1 = (6n-3)+2 = 3(2n-1)+2$. Therefore, the number of moves in this case is at least $\frac{2}{3}(6n-3)+2+2 = 4n+2$.

Case 2: Dominator responds by playing the vertex w where $w \neq u_1$.

Then, Staller plays the vertex u_1 in his next turn, and after that, he will play the leftmost vertex that can dominate one new vertex. Note that, in each Dominator's turn, he can dominate at most two new vertices. Therefore, excluding the first three moves, on average, a move by Dominator and a move by Staller dominate at most three new vertices. After Staller plays u_1, there are at least $(12n+2) - 2(3n+1) - 2 = 6n-2$ undominated vertices. Notice that $6n-2 = (6n-3)+1 = 3(2n-1)+1$. Hence, the number of moves in this case is at least $\frac{2}{3}(6n-3)+1+3 = 4n+2$.

From the above cases, $\gamma'_g(G) \geq 4n+2$. Consequently, $\gamma'_g(G) = 4n+2$.

We next show that $I_s(G) = 3n+2$. To show that $I_s(G) \leq 3n+2$, we present a strategy for Dominator.

Case 1: Staller starts the game by playing a vertex u_i for some $1 \leq i \leq 3n+1$.

Dominator responds by playing the vertex x_i. After that, Staller and Dominator cannot play any vertex u_j or x_j for all $1 \leq j \leq 3n+1$. For the remainder of the game, Staller and Dominator can only play either y_j or z_j for some $1 \leq j \leq 3n$. After the game ends, for each j, exactly one of y_j or z_j is played for $1 \leq j \leq 3n$. Therefore, $3n+2$ vertices are played and so the number of moves in this case is at most $3n+2$.

Case 2: Staller starts the game by playing a vertex x_i for some $1 \leq i \leq 3n+1$.

Dominator responses by playing the vertex u_i. By similar arguments as in case 1, we have that $3n+2$ vertices are played and so the number of moves in this case is at most $3n+2$.

Case 3: Staller starts the game by playing either y_i or z_i for some $1 \leq i \leq 3n$.

Dominator responds by playing the vertex u_i if Staller played y_i; otherwise, he plays u_{i+1}. By similar arguments as in case 1, we have that $(3n-1)+2 = 3n+1$ vertices are played. Therefore, the number of moves in this case is at most $3n+1$.

From the above cases, $I_s(G) \leq 3n+2$.

To show that $I_s(G) \geq 3n+2$, we present a strategy for Staller. Staller starts the game by playing the vertex x_1.

Case 1: Dominator responds by playing the vertex u_1.

Then, the vertices x_j and u_j are dominated for $1 \leq j \leq 3n+1$. Thus, after the game ends, for each $1 \leq j \leq 3n$, exactly one of y_j or z_j is played. Therefore, $3n+2$ vertices are played. Consequently, the number of moves in this case is at least $3n+2$.

Case 2: Dominator responds by playing the vertex y_i or z_i for some $1 \leq i \leq 3n$.

Without loss of generality, we may assume that Dominator responds by playing the vertex y_i for some $1 \leq i \leq 3n$. Then, Staller plays his next turn at the vertex u_1. At this point, the vertices x_j and u_j are dominated for $1 \leq j \leq 3n+1$, and y_i and z_i are also dominated. Thus, after the game ends, for each $1 \leq k \leq 3n$, exactly one of y_k or z_k is played. Hence, $3n+2$ vertices are played. Consequently, the number of moves in this case is at least $3n+2$.

Case 3: Dominator responds by playing the vertex x_i for some $2 \leq i \leq 3n+1$.

Then, each vertex u_j is dominated for $1 \leq j \leq 3n+1$. Note that, after Dominator plays x_i, there are at least $(12n+2) - 3n - 2 - 4 = 9n - 4$ undominated vertices. In each of Staller's turns, he plays to dominate at most two new undominated vertices. Notice that, in each turn of Dominator, he can dominate at most three new undominated vertices. Therefore, the number of moves is at least $\frac{2}{5}(9n-4) + 2 = 3n + \frac{3}{5}n + \frac{2}{5}$. If $n \geq 2$, then the number of moves is at least $3n+2$. For $n = 1$, it can be checked that the number of moves is at least 5.

From the above cases, $I_s(G) \geq 3n+2$. Consequently, $I_s(G) = 3n+2$.

Therefore, $\gamma'_g(G) - I_s(G) = 4n+2-(3n+2) = n$. □

By the definitions, $i(G) \leq I_d(G) \leq \alpha(G)$ and $i(G) \leq I_s(G) \leq \alpha(G)$. We show that the difference between the competition-independence numbers and independent domination numbers can be arbitrarily large.

Theorem 10. *For a nonnegative integer n, there is a graph G such that* $I_d(G) - i(G) = n$.

Proof. Let G be the graph obtained from $K_{n+2,n+2}$ by deleting a perfect matching. By Lemma 5, we have $I_d(G) = n+2$. Note that $i(G) = 2$. Thus $I_d(G) - i(G) = n$. □

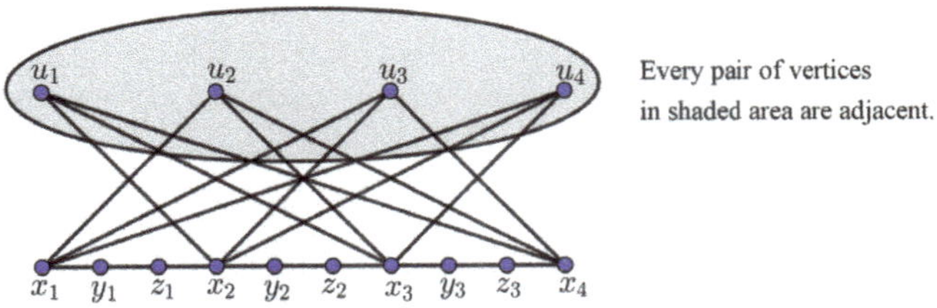

Figure 2. The graph G in the proof of Theorem 9, where $n = 1$.

Theorem 11. *For a nonnegative integer n, there is a graph G such that $\alpha(G) - I_d(G) = n$.*

Proof. Consider $K_{1,n+1}$. Note that $\alpha(K_{1,n+1}) = n+1$ and $I_d(K_{1,n+1}) = 1$. Thus $\alpha(K_{1,n+1}) - I_d(K_{1,n+1}) = n$. □

Theorem 12. *For a nonnegative integer n, there is a graph G such that $I_s(G) - i(G) = n$.*

Proof. Consider $K_{1,n+1}$. Note that $I_s(K_{1,n+1}) = n+1$ and $i(K_{1,n+1}) = 1$. Thus $I_s(K_{1,n+1}) - i(K_{1,n+1}) = n$. □

Theorem 13. *For a nonnegative integer n, there is a graph G such that $\alpha(G) - I_s(G) = n$.*

Proof. Let G be the graph obtained from $K_{n+2,n+2}$ by deleting a perfect matching. Note that $\alpha(G) = n+2$ and $I_s(G) = 2$. Thus, $\alpha(G) - I_s(G) = n$. □

Next, we show a family of graphs such that many parameters are equal.

The *corona product* of graphs G and H, denoted by $G \circ H$, is a graph obtained by taking one copy of G and $|V(G)|$ copies of H and by joining each vertex of the ith copy of H to the ith vertices of G, where $1 \leq i \leq |V(G)|$.

Theorem 14. *There exists a graph G such that $I_d(G) = I_s(G) = \gamma_g(G) = \gamma'_g(G) = \alpha(G) = i(G)$.*

Proof. Let G be a graph $K_n \circ K_1$. Note that $I_d(G) = I_s(G) = \gamma_g(G) = \gamma'_g(G) = \alpha(G) = i(G) = n$. □

3. Realization of the Competition-Independence Numbers

For a pair (a,b) of positive integers, we say G *realizes* (a,b) if $I_d(G) = a$ and $I_s(G) = b$.

In this section, we show that any pair (a,b) of positive integers can be realized.

Theorem 15. *For positive integers a and b, there is a connected graph G such that $I_d(G) = a$ and $I_s(G) = b$.*

Proof. Let a and b be positive integers.

Case 1: $a \leq b$.

Let G be the complete bipartite graph $K_{a,b}$. By Lemma 4, we have $I_d(G) = \min\{a,b\} = a$ and $I_s(G) = \max\{a,b\} = b$.

Case 2: $a > b$.

Let H be the complete b-partite graph such that the size of each partite set is a i.e., $H = K_{a,a,\ldots,a}$. Suppose that $V_1, V_2, \ldots, V_b$ are the partite sets of H and that $V_i = \{v_{i,1}, v_{i,2}, \ldots, v_{i,a}\}$ for $1 \leq i \leq b$. Let G be the graph $H - \{v_{i,j}v_{k,j} | 1 \leq i < k \leq b \text{ and } 1 \leq j \leq a\}$.

Let $T := \{v_{k,1} | 2 \leq k \leq b\}$. We first show that $I_d(G) = a$.

Without loss of generality, we may assume that Dominator starts the game by playing vertex $v_{1,1}$. Then, the set $(V_1 - \{v_{1,1}\}) \cup T$ is the set of all undominated vertices. Since $a > b$, Staller responds by playing a vertex in $V_1 - \{v_{1,1}\}$, and then, both players must alternately choose undominated vertices in the set V_1. Therefore, the game will end in a moves. Hence, $I_d(G) = a$.

We next show that $I_s(G) = b$.

Without loss of generality, we may assume that Staller starts the game by playing vertex $v_{1,1}$. Then, the set $(V_1 - \{v_{1,1}\}) \cup T$ is the set of all undominated vertices. Since $a > b$, Dominator responds by playing a vertex in the set T, and then, both players must alternately choose undominated vertices in the set T. Therefore, the game will end in b moves. Hence, $I_s(G) = b$. $\square$

In particular, Theorem 15 shows that the difference between the two competition-independence numbers of a graph can be arbitrarily large. This is very different from the domination game and the other variations where the difference is at most one [1,2,16].

4. Conclusions

In this paper, we studied the domination game such that the set of played vertices is independent. This game is known as the competition-independence game. We showed that the Continuation Principle for the domination game does not hold for the competition-independence game. We proved that the difference between the competition-independence numbers and the game domination numbers of a graph can be arbitrarily large. Also, we gave a family of graphs such that many parameters are equal. Furthermore, we showed that any pair of positive integers can be realized as the competition-independence numbers of some graph.

Author Contributions: Investigation, W.R.; writing—original draft preparation, W.R.; writing—review and editing, C.W.; supervision, C.W. All authors have read and agreed to the published version of the manuscript.

Funding: This research was funded by The Thailand Research Fund (grant number MRG6180301).

Acknowledgments: The authors would like to thank the reviewers for their suggestions and comments. The second author was supported by The Thailand Research Fund (grant number MRG6180301). The author takes this oppotunity to thank The Thailand Research Fund for the support.

Conflicts of Interest: The authors declare no conflict of interest.

References

1. Brešar, B.; Klavžar, S.; Rall, D.F. Domination game and an imagination strategy. *SIAM J. Discret. Math.* **2010**, *24*, 979–991. [CrossRef]
2. Kinnersley, W.B.; West, D.B.; Zamani, R. Extremal problems for game domination number. *SIAM J. Discret. Math.* **2013**, *27*, 2090–2107. [CrossRef]
3. Brešar, B.; Klavžar, S.; Košmrlj G.; Rall, D.F. Domination game: Extremal families of graphs for the 3/5-conjectures. *Discret. Appl. Math.* **2013**, *161*, 1308–1316. [CrossRef]
4. Bujtas, C. Domination game on forests. *Discret. Math.* **2015**, *338*, 2220–2228. [CrossRef]
5. Bujtas, C. On the game domination number of graphs with given minimum degree. *Electron. J. Combin.* **2015**, *22*, 3.29. [CrossRef]
6. Henning, M.A.; Kinnersley, W.B. Domination game: A proof of the 3/5-conjecture for graphs with minimum degree at least two. *SIAM J. Discret. Math.* **2016**, *30*, 20–35. [CrossRef]
7. Henning, M.A.; Löwenstein, C. Domination game: Extremal families for the 3/5-conjecture for forests. *Discuss. Math. Graph Theory* **2017**, *37*, 369–381. [CrossRef]
8. Dorbec, P.; Košmrlj, G.; Renault, G. The domination game played on unions of graphs. *Discret. Math.* **2015**, *338*, 71–79. [CrossRef]
9. James, T.; Klavžar, S.; Vijayakumar, A. The domination game on split graphs, *Bull. Aust. Math. Soc.* **2019**, *99*, 327–337. [CrossRef]
10. Ruksasakchai, W.; Onphaeng, K.; Worawannotai, C. Game domination numbers of a disjoint union of paths and cycles. *Quaest. Math.* **2019**, *42*, 1357–1372. [CrossRef]
11. Zamani, R. Hamiltonian cycles through specified edges in bipartite graphs, domination game, and the game of revolutionaries and spies. Ph.D. Thesis, University of Illinois at Urbana-Champaign, IL, USA.
12. Henning, M.A.; Klavžar, S.; Rall, D.F. Total version of the domination game. *Graphs Combin.* **2015**, *31*, 1453–1462. [CrossRef]
13. Bujtas, C. On the game total domination number. *Graphs Comb.* **2018**, *34*, 415–425. [CrossRef]

14. Dorbec, P.; Henning, M.A. Game total domination for cycles and paths. *Discret. Appl. Math.* **2016**, *208*, 7–18. [CrossRef]
15. Henning, M.A.; Klavžar, S.; Rall, D.F. The 4/5 upper bound on the game total domination number. *Combinatorica* **2017**, *37*, 223–251. [CrossRef]
16. Brešar, B.; Bujtas, C.; Gologranc, T.; Klavžar, S.; Košmrlj, G.; Marc, T.; Patkos, B.; Tuza, Z.; Vizer, M. The variety of domination games. *Aequ. Math.* **2019**, *93*, 1085–1109. [CrossRef]
17. Phillips, J.B.; Slater, P.J. An introduction to graph competition independence and enclaveless parameters. *Graph. Theory Notes N.Y.* **2001**, *41*, 37–41.
18. Phillips, J.B.; Slater, P.J. Graph competition independence and enclaveless parameters. *Congr. Numer.* **2002**, *154*, 79–100.
19. Goddard, W.; Henning, M.A. The competition-independence game in trees. *J. Combin. Math. Combin. Comput.* **2018**, *104*, 161–170.

MDPI
St. Alban-Anlage 66
4052 Basel
Switzerland
Tel. +41 61 683 77 34
Fax +41 61 302 89 18
www.mdpi.com

Mathematics Editorial Office
E-mail: mathematics@mdpi.com
www.mdpi.com/journal/mathematics

www.ingramcontent.com/pod-product-compliance
Lightning Source LLC
Chambersburg PA
CBHW061816210926
51850CB00006B/721
9783039287987